Advances in postharvest management of horticultural produce

It is widely recognised that agriculture is a significant contributor to global warming and climate change. Agriculture needs to reduce its environmental impact and adapt to current climate change whilst still feeding a growing population, i.e. become more 'climate-smart'. Burleigh Dodds Science Publishing is playing its part in achieving this by bringing together key research on making the production of the world's most important crops and livestock products more sustainable.

Based on extensive research, our publications specifically target the challenge of climate-smart agriculture. In this way we are using 'smart publishing' to help achieve climate-smart agriculture.

Burleigh Dodds Science Publishing is an independent and innovative publisher delivering high quality customer-focused agricultural science content in both print and online formats for the academic and research communities. Our aim is to build a foundation of knowledge on which researchers can build to meet the challenge of climate-smart agriculture.

For more information about Burleigh Dodds Science Publishing simply call us on +44 (0) 1223 839365, email info@bdspublishing.com or alternatively please visit our website at www.bdspublishing.com.

Related titles:

Achieving sustainable cultivation of apples
Print (ISBN 978-1-78676-032-6); Online (ISBN 978-1-78676-034-0, 978-1-78676-035-7)

Achieving sustainable cultivation of bananas Volume 1: Cultivation techniques
Print (ISBN 978-1-78676-156-9); Online (ISBN 978-1-78676-158-3, 978-1-78676-159-0)

Achieving sustainable cultivation of mangoes
Print (ISBN 978-1-78676-132-3); Online (ISBN 978-1-78676-134-7, 978-1-78676-135-4)

Achieving sustainable cultivation of temperate zone tree fruits and berries Volume 1: Physiology, genetics and cultivation
Print (ISBN 978-1-78676-208-5); Online (ISBN 978-1-78676-210-8, 978-1-78676-211-5)

Achieving sustainable cultivation of temperate zone tree fruits and berries Volume 2: Case studies
Print (ISBN 978-1-78676-212-2); Online (ISBN 978-1-78676-214-6, 978-1-78676-215-3)

Achieving sustainable cultivation of tomatoes
Print (ISBN 978-1-78676-040-1); Online (ISBN 978-1-78676-042-5, 978-1-78676-043-2)

Achieving sustainable cultivation of tropical fruits
Print (ISBN 978-1-78676-284-9); Online (ISBN 978-1-78676-286-3, 978-1-78676-287-0)

Achieving sustainable cultivation of vegetables
Print (ISBN 978-1-78676-236-8); Online (ISBN 978-1-78676-238-2, 978-1-78676-239-9)

Achieving sustainable greenhouse cultivation
Print (ISBN 9781-78676-280-1); Online (ISBN 978-1-78676-282-5, 978-1-78676-283-2)

Chapters are available individually from our online bookshop:
https://shop.bdspublishing.com

climate•SMART•publishing
IN AGRICULTURAL SCIENCE

BURLEIGH DODDS SERIES IN AGRICULTURAL SCIENCE
NUMBER 66

Advances in postharvest management of horticultural produce

Edited by Professor Chris Watkins, Cornell University, USA

burleigh dodds
SCIENCE PUBLISHING

Published by Burleigh Dodds Science Publishing Limited
82 High Street, Sawston, Cambridge CB22 3HJ, UK
www.bdspublishing.com

Burleigh Dodds Science Publishing, 1518 Walnut Street, Suite 900, Philadelphia, PA 19102-3406, USA

First published 2020 by Burleigh Dodds Science Publishing Limited

Library of Congress Control Number: 2019951729

British Library Cataloguing in Publication Data
A catalogue record for this book is available from the British Library

ISBN 978-1-78676-288-7 (Print)
ISBN 978-1-78676-291-7 (PDF)
ISBN 978-1-78676-290-0 (ePub)
ISSN 2059-6936 (print)
ISSN 2059-6944 (online)

DOI 10.19103/AS.2019.0055

Typeset by Deanta Global Publishing Services, Dublin, Ireland

Contents

Series list

Title	Series number
Achieving sustainable cultivation of maize - Vol 1 From improved varieties to local applications Edited by: Dr Dave Watson, CGIAR Maize Research Program Manager, CIMMYT, Mexico	001
Achieving sustainable cultivation of maize - Vol 2 Cultivation techniques, pest and disease control Edited by: Dr Dave Watson, CGIAR Maize Research Program Manager, CIMMYT, Mexico	002
Achieving sustainable cultivation of rice - Vol 1 Breeding for higher yield and quality Edited by: Prof. Takuji Sasaki, Tokyo University of Agriculture, Japan	003
Achieving sustainable cultivation of rice - Vol 2 Cultivation, pest and disease management Edited by: Prof. Takuji Sasaki, Tokyo University of Agriculture, Japan	004
Achieving sustainable cultivation of wheat - Vol 1 Breeding, quality traits, pests and diseases Edited by: Prof. Peter Langridge, The University of Adelaide, Australia	005
Achieving sustainable cultivation of wheat - Vol 2 Cultivation techniques Edited by: Prof. Peter Langridge, The University of Adelaide, Australia	006
Achieving sustainable cultivation of tomatoes Edited by: Dr Autar Mattoo, USDA-ARS, USA & Prof. Avtar Handa, Purdue University, USA	007
Achieving sustainable production of milk - Vol 1 Milk composition, genetics and breeding Edited by: Dr Nico van Belzen, International Dairy Federation (IDF), Belgium	008
Achieving sustainable production of milk - Vol 2 Safety, quality and sustainability Edited by: Dr Nico van Belzen, International Dairy Federation (IDF), Belgium	009
Achieving sustainable production of milk - Vol 3 Dairy herd management and welfare Edited by: Prof. John Webster, University of Bristol, UK	010
Ensuring safety and quality in the production of beef - Vol 1 Safety Edited by: Prof. Gary Acuff, Texas A&M University, USA & Prof. James Dickson, Iowa State University, USA	011
Ensuring safety and quality in the production of beef - Vol 2 Quality Edited by: Prof. Michael Dikeman, Kansas State University, USA	012
Achieving sustainable production of poultry meat - Vol 1 Safety, quality and sustainability Edited by: Prof. Steven C. Ricke, University of Arkansas, USA	013
Achieving sustainable production of poultry meat - Vol 2 Breeding and nutrition Edited by: Prof. Todd Applegate, University of Georgia, USA	014

Introduction

This book provides a comprehensive review of key advances in preservation techniques for fresh fruit and vegetables. It reviews the strengths and weaknesses of different disinfection techniques and focuses on advances in preservation techniques.

The first part of the book summarises developments and improvements in preservation technologies such as cooling, controlled atmosphere storage, modified atmosphere and active packaging and barrier coatings. The second part of the book focuses on postharvest safety management and disinfection. Chapters cover current research on mechanisms of pathogen contamination of fresh produce, improvements in sanitising regimes and disinfection techniques using heat, irradiation and plasma, ozone and natural microbials.

The final part of the book surveys advances in monitoring postharvest quality of fresh produce, the handling of organically produce specialty crops and smart distribution systems to maintain the quality of horticultural produce.

Part 1 Preservation techniques

Chapter 1 discusses advances in cooling technologies to preserve horticultural produce. Fruit and vegetables are living organisms that continue to respire after harvest, deteriorating over time, in a process known as senescence. Although it is impossible to halt senescence, removing the field-heat, and reducing the temperature of the product at the time of harvest, has substantial effects on maintaining product integrity. Cooling is the first 'cool' stage of the horticultural cool-chain, and is designed to rapidly and efficiently reduce the product temperature from the field temperature (often 20-30°C, depending on local conditions) to storage temperature (usually between 0-15°C, depending on the specific product. The chapter begins by discussing the main cooling methods applied to cool horticultural products, such as static cooling, forced-air cooling, hydro-cooling, vacuum cooling and icing. Many of these technologies are focused on managing and improving cooling outcomes and minimising costs. These factors are also discussed, along with a case study on the management of cooling operations for the kiwifruit industry in New Zealand.

The next chapter reviews advances in controlled atmosphere storage of horticultural produce, focusing on the advances of the last two decades. It emphasises the connection between technical/managerial aspects of the controlled atmosphere regime and retention of product quality during and after the storage period. Chapter 2 particularly focuses on: initial low O2 stress (ILOS); delayed CA storage; anti-ethylene compounds (e.g. 1-MCP); dynamic controlled atmosphere (DCA) systems based on chlorophyll fluorescence (CF), ethanol detection and respiratory quotient (RQ) determination for identifying the lower oxygen limit (LOL); and strategies for reducing the electrical energy demands of the storage room. The chapter concludes with a reflection upon likely future advancements in controlled atmosphere storage science over the next 10-20 years, including: refining and broadening DCA system and 1-MCP application (singly and together), more metabolomic research to understand the biochemical and molecular effects of hypoxic stress, and new protocols for reducing storage room-related electricity consumption.

Chapter 3 focuses on advances in modified atmosphere and active packaging of horticultural produce. The science and technology of modified atmosphere packaging (MAP), continues to expand not only within the industry but also in terms of the technologies that are becoming available. MAP has transitioned from a standalone technology to a platform where unique science and technologies can be incorporated directly into the system to address a wide range of specific needs within the fresh produce industry. MAP is a platform technology that continues to expand its role in the extension and optimization of fresh produce quality by allowing management of factors such as ethylene, moisture and antimicrobial action. However, the amount of improperly designed commercial packaging highlights the need for knowledge about the fundamentals and importance of MAP. The chapter reviews the fundamentals of MAP design, and a substantive update on the advancements of converting and additive technologies is also provided.

The final chapter of Part 1 discusses the advances in the use of barrier coating and additives in the preservation of fresh horticultural produce. Chapter 4 reviews developments in edible coatings and surface treatments to preserve whole and fresh-cut fruits and vegetables. It also highlights the different purposes and properties of coatings and their applications in such areas as maintaining appearance, preventing water loss and decay as well as the use of semipermeable coatings to create an internal modified atmosphere to preserve produce quality. The chapter also discusses additives and their applications in such areas as prevention of browning, maintaining firmness and the use of antimicrobials to extend shelf-life. It concludes by looking at how coatings and additives can be combined synergistically to optimize the quality of fresh horticultural produce.

Part 2 Safety management and disinfection techniques

The second part of the book begins with a discussion of postharvest risk management of biological hazards encountered in horticultural produce. Fresh produce remains the main cause of foodborne illness outbreaks implicating a diverse range of enteric pathogens. The primary source of pathogens is pre-harvest contamination, which then become disseminated during processing. Chapter 5 provides an overview of fresh produce related foodborne illness outbreaks and the characteristics of implicated pathogens. The chapter describes the routes by which pathogens can be introduced and distributed at different post-harvest stages, from the field through to food service. It introduces a range of risk management strategies based on antimicrobial coatings, antimicrobial gas treatments, irradiation, UV-C and those based on advanced oxidation process. The chapter explains application of Response Surface Modelling (RSM) to balance microbial reduction with changes in produce quality, providing a detailed case study. Finally, the chapter looks ahead to future research trends and provides guidance on further reading on the subject.

Chapter 6 reviews advances in understanding pathogens contaminating horticultural produce. There is an overall trend in most of the developed countries that confirms the rise in the number of microbial contamination outbreaks associated with fresh products. Major contributing factors are global consumer demand, increasing worldwide trade and the development of new molecular tools that allow the identification of the causative agents and the outbreak traceability. The chapter examines the surveillance, source attribution and microbial risk assessment of foodborne pathogens. It also identifies the main sources for pathogen contamination, including primary production, processing, distribution, and cold chain. The chapter deals with the routes of pathogens contaminating fresh fruits and vegetables (FFV), as well as the factors affecting the survival or inactivation of pathogens in horticultural products. Finally, the chapter looks ahead to future research trends in this area.

Moving on to Chapter 7, the chapter highlights advances in postharvest sanitising regimes for horticultural produce. Fresh and fresh cut fruits and vegetables are occasionally contaminated with foodborne pathogens, which has led to numerous recalls and outbreaks of foodborne illness, affecting consumer confidence. Leafy greens are more difficult to decontaminate from foodborne pathogens than other types of fresh produce, and the internalization of these pathogens into the tissue of fresh produce also impedes the efficacy of sanitizers. The chapter describes the use of sodium hypochlorite, peracetic acid, chlorine dioxide, electrolytic oxidizing water, acidified sodium chlorite and ozone as sanitizers of leafy greens. It describes their modes of action as well as their effectiveness and potential drawbacks in inactivating foodborne pathogens or native background microbiota.

The subject of Chapter 8 is advances in using heat for disinfection/disinfestation of horticultural produce. Research into the development of pre-storage heat treatments is continuing for both disinfection (postharvest decay) and disinfestation (insect control) outcomes. This is driven by loss in availability or efficacy of postharvest fungicides, or treatments for insects such as methyl bromide. The chapter summarises the range of methods of heat treatments, current research, the mode-of-action based on the product, pest and pathogen responses, and information on commercial use of heat treatments. The main commercial uses are hot air treatments of tropical fruit for fruit fly control, and control of decay by hot water treatments for citrus, organic apples, and tropical crops. There is increasing use of short duration treatments (e.g. hot water brushing or dips) for decay control and a broad range of heat treatments for bulb crops have been commercialized in the Netherlands. Cost of treatment is a key limitation, and technical innovations would improve commercial uptake.

Chapter 9 reviews advances in the use of irradiation for the market access of fresh horticultural produce. Commercial and research interest in the irradiation of fresh fruit and vegetables as a market access treatment is growing. Irradiation has been approved for use in many markets as an approved phytosanitary/quarantine treatment to allow international trade. The chapter focuses on the effects of irradiation on key aspects of citrus fruit quality (internal and external quality and vitamin C content). It also considers the safety of irradiated food, uses of irradiation in postharvest horticulture, and the current state of trade in irradiated horticultural produce. The chapter reviews the effects of irradiation on citrus fruit quality. Finally, the chapter looks ahead to future research trends in this area and suggests further reading on the subject.

Moving on to Chapter 10, the chapter discusses advances in the potential use of non-thermal plasma in postharvest treatment of fresh horticultural produce. Fresh horticultural produce related foodborne illnesses outbreaks have increased around the world, and despite the adoption of good on-farm and postharvest practices, microbial contamination of fresh produce can still occur. Non-thermal plasma (NTP) has recently emerged as an option for killing/inactivating a broad range of microorganisms such as bacteria, moulds, yeasts and viruses. The reactive oxidizing species present in the NTP can inactivate and/or eliminate microbial pathogens on the produce surfaces leading to improved food safety and decay control. The chapter provides an overview of the NTP technology, its potential applications in postharvest horticulture and barriers to commercialization. The technology has great potential to be a rapid, chemical-free, dry, environmentally friendly, and effective antimicrobial solution for food safety and decay control. Adoption of this technology will minimize microbial contamination, mitigate postharvest losses and reduce the amount of postharvest chemicals and water used, resulting in improved environmental health and sustainability.

The next chapter examines advances in the use of ozone in the disinfection of horticultural produce. Fruit and vegetables are recognised for their health-promoting properties, while at the same time being associated with numerous bacterial illnesses caused by foodborne pathogens. A sanitation step is crucial to ensure that the produce is safe for consumption. Ozone has been identified as a feasible solution for the fresh produce industry, and this chapter reviews the most up to date in-formation on the practical use of ozone against microorganisms (bacteria and fungi), and its effect on the fresh produce characteristics. The chapter describes ideas for future studies on the post-harvest use of ozone. Finally, the chapter directs those interested in the subject where to look for further information regarding ozone.

Chapter 12, the final chapter of Part 2, focuses on advances in the use of biological control agents in the disinfection of horticultural produce. The chapter discusses plant extracts and animal-based materials as postharvest treatments, the discovery of microbial agents with non-chemical postharvest treatments as well as the development and commercial application of postharvest biological control products.

Part 3 Monitoring and management

Chapter 13 discusses the monitoring of postharvest attributes, focusing primarily on instrumental techniques for measuring harvest maturity/fruit quality. The chapter begins by reviewing measurement technologies relevant to postharvest management, both in the context of technologies in current industry use and those with apparent promise for future use. The chapter also provides a brief description of these techniques in relation to their theory of operation, their limitations and potential alternative methodologies. The use of these measurements technologies is discussed in a wider context, with attention given to fruit and vegetable specifications, the point within the value chain at which the various measurements can be made, sampling and sorting statistics, communication options for remote logging, and labelling options. This information should be of practical use to those working in postharvest chains as well as to students and researchers involved in postharvest measurements.

The next chapter examines postharvest handling of organically produced specialty crops. Worldwide demand for organically produced specialty crops continues to increase. Both conventional and organic postharvest handling practices require a comprehensive approach from selection of plant material through production, harvest, storage, and marketing. Chapter 14 discusses the differences in organic and non-organic postharvest approaches. It provides a comparison of organic postharvest regulations worldwide as well as the different practices that are used. The chapter concludes with an overall summary of how postharvest handling of organically produced specialty crops needs to be improved as well as highlighting potential future research trends.

The final chapter of the book focuses on smart distribution to maintain shelf life of horticultural produce. Most fresh horticultural crops in developed countries that are harvested but not consumed (40 to 50%) are wasted because the quality is inadequate for marketing or does not meet consumer quality requirements. Reducing quality deterioration can both extend shelf life and reduce waste of fresh horticultural crops in the food distribution system. Monitoring quality and environmental conditions during food distribution and using the collected data to predict the remaining shelf life of produce at any point in the cold chain enables use of smart distribution management. Monitoring environmental conditions in order to calculate shelf life can also help identify where the cold chain can be improved. Chapter 15 highlights how optimising management of produce distribution systems can reduce waste and loss and also extend shelf life by minimising quality deterioration. It focuses on the logistics and supply chain management of fresh horticultural produce, and also reviews shelf life modelling of fresh produce and remote sensing of food quality and the environmental factors that influence food quality.

Part 1

Preservation techniques

Chapter 1

Advances in cooling technologies to preserve horticultural produce

J. R. Olatunji and A. R. East, Massey University, New Zealand

1 Introduction

Fruit and vegetables are living organisms that continue to respire after harvest, deteriorating over time, in processes known as ripening and senescence. Although it is impossible to halt senescence, removing the field-heat and reducing the temperature of the product at the time of harvest has substantial effects on maintaining product integrity (Boyette et al., 1989).

A supply-chain that incorporates temperature control is often referred to as the 'cool-chain', and implies a supply chain that consists of a series of unbroken storage and transportation stages maintaining the same low temperature conditions (Hundy et al., 2008). Cooling is the first stage of the horticultural cool-chain, and is designed to rapidly and efficiently reduce the product temperature from the field temperature (often 20–30°C, depending on local conditions) to storage temperature (usually between 0°C and 15°C, depending

http://dx.doi.org/10.19103/AS.2019.0055.01

on the specific product; Fraser, 1998; Brosnan and Sun, 2001a; Pathare and Opara, 2014). In general there are five methods applied to cool horticultural products:

- Static cooling;
- Forced-air cooling;
- Hydro-cooling;
- Vacuum cooling; and
- Icing.

Each of these cooling technologies will be discussed in this chapter. Choice of a suitable technology is a balance between:

- desired speed of cooling;
- availability of potable water and ability of the product to withstand contact with water (or ice);
- homogeneity of cooling;
- capital and operational cost - including cost of not cooling rapidly; and
- consideration of the volume of product to be processed.

Many of the advances in cooling technologies are focussed on managing and improving cooling outcomes and minimising costs. Each of these factors will be discussed in this chapter, along with a case study on the management of cooling operations for the kiwifruit industry in New Zealand.

2 Cooling methods and choices

Pre-cooling is a dedicated unit operation designed to rapidly and efficiently reduce the temperature of freshly harvested produce to the storage temperature (Brosnan and Sun, 2001a; Thompson et al., 2008). Concentrating capital and operational expenses into this first stage of the cool chain enables portable refrigeration systems - such as shipping containers and trucks - to operate successfully, allowing their physical size and refrigeration capacity to be reduced (Tassou et al., 2009). Appropriate pre-cooling can also reduce the size of cool storage refrigeration capacity by one-quarter to one-third (ASHRAE, 2010). Small improvements in energy efficiency translate into large overall savings, as pre-cooling is an energy intensive stage of the cool-chain (James and James, 2010), representing approximately 40% of the entire cool-chain energy usage (McCormick et al., 2010).

Pre-coolers are designed around the principle of heat transfer through contact between the warm product and a refrigerated fluid - typically air or water. The specifics of each different pre-cooler design is based on the needs of the product being cooled, the tolerance to contact with water, and the capital and operating costs (Thompson et al., 2008). Few studies compare all the options for the same product, with Cortbaoui et al. (2005) being an exception. Pre-cooler performance is reliant on the thermal properties of the chosen cooling fluid, and in particular on how it is distributed to effectively interact with the warm product. Therefore, advancements in pre-cooling technology have come from modelling and monitoring air and water velocity profiles and product temperatures, so that a preferable fluid distribution can be imposed. A large proportion of the research effort has focussed on improving air cooled systems due to their commercial popularity, large energy demand and comparatively slow cooling rates. Consequently, the focus has been on obtaining performance gains through both reducing cooling times and energy costs of operating air cooled systems often via improving air distribution through horticultural package design (Ferrua and Singh, 2009a; Ngcobo et al., 2011; Defraeye et al., 2014, Berry et al., 2016; O'Sullivan et al., 2016; Han et al., 2017).

All packaging materials and horticultural products can withstand contact with air, making air cooling systems the most common type. The simplest air cooler is a static cooler, also known as room cooling. In many ways, static cooling is identical to a standard cool storage operation: pallets and bins of product are placed into a refrigerated room, and fans circulate cold air throughout the space (Fig. 1a). Rooms used as static coolers may be required to have larger refrigeration capacity than normal cold storage rooms in order to achieve the initial reduction in temperature of the warm and rapidly respiring product. An additional engineering strategy is to have fans with a large range of speed control, so that a higher air velocity can be used during initial cooling, with lower speeds used once product has reached storage temperature (Yost, 1984;

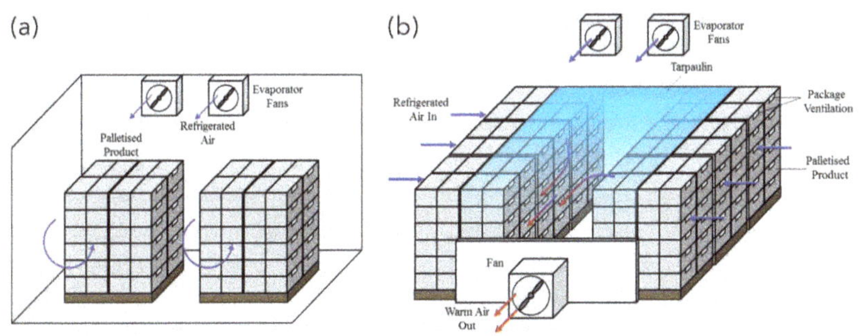

Figure 1 (a) Static or room cooler for horticultural produce. Fans circulate refrigerated air around the room, exchanging heat with pallets stacked inside; (b) A tunnel cooler, a common forced-air cooling set up. Pallets of horticultural produce are stacked into two rows and covered with a tarpaulin, where a fan is used to create a vacuum in the cavity to draw refrigerated air through the palletised produce.

Koca and Hellickson, 1993; East et al., 2013; Ambaw et al., 2016; Gruyters et al., 2018a). As there is little direct contact between the product and the refrigerated air, static cooling is the slowest method, with typical cooling times on the order of weeks (Sargent et al., 2007). Berry et al. (2013) showed that levels of cooling heterogeneity are high in static cooled produce, because fruit nearest to the pallet surface cools more rapidly than fruit in the centre of the pallet. Hoang et al. (2000) and Delele et al. (2009) used computational fluid dynamics (CFD) to determine airflow patterns inside cool storage rooms. Air leaves the evaporator fan along the ceiling and returns along the floor in the opposite direction, a pattern that favours pallets further away from the evaporator fan, to the detriment of pallets nearest to the evaporator.

A more advanced air cooler is forced-draft cooling, also known as forced-air cooling. This operation uses a combination of fans and structures to create a pressure drop across a pallet or bin, forcing refrigerated air to flow through the packaging structure and directly over the warm product (Fig. 1b), promoting faster and more uniform cooling. The universality of air cooling systems coupled with the improved cooling time makes forced-air cooling a popular pre-cooling method in the horticulture industry. An additional advantage of applying this method is that it can be applied after packing and grading, meaning that the energy invested in cooling is only expended on product that is known to be of saleable quality. However, forced-air cooling requires considerably higher handling and electricity costs (in comparison to room cooling) that must be balanced against product quality losses associated with slower cooling (Thompson et al., 2008).

Direct interaction of higher airflow velocities with the product may result in greater water loss from the product, and a higher potential for shrivelling or wilt. Even though the rate of moisture transfer is higher than in static cooling, there is a considerable reduction in the time in which the difference in partial pressure of water inside the fruit and the environment is large (Talbot and Baird, 1991), so that the total amount of moisture transfer is less (O'Sullivan et al., 2016; Huang et al., 2017). Some products require additional protection from moisture loss either during cooling or in the remainder of the cool chain, resulting in many products often being packaged in a polyliner bag. In other cases, significant components of packaging are utilised for the purposes of preparation for consumer presentation and sale (e.g. berryfruit). In either case, these additional packaging components can have large influences on cooling performance (Ferrua and Singh, 2011; Delele et al., 2012; O'Sullivan et al., 2016; Ambaw et al., 2017). The pattern of refrigerated airflow through the pallet is determined by the design of the packaging - specifically the size, shape, placement and alignment of the ventilation (Berry et al., 2016; Defraeye et al., 2014). As a result, package design has a large impact on the performance of forced-air coolers (Fadiji et al., 2016; Berry et al., 2017; Fadiji et al., 2019). As the most common and spatially complex cooling method, forced-draft cooling as a technique dominates the research landscape, with more details discussed later in this chapter.

Water cooled systems, also known as hydro-cooling, attentively utilise water as the cooling fluid, by either submerging or spraying the product (Bachmann and Earles, 2000; Sargent et al., 2007; Thompson et al., 2008). Water has favourable thermal properties compared with air (in terms of thermal conductivity and heat transfer coefficients), and as a result cools product at rates near to the theoretical maximal rate. Exposure to moisture can however stimulate decay development in many fresh produce and/or the water can become an effective medium for spreading infection, meaning that application of this method can be unsuitable for achieving optimal storage life for many products. Icing is also an alternative hydro-cooling option, for the few products that can withstand direct exposure to ice without resulting in freezing damage or inducing rot development (as also occurs with traditional hydro-cooling), such as broccoli, sweet corn, carrots, brussel sprouts and cantaloupes (Vigneault and Goyette, 2001). Icing is achieved by directly pumping slush ice into the packaging system containing the product, which remains and drains from the product when it enters the supply chain. This method hence introduces other issues, including the need to use packaging that maintains mechanical strength when wet, the requirement to safely drain and discard the melt water, and significantly increasing the mass of the packaging system. Where hydro-cooling or icing is a suitable choice, these systems have seen little advancement, with innovations instead addressing issues tangential to cooling performance, such as appropriate sterilisation of recirculated water for water cooling (Vigneault et al., 2000) or using the water as a vector for other chemical treatments (Rivera et al., 2006) and the development of new packaging that can withstand submersion or ice (Vigneault and Goyette, 2001; Vigneault et al., 2004).

The remaining cooling operation of note is vacuum cooling. This operation is distinct from the others as it does not rely on a cooling fluid, instead a vacuum cooler lowers the pressure surrounding the product to a pressure where the surface water evaporates, removing the latent heat required for the phase change from the product. This technique is less common for horticultural products but has potential to be rapid, homogeneous and cost effective - as such, advances in this area are explored in further detail later in this chapter.

Deciding which pre-cooling method to use is a balance of many different factors. Air coolers are suitable in almost all cases, with the static cooler being the most cost effective operationally, having low electricity and handling costs. However because of the extended cooling time, they can require a substantially larger capital and land requirement to house and cool the same volume of product (Boyette et al., 1990) and result in higher levels of fruit loss due to senescence. Hydro-cooling and icing are very rapid in terms of cooling but can only be applied to products that can withstand contact with free moisture without inducing decay development. Vacuum cooling requires expensive specialist equipment, and introduces handling costs as a result of the batch processing nature of the operation. For some products, such as most leafy vegetables, cooling through this operation can be economically and operationally viable.

3 Advances in package design to influence cooling

Packaging is essential for the success of the horticultural supply chain, as it creates a physical barrier between the vulnerable product and the external environment, protecting it from mechanical injury. Packaging also facilitates the mass scale transport of product through palletisation. The design of packaging can impact the performance of postharvest cooling operations because it also acts as a physical barrier between the product and the cooling medium. For example, in forced-draft cooling, the nature of the package ventilation determines the airflow pattern inside of the pallet structure (Ferrua and Singh, 2009b), and different airflow patterns result in different global and local rates of convective heat transfer (O'Sullivan et al., 2016). Since forced-air cooling is the most common horticultural pre-cooling process, a significant proportion of the research effort in recent years has focussed on how to optimise this process through package design.

The link between packaging and cooling performance has been studied experimentally as early as 1937 (Smith, 1937). Optimising the package design from a purely experimental approach has proven to be difficult, as the size and cost of manufacturing and testing many different package design iterations is prohibitive. This approach also does not allow a thorough analysis of the physical mechanisms, such as conduction, convection, evaporation and radiation, that combine to create a specific cooling rate. As a result, designing packaging through physical experimentation of potential designs becomes a 'black box' process, where the performance of a new design can be compared to previous iterations, yet understanding as to why one design is superior to another relies on theory and rules of thumb. One such key finding was the existence of a maximum vent size that enables efficient cooling. While increasing the total opening area (TOA) allows more refrigerated air into the system, this sacrifices the structural integrity of the box. However, de Castro et al. (2004) and Pathare et al. (2012) showed that cooling benefits were approximately maximised between 7–14% TOA, depending on the product; though the mechanism behind this phenomenon was speculative.

Improvements in computational power has stimulated increasing use of mathematical modelling as a design tool. A large number of CFD and finite element models (FEM) accurately predict global and local cooling rates (Fig. 2; Zou et al., 2006; Ferrua and Singh, 2009b; Delele et al., 2012, 2013; Han et al., 2015; Berry et al., 2016; O'Sullivan et al., 2016); as well as predict package strength (Han and Park, 2007; Fadiji et al., 2016, 2018). Mathematical models have advantages over physical testing in that a validated model can test an endless number of design iterations in a digital environment, where material costs are free; and many different designs can be investigated simultaneously. Some of the most notable work in which this ability to assess many package designs is demonstrated in Delele et al. (2013), Berry et al. (2016) and Han et al. (2017).

Modelling provides insights into the physical mechanisms of forced-air cooling. Ferrua and Singh (2009b) and Delele et al. (2013) used conjugated CFD models to propose a mechanistic reason for the existence of a maximum vent size. Although a larger amount of air passes through the system at higher vent openings, a high percentage bypasses the product; and air that does not

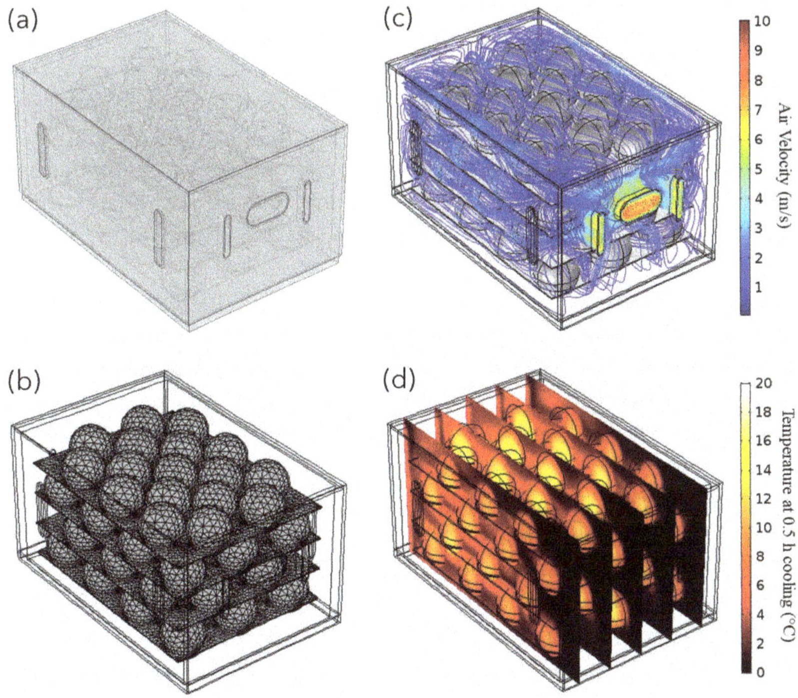

Figure 2 Example of a conjugated momentum and heat transfer mathematical model for forced-air cooling for apples in trays and packaging. (a) model geometry of apples and packaging; (b) finite element mesh of model geometry; (c) predicted airflow velocity streamlines; (d) predicted temperature profile at 0.5 h cooling. Model created using COMSOL Multiphysics.

bypass is warmed as it exchanges heat with the fruit. Warmer air then interacts with fruit at the back of the box or pallet, resulting in a relatively reduced temperature gradient between the air and fruit (in comparison to the front of the pallet), resulting in no significant increase in rate of heat transfer (O'Sullivan et al., 2016). Redding et al. (2016) also demonstrated with a simple analysis that as air velocity (and hence heat transfer coefficient) increases, the Biot number becomes greater than 1, meaning that the major resistance to cooling becomes the conduction within the fruit rather than the removal of the heat from the surface of the fruit. As a result, increasing air velocity becomes a diminishing return.

Ferrua and Singh (2009c) also used modelling to demonstrate the importance of evaporation as a cooling mechanism. Even as little as 0.5% moisture loss contributed up to 27% of the total cooling for strawberries in clamshells, a phenomenon that is difficult to experimentally measure, as the quantities of mass transfer are small. Through CFD modelling of the forced-air cooling process, Ferrua and Singh (2009a) and Defraeye et al. (2013) demonstrated the importance of bypass airflow. Previously it was thought that bypass air should be minimised as much as possible, but it was shown that some level of bypass was an important mechanism for delivering fresh refrigerated air deeper into the packaging structure, promoting faster and more uniform cooling. Natural convection is commonly excluded from the modelling process as forced convection is much more significant, however O'Sullivan et al. (2016) showed numerically that it was a vital heat transport phenomenon in systems that encapsulate air (i.e. those with multiple packaging components).

While mathematical modelling has proven to be integral to the optimisation of packaging for horticultural cooling processes, a number of significant issues must be addressed for it to continue to be a successful design tool. Firstly, constructing models (e.g. Fig. 2a and b) is still a very time-consuming and arduous task. It is notable that the first authors of most recent publications are PhD students, suggesting that at least a year of time investment is required. Even once created, models are often case specific, so that results from a model of an apple box cannot be directly applied to a box of grapes, leading to the need for yet another model to be developed. Once a validated model has been developed, it can be time consuming to adapt the model to new solutions (designs), and can be time consuming with simulation times on the scale of days or weeks. This means the usefulness of mathematical models as an iterative design tool remain limited at this time. A focus on developing simplified models with a lower level of fidelity, but not below the threshold of providing useful results, is an alternative strategy. For example, Tanner et al. (2002) and Olatunji et al. (2016) created a generalised zonal model for pre-cooling operations that provided rapid solutions; and Dehghannya et al. (2011)

used a simplified 2-D CFD model to demonstrate the impact of vent number and location on cooling uniformity.

The impact of radiative heat transfer is still unexplored because of the computational load, though some groups have shown it to be insignificant (Ferrua and Singh, 2009b; Delele et al., 2013; Defraeye et al., 2014). As computational power continues to increase, radiation should start to be included in conjugated heat and momentum transfer models.

4 Advances in assessing homogeneity

The performance or efficiency of different cooling operations are typically compared by reporting the average cooling rate, using metrics such as the half-cooling time (HCT) or seven-eighths cooling time (SECT; Brosnan and Sun, 2001a). While important, when reported in isolation it ignores other factors that impact efficacy, such as the cooling heterogeneity. Differences in cooling rates and hence temperatures between individual products can potentially perpetuate product variability in the batch. For example, products that cool much faster than the average cooling rate may have an increase risk to developing chilling injury (Zhao et al., 2014), while other products that cool slower than average may have a reduced shelf life or quality due to advanced senescence (Nunes et al., 1995; Jackson et al., 1999; Paniagua et al., 2013).

Development of new advanced methods for reporting heterogeneity have been aided in part by an increasing capability to measure a wide spatial sample of temperatures - thanks to an increase in the quality and reduction in price of temperature measurement equipment, and an increased use of mathematical models that can predict time-temperature data at very minute levels of detail.

There have recently been advances in how heterogeneity is reported and compared beyond the legacy method of using the relative standard deviation (Olatunji et al., 2017). 'Defraeye's method' was introduced in Defraeye et al. (2013). This method utilises CFD modelling outputs to calculate the effective convective heat transfer coefficient (CHTC) on the bulk surface of the product during forced-air cooling at time $t = 0$:

$$h = \frac{q_{c,w}}{T_w - T_{ref}} \tag{1}$$

Where h is the CHTC ($W \cdot m^{-2} \cdot K^{-1}$), $q_{c,w}$ is the convective heat flux at the air-fruit interface ($J \cdot s^{-1} \cdot m^{-2}$) at time $t = 0$, T_w is the initial fruit temperature (°C) and T_{ref} the initial refrigerated air temperature (°C). Plotting a histogram of the predicted CHTC reveals the evenness (or otherwise) of the cooling process. Plotting multiple histograms of different package designs or airflow velocities and comparing the shape of the different distributions of the CHTC allows conclusions to be drawn about the relative levels of cooling heterogeneity. Defraeye's method is very detailed, with minute differences in the levels of heterogeneity appearing in the histograms of the CHTC between processes. As such, it is increasingly being used in the literature (Berry et al., 2016; Gruyters et al., 2018b).

Another new heterogeneity metric, the overall heterogeneity index (OHI) was proposed in Olatunji et al. (2017). Using fruit time-temperature data, the OHI reports the variability of cooling rates throughout the entire cooling process as a dimensionless number. The OHI is calculated by considering the range of the fractional unaccomplished temperature change of each data location to the average:

$$OHI = \int_{0}^{1} \int_{0.125}^{1} |\Delta Y| \, d\bar{Y} dF(\Delta Y) \quad (2)$$

Where ΔY is the fractional unaccomplished temperature difference and $\bar{Y}$ is the average fractional unaccomplished temperature change. The OHI has a lower limit of 0 (perfect homogeneity) and no upper limit. Therefore, when reporting the OHI of two or more systems, the lowest value is a more uniform process and can be reported as superior in terms of cooling homogeneity. Fruit time-temperature data can be supplied from experimental cooling data or from a mathematical model. This method has been used by several groups (Mercier et al., 2017a,b; Han et al., 2017; Defraeye et al., 2017; Gruyters et al., 2018b).

5 Fruit simulators

Measuring the performance of horticultural cooling operations are typically performed using real products, to create a close representation of the industrially relevant cooling process. It has been recently suggested that use of real horticultural produce as the thermal mass can limit the speed and quality of research outputs (Redding et al., 2016; Defraeye et al., 2017; Huang et al., 2017). Horticultural products are variable in size and shape (Moreda et al., 2012), and fruit of the same cultivar can have variable thermal properties (specific heat capacity, thermal conductivity and density) due to environmental factors during fruit growth (Alcobendas et al., 2013). The difference between two cooling operations could be a result of these physiological differences, rather than the intervention being studied by the researcher.

Horticultural products are also perishable. Therefore the time during which experiments can be performed is limited to a finite time window every season, therefore restricting researchers to performing only a few cooling experiments per year. Perishability also potentially impacts the cooling rate - fruit closer to the end of their shelf life may cool at different rates than freshly picked fruit due to differences in the levels of moisture loss (ASHRAE, 2010).

Several researchers have used 'fruit simulators' in lieu of real fruit. A fruit simulator is a shelf-stable object that mimics the shape, size and thermal properties of the fruit. De Castro (2004) used plastic spheres filled with agar gel and stacked them into a ball matrix inside of a package to simulate a packed bed. The impact of a wide variety of package ventilation orientations were able to be empirically measured on a single box scale using this method. Defraeye et al. (2017) engineered a more realistic fruit simulator for apples. The fruit simulator consisted of a shell of polyamide, moulded to the precise shape of a Braeburn apple, appropriated via CT scanning. The shell consisted of two halves, so that temperature monitoring equipment could be placed inside the fruit simulator; and the shell was hollow, so that it could be filled with a substance with a precise thermal mass (in this case, a water-carbohydrate-air gel mixture). The thermal response of the fruit simulator was to within 3% of real apples.

Redding et al. (2016) concluded that the most important thermal properties to mimic in a fruit simulator are the shape and thermal conductivity. Even if the fruit simulator has a different heat capacity and/or density, and hence has a different thermal response to real fruit, the cooling rate can be adjusted via dimensionless analysis to be compared with parity to real fruit, so long as the thermal conductivity and shape are comparable. This is particularly important when selecting the material the simulator will be manufactured from, as a perfect match of all thermal properties may be difficult.

Large scale industrial cooling trials consisting entirely of fruit simulators could be explored. Though the initial cost of using fruit simulators is much higher than real fruit, the shelf stability means the same batch of simulators can be used without time restriction for many years, so there is a reasonable timeframe for return on investment, with the data likely being more reliable, consistent and comparable.

6 Cooling impact on long-term product outcomes

Cooling is merely a means to an ends. The true goal of postharvest cooling is the preservation of produce quality in preparation for transport through the cool-chain. Typically it is considered that more rapid cooling results in better product outcomes. A slow cooling process is considered detrimental to product quality, as extended periods of time at elevated temperatures increases degradation due to the elevated rates of respiration, and encourages pathogenic micro-organisms to develop. This is supported by numerous studies demonstrating impacts of delays to start cooling on long-term product outcomes (Nunes et al., 1995; Liang et al., 2013; Toivonen, 2015).

Cool temperatures are also however known to induce physiological disorders in fresh produce known as 'chilling injuries', that result in disfiguring browning, blemishes or unusual changes in firmness that lead to the product become unsaleable (Biswas et al., 2016). Although the general principle of cooling as rapidly as possible remains, a growing body of evidence suggests a more complex relationship between timing of cooling initiation, cooling rates and the consequent development of chilling injuries in storage. In particular, for products that are generally known to store well for long periods at close to freezing temperature (e.g. apples, pears and kiwifruit), incidences of physiological disorders observed are being more commonly linked not only to temperature and time in storage but also the means in which cooling was performed. For example, Watkins et al. (2005) reported that a 7-day delay at 10°C reduced subsequent development of soft scald and low temperature breakdown in 'Honeycrisp' apples. For kiwifruit, a number of studies have demonstrated that gradual cooling results in a lower propensity for developing chilling injury after subsequent storage (Yang et al., 2012; Zhao et al., 2014). Similar responses have been reported for tomatoes (Galvez et al., 2010). The potential exists for any crop to have a 'goldilocks' cooling zone, where the best possible after storage quality is achieved by not cooling too quickly or too slowly either.

As a result of the observed rapid cooling effects on chilling injury development, 'step down cooling' or 'low temperature conditioning' treatments, in which fruit are exposed to cool temperatures different from the storage temperature remain an active part of postharvest research. Examples of work in which treatments have been demonstrated to be beneficial include for avocado (Woolf et al., 2003), eggplant (Shi et al., 2018) and grapefruit (Chaudhary et al., 2014). These treatments do create additional challenges in achieving the multistage cool down in an industrial setting. The vast majority of this work focuses on staged cooling, where a question remains over whether the speed of transition (between planned temperatures) may also contribute to chilling injury development.

7 Advances in vacuum cooling

Vacuum cooling is the technique of pulling a vacuum around the produce to a pressure that induces boiling of the water from the surface of the product. In the act of boiling, latent heat is removed from the product surface, cooling the product and requiring a further pressure reduction to induce further cooling. Vacuum cooling has long been known as the most energy efficient (Thompson et al., 2008) and rapid pre-cooling process (Sun and Zheng, 2006). However, it remains relatively rare in an industrial setting due to the high capital and handling costs, comparably smaller batch sizes (Brosnan and Sun, 2001a) and risk of product damage due to water loss.

Given that the primary mechanism of cooling is phase change on the surface of the product, this technique is favoured for products with a high surface area to volume ratio - most notably lettuce (Kongwong et al., 2019), but also other leafy vegetables and cut flowers (Brosnan and Sun, 2001b). Assuming a product specific heat equal to water, for every 1°C change, approximately 0.17% loss of weight is required (via water evaporation), exposing the product to a significant risk of weight loss issues, such as shrivel or wilt.

Zhu et al. (2019) reviewed the recent advances in vacuum cooling technology, as well as improvements in operational conditions to enhance the performance of the process. One such advance is water spraying. This technique sprays water onto the surface of the product prior to being placed into the cooling chamber (Brosnan and Sun, 2001b) or incorporates spray nozzles into the cooling chamber itself (Ding et al., 2016), so that this free moisture evaporates during cooling instead of the moisture inside the product. Ding et al. (2016) reported that this technique lowered total moisture loss for vacuum cooling to a degree comparable to other pre-cooling operations for broccoli; and Brosnan and Sun (2001b) eliminated weight loss in freshly cut flowers.

Another notable advance highlighted by Zhu et al. (2019) is modified atmosphere vacuum cooling (MAVC). During the pressure recovery stage when atmospheric pressure is restored, rather than re-filling the chamber with air, different gas mixture combinations are used instead. This allows gas treatments that were previously a separate operation to be integrated into the cooling process. For example, Vurma et al. (2009) re-pressurised vacuum cooled spinach leaves with ozone, the intention being that O_3 is a disinfectant that naturally decays into non-toxic O_2, combining the pre-cooling and sanitation stages. Ozone treatment reduced the population of inoculated *E. coli* O157:H7 by approximately 2 log CFU/g compared with the untreated control samples. Zhu et al. (2018) re-pressurised vacuum cooled cabbages using air with a higher CO_2 concentration, and maintained the modified atmosphere for 21 days, and an improvement in the shelf life and sensory quality was observed compared with control samples.

Analytical optimisation techniques have also helped in the advance of vacuum cooling systems. Tian et al. (2014) showed a wide variability in weight loss and final cooling temperature for vacuum cooled broccoli when changing the operational variables of product weight, vacuum pressure, water-spraying volume and processing time. An optimisation algorithm then chose the best combination of operational parameters to minimise moisture loss and final product temperature. Santana et al. (2018) used a similar experimental approach for broccoli, but found the optimal processing conditions using a genetic algorithm.

As highlighted in section 3, mathematical models of pre-cooling processes can be useful in the optimisation of pre-cooling processes. So far, mathematical models for vacuum cooling of horticultural produce have been scarce, exceptions being He and Li (2003) who produced a cooling model for a singular spherical product and Sun and Hu (2002) who produced a CFD model predicting the temperature profile inside of vacuum cooled meat. Therefore, there is a potential for further advances to be discovered through the development and use of FEM that use realistic fruit or vegetable geometries and have an industry relevant scale.

8 Horticultural cooling in developing countries

Cold chain technology had developed in large part in the United States and Europe for horticultural products since the early 1950s (Kitinoja, 2013). Despite the rapid growth of the developing world since then, and the subsequent increase in food production (Szirmai and Verspagen, 2015), cold chains remain under developed in most developing countries (Gligor et al., 2018), with over 55% losses in some markets (Jaspreet et al., 2013). A lack of refrigeration (and hence cooling capability) limits shelf life, restricting access of the product to local markets. This represents an economic bottleneck and also risks the nutritional wellbeing of those at the very bottom of the socioeconomic global ladder.

Gligor et al. (2018) studied the cold chain in the developing nation Vietnam, and found that one of the major roadblocks was a lack of expertise. Access to higher education is limited across the developing world, so that there are few engineers with the required knowledge to design cold chains effectively. They suggested that international cooperation between the first and third world could bridge this gap. Capital expenses are a significant barrier: a cooling facility capable of storing 1000 pallets of produce can cost USD $670 000, and a single refrigerated truck can cost USD $40 000. Comparing this to the Vietnamese GDP per capita of USD $2343 in 2017 (World Bank, 2019) these costs can be prohibitive.

Producing effective refrigeration in developing countries is an active area of research and has produced many innovative solutions to date. These solutions have low start-up costs, utilise natural phenomenon to induce cooling (such as evaporation) instead of electricity, and adapt equipment typically used in other industries to become useful to the horticultural sector. One such innovation is Coolbot™ (www.storeitcold.com). This promising new device converts a residential air conditioner into a commercial refrigeration unit for horticultural produce. A Coolbot™ is a small electronic controller that is attached to an air conditioner. While air conditioners typically do not cool below 18°C, the Coolbot™ can force the air conditioner to produce air temperatures as low as

0°C. It has a sensor which is placed on the air fins to detect and prevent ice build-up. Within small insulated rooms, a standard air conditioner combined with a CoolbotTM can produce temperatures much lower than ambient for a fraction of the price of traditional refrigerated cool stores; for example, Karithi (2016) produced a 10°C storage temperature in just 6 h, and maintained this lower temperature for mangos in Kenya (ambient temperature of 30°C) for 40 days, improving the shelf life and quality outcomes of the fruit.

Another promising refrigeration technology that is gaining traction in developing countries is passive evaporative refrigeration. This technique is based around the physical phenomenon in which the evaporation of liquid water cools the surrounding air, as evaporation requires a large degree of energy to complete phase change. By placing a material that can hold or absorb water - such as sand, porous ceramics or cloth - next to or over an enclosed space, as ambient warm air or radiation from the sun evaporates the water, air inside the enclosure is cooled, along with the horticultural products stored inside (Ndukwu and Manuwa, 2014). Passive evaporative coolers have been used for thousands of years for a variety of purposes, however interest in their use in the horticultural industry has grown recently (Kitinoja and Thompson, 2010), with modern technology and engineering working to increase the efficiency of passive evaporative cooling systems to a level where it can be used effectively at an industrial scale (lal Basediya et al., 2013). Systems such as these operate at peak efficiency when the climate is hot and dry, the thermodynamic limit of performance being the differential between the wet and dry bulb temperatures. Passive evaporative coolers can produce storage temperatures of 15–20°C with no electrical cost (Ndukwu and Manuwa, 2014).

There are opportunities to develop new low-cost cooling technologies for the developing world that extend beyond altruism. As the developing world contains a majority of the human population (United Nations, 2010), there is a high demand for innovative solutions to their horticultural storability problems, which should be explored by entrepreneurs and philanthropists in industry and academia.

9 Case study: Cooling of the New Zealand Hayward Kiwifruit

Kiwifruit is the largest fresh horticultural export from New Zealand. A case study and a discussion of the considerations around commercial pre-cooling for this product are provided. A large proportion of New Zealand's NZD $1.6 billion in exported kiwifruit in 2017 were 'Hayward' green fruit packed in a form referred to as modular bulk packs. Modular bulk packs are made of corrugated fibreboard and contain approximately 10 kg of fruit per box. The fruit are wrapped in polyliners (thin polyethylene bags) to prevent moisture loss and shrivelling in storage. Full details about the modular bulk pack for kiwifruit are provided in O'Sullivan et al. (2016). Fruits are packed at their field temperature (15–30°C) and stacked into pallets weighing approximately 1 tonne (100 boxes per pallet).

To fairly compare the cooling rates of different methods, the fractional unaccomplished temperature change (FUTC) is commonly used, a dimensionless temperature term that removes the bias of the starting and refrigeration temperatures (Brosnan and Sun, 2001a). The FUTC is calculated using Eq. (3):

$$Y_{t,p} = \frac{T_{t,p} - T_a}{T_{i,p} - T_a} \tag{3}$$

Where Y is the FUTC, T is the temperature at time t and position p, T_a is the refrigeration temperature, and T_i is the initial temperature at position p. The cooling rate of different cooling methods can be compared with the HCT (when $Y = 0.5$), or alternatively, if Newton's law of cooling is assumed to apply, then the profile of the average temperature can be modelled as an exponential decay function:

$$Y = e^{-Ct} \tag{4}$$

Where C is the cooling coefficient (h^{-1}), and t is the time (h). By determining the best C to fit the experimental data, the cooling performance can be assessed by comparing the different cooling coefficient values.

Cooling uniformity is also an important performance metric, as an uneven cooling profile represents a form of inefficiency. Therefore temperature heterogeneity is compared using heterogeneity plots developed by Olatunji et al. (2017).

As stated previously in Section 2, the possible pre-cooling methods are: room cooling, forced-air cooling, hydro-cooling, icing and vacuum cooling. Icing must be discounted, as kiwifruit cannot withstand freezing temperatures without significantly inducing chilling injury (Gerasopoulos et al., 2006). Hydro-cooling must also be discounted as fibreboard packaging loses structural integrity when submerged, or otherwise exposed to high levels of humidity for prolonged periods. The packaging could be replaced with wax coated fibreboard, plastic or wood - materials that can better withstand contact with water - but hydro-cooling would remain unsuitable as free moisture induces fungal spores to germinate in kiwifruit. In particular kiwifruit are susceptible to botrytis rot (Michailides and Elmer, 2000).

The cooling rates and levels of cooling uniformity of room cooling, forced-air cooling and vacuum cooling were measured and compared for pallets of Hayward kiwifruit. Room cooling was performed in an industrial setting in

Tauranga, New Zealand in 2013 (unpublished data); forced-air cooling was performed in a laboratory setting at Massey University in Palmerston North, New Zealand, with a volumetric airflow rate of 0.38 $m^3 \cdot s^{-1}$ in 2014 (details in O'Sullivan et al., 2016); and vacuum cooling was performed at the Animal and Veterinary Authority of Singapore in Singapore in 2013, where the pressure was lowered from ambient to 1.32 kPa for 4 h (unpublished data). Temperatures were recorded in all experiments by inserting thermocouples directly into the centre of a sample of fruit, with multiple (at least 5 fruit) measured in each box. Average cooling rates for each operation are shown in Fig. 3a, and the temperature difference of individual boxes is shown in Fig. 3b–d.

Vacuum cooling was observed to be the fastest cooling method, with a HCT of 1.96 h and a cooling coefficient of $C = 0.32\ h^{-1}$ (though the Newton's law cooling model appears not to apply in this scenario due to the low R^2 value); just ahead of forced-air cooling, with a HCT of 3.12 h and $C = 0.21\ h^{-1}$. The static cooler was significantly slower, with a HCT of 29 h and $C = 0.023\ h^{-1}$. These findings are consistent with Thompson et al. (2008). Though vacuum cooling showed a small advantage in terms of cooling rate, its smaller batch size, larger capital costs and increase in operational complexity makes it impractical for an industry where approximately half a million pallets of kiwifruit need to be processed in the span of just 3 months. Comparably, forced-air cooling is a much easier operation to implement, with only a small compromise in cooling rate.

All three processes were highly variable in terms of cooling heterogeneity, with a large separation in temperatures between individual boxes, even on the same pallet, throughout the cooling period (Fig. 3b–d). Vacuum cooling was particularly heterogeneous, with Box 4 at 0.2 FUTC units below average while at the same moment Box 1 was 0.15 FUTC units above average. Box position was correlated to heterogeneity in the case of forced-air cooling, with boxes nearer to the inlet of refrigerated air (Boxes 1 and 5) cooling disproportionately quickly, while boxes furthest away (Boxes 4 and 7) cooling slowly. There was not as clear a spatial cooling trend for room cooling, however. It might be expected Box 1 and Box 6 cooled at a comparable rate, as they are both externally facing and in contact with refrigerated air. However, Box 1 cooled quickly, while Box 6 lagged behind. Natural convection currents that form in static coolers, as described by Delele et al. (2009), are the likely culprit, pushing air in the direction of Box 1; this is further evidenced by Box 5 also cooling rapidly compared to the average. Vacuum cooling showed an even less clear spatial trend. Considering the only mechanism for cooling is evaporation, the fruit with the highest surface moisture content and the lowest total resistance to vapour transfer would cool faster, and there is currently no evidence to suggest this is correlated to box position on the pallet.

Comparing forced-air cooling to static cooling, recent studies have shown that very rapid cooling has the potential to increase the probability of chilling

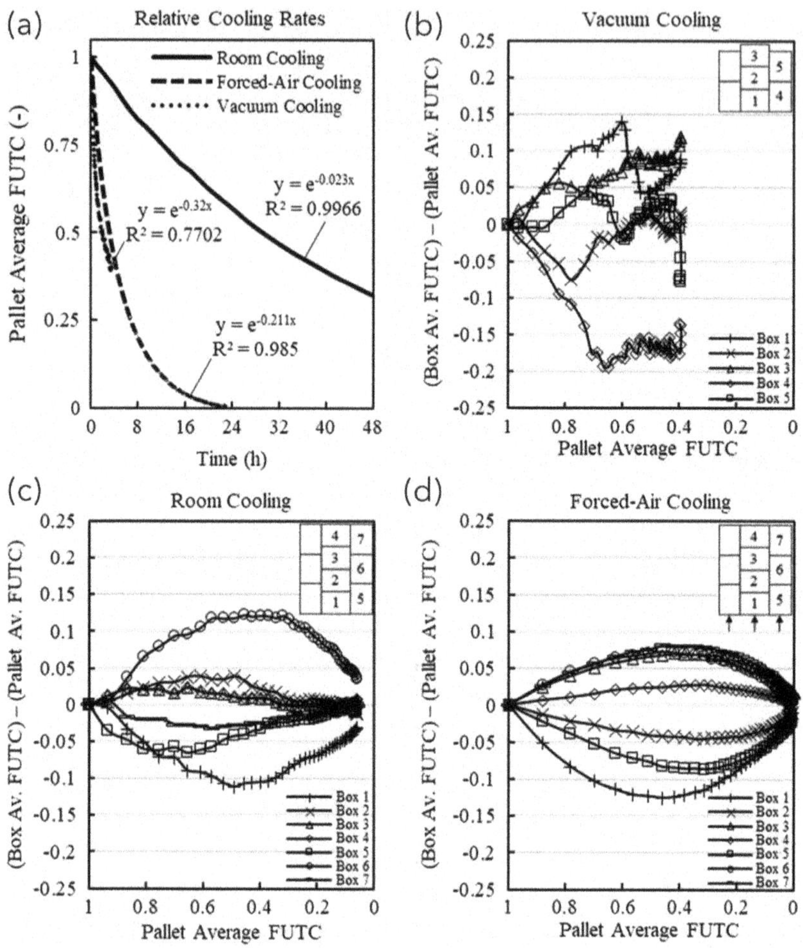

Figure 3 Relative performance of key pre-cooling operations for pallets of Hayward kiwifruit; (a) Pallet average cooling rates, with fitted cooling coefficients; temperature heterogeneity plots for (b) Vacuum cooling, (c) Room cooling and (d) Forced-air cooling.

injury (Lallu and Webb, 1997; Yang et al., 2012), while Zhao et al. (2014) also showed that a delay in cooling may reduce future rot incidence in storage. As a result, a delay at ambient conditions is designed and applied to the kiwifruit supply chain, where immediate cooling is stalled for a period of 12–24 h. This is referred to as curing, and is where fruit are left overnight in bins, prior to packaging (Pennycook and Manning, 1992; Bautista-Baños et al., 1997; East et al., 2018). Curing has the benefit of both reducing decay development and SBD development after long-term storage. Harvey and Harris (1986) demonstrated that slower cooling contributes to an increase in softening during the delay, which can be alarming for operators given that fruit firmness can be a key product attribute that limits shelf life. However, this same study also demonstrated that the difference generated during the delay in cooling diminishes during storage, and hence should not be of concern.

Given the results presented above and the discussion pertaining to the most recent research on Hayward kiwifruit, a forced-air cooling pre-cooling operation combined with curing is the best option. This strategy mitigates the chilling injury effects of rapid cooling and produces long-term storage performance. The focus of the kiwifruit industry now is on packaging design that is tailored to maximise the performance of forced-air cooling, given that this is the preferred pre-cooling method - a package design that produces the optimum airflow pathway through the pallet structure to produce rapid and uniform cooling - hence the work of O'Sullivan et al. (2016) and others (see Section 3).

10 Conclusion

Refrigeration remains a vital component of the supply of horticultural produce. Efforts to improve cooling technologies as applied to the horticultural supply chain have mainly focused on optimisation of air pre-coolers by designing new packages that produce beneficial airflow patterns for faster and more efficient cooling. The increased use of mathematical modelling has proved to be a useful design tool, allowing engineers and operators to investigate the impact of many different iterations of design parameters in a simulation environment. Mathematical modelling has also provided insight into the importance of bypass air and evaporation as significant modes contributing to accelerated heat transfer.

Considering low temperature disorders and chilling injuries, an emerging body of evidence suggests that most products have a 'goldilocks' cooling zone. Suggested solutions include multistage cool down operations and the curing of fruit before cooling. Further investigation of the existence of a goldilocks cooling zone for other horticultural products is warranted.

There is still potential for underutilised cooling technologies to supplant current methods. Vacuum cooling is often considered more rapid and uniform than air coolers, but it is not a feasible replacement until the initial set up and operations costs are reduced and moisture loss concerns are mitigated. In developing countries where access to capital is low and electricity may be scarce, innovative technologies such as Coolbot™ and passive evaporative cooling systems have potential to be the preferred refrigerated cooling and storage solution. There is an opportunity for researchers, entrepreneurs and philanthropists to invest in developing these technologies to increase their efficiency at an industrially relevant scale.

11 Acknowledgements

The authors gratefully acknowledge the New Zealand Ministry of Business, Innovation and Employment Zespri International and Oji Fibre Solutions for funding the work (Fibreboard Packaging Design Project, MAUX1302); that has led to the authors' expertise. We also acknowledge the help of Dr. Justin O'Sullivan and Dr. Young-Min Shim for planning and assisting in experimental work; and students Alicia Tan, Lyall McDonald, Julia Zhou, Angela Yang and Tim Cook for their help collecting the experimental data.

12 Where to look for further information

- Pieter Verboven and Bart Nicolai, KU Leuven, Leuven, Belgium
- Linus Opara, Stellenbosh University, South Africa
- Andrew East, Massey University, New Zealand
- Jia-Wei Han, National Engineering Research Center for Information Technology in Agriculture, Beijing, China
- Ismail Uysal, University of South Florida, Tampa, USA

13 References

Alcobendas, R., Mirás-Avalos, J. M., Alarcón, J. J. and Nicolás, E. 2013. Effects of irrigation and fruit position on size, colour, firmness and sugar contents of fruits in a mid-late maturing peach cultivar. *Scientia Horticulturae* 164, 340–7. doi:10.1016/j.scienta.2013.09.048.

Ambaw, A., Bessemans, N., Gruyters, W., Gwanpua, S. G., Schenk, A., De Roeck, A., Delele, M. A., Verboven, P. and Nicolai, B. M. 2016. Analysis of the spatiotemporal temperature fluctuations inside an apple cool store in response to energy use concerns. *International Journal of Refrigeration* 66, 156–68. doi:10.1016/j.ijrefrig.2016.02.004.

Ambaw, A., Mukama, M. and Opara, U. L. 2017. Analysis of the effects of package design on the rate and uniformity of cooling of stacked pomegranates: numerical and experimental studies. *Computers and Electronics in Agriculture* 136, 13–24. doi:10.1016/j.compag.2017.02.015.

ASHRAE. 2010. *ASHRAE Handbook: Refrigeration*. American Society of Heating, Refrigeration and Air-Conditioning Engineers, Atlanta, GA.

Bachmann, J. and Earles, R. 2000. Postharvest handling of fruits and vegetables. Horticulture technical note. Appropriate Technology Transfer for Rural Areas.

Bautista-Baños, S., Long, P. G. and Ganesh, S. 1997. Curing of kiwifruit for control of postharvest infection by *Botrytis cinerea*. *Postharvest Biology and Technology* 12(2), 137–45. doi:10.1016/S0925-5214(97)00043-4.

Berry, T., Delele, M. A., Opara, U. L. and Cronjé, P. J. R. 2013. Spatial variation of fruit cooling rate and relative humidity inside 'super-vent' packaging for citrus during static cooling. *Acta Horticulturae* 1007(1007), 207–13. doi:10.17660/ActaHortic.2013.1007.20.

Berry, T. M., Defraeye, T., Nicolaï, B. M. and Opara, U. L. 2016. Multiparameter analysis of cooling efficiency of ventilated fruit cartons using CFD: impact of vent hole design and internal packaging. *Food and Bioprocess Technology* 9(9), 1481-93. doi:10.1007/s11947-016-1733-y.

Berry, T. M., Fadiji, T. S., Defraeye, T. and Opara, U. L. 2017. The role of horticultural carton vent hole design on cooling efficiency and compression strength: a multi-parameter approach. *Postharvest Biology and Technology* 124, 62-74. doi:10.1016/j.postharvbio.2016.10.005.

Biswas, P., East, A. R., Hewett, E. W. and Heyes, J. A. 2016. Chilling injury in tomato fruit. *Horticultural Reviews* 44, 229-78.

Boyette, M. D., Wilson, L. G. and Estes, E. A. 1989. Forced-air cooling. North Carolina Agricultural Extension Service.

Boyette, M. D., Wilson, L. G. and Estes, E. A. 1990. Postharvest cooling and handling of apples. North Carolina Agricultural Extension Service.

Brosnan, T. and Sun, D. W. 2001a. Precooling techniques and applications for horticultural products – a review. *International Journal of Refrigeration* 24(2), 154-70. doi:10.1016/S0140-7007(00)00017-7.

Brosnan, T. and Sun, D. W. 2001b. Compensation for water loss in vacuum-precooled cut lily flowers. *Journal of Agricultural Engineering Research* 79(3), 299-305. doi:10.1006/jaer.2001.0704.

Chaudhary, P. R., Jayaprakasha, G. K., Porat, R. and Patil, B. S. 2014. Low temperature conditioning reduces chilling injury while maintaining quality and certain bioactive compounds of 'Star Ruby' grapefruit. *Food Chemistry* 153, 243-9. doi:10.1016/j.foodchem.2013.12.043.

Cortbaoui, P., Vigneault, C., Gariepy, Y., Goyette, B., Raghavan, V. G. S. and Charles, M. T. 2005. Assessment of precooling techniques for sweet corn. *ASAE Annual Meeting*. American Society of Agricultural and Biological Engineers. Paper 056075.

de Castro, L. R., Vigneault, C. and Cortez, L. A. B. 2004. Container opening design for horticultural produce cooling efficiency. *Food, Agriculture and Environment* 2, 135-40.

Defraeye, T., Lambrecht, R., Tsige, A. A., Delele, M. A., Opara, U. L., Cronjé, P., Verboven, P. and Nicolai, B. 2013. Forced-convective cooling of citrus fruit: package design. *Journal of Food Engineering* 118(1), 8-18. doi:10.1016/j.jfoodeng.2013.03.026.

Defraeye, T., Lambrecht, R., Delele, M. A., Tsige, A. A., Opara, U. L., Cronjé, P., Verboven, P. and Nicolai, B. 2014. Forced-convective cooling of citrus fruit: cooling conditions and energy consumption in relation to package design. *Journal of Food Engineering* 121, 118-27. doi:10.1016/j.jfoodeng.2013.08.021.

Defraeye, T., Wu, W., Prawiranto, K., Fortunato, G., Kemp, S., Hartmann, S., Cronje, P., Verboven, P. and Nicolai, B. 2017. Artificial fruit for monitoring the thermal history of horticultural produce in the cold chain. *Journal of Food Engineering* 215, 51-60. doi:10.1016/j.jfoodeng.2017.07.012.

Dehghannya, J., Ngadi, M. and Vigneault, C. 2011. Mathematical modeling of airflow and heat transfer during forced convection cooling of produce considering various package vent areas. *Food Control* 22(8), 1393-9. doi:10.1016/j.foodcont.2011.02.019.

Delele, M. A., Schenk, A., Ramon, H., Nicolai, B. M. and Verboven, P. 2009. Evaluation of chicory root cold store humidification system using computational fluid dynamics. *Journal of Food Engineering* 94(1), 110-21. doi:10.1016/j.jfoodeng.2009.03.004.

Delele, M. A., Ngcobo, M. E. K., Opara, U. L. and Meyer, C. J. 2012. Investigating the effects of table grape package components and stacking on airflow, heat and mass transfer using 3-D CFD modelling. *Food and Bioprocess Technology* 6(9), 2571–85. doi:10.1007/s11947-012-0895-5.

Delele, M. A., Ngcobo, M. E. K., Getahun, S. T., Chen, L., Mellmann, J. and Opara, U. L. 2013. Studying airflow and heat transfer characteristics of a horticultural produce packaging system using a 3-D CFD model. Part II: effect of package design. *Postharvest Biology and Technology* 86, 546–55. doi:10.1016/j.postharvbio.2013.08.015.

Ding, T., Liu, F., Ling, J. G., Kang, M. L., Yu, J. F., Ye, X. Q. and Liu, D. H. 2016. Comparison of different cooling methods for extending shelf life of postharvest broccoli. *International Journal of Agriculture and Biological Engineering* 9, 178–85.

East, A. R., Smale, N. J. and Trujillo, F. J. 2013. Potential for energy cost savings by utilising alternative temperature control strategies for controlled atmosphere stored apples. *International Journal of Refrigeration* 36(3), 1109–17. doi:10.1016/j.ijrefrig.2012.10.028.

East, A. R., Doleh, L., Jeffery, P. B., Tapia, N., Gwanpua, S. G. and Woolf, A. F. 2018. What happens in a full-scale kiwifruit curing stack? *Acta Horticulturae* 1218, 525–32. doi:10.17660/ActaHortic.2018.1218.72.

Fadiji, T., Coetzee, C. and Opara, U. L. 2016. Compression strength of ventilated corrugated paperboard packages: numerical modelling, experimental validation and effects of vent geometric design. *Biosystems Engineering* 151, 231–47. doi:10.1016/j.biosystemseng.2016.09.010.

Fadiji, T., Ambaw, A., Coetzee, C. J., Berry, T. M. and Opara, U. L. 2018. Application of finite element analysis to predict the mechanical strength of ventilated corrugated paperboard packaging for handling fresh produce. *Biosystems Engineering* 174, 260–81. doi:10.1016/j.biosystemseng.2018.07.014.

Fadiji, T., Coetzee, C. J., Berry, T. M. and Opara, U. L. 2019. Investigating the role of geometrical configurations of ventilated fresh produce packaging to improve the mechanical strength–Experimental and numerical approaches. *Food Packaging and Shelf Life* 20. doi:10.1016/j.fpsl.2019.100312.

Ferrua, M. J. and Singh, R. P. 2009a. Design guidelines for the forced-air cooling process of strawberries. *International Journal of Refrigeration* 32(8), 1932–43. doi:10.1016/j.ijrefrig.2009.07.011.

Ferrua, M. J. and Singh, R. P. 2009b. Modeling the forced-air cooling process of fresh strawberry packages, Part I: numerical model. *International Journal of Refrigeration* 32(2), 335–48. doi:10.1016/j.ijrefrig.2008.04.010.

Ferrua, M. J. and Singh, R. P. 2009c. Modeling the forced-air cooling process of fresh strawberry packages, Part II: experimental validation of the flow model. *International Journal of Refrigeration* 32(2), 349–58. doi:10.1016/j.ijrefrig.2008.04.009.

Ferrua, M. J. and Singh, R. P. 2011. Improved airflow method and packaging system for forced-air cooling of strawberries. *International Journal of Refrigeration* 34(4), 1162–73. doi:10.1016/j.ijrefrig.2011.01.018.

Fraser, H. F. 1998. *Tunnel Forced-Air Coolers for Fresh Fruits & Vegetables*. Ontario Ministry of Agriculture, Food and Rural Affairs.

Galvez, A. B., Garcia, M. V., Corrales, J. C., Lopez, A. C. and Valenzuela, J. A. L. 2010. Effect of gradual cooling storage on chilling injury and phenylalanine ammonia-lyase activity in tomato fruit. *Journal of Food Biochemistry* 34(2), 295–307. doi:10.1111/j.1745-4514.2009.00279.x.

Gerasopoulos, D., Chlioumis, G. and Sfakiotakis, E. 2006. Non-freezing points below zero induce low-temperature breakdown of kiwifruit at harvest. *Journal of the Science of Food and Agriculture* 86(6), 886–90. doi:10.1002/jsfa.2429.

Gligor, D., Tan, A. and Nguyen, T. N. T. 2018. The obstacles to cold chain implementation in developing countries: insights from Vietnam. *The International Journal of Logistics Management* 29(3), 942–58. doi:10.1108/IJLM-02-2017-0026.

Gruyters, W., Verboven, P., Delele, M., Gwanpua, S. G., Schenk, A. and Nicolaï, B. 2018a. A numerical evaluation of adaptive on-off cooling strategies for energy savings during long-term storage of apples. *International Journal of Refrigeration* 85, 431–40. doi:10.1016/j.ijrefrig.2017.10.018.

Gruyters, W., Verboven, P., Diels, E., Rogge, S., Smeets, B., Ramon, H., Defraeye, T. and Nicolaï, B. M. 2018b. Modelling cooling of packaged fruit using 3D shape models. *Food and Bioprocess Technology* 11(11), 2008–20. doi:10.1007/s11947-018-2163-9.

Han, J. and Park, J. M. 2007. Finite element analysis of vent/hand hole designs for corrugated fibreboard boxes. *Packaging Technology and Science: An International Journal* 20, 39–47.

Han, J. W., Zhao, C. J., Yang, X. T., Qian, J. P. and Fan, B. L. 2015. Computational modeling of airflow and heat transfer in a vented box during cooling: optimal package design. *Applied Thermal Engineering* 91, 883–93. doi:10.1016/j.applthermaleng.2015.08.060.

Han, J. W., Qian, J. P., Zhao, C. J., Yang, X. T. and Fan, B. L. 2017. Mathematical modelling of cooling efficiency of ventilated packaging: integral performance evaluation. *International Journal of Heat and Mass Transfer* 111, 386–97. doi:10.1016/j.ijheatmasstransfer.2017.04.015.

Harvey, J. M. and Harris, C. M. 1986. In-storage softening of kiwi fruit: effects of delayed cooling. *International Journal of Refrigeration* 9(6), 352–6. doi:10.1016/0140-7007(86)90007-1.

He, S. and Li, Y. 2003. Theoretical simulation of vacuum cooling of spherical foods. *Applied Thermal Engineering* 23(12), 1489–501. doi:10.1016/S1359-4311(03)00085-1.

Hoang, M. L., Verboven, P., De Baerdemaeker, J. and Nicolai, B. M. 2000. Analysis of the air flow in a cold store by means of computational fluid dynamics. *International Journal of Refrigeration* 23(2), 127–40. doi:10.1016/S0140-7007(99)00043-2.

Huang, H., Tunnicliffe, M., Shim, Y. M. and Bronlund, J. E. 2017. Model based development of fruit simulators. *Proceedings from 20th International ESA form Conference on Material Forming*, Dublin, Ireland.

Hundy, G. F., Trott, A. R., Welch, T. and Trott, A. R. 2008. Refrigeration and air-conditioning. Amsterdam: Butterworth-Heinemann/Elsevier. *Journal of Refrigeration* 23, 127–40.

Jackson, E. D., Sanford, K. A., Lawrence, R. A., McRae, K. B. and Stark, R. 1999. Lowbush blueberry quality changes in response to prepacking delays and holding temperatures. *Postharvest Biology and Technology* 15(2), 117–26. doi:10.1016/S0925-5214(98)00075-1.

James, S. J. and James, C. 2010. The food cold chain and climate change. *Food Research International* 43(7), 1944–56.

Jaspreet, A., Regmi, A., Joan, R. F. and Corinne, A. 2013. Estimating post harvest food losses: developing a consistent global estimation framework. *First International Conference on Food Security, Agricultural and Applied Economics Association*, 29th September – 2nd October 2013, Noordwijkerhout, the Netherlands.

Karithi, E. M. 2016. Evaluation of the efficacy of Coolbot™ cold storage technology to preserve quality and extend shelf life of mango fruits. Master's Thesis. Department of Plant Science and Crop Protection, Faculty of Agriculture, University of Nairobi, Kenya.

Kitinoja, L. 2013. Use of cold chains for reducing food losses in developing countries. *Population* 6, 5–60.

Kitinoja, L. and Thompson, J. F. 2010. Pre-cooling systems for small-scale producers. *Stewart Postharvest Review* 6, 1–14.

Koca, R. W. and Hellickson, M. L. 1993. Energy savings in evaporator fan-cycled apple storages. *Applied Engineering in Agriculture* 9(6), 553–60. doi:10.13031/2013.26021.

Kongwong, P., Boonyakiat, D. and Poonlarp, P. 2019. Extending shelf life and qualities of baby cos lettuce using commercial pre-cooling systems. *Postharvest Biology and Technology* 150, 60–70. doi:10.1016/j.postharvbio.2018.12.012.

lal Basediya, A., Samuel, D. V. K. and Beera, V. 2013. Evaporative cooling system for storage of fruits and vegetables – a review. *Journal of Food Science and Technology* 50(3), 429–42. doi:10.1007/s13197-011-0311-6.

Lallu, N. and Webb, D. J. 1997. Physiological and economic analysis of precooling kiwifruit. *Acta Horticulturae* 444(444), 691–8. doi:10.17660/ActaHortic.1997.444.106.

Liang, Y. S., Wongmetha, O., Wu, P. S. and Ke, L. S. 2013. Influence of hydrocooling on browning and quality of litchi cultivar Feizixiao during storage. *International Journal of Refrigeration* 36(3), 1173–9. doi:10.1016/j.ijrefrig.2012.11.007.

McCormick, R., Neuwald, D. A. and Streif, J. 2010. A case study: potential energy savings using 1-Mcp with 'gala' apples in commercial CA storage. *Acta Horticulturae* 877(877), 323–6. doi:10.17660/ActaHortic.2010.877.39.

Mercier, S., Marcos, B. and Uysal, I. 2017a. Identification of the best temperature measurement position inside a food pallet for the prediction of its temperature distribution. *International Journal of Refrigeration* 76, 147–59. doi:10.1016/j.ijrefrig.2017.01.025.

Mercier, S., Villeneuve, S., Mondor, M. and Uysal, I. 2017b. Time–temperature management along the food cold chain: a review of recent developments. *Comprehensive Reviews in Food Science and Food Safety* 16(4), 647–67. doi:10.1111/1541-4337.12269.

Michailides, T. J. and Elmer, P. A. G. 2000. Botrytis gray mold of kiwifruit caused by *Botrytis cinerea* in the United States and New Zealand. *Plant Disease* 84(3), 208–23. doi:10.1094/PDIS.2000.84.3.208.

Moreda, G. P., Muñoz, M. A., Ruiz-Altisent, M. and Perdigones, A. 2012. Shape determination of horticultural produce using two-dimensional computer vision – a review. *Journal of Food Engineering* 108(2), 245–61. doi:10.1016/j.jfoodeng.2011.08.011.

Ndukwu, M. C. and Manuwa, S. I. 2014. Review of research and application of evaporative cooling in preservation of fresh agricultural produce. *International Journal of Agriculture and Biological Engineering* 7, 85–102.

Ngcobo, M., Opara, U. and Thiart, G. 2011. Effects of packaging liners on cooling rate and quality attributes of table grape (cv. Regal Seedless). *Packaging Technology and Science* 25, 73–84.

Nunes, M. C. N., Brecht, J. K., Morais, A. M. M. B. and Sargent, S. A. 1995. Physical and chemical quality characteristics of strawberries after storage are reduced by a short delay to cooling. *Postharvest Biology and Technology* 6(1–2), 17–28. doi:10.1016/0925-5214(94)00048-W.

Olatunji, J. R., Shim, Y. M., Love, R. J., Ferrua, M. J. and East, A. R. 2016. Toward a fast and flexible transfer model for horticultural products packaged into boxes. *4th IIR International Conference on Sustainability and the Cold Chain*. Proceedings: Auckland, New Zealand, 7-9 April 2016.

Olatunji, J. R., Love, R. J., Shim, Y. M., Ferrua, M. J. and East, A. R. 2017. Quantifying and visualising variation in batch operations: a new heterogeneity index. *Journal of Food Engineering* 196, 81-93. doi:10.1016/j.jfoodeng.2016.10.004.

O'Sullivan, J., Ferrua, M. J., Love, R., Verboven, P., Nicolaï, B. and East, A. R. 2016. Modelling the forced-air cooling mechanisms and performance of polylined horticultural produce. *Postharvest Biology and Technology* 120, 23-35. doi:10.1016/j.postharvbio.2016.05.008.

Paniagua, A. C., East, A. R. and Heyes, J. A. 2013. Effects of delays in cooling on blueberry quality outcomes. *Acta Horticulturae* 1012(1012), 1493-8. doi:10.17660/ActaHortic.2013.1012.202.

Pathare, P. B. and Opara, U. L. 2014. Structural design of corrugated boxes for horticultural produce: a review. *Biosystems Engineering* 125, 128-40. doi:10.1016/j.biosystemseng.2014.06.021.

Pathare, P. B., Opara, U. L., Vigneault, C., Delele, M. A. and Al-Said, F. A. J. 2012. Design of packaging vents for cooling fresh horticultural produce. *Food and Bioprocess Technology* 5(6), 2031-45. doi:10.1007/s11947-012-0883-9.

Pennycook, S. R. and Manning, M. A. 1992. Picking wound curing to reduce Botrytis storage rot of kiwifruit. *New Zealand Journal of Crop and Horticultural Science* 20(3), 357-60. doi:10.1080/01140671.1992.10421779.

Redding, G. P., Yang, A., Shim, Y. M., Olatunji, J. R. and East, A. R. 2016. A review of the use and design of produce simulators for horticultural forced-air cooling studies. *Journal of Food Engineering* 190, 80-93. doi:10.1016/j.jfoodeng.2016.06.014.

Rivera, J. R. E., Stone, M. B., Stushnoff, C., Pilon-Smits, E. and Kendall, P. A. 2006. Effects of ascorbic acid applied by two hydrocooling methods on physical and chemical properties of green leaf lettuce stored at 5°C. *Journal of Food Science* 71(3), S270-6. doi:10.1111/j.1365-2621.2006.tb15653.x.

Santana, J. C. C., Araújo, S. A., Alves, W. A. L., Belan, P. A., Jiangang, L., Jianchu, C. and Donghong, L. 2018. Optimization of vacuum cooling treatment of postharvest broccoli using response surface methodology combined with genetic algorithm technique. *Computers and Electronics in Agriculture* 144, 209-15. doi:10.1016/j.compag.2017.12.010.

Sargent, S. A., Ritenour, M. A., Brecht, J. K. and Bartz, J. A. 2007. *Handling, Cooling and Sanitation Techniques for Maintaining Postharvest Quality*. University of Florida.

Shi, J., Zuo, J., Zhou, F., Gao, L., Wang, Q. and Jiang, A. 2018. Low-temperature conditioning enhances chilling tolerance and reduces damage in cold-stored eggplant (*Solanum melongena* L.) fruit. *Postharvest Biology and Technology* 141, 33-8. doi:10.1016/j.postharvbio.2018.03.007.

Smith, A. J. M. 1937. *The Pre-Cooling of Fruits in South Africa*. Government Printer, Pretoria, p. 84.

Sun, D. and Hu, Z. 2002. CFD predicting the effects of various parameters on core temperature and weight loss profiles of cooked meat during vacuum cooling. *Computers and Electronics in Agriculture* 34(1-3), 111-27. doi:10.1016/S0168-1699(01)00182-X.

Sun, D. and Zheng, L. 2006. Vacuum cooling technology for the agri-food industry: past, present and future. *Journal of Food Engineering* 77(2), 203–14. doi:10.1016/j.jfoodeng.2005.06.023.

Szirmai, A. and Verspagen, B. 2015. Manufacturing and economic growth in developing countries, 1950–2005. *Structural Change and Economic Dynamics* 34, 46–59. doi:10.1016/j.strueco.2015.06.002.

Talbot, M. T. and Baird, C. D. 1991. Evaluating commercial forced-air precoolers. Paper-American Society of Agricultural Engineers (USA).

Tanner, D. J., Cleland, A. C., Opara, L. U. and Robertson, T. R. 2002. A generalized mathematical modelling methodology for design of horticultural food packages exposed to refrigerated conditions. *International Journal of Refrigeration* 25(1), 33–42.

Tassou, S. A., De-Lille, G. and Ge., Y. T. 2009. Food transport refrigeration – approaches to reduce energy consumption and environmental impacts of road transport. *Applied Thermal Engineering* 29(8–9), 1467–77. doi:10.1016/j.applthermaleng.2008.06.027.

Thompson, J. F., Mitchell, F. G., Rumsey, T. R., Kasmire, R. F. and Crisosto, C. H. 2008. *Commercial Cooling of Fruits, Vegetables, and Flowers*. University of California, Division of Agriculture and Natural Resources, Oakland, CA, p. 11.

Tian, D., Alves, W., Araújo, S., Santana, J., Jiangang, L., Jianchu, C. and Donghong, L. 2014. Simulation approach for optimal design of vacuum cooling on broccoli by simulated annealing technique. *International Journal of Agricultural and Biological Engineering* 7, 111–5.

Toivonen, P. M. A. 2015. Integrated analysis for improving export of sweet cherries and how a small industry can comete by focussing on premium quality. *Acta Horticulturae* 1079, 75–82.

United Nations. 2010. Population facts. Report of the Department of Economic and Social Affairs, Population Division, 5, pp. 1–4.

Vigneault, C. and Goyette, B. 2001. Loss of ice through container openings during liquid-ice cooling of horticultural crops. *Canadian Biosystems Engineering* 43, 3–45.

Vigneault, C., Bartz, J. A. and Sargent, S. A. 2000. Postharvest decay risk associated with hydrocooling tomatoes. *Plant Disease* 84(12), 1314–8. doi:10.1094/PDIS.2000.84.12.1314.

Vigneault, C., Goyette, B., Markarian, N. R., Hui, C. K. P., Cote, S., Charles, M. T. and Emond, J. P. 2004. Plastic container opening area for optimum hydrocooling. *Canadian Biosystems Engineering* 46, 41–4.

Vurma, M., Pandit, R. B., Sastry, S. K. and Yousef, A. E. 2009. Inactiviation of *Escherichia coli* O157:H7 and natural microbiota on spinach leaves using gaseous ozone during vacuum cooling and simulated transportation. *Journal of Food Protection* 72(7), 1538–46. doi:10.4315/0362-028x-72.7.1538.

Watkins, C. B., Erkan, M., Nock, J. F., Iungerman, K. A., Beaudry, R. M. and Moran, R. E. 2005. Harvest date effects on maturity, quality, and storage disorders of "Honeycrisp" apples. *Horticultural Science* 40(1), 164–9. doi:10.21273/HORTSCI.40.1.164.

Woolf, A. B., Cox, K. A., White, A. and Ferguson, I. B. 2003. Low temperature conditioning treatments reduce external chilling injury of 'Hass' avocados. *Postharvest Biology and Technology* 28(1), 113–22. doi:10.1016/S0925-5214(02)00178-3.

World Bank. 2019. GDP per capita (current US$) | Data. *Data.worldbank.org*. Retrieved on 23 February 2019, from www.data.worldbank.org/indicator/Ny.gDP.PCAP.CD?locations=VN.

Yang, Q., Rao, J., Yi, S., Meng, K., Wu, J. and Hou, Y. 2012. Antioxidant enzyme activity and chilling injury during low-temperature storage of Kiwifruit cv. Hongyang exposed to gradual postharvest cooling. *Horticulture, Environment, and Biotechnology* 53(6), 505–12. doi:10.1007/s13580-012-0101-8.

Yost, G. E. 1984. Energy savings through the use of fan and refrigeration cycling in apple cold storage. *Transactions of the American Society of Agricultural Engineers* 27, 497–501.

Zhao, J. M., Bronlund, J. E. and East, A. R. 2014. Effect of cooling rates on kiwifruit firmness and rot incidence in subsequent storage. *Proceedings of the 5th Postharvest Unlimited Conference*, Cyprus, pp. 313–8. doi:10.17660/ActaHortic.2015.1079.38.

Zhu, Z., Wu, X., Geng, Y., Sun, D., Chen, H., Zhao, Y., Zhou, W., Li, X. and Pan, H. 2018. Effects of modified atmosphere vacuum cooling (MAVC) on the quality of three different leafy cabbages. *LWT* 94, 190–7.

Zhu, Z., Geng, Y. and Sun, D. 2019. Effects of operation processes and conditions on enhancing performances of vacuum cooling of foods: a review. *Trends in Food Science and Technology* 85, 67–77. doi:10.1016/j.tifs.2018.12.011.

Zou, Q., Opara, L. U. and McKibbin, R. 2006. A CFD modeling system for airflow and heat transfer in ventilated packaging for fresh foods: I. Initial analysis and development of mathematical models. *Journal of Food Engineering* 77(4), 1037–47. doi:10.1016/j.jfoodeng.2005.08.042.

Chapter 2

Advances in controlled atmosphere storage of horticultural produce

John DeLong, Agriculture and Agri-Food Canada, Canada; and Robert Prange, Dalhousie University, Canada

1 Introduction

A controlled atmosphere storage (CAS) environment is an enclosed system (i.e. a chamber/room sealed from the external ambient environment) used for storing plant organs in which the enclosure's biological active gases (particularly O_2 and CO_2) and temperature are continuously monitored and controlled at static settings or within a limited dynamic range (Dilley, 2006; DeLong, 2017). In a CAS room, the gas and temperature settings of the store markedly retard active catabolism thereby reducing the rate of cellular senescence. Compared with a non-CAS environment (e.g. refrigerated air (RA) storage at 21 kPa O_2, 0.03 kPa CO_2), the general outcome is a significant extension of the length of time in which the crop can be stored without quality loss from advancing senescence.

Although ancient civilizations routinely practiced storing fruits and vegetables in sealed pots, underground pits, and chambers, the development of modern CAS has a more recent history, with the majority of technological

http://dx.doi.org/10.19103/AS.2019.0055.02
Published by Burleigh Dodds Science Publishing Limited, 2020.

advances occurring in the twentieth century (Dilley, 2006; Hoehn et al., 2009; DeLong, 2017). The cumulative outcome of these advances is the capacity to retain a very high degree of quality of perishable plant organs over increasingly long periods of time. While defining a 'high degree of quality' can be complex, it may be envisioned as retaining the sensory (taste, smell, visual) and nutritional state of the organ as at the time of harvest (or nearly so). From a physiological perspective, retention of quality can be generally thought of as mitigating the loss of cellular function/integrity associated with cellular senescence. During the last two decades (i.e. late 1990s to present), progress in CAS resides less in the development of new standards for storage room construction, refrigeration or gas monitoring technologies, or in novel discoveries in biology or chemistry (with perhaps the exception of the discovery and commercialization of the anti-ethylene compounds 1-methylcyclopropene (1-MCP) and aminoethoxyvinylglycine (AVG) (see Section 6)), but more in the way the harvested crop is managed for quality retention within the storage room. In this light, this chapter will highlight advances in CA storage science of horticultural crops of the last two decades with emphasis on obtaining the ideal or optimum CA storage regime for a particular commodity. As the majority of the world's CA storages are employed for holding apple and pear fruit (the latter to a lesser extent), this chapter will largely focus on advances associated with apple and pear storage by highlighting developments in: initial low O_2 stress, delayed CAS; the anti-ethylene compounds 1-MCP and AVG; dynamic controlled atmosphere (DCA) storage based on chlorophyll fluorescence (CF), ethanol detection, and respiratory quotient (RQ) determination; and strategies for reducing electrical energy consumption during longer-term storage periods. This chapter also includes a case study, based on a research article by Wright et al. (2010), which explains several important aspects of DCA-CF storage technology.

2 Initial low oxygen stress (ILOS)

Applying an initial low oxygen stress (ILOS) treatment involves subjecting fruit at the very beginning of the CA storage to O_2 levels that are near or below the lower oxygen limit ((LOL); see Section 5 for description of the LOL) for a relatively short time period (e.g., 0.25–0.50 kPa O_2 for 2 weeks). Following the ILOS period, storage room O_2 concentrations are raised to safe (non-injurious) levels above the LOL (e.g., 1.5–2.0 kPa O_2). The impetus for developing ILOS protocols is largely connected to reduction of the epidermal storage disorder superficial scald (Little et al., 1982; Wang and Dilley, 2000), which has historically been controlled on susceptible cultivars by the antioxidants diphenylamine (DPA) or ethoxyquin (Smock, 1955; Huelin and Coggiola,

1970; Hansen and Mellenthin, 1979). However, with the present emphasis on reducing chemical residues in fruits and vegetables, and with the current ban on DPA usage in many European countries (Commission Implementing Regulation 578/2012, 2012), alternate superficial control measures on apple and pear fruit in particular have received considerable research effort (Lurie and Watkins, 2012; Zanella and Stürz, 2013).

Under an ILOS regime, the hypoxic stress experienced by the fruit tissue often results in marked reduction of superficial scald in susceptible fruit along with better retention of quality attributes like firmness and acidity (Truter et al., 1994; Wang and Dilley, 2000; Zanella, 2003; Mattè et al., 2005). At the cellular level, it is thought that an increase in fermentative ethanol production while the fruit is held at low O_2 (i.e. below the LOL) inhibits the key biochemical oxidation reactions in scald development (Wang et al., 1997; Ghahramani and Scott, 1998). Wang and Dilley (2000) showed that an ILOS treatment of 0.25 kPa O_2 effectively controlled superficial scald expression in 'Granny Smith', 'Law Rome', and 'Delicious' fruit as well as inhibited the biochemical scald precursors α-farnesene and 6-methyl-5-hepten-2-one. Zanella (2003) observed that an ILOS treatment of 0.4 kPa O_2 for 2 weeks followed by 4 or 6 months at 1.0 kPa controlled superficial scald of 'Granny Smith' fruit more effectively than a constant regime of 1.0 kPa O_2. In some cases, an ILOS regime does not confer commercially relevant superficial scald control as was observed on 'Delicious' fruit after an ILOS treatment of 0.4 kPa for 10 days followed by 0.7 or 1.0 kPa O_2 for 6–9 months (Lau, 1997).

For organically managed fruit, the ILOS protocol may be an effective two-stage static strategy for controlling economically damaging oxidation reactions that manifest during storage, where stage one is the ILOS treatment (e.g. 0.4 kPa O_2 for 10–14 days) and stage two is the long-term storage regime (e.g. 1.5–2.0 kPa O_2 for 8 months). However, in order to achieve the very low O_2 levels required for an ILOS treatment, several requirements are needed, such as room air-tightness, adequate nitrogen generation capacity (for displacing O_2), and highly accurate O_2 sensors for atmosphere O_2 measurement. However effective in controlling superficial scald and other oxidative disorders, a widespread commercial adoption of ILOS has not occurred, largely due to the introduction of 1-MCP during the early 2000s (see Section 6), the risk of inducing low-O_2 injury (e.g. fermentative off-flavors, skin purpling), and the development of dynamic CA systems (see Section 5). Nonetheless, for those managers who do not want to treat their produce with postharvest antioxidants or anti-ethylene chemicals and who have the technical capacity to apply the initial low O_2 regime, the ILOS protocol can be an effective, chemical-free method for long-term quality retention by exposing produce to very low O_2 partial pressures for a controlled period of time.

3 Anti-ethylene compounds

The gaseous plant hormone ethylene affects many processes associated with fruit maturation including ripening, senescence, and respiration in general, chlorophyll degradation (yellowing), tissue softening, and organ abscission from the parent plant (Saltveit, 1999). For most fruit, quality retention after harvest involves minimizing or eliminating the progress in ripening that ethylene promotes. In the last 20–25 years, two anti-ethylene products with differing modes of activity have played a significant role in how the worldwide horticultural industry (particularly for climacteric tree fruits and especially apples) manages the effect of endogenous ethylene production in fruit crops.

Aminoethoxyvinylglycine (AVG) impedes ethylene synthesis by inhibiting the conversion of S-adenosylmethionine to aminocyclopropanecarboxylic acid by its interaction with the enzyme aminocyclopropanecarboxylic acid synthase in the ethylene biosynthetic pathway (Yang, 1985). The commercial AVG product is known as ReTain® and is sold by Valent Biosciences in the United States and by Valent in Canada (Valent, 2017). AVG is largely used as a harvest management tool, being applied as a preharvest spray 1–4 weeks prior to anticipated harvest to prevent preharvest fruit drop in susceptible cultivars, to generally widen the harvest window and to promote varietal color development by delaying and extending harvest (Greene and Schupp, 2004; Robinson et al., 2006; Watkins, 2010). Storage and/or post-storage benefits are less noted but can include better firmness retention, slowing of tissue softening, and improving fruit response to 1-MCP application (Robinson et al., 2006; Watkins, 2010).

1-Methylcycloproprene (1-MCP) is a gaseous, non-toxic cyclopropene (C_4H_6) derivative that competitively inhibits ethylene action by irreversibly binding to several ethylene receptors (e.g. ETR1, ETR2, EIN4, ERS1, and ERS2) in the plant cell (Binder and Bleecker, 2003). Once bound to the receptor(s), the typical effects of ethylene in accelerating ripening and senescence chemistry do not occur regardless of the subsequent concentration of ethylene surrounding the bound receptor complex (Sisler and Serek, 2003; Prange and DeLong, 2003). The discovery of the ethylene-inhibiting properties of 1-MCP and its ensuing commercialization is a major advance in horticultural science affecting broad commodity categories (especially for climacteric fruit like apple) as well as being an invaluable tool for researchers investigating the role of ethylene in an array of physiological responses at basic and applied levels (Watkins, 2010, 2018). Since being commercialized in 2002 for edible food products, the literature records 1-MCP usage on at least 33 climacteric and 17 non-climacteric fruits, as well as on about 33 vegetables (Watkins, 2018). Commercially, 1-MCP is sold as SmartFresh™ for postharvest treatment on a variety of crops, or as Harvista™ for preharvest field application on apples, pears, and cherries (AgroFresh, 2019). In addition, new 1-MCP products like Fysium® (patented

1-MCP generator system) (Fysium®:Pace, 2019) and Hazel® (1-MCP-containing sachet) (Hazel Technologies Inc., 2019) are currently available, adding unique product delivery options for controlling ethylene-related senescence for a number of fruit crops.

4 Delayed controlled atmosphere storage

Cooling newly harvested crops to remove field heat is normally the first stage in ensuring quality retention during storage for most fruits and vegetables. Once cooled (usually within several days of harvest at temperatures ranging from 1 to 5°C for apples), the crop enters the CA establishment phase - the process of acclimating over time to an increased reduction of O_2 along with an increase in CO_2 partial pressures in the storage room, eventually reaching a static and safe (non-injurious) gas level for the duration of the holding period. Establishing this gaseous environment usually occurs over a 10–14 day period for most apple and pear cultivars, and involves further temperature reduction as well reducing the O_2 content of the storage room air volume with CA control technology via nitrogen gas flushing. A concurrent goal during the O_2 reduction period is to minimize low O_2 cellular shock (i.e. hypoxia- or anoxia-induced) to the plant organ so that the tissue does not incur economic injury.

For many apple cultivars (e.g. 'Delicious', 'Elstar', 'Gala', 'Golden Delicious', 'Jonagold'), it is common practice to establish the CAS as rapidly as possible after transporting harvested fruit to the storage facility in order to minimize respiration-related losses. However, for other cultivars (e.g. 'Empire', 'Braeburn', 'Fuji', 'Honeycrisp'), initiating a CAS regime immediately following harvest may exacerbate chilling injury, internal flesh browning, and CO_2-related disorders (Saquet et al., 2003a; Argenta et al., 2000; de Castro et al., 2007; DeEll and Ehsani-Moghaddam, 2012). To reduce the potential for disorder development, storage operators can delay the onset of the CAS regime for several weeks (e.g., 2–8) and then establish the CA environment for the remainder of the storage period (Saquet et al., 2003a,b; de Castro et al., 2007; DeEll et al., 2016, Watkins and Nock, 2012a).

Pre-storage application of 1-MCP followed by rapid establishment of a CA environment can also increase the risk of developing internal flesh browning and CO_2-related disorders in apple fruit (DeEll and Ehsani-Moghaddam, 2012; Watkins and Nock, 2012b). Several groups have shown that applying 1-MCP immediately following harvest, coupled with delaying the establishment of the CA regime by several weeks or more, is a viable solution for maintaining fruit cortical firmness and acidity as well as reducing browning and CO_2-induced problems. DeEll and Ehsani-Moghaddam (2012) treated 'Empire' apples within 24 hours after harvest with 1-MCP and then either placed the fruit immediately in standard CAS (2.5 kPa O_2, 2.0 kPa CO_2, 3°C) for 9 months, or in air at 3°C for one or two months plus 7 or 8 months of standard CAS regime. Delaying the CA environment reduced external CO_2 injury and flesh browning, while application of 1-MCP immediately after harvest helped retain flesh firmness and acidity over the 9-month storage. Using this basic approach, similar quality retention results have been demonstrated for the cultivars 'Honeycrisp' (DeEll et al., 2016; Watkins and Nock, 2012a), 'Empire', and 'McIntosh' (Watkins and Nock, 2012b).

Another issue that 'delayed CA' may help solve is the inability of commercial operators to fill large storage rooms rapidly enough to establish the CA regime within a few days of harvest. This may be due to many reasons, including several-to-many growers, difficulties managing harvest labor, the multiple pick nature of most apple cultivars, and the need to store several cultivars with differing maturities in the same storage room. If the fruit can be treated with 1-MCP promptly after harvest, the operator then has a longer period of time to decide when to establish the CA regime. In order to minimize the loss of fruit quality within the delayed CA system, however, the operator needs to focus on the idiosyncrasies of the cultivar being stored, for example, the optimal stage of fruit maturity for harvest and the innate susceptibility of the cultivar(s) to storage disorders.

5 Dynamic controlled atmosphere (DCA) storage

Saltveit (2003) presciently expressed the concern that a truly optimal CA storage environment for a horticultural crop would not be obtainable because the science of establishing it was a static exercise. He recognized that ideal storage settings must take into account all causes of seasonal and postharvest influence which could alter the CA condition and not provide the optimal storage environment:

> ...the natural variability in the raw material and its dynamic response to processing and storage conditions may render it impossible to identify a truly optimal storage atmosphere (Saltveit, 2003).

And,

> Truly significant advances in the use of CA and modified atmospheres may require the development of mathematical models that incorporate some measure of the commodity's dynamic response to the storage environment (Saltveit, 2003).

He recognized that until the CA conditions (largely O_2 settings) were dynamically attuned to the seasonal 'in-the-field' influences as well as to the acclimation-related changes of the crop during storage, creating an optimal CA environment in real time was not possible to achieve.

Since Saltveit published that article, several dynamic CA technologies have been developed and are now commercially available for pome fruit (i.e. apples and pears) storage. The basic operational principles for the three main technologies include: chlorophyll fluorescence emission (DeLong et al., 2004; Prange et al., 2013; Zanella and Stürz, 2015), ethanol accumulation (Schouten et al., 1998; Veltman et al., 2003), and RQ calculation (Gasser et al., 2010; Van Schaik et al., 2015; Bessemans et al., 2016) to identify the LOL.

The LOL is a key and integral concept for the operational success of DCA systems and low O_2 storage in general. It can be thought of as the metabolic threshold between aerobic and anaerobic respiration, where the former is minimal and the latter is not invoked. Once identified, an optimum storage O_2 atmosphere would occur slightly above the LOL (e.g. 0.1 to 0.3 kPa) resulting in the lowest possible O_2 setting which does not induce hypoxic or anoxic injury to the crop. Other terms such as the anaerobic compensation point (ACP) or the RQ or RQ breakpoint (RQB) are commonly used when describing states of biological hypoxia or anoxia. The ACP is defined as the oxygen concentration at which CO_2 production is minimal (Boersig et al., 1988). As the O_2 concentration declines, cells respond by slowing the rate of aerobic respiration which concomitantly decreases the amount of CO_2 generated. Below a certain oxygen concentration, an increase in low O_2-induced fermentation results in an increase in overall CO_2 production (Boersig et al., 1988). The ACP is therefore the point at which a decrease in aerobic, mitochondrial CO_2 production is exceeded by the increase in CO_2 due to fermentation. The RQ is the ratio of the moles of CO_2 evolved divided by the moles of O_2 consumed. As respiration slows due to declining O_2 levels, the RQ ratio remains relatively constant at about 1.0 in apple. However, the RQ rises distinctly below a certain O_2 level (termed the RQB) which results from the decarboxylation of pyruvate by pyruvate decarboxylase. This increase in the rate of glycolysis is historically known as the Pasteur effect (Boersig et al., 1988) and signifies the onset of fermentation.

Storage room O_2 and CO_2 levels for fruits and vegetables (as well as temperature) have been historically established by years of trial until a safe but effective product 'quality retention' level is validated. These recommended CA storage settings were contingent upon several considerations, including the commodity type, stage of ripeness or maturity, growing season conditions, the general climate where the crop was grown (e.g. maritime vs. continental) – as well as the CA-related technology available (Dilley, 2010; Wright et al., 2015). In general, the CA recommendations for a specific fruit or vegetable were set conservatively high to avoid any undesirable changes or injury occurring if the O_2 level (for example) fell too low. While safe (and sensible), this practice could not take advantage of the potential for greater CA storage benefits if the O_2 level had been set lower. With the three main DCA systems currently available (see succeeding sections below), the LOL can be identified resulting in a recommendation for the ideal storage room O_2 setting - generally set 0.1–0.3 kPa above the LOL to avoid hypoxic/anoxic injury to the crop over the storage duration. As the produce usually acclimates to some degree during storage, periodically identifying the potentially changing LOL (e.g. once per month) and consequently adjusting the CA room's O_2 levels, will ensure the lowest possible O_2 setting over the entirety of the CAS period.

5.1 DCS-ethanol

Although using ethanol to identify safe O_2 levels for apple fruit can be traced back to Blackman (1928), it was not until the mid-to-late 1990s that work by Schouten et al. (1997, 1998) demonstrated CAS O_2 control was possible through headspace ethanol measurements resulting in crop quality benefits. Veltman et al. (2003) introduced a patented technology called the 'Dynamic Control System (DCS)' where an ethanol sensor measured headspace ethanol emission from 'Elstar' fruit under increasingly low O_2 storage conditions (e.g. <1.2 kPa O_2). The LOL is associated with the accumulation of ethanol above a critical threshold level (e.g. 500 nl L^{-1}) due to fermentative cellular metabolism as hypoxic or anoxic conditions occur. At this point, the lowest safe O_2 setting would be attained by slightly increasing the room O_2 content just above where the LOL was identified (e.g. in 0.1 kPa O_2 increments) concomitant with the ethanol threshold value decreasing to an acceptably safe concentration (e.g. <500 nl L^{-1}). 'Elstar' fruit managed by the DCS-ethanol technology have higher firmness, acidity, and color retention as well as some disorder reduction for longer post-storage periods compared with fruit stored under conventional CAS (Veltman et al., 2003). The DCS-ethanol system is currently available commercially as DCS™ which is sold by Storex, Gravendeel, The Netherlands (http://www.storex.nl/applications-fruitstorage.html, accessed 24 October 2019).

5.2 DCA-respiratory quotient (DCA-RQ)

The use of the RQ - defined as the ratio of the number of moles of CO_2 evolved divided by the moles of O_2 consumed by the produce, is key to this DCA protocol. In apple fruit, the RQ is around 1 when the crop is under no hypoxic or anoxic stress. As the DCA environment is imposed on a stored crop, there will be a relatively constant RQ until the LOL is reached, at which point an increase in the RQ value is noted. This RQ rise is caused by increased CO_2 production by the plant tissue due to the decarboxylation of pyruvate (3 carbon) to acetaldehyde (2 carbon) by the enzyme pyruvate decarboxylase during anaerobic fermentation (Taiz and Zeiger, 2010). At this point, the operator will turn the O_2 level of the room up 0.1–0.3 kPa or more until the CO_2 production declines and the RQ drops to pre-stressed levels indicating that the

O_2 setting is above the LOL. In this way, the lowest possible storage room O_2 level is achieved without injuring the fruit.

Van Schaik et al. (2015) used the DCA-RQ approach to dynamically control the O_2 level of stored 'Elstar' fruit and found that RQ and ethanol increases at 0.3 kPa O_2 appeared to be reliable markers of the onset of fermentation. The decrease in respiratory metabolism of the stored fruit and the associated decrease in catabolic heat, suggested that room cooling costs may also be lessened with the DCA-RQ system (Van Schaik et al., 2015). Bessemans et al. (2016, 2018) observed that 'Granny Smith' and 'Golden Delicious' apples stored under DCA-RQ regimes maintained higher post-storage flesh firmness and color retention compared with conventional CA-stored fruit. In these experiments, the storage O_2 level was set at the RQB which ranged from 0.2–0.45 kPa. Low fruit ethanol measurements also indicated that the DCA-RQ regime did not promote a state of cellular anaerobiosis. In addition, 'Granny Smith' superficial scald was completely controlled for up to 14 days after storage removal. In general, the quality retention of DCA-RQ-stored 'Granny Smith' fruit was between that of CA and CA+1-MCP after nine storage months and a 14-day shelf life at 18°C. These results demonstrate that while the DCA-RQ treatment was somewhat less effective for maintaining firmness and controlling superficial scald compared with 1-MCP, 'Granny Smith' ripening did occur in the DCA-RQ fruit after removal, which in some cases, can be a problem in 1-MCP-treated apples (Aubert et al., 2015; Cliff and Toivonen, 2017).

A DCA-RQ-based system which measures the O_2 consumed and CO_2 generated by stored fruit is commercially available by the patented name of SafePod™ and is sold by Storage Control Systems, Sparta, Michigan, USA (http ://www.storagecontrol.ltd/products/safepod-system/, accessed 24 October 2019). An enclosed container is placed in the CA storage room and holding up to 70–80 kg of sample fruit, has the same temperature, humidity and gas composition as the CA store (Fig. 1). At the operator's initiative, the enclosure is sealed from the CA room by valve closure, the O_2 and CO_2 are measured and the RQ then calculated. Once the RQ is determined, the storage room can be adjusted for optimal O_2 and CO_2 concentrations. Currently, there are approximately 70 and 150 users of SafePod™ in the United Kingdom and the United States, respectively (Prange, 2018).

Recently, DeEll and Lum (2017) using a SafePod™ system, reported that 'Empire' fruit stored at 0.6 kPa O2 and 0.5 kPa CO_2 for 8 months at 3.0°C had higher firmness and negligible CO_2 injury, flesh or core browning compared with fruit stored at 2.5 kPa O_2 and 2 kPa CO_2 or 1.5 kPa O_2 and 1.2 kPa CO_2 and held at 1.5°C and 3.0°C. Interestingly, they also observed that application of 1-MCP at 3°C in ambient atmosphere after 8 months of 0.6 kPa O_2 in the SafePod™ (DCA-RQ) system, conferred firmness benefits to 'Empire' fruit after a shelf life of 14 days at room temperature. Although based only on a single

Figure 1 The Safepod™ unit, containing apple fruit on a pallet placed in a CA storage room. Source: reproduced with permission from Storage Control Systems Inc., (Sparta, Michigan).

year of data, this 'post-storage' firmness retention demonstrates an interesting interaction between 1-MCP and an ultra-low O_2 storage regime and may lead to more research to determine which cultivars under what low O_2 environments favorably respond to 1-MCP application after fruit are removed from DCA storage.

5.3 DCA-chlorophyll fluorescence (DCA-CF)

The use of CF as a dynamic indicator of the LOL in fruit and vegetable tissue was first introduced at the ISHS CA symposium in Rotterdam (the Netherlands) (Prange et al., 2001) and was then initially adopted commercially in the 2003-2004 apple storage season by Washington State (USA) and the South Tyrol region (Italy) (Prange et al., 2013; Zanella and Strüz, 2013, 2015). Since then, its commercial form, known as HarvestWatch™ (Fig. 2), has been used worldwide, but is especially popular in the apple growing areas of Europe, particularly, in Northern Italy. It is also used extensively for optimizing CAS conditions for pear, including control of superficial scald (Prange et al., 2011; Vanoli et al., 2015, 2016; Deuchande et al., 2016).

The fundamental operating principal of the DCA-CF technology is the sensor-based generation of a frequency modulated red light pulse emitted for 60 seconds once per one or two hour(s) on a chlorophyll-containing plant organ (e.g. apple) in a dark adapted state. During the CF measurement cycle, the far-red light LED duty cycle is increased resulting in an increasingly more intense photon flux. The subsequent CF signal emitted by the organ's chloroplasts is detected by the system's sensor and then subjected to an algorithm which calculates the fluorescence at theoretically zero irradiance (deemed Fα), which is similar to the initial fluorescence parameter, Fo, of the pulse amplitude

Figure 2 A DCA-CF (HarvestWatch™) system comprised of a fluorescence interactive response monitor (FIRM) sensor, each mounted within a kennel containing fruit and connected using CAT 6/7 cables to a central hub and then to a computer running the HarvestWatch™ software.

modulation technology (Wright et al., 2008). When placed in close proximity to fruit under CA conditions (i.e. within a plastic kennel placed in a bin of fruit, Fig. 3), the Fα signal detected by the sensor over the irradiance period is then displayed on the system's computer monitor in real time. If the O_2 level in the CA storage is above the LOL, Fα is flat (= non-stressed); if the room O_2 level goes below the LOL, an abrupt increase in Fα occurs which indicates that the produce is in a low-O_2 state of stress, alerting the operator to increase the O_2 settings for the CA room until Fα becomes flat again (non-stressed state) (Fig. 4). In identifying the LOL by the abrupt increase in Fα, the theoretically lowest possible CA room O_2 setting is obtained by adjusting the room's O_2 concentration about 0.1–0.3 kPa above the LOL - at which point the Fα signal declines to pre-stress levels. Thus, adjusting the storage room's O_2 level as directed by the product's fluorescence response to the O_2 setting represents the essential 'dynamic' nature of the technology.

Figure 3 A common arrangement of six DCA-CF (HarvestWatch™) FIRM kennels containing apple fruit connected to one hub in a commercial CAS room.

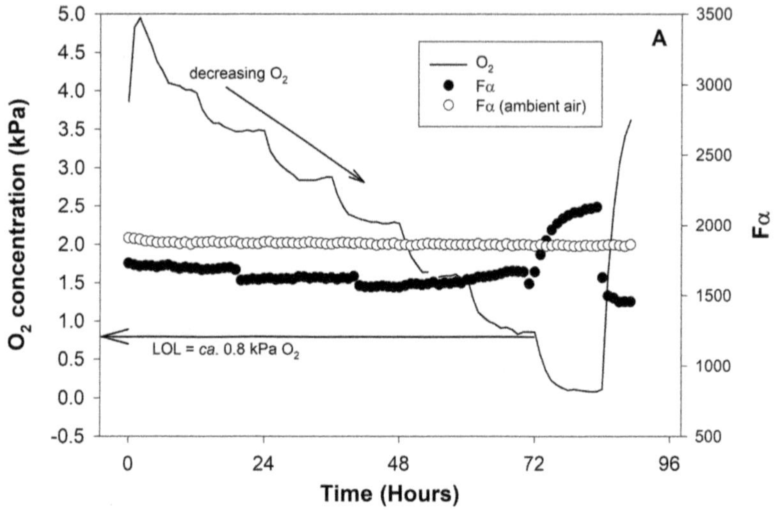

Figure 4 Using HarvestWatch™ to determine the LOL in an apple sample. The O_2 is gradually reduced and the chlorophyll fluorescence signal (Fα, solid dots) suddenly spikes at ca. 72 h. The LOL is the O_2 value (0.8 kPa) where the Fα increase (indicates hypoxic stress) begins. The O_2 is then slightly increased above the LOL at ca. 84 h and the Fα signal immediately returns to its non-stressed baseline level (Prange et al., 2003; Prange, 2018).

Of the commercially available DCA options, the CF-based HarvestWatch™ system is the most extensively researched and used industrially (Prange et al., 2015; Prange, 2018). A recent conservative 2017 estimate of the worldwide usage of HarvestWatch™ indicates approximately 1980 commercial storage rooms involving more than 20 cultivars of apples and pears in 26 countries; it is also used in about 20 research institutions (Prange, 2018). From early 2000 until about 2013, DCA-CF was predominantly used in Italy; since then, usage in other countries has collectively surpassed Italy's employ of the technology. Presently, the HarvestWatch™ system is found in many European countries, as well as in Australia, Argentina, Brazil, Chile, Israel, New Zealand, Russia, the Republic of South Africa, Turkey, and in the United States. Comparing DCA-type technologies in current use in the worldwide fruit storage industry, it is estimated that 85–90% employ DCA-CF (e.g. HarvestWatch™), while 10–15% use ethanol (e.g. DCS™) or RQ-based systems (Prange, 2018). The HarvestWatch™ technology is sold by Isolcell S. P. A. (https://isolcell.com/en/dca-isostore/, accessed 4 April 2019), located in Bolzano, Italy.

Over the last 14 or 15 years, the HarvestWatch™ system has been primarily used for pome fruit storage, with articles reporting at least 19 and 9 cultivars of apple and pear fruit, respectively; reports of HarvestWatch™ storage of avocado and kiwifruit have also appeared (Table 1). Although it has also been

Table 1 List of publications on DCA use in fruits and vegetables

Commodity	Cultivar	Publication
Fruits		
Apple	Braeburn	Bessemans et al. (2018), Gasser et al. (2008), Lafer (2008, 2009), Poldervaart (2010a), Prange et al. (2012), Withnall (2008), Zanella and Rossi (2015)
	Cortland	DeLong et al. (2004, 2007), Prange et al. (2012), Wright et al. (2012)
	Cripp's Pink (Pink Lady)	Aubert et al. (2015), Withnall (2008), Prange et al. (2012), Zanella and Rossi (2015)
	Delicious (Red)	Brizzolara et al. (2017), DeLong et al. (2004, 2007), Lafer (2009), Prange et al. (2012), Stephens and Tanner (2005), Withnall (2008), Wright et al. (2012), Zanella and Rossi (2015), Zanella and Stürz (2013)
	Elstar	Gasser et al. (2008), Köpcke (2009, 2015), Poldervaart (2010a), Maxin and Weber (2011), Prange et al. (2012)
	Empire	DeEll and Lum (2017)
	Fuji	Thewes et al. (2017b, 2018), Weber et al. (2017), Withnall (2008), Zanella and Rossi (2015), Zanella et al. (2008)
	Gala (Galaxy)	Both et al. (2017, 2018), Brackmann et al. (2015), de Oliveira Anese et al. (2018), Nock et al. (2019), Stanger et al. (2018), Thewes et al. (2015, 2017a), Withnall (2008), Zanella and Rossi (2015), Zanella et al. (2008)
	Golden Delicious	DeLong et al. (2004), Gabioud Rebeaud and Gasser (2015), Gasser et al. (2008), Kittemann et al. (2015), Poldervaart (2010a), Prange et al. (2012), Withnall (2008), Zanella and Rossi (2015), Zanella et al. (2008)
	Honeycrisp	DeLong et al. (2004), Prange et al. (2010, 2012), Wright et al. (2008, 2010, 2011, 2012)
	Gloster	Köpcke (2015)
	Granny Smith	Bessemans et al. (2016, 2018), Brizzolara et al. (2017), Calhan et al. (2016), Eren et al. (2015), Lafer (2009), Prange et al. (2012), Torres and Hernández (2015), Withnall (2008), Wright et al. (2012), Zanella et al. (2005)
	Idared	Poldervaart (2010a), Prange et al. (2012)
	Jonagold	DeLong et al. (2004), Kittemann et al. (2015), Köpcke (2015), Prange et al. (2012)
	Maigold	Gasser et al. (2008), Poldervaart (2010a)
	McIntosh	DeLong et al. (2004), Prange et al. (2002, 2003)
	Northern Spy	Prange et al. (2005)

(*Continued*)

Table 1 (*Continued*)

Commodity	Cultivar	Publication
	Pinova (Piñata)	Kittemann et al. (2015), Raffo et al. (2009)
	Topaz	Gasser and von Arx (2015), Lafer (2009), Poldervaart (2010a,b)
Avocado	-	Prange et al. (2003), Yearsley et al. (2003)
	Hass	Burdon (2009), Burdon and Lallu (2008), Burdon et al. (2008, 2010), Prange et al. (2002), Yearsley et al. (2003)
Banana	-	Prange et al. (2002, 2003)
Kiwifruit	-	Prange et al. (2002, 2003)
	Hayward	Lallu and Burdon (2007)
Mango	-	Prange et al. (2002, 2003)
Pear	-	Prange et al. (2003)
	Abbé Fétel (Abaté Fétel)	Folchi et al. (2015), Rizzolo et al. (2008, 2014, 2015b), Vanoli et al. (2007, 2010, 2015, 2016), Zerbini and Grassi (2010)
	Bartlett (Williams Bon Chrétien)	Mattheis (2007), Prange et al. (2002, 2011, 2012)
	Conference	Rizzolo et al. (2015a), Zerbini and Grasso (2010)
	d'Anjou	Mattheis (2007), Mattheis and Ruddell (2011), Prange et al. (2011), 2012
	Forelle	Prange et al. (2011)
	Packham's Triumph	Prange et al. (2011, 2012)
	Rocha	Deuchande et al. (2016)
	Spadona	Prange et al. (2011)
	Uta	Lafer (2011)
Vegetables		
Asparagus	-	Prange et al. (2005)
Broccoli	-	Prange et al. (2005)
Cabbage	-	Prange et al. (2003)
Pepper (Green)	-	Prange et al. (2003)
Lettuce	Iceberg	Prange et al. (2003, 2005)
	Romaine	Prange et al. (2003)
Spinach	-	Prange et al. (2012), Wright et al. (2011, 2012)

shown experimentally to markedly increase the 'green life' of asparagus, broccoli, cabbage, green pepper, lettuce, and spinach during storage (Table 1), DCA-CF is not yet used commercially for these crops.

Compared to a standard CAS regime, DCA-CF benefits for apples includes higher retention of flesh firmness (DeLong et al., 2004; Lafer, 2008; Gabioud

Rebeaud and Gasser, 2015; Köpcke, 2015; Zanella and Rossi, 2015; Calhan et al., 2016; Thewes et al., 2017b) and acidity (Gabioud Rebeaud and Gasser, 2015; Calhan et al., 2016), lower internal ethylene (Thewes et al., 2015; Calhan et al., 2016; Weber et al., 2017) as well as the reduction of disorders like superficial scald (DeLong et al., 2004; Zanella et al., 2005; Calhan et al., 2016), cortical or stem end flesh browning (Lafer, 2008; Köpcke, 2015; Nock et al., 2019) and core flush (Zanella et al., 2005; Köpcke, 2015).

When compared with conventional CAS, DCA-CF storage for pears generally retains higher cortical firmness (Lafer, 2011; Folchi et al., 2015) and reduces superficial scald at removal and/or during a shelf life period (Vanoli et al., 2010, 2016; Folchi et al., 2015; Mattheis and Ruddel, 2011). In some cultivars (e.g. 'Abbé Fétel'), a DCA-CF environment may exacerbate soft scald and/or internal breakdown and browning (Vanoli et al., 2010, 2015, 2016; Folchi et al., 2015), while for others, (e.g. 'Uta' and 'Rocha') it reduces or eliminates the same disorders (Lafer, 2011; Deuchande et al., 2016; Saquet et al., 2017). Interestingly, Vanoli et al. (2016) observed that disorders were reduced in 'Abbé Fétel' fruit held for 20–28 weeks in a DCA-CF regime (0.8 kPa O_2, 0.45 kPa CO_2) at −0.5°C when compared with 1.0°C. Thus, optimal DCA-CF pear storage requires testing with temperature and low O_2 combinations in order to establish the most favorable environment for quality retention.

When a pre-storage application of 1-MCP is combined with a DCA-CF storage regime, a high degree of firmness retention and marked disorder reduction after removal is frequently observed (Vanoli et al., 2010, 2016). Unfortunately, 1-MCP treatment is often associated with the inability of fruit to fully ripening following storage removal (Rizzolo et al., 2014, 2015a,b; Vanoli et al., 2015, 2016) which precludes its common use in the pear industry. However, implementing an ultra-low O_2 DCA regime can confer quality retention benefits similar to that of 1-MCP and allow the post-storage ripening needed to develop the sensory attributes characteristic of optimally ripe pears (Vanoli et al., 2016; Rizzolo et al., 2014, 2015a,b).

Several papers have appeared in recent years comparing the long-term apple storage quality retention capacity of DCA-RQ and DCA-CF systems. In most reports, the two DCA methods similarly maintain higher fruit firmness and reduce disorder incidence when compared with a conventional CA regime for the cultivar tested (Both et al., 2017; Weber et al., 2017). Interestingly, 'Fuji' fruit under DCA conditions combined with 1-MCP develop internal cortical cavities likely due to heightened 1-MCP-induced CO_2 sensitivity (Thewes et al., 2017b; Lafer, 2008). This 1-MCP-related problem has been observed in both conventional CAS and DCA environments among various cultivars including 'Empire' (Jung and Watkins, 2011; Fawbush et al., 2008; DeEll et al., 2003), 'Braeburn' (Lafer, 2008; Argenta et al., 2000), 'Honeycrisp' (Watkins and Nock, 2012a; DeEll et al., 2016) and 'McIntosh' (DeEll et al., 2003; Nock and Watkins, 2013). Thus, understanding the potential exacerbation of CO_2 sensitivity of the cultivar(s) being stored to both 1-MCP and ultra-low O_2 exposure is critical for avoiding the development of economically devastating disorders.

6 1-Methylcyclopropene (1-MCP) and DCA technologies

Since commercialization for use on fruit crops in the early 2000s, DCA (see Section 5) and 1-MCP technologies have globally transformed the postharvest management of tree fruit crops (particularly 1-MCP). The choice of which to employ is influenced by several factors, including: product and application costs, the degree of chemical-free management desired, capacity to integrate the technology within existing storage infrastructure, specific cultivar sensitivity to 1-MCP-exacerbated disorders, regional familiarity with the technology, as well as the availability of technical advice and support (Watkins, 2010; Prange et al., 2013; Gabioud Rebeaud and Gasser, 2015).

In research comparing 1-MCP vs DCA technology over longer-term apple storage (e.g. >5 months), higher fruit quality retention can be conferred by either system compared with a conventional CAS regime. After storing several cultivars for five-seven months, Zanella and Rossi (2015) observed that 'Golden Delicious', 'Delicious', 'Cripps Pink' and 'Braeburn' fruit maintained higher firmness in DCA-CF storage or with1-MCP treatment compared with those held in standard CA (1.5–1.8 kPa O_2) or ULO (1.0 kPa O_2) storage at removal or after a shelf life period.

Where SCA (1.5–2.5 kPa O_2)-stored fruit are compared with those receiving a DCA (<1 kPa O_2) or separate 1-MCP treatment, it is commonly reported that immediately after storage removal, the DCA- and 1-MCP-treated samples have higher quality retention (i.e. firmness or acidity) than fruit in conventional CA (or in ULO conditions, in some cases) (Lafer, 2008; Calhan et al., 2016; Köpcke, 2015; DeEll and Lum, 2017). During a subsequent shelf life period (e.g. 7 days at 20–25°C), DCA fruit usually maintain a higher degree of firmness and/or acidity than those held in standard CA, while the 1-MCP fruit are usually more firm and retain more acid than those held in a DCA or conventional CAS regime. Thus, the DCA treatment generally appears to be less effective than 1-MCP in maintaining firmness the longer the fruit are removed from the storage environment (Zanella et al., 2008; Köpcke, 2015; Bessemans et al., 2016; Watkins, 2018) although exceptions to this have been reported (Zanella et al., 2008; Kitteman et al., 2015).

While higher post-storage firmness and acidity retention is generally a positive 1-MCP effect, the inability of 1-MCP-treated fruit to fully ripen after storage removal limits its use on some tree fruits such as pears, which undergo considerable post-storage ripening before consumption (Villalobos-Acuña et al., 2011; Rizzolo et al., 2014; Vanoli et al, 2010, 2015). In addition, preference for non-1-MCP-treated fruit has been documented due to 1-MCP-related suppression of attributes that are important contributors to the sensory perception (e.g. taste, smell) of an ideally mature eating apple (Marin et al., 2009; Juhnevica et al., 2013, 2016; Aubert et al., 2015). In contrast, DCA-stored fruit generally ripen fully once they are removed from low O_2 storage, eventually developing the cultivar-specific sensory traits that are characteristic of fruit in an optimally ripened state (Raffo et al., 2009; Rizzolo et al., 2014; Vanoli et al., 2015).

There is little research on the effects of combining DCA storage and 1-MCP for fruit quality maintenance and disorder control over long-term storage periods. In part, this is due to the high cost of simultaneously using both technologies on large volumes of industrially stored fruit, the lack of information on optimal 1-MCP concentrations, as well the scarcity of data on the type and duration of DCA technology promoting higher crop quality retention. Recently, however, some studies on DCA and 1-MCP combinations have appeared and show interesting outcomes when fruit are evaluated post-storage. Thewes et al. (2018) observed that 1-MCP application suppressed anaerobic metabolism products (i.e. ethanol, ethyl acetate, acetaldehyde) when combined with 9 months of DCA-RQ (1.5) storage and a 7-day shelf life in 'Gala Galaxy', 'Fuji' and 'Pink Lady' fruit. Köpcke (2015) noted that the combination of 1-MCP and 5-6 months of DCA-CF storage generally maintained higher quality and reduced some physiological disorders in 'Elstar', 'Jonagold' and 'Gloster' fruit compared with either treatment alone, especially at higher storage temperatures (i.e. 2 vs. 3.5 and 10°C). Weber et al. (2017) studied the effects of 9 months of CA, DCA-RQ and DCA-CF storage with and without the addition of 1-MCP on 'Fuji Suprema'. Their findings show that the two DCA treatments maintained fruit quality similarly, and while 1-MCP helped to retain flesh firmness and reduced anaerobic volatile production, it could not be recommended for 'Fuji Suprema' under DCA conditions due to an increase in cortical cavities.

Recent work on pears show that 1-MCP confers control of economically significant disorders as well as a high degree of firmness retention, but the fruit generally do not soften when removed from the storage room (Rizzolo et al., 2015a; Vanoli et al., 2016). A DCA environment can maintain similar pear quality as 1-MCP, but for shorter storage durations with temperature control being a vital component (Vanoli et al., 2016). Although quality retention results are not as consistent as with 1-MCP, DCA-stored pears generally ripen adequately for the consumer (Rizzolo et al., 2014; Vanoli et al., 2016). If using both DCA and 1-MCP technologies can be financially justified and technically optimized for the crop and cultivar of interest, it may result in maximization of the quality of fruit delivered to the marketplace.

7 Storage energy reduction strategies

Storing produce (e.g. apple, pear) at low, non-injurious temperatures (usually >−1°C, depending on the crop) is a fundamental horticultural practice for reducing the rate of catabolic processes associated with cellular senescence. Selection of the lowest temperature for a particular commodity is usually based on years of experience with the goal of avoiding freezing and low-temperature/chilling injury, while simultaneously facilitating the highest possible degree of fruit quality retention (e.g. firmness) upon removal (DeLong, 2017). The current industry practise of establishing and maintaining large refrigerated rooms to continuously maintain low storage temperatures, however, represents a major annual electrical expense for the storage operation (Wilcox, 2001; East at al., 2013).

In light of the high electricity costs incurred when operating large refrigerated storages, several strategies have evolved over the last decade or so to reduce electrical power demand, including: (i) using constant, but higher-than-conventional temperatures [e.g. 2-4C above standard settings (Brown, 2011; McCormick et al., 2010, 2012)]; (ii) on/off cycling of compressors, fans and other equipment (Koca and Hellickson, 1993; East et al., 2013); and (iii) load-shift cooling (using electricity when daily regional demand is lowest) where utility companies permit this option. In all of these scenarios, the average air temperature of the storage room increases, which must then be managed to diminish any negative fruit quality outcomes at removal.

Several groups have demonstrated that ULO storage (e.g. 1–1.5 kPa O_2) coupled with the use of 1-MCP, or establishing a DCA environment (<1 kPa O_2) enables apple fruit to be held at higher relative temperatures, resulting in significant reduction of refrigeration-related electrical demand during long-term storage periods. For example, Kittemann et al. (2015) held 'Jonagold', 'Golden Delicious' and 'Pinova' fruit for 7 months in ULO (1 kPa O_2, 2.5 kPa CO_2)+1-MCP at 5°C and in DCA (0.7 kPa O_2, 1.5 kPa CO_2) at 1°C resulting in electrical cost savings of 70% and 20%, respectively, compared with ULO (1 kPa O_2, 2.5 kPa CO_2) at 1°C. The bulk of the electricity savings were due to lessened demand for the refrigeration compressor and ventilation fans. (Interestingly, DCA-related savings were due to less cooling and shorter ventilation demands as the 1 and 5°C rooms were located on each side of DCA storage room providing substantial cooling on two of its four walls). Fruit quality in their study was not compromised in the DCA environment or in fruit treated with 1-MCP and held at the higher temperatures. By using the combination of higher storage room temperatures and 1-MCP, other groups have reported electrical cost savings over long-term storage periods of ≥35% for 'Gala' (McCormick et al., 2010; Mazzurana et al., 2016), 30% for 'Pink Lady' (Brown, 2011), and 40% for 'Cripps Pink' (East et al., 2013) without compromising fruit quality. For 'Royal Gala' and 'Pink Lady', electrical cost savings of 35% and 15% were observed during cooling and storage, respectively, by using DCA-CF and a higher storage temperature of 5°C (Prange et al., 2013). In some of these studies, taste panels had a preference for the 1-MCP-treated apples held at the higher temperature, or for fruit held in DCA conditions (McCormick et al., 2010; Brown, 2011; Kittemann et al., 2015). As the electricity-saving benefits of combing 1-MCP and higher temperature storage, or establishing a DCA are realized, the conundrum for storage managers will then be to decide if the expense for the 1-MCP treatment or the DCA technology is offset by the potential savings in electricity.

8 Case study: DCA-chlorophyll fluorescence (CF)

The influences of temperature, scan interval during measurement, and rate of O_2 decline on the CF signatures of 'Honeycrisp' apple fruit in DCA-CF storage were investigated in this article. This work helps resolve the issue of how base level (non-stressed) and spiking (indicating hypoxic stress) fluorescence signals change when two important DCA storage environment factors (i.e. temperature, O_2 reduction) as well the time period between measurements, are altered. The results of this study help researchers understand some of the practical idiosyncrasies of the DCA-CF system and may lead to insights into how chloroplasts and other organelles acclimate under ultra-low O_2 storage conditions. For practical purposes, these findings are useful for optimizing the DCA-CF technology currently used in fruit storage facilities worldwide (Wright et al., 2010).

Four, 4-L plastic jars each received five or six 'Honeycrisp' apple fruit (~1 kg), were placed in the dark and then had CF (i.e. Fα) monitored on each jar by a pulse frequency modulated (PFM) CF system (HarvestWatch™). Intervals between CF scans ranged from 0.25, 0.5, 1, 2 and 4 hours with Fα being measured at 3.5°C. A sub-LOL event (<1 kPa O_2, causing a Fα spike) was induced at each scan interval-temperature combination which resulted in the novel observation that sub-LOL-induced Fα spike intensities were higher the longer the scan interval; the Fα response was strongly quenched at shorter intervals (Fig. 5). The practical implication of this finding is that a dark adaptation period of about 2 hours between measurement cycles is ideal for 'Honeycrisp' (and likely other apple fruit) for producing an optimal Fα spike during a sub-LOL event (i.e. hypoxic stress event).

In another experiment, about 1 kg of 'Honeycrisp' fruit were put into each of twelve, 4-L plastic jars, were then placed in a dark room with the temperature adjusted to 20, 10, 3.5 and 0°C and held at each level for three days. On the third day, a sub-LOL event occurred by N_2 gas flushing which induced a Fα spike as measured by the HarvestWatch™ system. The data show that both the

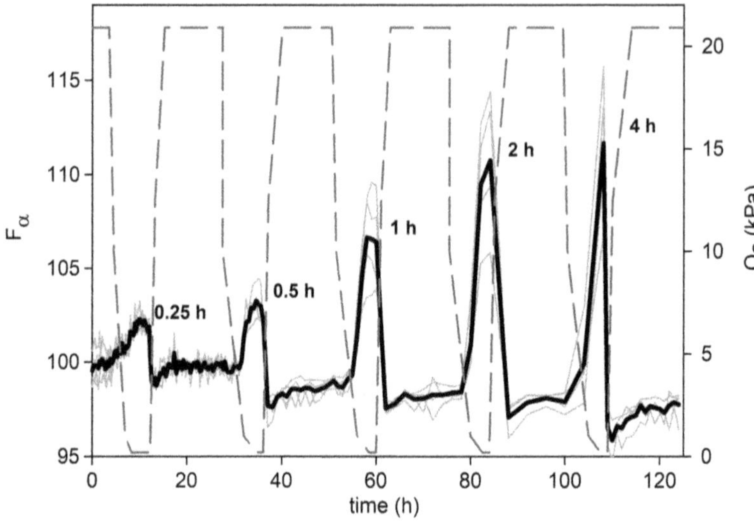

Figure 5 A representative plot showing the effect of pulse frequency modulated (PFM) scan intervals (0.25, 0.5, 1, 2 and 4 h scan intervals which were each held for 24 h) on the low-O_2 (- -) fluorescence response (Fα) (generated using ≈10 h nitrogen purge) from apples in jars held at 3.5C under ambient atmosphere. Individual scans are shown as solid grey lines and the averaged response (n = 4) is solid black.

LOL and Fα spike height increased with warmer temperatures, but decreased at lower temperatures; the Fα spike apex values, however, were largely similar irrespective of temperature (Fig. 6).

In another experiment conducted at 20 and 3.5°C, five plastic jars were each filled with approximately 1 kg of 'Honeycrisp' fruit. The O_2 levels in the fruit jars were then reduced from ambient (21 kPa O_2) to ~2 kPa O_2 over 24 hours and then lowered to 0.1 kPa O_2 over several days (gradual decline). In other fruit-filled jars, the O_2 decline to sub-LOL levels occurred within 8 hours (rapid decline). The data show that 'Honeycrisp' LOL values were highest under the slow (several days) and then at the fast O_2 decline rate (~8 hours) at the higher temperature (Table 2). Interestingly, at the same temperature (20C), the ΔFα spike intensity was lower for the slow O_2 decline compared with those fruit receiving the faster O_2 reduction. In addition, fruit held at 3.5C displayed the lowest ΔFα spike signals. Although it is unknown why the rate of O_2 decline alters both the LOL and Fα spike intensities, it may be that a more gradual onset of hypoxic/anoxic stress results in metabolic changes that induce a type of acclimation as O_2 is more slowly depleted from the immediate fruit cell environment. The LOL increase at the higher temperature could be related to O_2 solubility properties within fruit cells (O_2 solubility in solution declines as

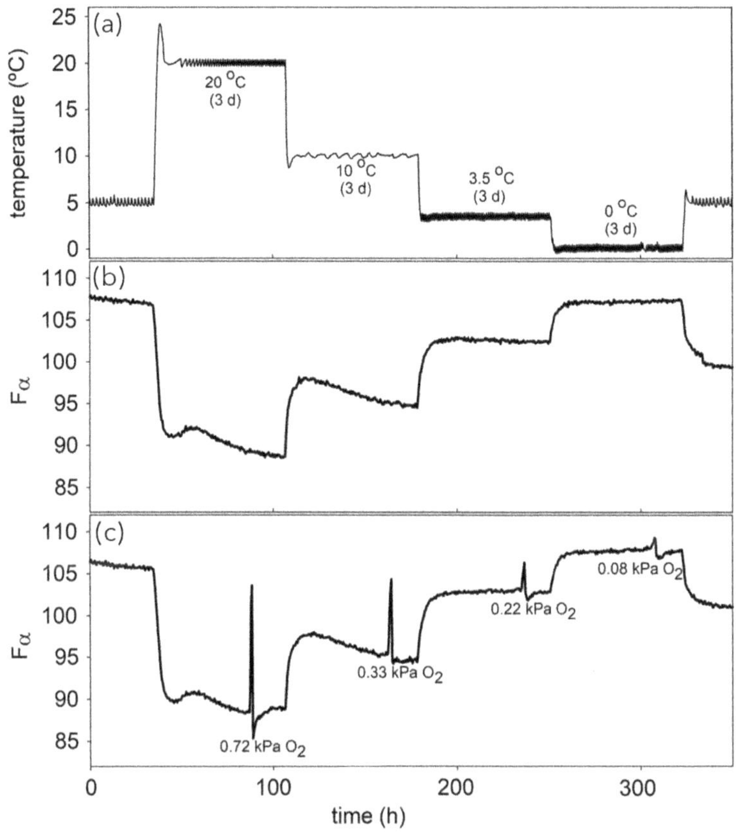

Figure 6 The effect of (a) temperature on (b) Fα and (c) Fα and the LOL (kPa O_2) during sub-LOL episodes on the third day following the temperature change in jars of apples. Plots (b) and (c) are each comprised of four individual scans.

Table 2 The effect of temperature and rate of O_2 decline on the average (n=8) LOL and fluorescence (Fα) spike intensity

O_2 decline	Temperature (°C)	LOL (kPa O_2)	Spike intensity (ΔFα)
Fast (~8 h)	20	0.72[b]	18.0[a]
	3.5	0.22[c]	3.9[c]
Slow (Days)	20	1.01[a]	10.5[b]
	3.5	0.27[d]	3.0[d]

Averages with similar letter groupings within a column are not significantly different ($P < 0.005$).

temperature increases), i.e. less dissolved O_2 in the cell is related to higher LOL thresholds; it may also be that higher metabolic rates at higher temperatures in general explain the LOL differences.

This study demonstrates the importance of selecting a Fα scan interval of sufficient duration so that Fα spikes indicating hypoxia/anoxia stress (i.e. O_2 levels below the LOL) can be accurately and easily recognized. Also, this work is the first to describe the relationship between temperature and the LOL using a DCA-CF system and also shows that the rate of O_2 reduction in the storage room can alter the magnitude of the LOL as well as the Fα spike height. The novel observation that the CF baseline is sensitive to temperature may explain why Fα signal movement can occur in commercial DCA-CF storages (barring sub-LOL events). However, the cellular mechanism(s) responsible for inducing temperature- and O_2-induced LOL and Fα changes is currently unknown.

9 Conclusion and future trends

The capacity to store agricultural food crops for long periods after harvest is foundational to the development and sustaining of modern societies. For perishable horticultural products, the twentieth century witnessed unprecedented technological innovation supported by marked advances in the knowledge of plant physiology which led to the discipline of postharvest storage science. This combination of knowledge and technical ability has been so successful that today perishable crops like apple can be held for 12 months with little loss of the nutritional and sensory attributes of fruit newly harvested. Essential to this ability is the development of CAS environments that use low, non-chilling, non-injury-inducing temperatures and O_2 and CO_2 concentrations that markedly arrest cellular respiration and senescence catabolism. Without the CAS capacity currently operational in the developed world, both the quantity and kinds of horticultural produce on the shelves of local supermarkets would look very different than that to which we have grown accustomed.

Future advances in CAS for horticultural crops will likely see more detailed procedures for establishing low O_2 regimes (e.g. ULO, DCA systems) on a crop-by-crop basis. This will enable those crops (and cultivars) that differ in response to ultra-low O_2 to be managed without incurring injury and economic loss (Prange et al., 2015; Wright et al., 2015; Table 1). This will be particularly important during the first 4–8 weeks of establishing the CA environment which includes the initial O_2 'pull-down' protocol, one or several mid-storage alterations in the low-O_2 setting as the crop acclimates to low O_2 partial pressures, and ascertaining how best to manage the low-O_2 regime when the crop is also treated with 1-MCP (Vanoli et al., 2016; DeEll and Lum, 2017).

The next decade or so will likely see more intricate protocols for applying 1-MCP to a wider diversity of horticultural crops, including: new or modified crop/cultivar-specific product concentration recommendations, unconventional applications (e.g. at mid- or post-storage removal), multiple applications (e.g. pre-storage and post-storage), coordination of preharvest and postharvest product application for high-value crops, and fine-tuning the timing of establishing a CA regime (e.g. delayed CA) when a crop has been 1-MCP-treated immediately after harvest. Since being commercialized in the early 2000s, 1-MCP, perhaps more than any other single technology, has revolutionized the management of perishable crops following harvest (particularly pome fruits), being valued for its capacity to maintain important 'at-harvest' quality attributes during and after a storage period with a single application. Although not without its problems (e.g. exacerbating CO_2-related tissue injury on susceptible apple and pear cultivars), the worldwide horticultural industry will seek to find increasingly creative ways to better manage 1-MCP and to further its use.

Another likely future development in CAS science will be the use of 1-MCP and/or DCA systems to reduce electrical energy consumption by elevating storage temperatures, with possible annual electricity savings ranging from 20–70% (McCormick et al., 2012; Kitteman et al., 2015; Neuwald et al., 2016). Benefits to the stored crop are due to the reduction in respiration and general senescence catabolism facilitated either by the ultra-low O_2 partial pressures occurring in a ULO or DCA environment, or to 1-MCP application (or both). If post-storage fruit quality is consistently high under ULO or DCA regimes, warmer storage temperatures may reduce the incidence of disorders such as internal flesh browning, skin spotting and watercore, as well as decay in apple cultivars (Köpcke, 2015; Neuwald and Kittemann, 2016; Both et al., 2018). In addition, 1-MCP-treated fruit stored or conditioned at higher temperatures may show less disorder expression upon removal and have a better taste profile compared with fruit held at lower, more conventional storage temperatures (Kittemann et al., 2015; Köpcke, 2015; Mazzurana et al., 2016).

Lastly, with the current capacity for high throughput metabolomic research, more effort is expected to be directed toward understanding the metabolic changes occurring in tissue under hypoxic stress induced by ULO or DCA storage environments. According to the findings of Brizzolara et al. (2017), some responses to low O_2 atmospheres are cultivar dependent, while others appear to be universal in apple cortical tissue (for example), especially those that point to metabolic accommodation for maintaining the energy status within cells undergoing hypoxic stress. Physiologists are only beginning to unravel the biochemical complexities that occur when a plant organ undergoes low O_2 stress for shorter or longer durations. The expectation is that during the next decade new knowledge in this area will help us sort out the major and minor cellular responses to imposed, harsh storage environments, that oddly enough, markedly contribute to the flourishing of human civilization.

10 Where to look for further information

Further reading

- Thompson, A. K., Prange, R. K., Bancroft, R. D. and Puttongsiri, T. (2018), *Controlled Atmosphere Storage of Fruit & Vegetables, 3rd Ed.*, Oxfordshire: CAB International.
- Watkins, C. B. (2018), New technologies for storage of horticultural products - there is more to adoption than availability!, *Acta Hort.*, 1194, 1233-44.
- Wills, R. B. H. and Golding, J. (2015), *Advances in Postharvest Fruit and Vegetable Technology*, Boca Raton: CRC Press.
- Wright, A. H., DeLong, J. M., Arul, J. and Prange, R. K. (2015), The trend toward lower oxygen levels during apple (*Malus x domestica* Borkh) storage - a review, *J. Hortic. Sci & Biotech.*, 90(1), 1-13.
- Yahia, E. M. (2009), *Modified and Controlled Atmospheres for the Storage, Transportation, and Packaging of Horticultural Commodities*, Boca Raton: CRC Press.

Key journals/conferences

- Postharvest Biology and Technology.
- Journal of the American Society for Horticultural Science, HortScience, HortTechnology (journals of the American Society for Horticultural Science).
- ASHS (American Society for Horticultural Science) annual conferences (https://ashs.org).
- ISHS (International Society for Horticultural Science) conferences (http://www.ishs.org).

11 References

AgroFresh. (2019), https://www.agrofresh.com/ (accessed 14 January 2019).

Argenta, L., Fan, X. and Mattheis, J. (2000), Delaying establishment of controlled atmosphere of CO_2 exposure reduces 'Fuji' apple CO_2 injury without excessive fruit quality loss, *Postharvest Biol. Technol.*, 20, 221-2.

Aubert, C., Mathieu-Hurtiger, V. and Vaysse, P. (2015), Effects of dynamic atmosphere on volatile compounds, polyphenolic content, overall fruit quality, and sensory evaluation of 'Pink-Lady®' apple, *Acta Hortic.*, 1071, 275-80.

Bessemans, N., Verboven, P., Verlinden, B. E. and Nicolai, B. M. (2016), A novel type of dynamic controlled storage based on the respiratory quotient (RQ-DCA), *Postharvest Biol. Technol.*, 115, 91-102.

Bessemans, N., Verboven, P., Verlinden, B. E. and Nicolai, B. M. (2018), RQ-based dynamic controlled atmosphere storage of apple fruit, *Acta Hortic.*, 1194, 681-7.

Published by Burleigh Dodds Science Publishing Limited, 2020.

Binder, B. M. and Bleecker, A. B. (2003), A model for ethylene receptor function and 1-methylcyclopropene action, *Acta Hortic.*, 628, 177–87.

Blackman, F. F. (1928), Analytic studies in plant respiration, *Proc. Royal Soc. London, Series B*, 103, 491–523.

Boersig, M. R., Kader, A. A. and Romani, R. J. (1988), Aerobic-anaerobic respiratory transition in pear fruit and cultured pear fruit cells, *J. Amer. Soc. Hortic. Sci.*, 113, 869–73.

Both, V., Thewes, F. R., Brackmann, A., de Oliveira Anese, R., de Freitas Ferreira, D. and Wagner, R. (2017), Effects of dynamic controlled atmosphere by respiratory quotient on some quality parameters and volatile of 'Royal Gala' apple after long-term storage, *Food Chem.*, 215, 483–92.

Both, V., Brackmann, A., Thewes, F. R., Weber, A., Schultz, E. E. and Ludwig, V. (2018), The influence of temperature and 1-MCP on quality of 'Galaxy' apple stored in controlled atmosphere and dynamic controlled atmosphere, *Food Pack. Shelf Life*, 16, 168–77.

Brackmann, A., Weber, A. and Both, V. (2015), CO_2 partial pressure for respiratory quotient and HarvestWatch™ dynamic controlled atmosphere for 'Galaxy' apples storage, *Acta Hortic.*, 1079, 435–40.

Brizzolara, S., Santucci, C., Tenori, L., Hertog, M., Nicolai, B., Stürtz, S., Zanella, A. and Tonutti, P. (2017), A metabolomic approach to elucidate apple fruit response to static and dynamic controlled atmosphere storage, *Postharvest Biol. Technol.*, 127, 76–87.

Brown, G. (2011), Saving apple storage costs, *Aust. Fruitgrower*, 5(1), 19–21.

Burdon, J. (2009), Dynamic CA storage of avocados: Technology for managing exports?, http://www.avocadosource.com/Journals/AUSNZ/AUSNZ_2009/BurdonJ2009.pdf (accessed 15 December 2018).

Burdon, J. and Lallu, N. (2008), Dynamic controlled atmosphere storage of avocados: Technology for managing exports? *N.Z. Avocado Grow. Assoc. Ann. Res. Report*, 8, 123–6.

Burdon, J., Lallu, N., Haynes, G., McDermott, K. and Billing, D. (2008), The effect of delays in establishment of a static or dynamic controlled atmosphere on the quality of 'Hass' avocado fruit, *Postharvest Biol. Technol.*, (49), 61–8.

Burdon, J., Lallu, N., Haynes, G., Pidakala, P., Billing, D. and McDermott, K. (2010), Dynamic controlled atmosphere storage of New Zealand-grown 'Hass' avocado fruit, *Acta Hortic.*, 876, 47–54.

Calhan, Ő., Onursal, C. E., Güneyli, A., Eren, İ. and Koyuncu, M. A. (2016), Effects of different storage techniques and 1-MCP application on quality of 'Granny Smith' apple, Acta Hortic., 1120, 123–9.

Cliff, M. A. and Toivonen, P. M. A. (2017), Sensory and quality characteristics of 'Ambrosia' apples in relation to harvest maturity for fruit stored up to eight months, *Postharvest Biol. Technol.*, 132, 145–53.

Commission Implementing Regulation (EU) No 578/2012. (2012), Commission Implementing Regulation (EU) No 578/2012 of 29 June 2012 concerning the non approval of the active substance diphenylamine, in accordance with Regulation (EC) No 1107/2009 of the European Parliament and of the Council concerning the placing of plant protection products on the market, *Off. J. Eur. Union*, L171/2.

de Castro, E., Biasi, B., Mitcham, E., Tustin, S., Tanner, D. and Jobling, J. (2007), Carbon dioxide-induced flesh browning in Pink Lady apples, *J. Amer. Hort. Sci.*, 132(5), 713–19.

DeEll, J. R. and Ehsani-Moghaddam, B. (2012), Delayed controlled atmosphere storage affects disorders of 'Empire' apples, Postharvest Biol. Technol., 67, 167–71.

DeEll, J. R. and Lum, G. B. (2017), Effects of low oxygen and 1-methylcyclopropene on storage disorders of 'Empire' apples, *HortScience*, 52(9), 1265–70.

DeEll, J. R., Murr, D. P., Len Wiley, L. and Porteous, M. D. (2003), 1-Methylcyclopropene (1-MCP) increases CO_2 injury in apples, *Acta Hortic.*, 600, 277–80.

DeEll, J. R., Lum, G. B. and Ehsani-Moghaddam, B. (2016), Effects of delayed controlled atmosphere storage on disorder development in 'Honeycrisp' apples, *Can. J. Plant Sci.*, 96, 621–9.

DeLong, J. M. (2017), Volume 1: Storage, in B. Thomas, D. J. Murphy and B. G. Murray (Eds), *Encyclopedia of Applied Plant Sciences*, pp. 340–50. Waltham, MA: Elsevier, Academic Press.

DeLong, J. M., Prange, R. K., Leyte, J. C. and Harrison, P. A. (2004), A new technology that determines low-oxygen thresholds in controlled-atmosphere-stored apples, *HortTechnol.*, 14, 262–6.

DeLong, J. M., Prange, R. K. and Harrison, P. A. (2007), Chlorophyll fluorescence-based low-O_2 CA storage of organic 'Cortland' and 'Delicious' apples, *Acta Hortic.*, 737, 31–7.

de Oliveira Anese, R., Brackmann, A., Wendt, L. M., Thewes, F. R., Schultz, E. E., Ludwig, V. and Berghetti, M. R. P. (2018), Interaction of 1-methylcyclopropene, temperature and dynamic controlled atmosphere by respiratory quotient of 'Galaxy' apples storage, *Food Pack. Shelf Life.*, 20, 100246. doi: 10.1016/j.fpsl.2018.07.004.

Deuchande, T., Carvalho, S. M. P., Gutteres, U., Fidalgo, F., Isidoro, N., Larrigaudière, C. and Vasconcelos, M. W. (2016), Dynamic controlled atmosphere for prevention of internal browning disorders in 'Rocha' pear, *Food Sci. Technol.*, 65, 725–30.

Dilley, D. R. (2006), Development of controlled atmosphere storage technologies, *Stewart Postharvest Rev.*, 6(5), 1–8.

Dilley, D. R. (2010), Controlled atmosphere storage - chronology and technology, *Acta Hortic.*, 857, 493–501.

East, A. R., Smale, N. J. and Trujillo, F. J. (2013), Potential for energy cost savings by utilising alternative temperature control strategies for controlled atmosphere stored apples, *Inter. J. Refrig.*, 36, 1109–17.

Eren, I., Calhan, Ö., Onursal, C. E. and Güneyli, A. (2015), Effects on controlled atmosphere, dynamic controlled atmosphere and 1-MCP of quality of 'Granny Smith' apples, *Acta Hortic.*, 1071, 495–502.

Fawbush, F., Nock, J. F. and Watkins, C. B. (2008), External carbon dioxide injury and 1-methylcyclopropene (1-MCP) in the 'Empire' apple, *Postharvest Biol. Technol.*, 48, 92–8.

Folchi, A., Bertolini, P. and Mazzoni, D. (2015), Influence of controlled atmosphere and dynamic controlled atmosphere on physiology, scald prevention and CA related disorders of 'Abbé Fétel' pears during long-term storage, *Acta Hortic.*, 1071, 195–202.

Fysium®:Pace. (2019), Fysium® 1-MCP, https://www.paceint.com/product/fysium (accessed 15 March 2019).

Gabioud Rebeaud, S. and Gasser, F. (2015), Fruit quality as affected by 1-MCP treatment and DCA storage - a comparison of the two methods, *Eur. J. Hortic. Sci.*, 80(1), 18–24.

Gasser, F. and von Arx, K. (2015), Dynamic CA storage of organic apple cultivars, *Acta Hortic.*, 1071, 527–32.

Gasser, F., Eppler, T., Naunheim, W., Gabioud, S. and Hoehn, E. (2008), Control of the critical oxygen level during dynamic CA storage of apples by monitoring respiration as well as chlorophyll fluorescence, *Acta Hortic.*, 796, 69-76.

Gasser, F., Eppler, T., Naunheim, W., Gabioud, S. and Bozzi Nising, A. (2010), Dynamic CA storage of apples: Monitoring of the critical oxygen concentration and adjustment of optimum conditions during oxygen reduction, *Acta Hortic.*, 876, 39-46.

Ghahramani, F. and Scott, K. J. (1998), Oxygen stress of 'Granny Smith' apples in relation to superficial scald, ethanol, α-farnesene and conjugated trienes, *Aust. J. Agric. Res.*, 49, 207-10.

Greene, D. W. and Schupp, J. R. (2004), Effect of aminoethoxyvinylglycine (AVG) on preharvest drop, fruit quality, and maturatioh of 'McIntosh' apples. II. Effect of timing and concentration relationships and spray volume, *HortScience*, 39(5), 1036-41.

Hansen, F. and Mellenthin, W. M. (1979), Commercial handling and storage practices for winter pears, *Agr. Expt. Sta., Oregon State Univ., Spec. Rpt.*, pp. 5-6.

Hazel Technologies Inc. (2019), Increase shelf life. Decrease waste, https://www.hazeltechnologies.com (accessed 15 March 2019).

Hoehn, E., Prange, R. K. and Vigneault, C. (2009), Chapter 2: Storage technology and applications, in E. M. Yahia (Ed.), *Modified and Controlled Atmospheres for the Storage, Transportation, and Packaging of Horticultural Commodities*, pp. 17-50. Baton Rouge: CRC Press.

Huelin, F. E. and Coggiola, I. M. (1970), Superficial scald, a functional disorder of stored apples. V. Oxidation of α-farnesene and its inhibition by diphenylamine, *J. Sci. Food Agric.*, 21, 44-8.

Juhnevica-Radenkova, K. and Radenkovs, V. (2016), Influence of 1-methylcyclopropene and ULO conditions on sensory characteristics of apple fruit grown in Latvia, *J. Hortic. Res.*, 24(1), 37-46.

Juhnevica -Radenkova, K., Skudra, L., Skrivele, M, Radenkovs, V., Seglina, D. and Stepanovs, A. (2013), Effect of 1-methylcyclopropene treatment on sensory characteristics of apple fruit, *Environ. Exp. Biol.*, 11, 99-105.

Jung, S. K. and Watkins, C. B. (2011), Involvement of ethylene in browning development of controlled atmosphere-stored 'Empire' apple fruit, *Postharvest Biol. Technol.* 59, 219-26.

Kittemann, D., McCormick, R. and Neuwald, D. A. (2015), Effect of high temperature and 1-MCP application or dynamic controlled atmosphere on energy savings during apple storage, *Eur. J. Hortic. Sci.*, 80(1), 33-8.

Koca, R. W. and Hellickson, M. L. (1993), Energy savings in evaporator fan-cycled apple storages, *Appl. Eng. Agric.*, 9(6), 553-60.

Köpcke, D. (2009), Einfluss der sauerstoff-konzentration im lager auf den schalenfleckenbefall bei 'Elstar', *Mitteilungen des Obstbauversuchsringes des Alten Landes*, 64, 303-9.

Köpcke, D. (2015), 1-Methylcyclopropene (1-MCP) and dynamic controlled atmosphere (DCA) applications under elevated storage temperatures: Effects on fruit quality of 'Elstar', 'Jonagold' and 'Gloster' apple (*Malus domestica* Borkh.), *Eur. J. Hortic. Sci.*, 80(1), 25-32.

Lafer, G. (2008), Storability and fruit quality of 'Braeburn' apples as affected by harvest date, 1-MCP treatment and different storage conditions, *Acta Hortic.*, 796, 179-83.

Lafer, G. (2009). Practical experience with new storage technologies in Austria – dynamic CA (DCA) storage and SmartFresh™, *Eur. Fruit Growers Mag.*,1(1), 24-7.

Lafer, G. (2011), Effect of different CA storage conditions on storability and fruit quality of organically grown 'Uta' pears, *Acta Hortic.*, 909, 757–60.

Lallu, N. and Burdon, J. (2007), Experiences with recent postharvest technologies in kiwifruit, *Acta Hortic.*, 753, 733–40.

Lau, O. L. (1997), Initial low oxygen stress offers no scald control benefits to 'Starkrimson Delicious' apples in 0.7 kPa O_2 storage, *HortScience*, 32(7), 1239–41.

Little, C. R., Faragher, J. D. and Taylor, H. J. (1982), Effects of initial oxygen stress treatments in low oxygen modified atmosphere storage of 'Granny Smith' apple, *J. Amer. Soc. Hortic. Sci.*, 107, 320–3.

Lurie, S. and Watkins, C. B. (2012), Superficial scald, its etiology and control, *Postharvest Biol. Technol.*, 65, 44–60.

Neuwald, D. A. and Kittemann, D. (2016), The incidence of Neofabraea spp. in `Pinova' apples can be reduced at elevated storage temperatures, *Acta Hortic.*, 1144, 231–236. doi. 10.17660/ActaHortic.2016.1144.34.

Neuwald, D. A., Spuhler, M., Wünsche, J. and Kittemann, D. (2016), New apple storage technologies can reduce energy usage and improve storage life, *17th International Conference on Organic Fruit-Growing*, Stuttgart, Germany, http://www.ecofruit.net/2016/29_Neuwald_184bis187.pdf (accessed 15 December 2018).

Nock, J. F. and Watkins, C. B. (2013), Repeated treatment of apple fruit with 1-methylcyclopropene (1-MCP) prior to controlled atmosphere storage, *Postharvest Biol. Technol.*, 79, 73–9.

Nock, J. F., Doerflinger, F. C., Sutanto, G., Al Shoffe, Y., Gunes, N., Zhang, Y., Wright, A. H., DeLong, J. and Watkins, C. B. (2019), Managing stem-end flesh browning, a physiological disorder of 'Gala' apple, *Acta Hortic* (in press).

Marin, A. B., Colonna, A. E., Kudo, K., Kuperfman, E. M. and Mattheis, J. (2009), Measuring consumer response to 'Gala' apples treated with 1-methylcyclopropene (1-MCP), *Postharvest Biol. Technol.*, 51, 73–9.

Mattè, P., Buglia, L., Boschetti, A., Fadanelli, L., Christè, C. and Zeni, F. (2005), ILOS+ULO as a practical technology for apple scald prevention, *Acta Hortic.*, 682, 1543–50.

Mattheis, J. (2007), Harvest and postharvest practices for optimum quality, Final Project Report. Project Number PR-04-433, Washington Tree Fruit Research Commission Research Reports, http://jenny.tfrec.wsu.edu/wtfrc/PDFfinalReports/2007FinalReports/MattheisPear.pdf (accessed 15 December 2018).

Mattheis, J. P. and Rudell, D. (2011), Responses of 'd'Anjou' pear (*Pyrus communis* L.) fruit to storage at low oxygen setpoints determined by monitoring fruit chlorophyll fluorescence, *Postharvest Biol. Technol.*, 60, 125–9.

Maxin, P. and Weber, R. W. S. (2011), Control of *Phacidiopycnis washingtonensis* storage rot of apples by hot-water treatments without the ethylene inhibitor 1-MCP, *J. Plant Dis. Protect.*, 118(6), 222–4.

Mazzurana, E. R., Argenta, L. C., Do Amarante, C. V. T. and Steffens, C. A. (2016), Potential benefits of temperature increase during storage under controlled atmosphere of 'Gala' apples treated with1-MCP, *Rev. Brasil. Frutic.*, 38(1), 43–52.

McCormick, R., Neuwald, D. A. and Streif, J. (2010), A case study: Potential energy savings using a 1-MCP with 'Gala' apples in commercial CA storage, *Acta Hortic.*, 877, 323–6.

McCormick, R., Neuwald, D. A. and Streif, J. (2012), Commercial apple CA storage temperature regimes with 1-MCP (SmartFresh™): Benefits and risks, *Acta Hortic.*, 877, 263–70.

Poldervaart, G. (2010a), KOB storage day 2010: DCA storage goes from strength to strength, *Eur. Fruit Mag.*, 2(8), 6-9.
Poldervaart, G. (2010b), DCA viable alternative for organic Topaz, *Eur. Fruit Mag.* 2(8), 15.
Prange, R. K. (2018), Chapter 9: Dynamic DA storage, in A. K., Thompson, R. K. Prange, R. D. Bancroft and T. Puttongsiri (Eds), *Controlled Atmosphere Storage of Fruit & Vegetables*, 3rd Ed., pp. 125-42. Oxfordshire: CAB International.
Prange, R. K. and DeLong, J. M. (2003), 1-Methylcyclopropene: The 'magic bullet' for horticultural products? *Chron. Hortic.*, 43(1), 11-14.
Prange, R. K., DeLong, J. M., Harrison, P. A., Leyte, J. C. and McLean, S. D. (2001), Oxygen concentration affects chlorophyll fluorescence in chlorophyll-containing fruits and vegetables, *8th International Controlled Atmosphere Research Conference (CA2001)*, Rotterdam, The Netherlands, 8-13 July 2001 (Abstract).
Prange, R. K., Delong, J. M., Leyte, J. C. and Harrison, P. A. (2002), Oxygen concentration affects chlorophyll fluorescence in chlorophyll-containing fruit, *Postharvest Biol. Technol.*, 24, 201-5.
Prange, R. K., DeLong, J. M., Harrison, P. A., Leyte, J. C. and McLean, S. D. (2003), Oxygen concentration affects chlorophyll fluorescence in chlorophyll-containing fruit and vegetables, *J. Amer. Soc. Hortic. Sci.*,128, 603-7.
Prange, R. K., DeLong, J. M. and Harrison, P. A. (2005), Quality management through respiration control: Is there a relationship between lowest acceptable respiration, chlorophyll fluorescence and cytoplasmic acidosis? *Acta Hortic.*, 682, 823-30.
Prange, R. K., DeLong, J. M. and Wright, A. H. (2010), Chlorophyll fluorescence: Applications in postharvest horticulture, *Chron. Hortic.*, 50(1), 13-16.
Prange, R. K., DeLong, J. M. and Wright, A. H. (2011), Storage of pears using dynamic controlled-atmosphere (DCA), a non-chemical method, *Acta Hortic.*, 909, 707-17.
Prange, R. K., DeLong, J. M. and Wright, A. H. (2012), Improving our understanding of storage stress using chlorophyll fluorescence, *Acta Hortic.*, 945, 89-96.
Prange, R. K., DeLong, J. M., Wright, A. H. and Zanella, A. (2013), History, current situation and future prospects for dynamic controlled atmosphere (DCA) storage of fruits and vegetables, using chlorophyll fluorescence, *Acta Hortic.*, 1012, 905-15.
Prange, R. K., Wright, A. H., DeLong, J. M. and Zanella, A. (2015), A review on the successful adoption of dynamic controlled-atmosphere (DCA) storage as a replacement for diphenylamine (DPA), the chemical used for control of superficial scald in apples and pears, *Acta Hortic.*, 1071, 389-96.
Raffo, A., Kelderer, M., Paoletti, F. and Zanella, A. (2009), Impact of innovative controlled atmosphere storage technologies and postharvest treatments on volatile compound production in Cv. Pinova apples, *J. Agric. Food Chem.*, 57, 915-23.
Rizzolo, A., Vanoli, M., Grassi, M. and Eccher Zerbini, P. (2008), Gas exchange in 1-methylcyclopropene treated 'Abbe Fetel' pears during storage in different atmospheres, *Acta Hortic.*, 796, 143-6.
Rizzolo, A., Grassi, M. and Vanoli, M. (2014), 1-Methylcyclopropene application, storage temperature and atmosphere modulate sensory quality changes in shelf-life of 'Abbé Fétel' pears, *Postharvest Biol. Technol.*, 92, 87-97.
Rizzolo, A., Buccheri, M., Bianchi, G., Grassi, M. and Vanoli, M. (2015a), Quality of 'Conference' pears as affected by initial low oxygen stress, dynamically controlled atmosphere and 1-MCP treatment, *Acta Hortic.*, 1079, 343-50.

Rizzolo, A., Grassi, M. and Vanoli, M. (2015b), Influence of storage (time, temperature, atmosphere) on ripening, ethylene production and texture of 1-MCP treated 'Abbé Fétel' pears, *Postharvest Biol. Technol.*, 109, 20-9.

Robinson, T. L., Watkins, C. B., Hoying, S. A., Nock, J. F. and Iungerman, K. I. (2006), Aminoethoxyvinylglycine and 1-methylcyclopropene effects on 'McIntosh' preharvest drop, fruit maturation and fruit quality after storage, *Acta Hortic.*, 727, 473-80.

Saltveit, M. E. (1999), Effect of ethylene on quality of fresh fruits and vegetables, *Postharvest Biol. Technol.*, 15, 279-92.

Saltveit, M. E. (2003), Is it possible to find an optimal controlled atmosphere? *Postharvest Biol. Technol.*, 27, 3-13.

Saquet, A. A., Streif, J. and Bangerth, F. (2003a), Energy metabolism and membrane lipid alterations in relation to brown heart development in 'Conference' pears during delay controlled atmosphere storage, *Postharvest Biol. Technol.*, 30, 1123-32.

Saquet, A. A., Streif, J. and Bangerth, F. (2003b), Reducing internal browning disorders in 'Braeburn' apples by delayed controlled atmosphere storage and some related physiological and biochemical changes, *Acta Hortic.* 628, 453-8.

Saquet, A. A., Streif, J. and Almeida, D. P. F. (2017), Responses of 'Rocha' pear to delayed controlled atmosphere storage depend on oxygen partial pressure, *Sci Hortic.*, 222, 17-21.

Schouten, S. P., Prange, R. K., Verschoor, J. Lammers, T. R. and Oosterhaven, J. (1997), Improvement of quality of Elstar apples by dynamic control of ULO conditions, *Postharvest Hortic. Ser.*, 16, 71-8.

Schouten, S. P., Prange, R. K., Verschoor, J. L., Lammers, T. R. and Oosterhaven, J. (1998), Improvement of quality of Elstar apples by dynamic control of ULO conditions, in I. Farkas (Ed.), *2nd International Workshop on Control Applications in Post-Harvest and Processing Technology*, pp. 25-9. Budapest, Hungary: Agrotechnological Research Institute, https://www.sciencedirect.com/journal/ifac-proceedings-volumes/vol/31/issue/9 (accessed 21 March 2019).

Sisler, E. C. and Serek, M. (2003), Compounds interacting with the ethylene receptor in plants, *Plant Biol.*, 5, 473-80.

Smock, R. M. (1955), A new method of scald control, *Amer. Fruit Grower*, 75, 20.

Stanger, M. C., Steffens, C. A., Soethe, C., Moreira, M. A., do Amarante, C. V. T., Both, V. and Brackmann, A. (2018), Phenolic compounds content and antioxidant activity of 'Galaxy' apples stored in dynamic controlled atmosphere and ultralow oxygen conditions, *Postharvest Biol. Technol.*, 144, 70-6.

Stephens, B. E. and Tanner, D. J. (2005), The Harvest Watch system - measuring fruit's healthy glow, *Acta Hortic.*, 687, 363-4.

Taiz, L. and Zeiger, E. (2010), *Plant Physiology, 5th Ed.*, Sunderland, MA: Sinauer Associates Inc.

Thewes, F. R., Both, V., Brackmann, A., Weber, A. and de Oliveira Anese, R. (2015), Dynamic controlled atmosphere and ultralow oxygen storage on 'Gala' mutants quality maintenance, *Food Chem.*, 188, 72-80.

Thewes, F. R., Brackmann, A., de Oliveira Anese, R., Bronzatto, E. S., Schultz, E. E. and Wager, R. (2017a), Dynamic controlled atmosphere storage suppresses metabolism and enhances volatile concentrations of 'Galaxy' apple harvested at three maturity stages, *Postharvest Biol. Technol.*, 127, 1-13.

Thewes, F. R., Brackmann, A., Both, V., Weber, A., de Oliveira Anese, R., dos Santos Ferrao, T. and Wager, R. (2017b), The different impacts of dynamic controlled atmosphere and controlled atmosphere storage in the quality attributes of 'Fuji Suprema' apples, *Postharvest Biol. Technol.*, 130, 7-20.

Thewes, F. R., Brackmann, A., de Oliveira Anese, R., Ludwig, V., Schultz, E. E. and Berghetti, M. R. P. (2018), 1-Methylcyclopropene suppresses anaerobic metabolism in apples stored under dynamic controlled atmosphere monitored by respiratory quotient, *Sci. Hortic.*, 227, 288-95.

Torres, C. A. and Hernández, O. (2015), Superficial scald assessment on 'Granny Smith' apples stored under dynamic controlled atmospheres in commercial operations in Chile, *Acta Hortic.*, 1079, 421-8.

Truter, A. B., Combrink, J. C. and Burger, S. A. (1994), Control of superficial scald in 'Granny Smith' apples by ultra-low and stress levels of oxygen as an alternative to diphenylamine, *J. Hortic. Sci.*, 69(3), 581-7.

Valent. (2017), ReTain plant growth regulator label, https://www.valent.ca/valentus/Data/Labels/2017-RET-0001-E.pdf (accessed 17 December 2018).

Vanoli, M., Rizzolo, A., Grassi, M. and Eccher Zerbini, P. (2007), Storage disorders and quality in 'Abbé Fétel' pears treated with 1-methylcyclopropene, *Proceedings of the International Congress COST Action 924: Novel Approaches for the Control of Postharvest Diseases and Disorders*, Bologna, pp. 269-77, https://www.cabdirect.org/cabdirect/abstract/20093269505?q=(Storage+disorders+and+quality+in+%27Abb%c3%a9+F%c3%a9tel%27+pears+treated+with+1-methylcyclopropene (accessed 3 April 2019).

Vanoli, M., Eccher Zerbini, P., Grassi, M. and Rizzolo, A. (2010), Ethylene production and quality in 1-methylcyclopropene treated 'Abbé Fétel' pears after storage in dynamically controlled atmosphere, *Acta Hortic.*, 876, 31-8.

Vanoli, M., Rizzolo, A. and Grassi, M. (2015), Fruit quality and sensory characteristics of 1-MCP treated 'Abbé Fétel' pears after storage under dynamic controlled atmosphere at different temperatures, *Acta Hortic.*, 1071, 437-45.

Vanoli, M., Grassi, M. and Rizzolo, A. (2016), Ripening behavior and physiological disorders of 'Abate Fetel' pears treated at harvest with 1-MCP and stored at different temperatures and atmospheres, *Postharvest Biol. Technol.*, 111, 275-85.

Van Schaik, A. C. R., Van de Geijn, F. G., Verschoor, J. A. and Veltman, R. H. (2015), A new interactive storage concept: Dynamic controlled of respiration, *Acta Hortic.*, 1071, 245-51.

Veltman, R. H., Vershoor, J. A. and van Dugteren, J. H. R. (2003), Dynamic control system (DCS) for apples (*Malus domestica* Borkh cv 'Elstar'): Optimal quality through storage based on product response, *Postharvest Biol. Technol.*, 27, 79-86.

Villalobos-Acuña, M. G., Biasi, W. V., Flores, S., Jiang, C.-Z., Reid, M. S., Willits, N. H. and Mitcham, E. J. (2011), Effect of maturity and cold storage on ethylene biosynthesis and ripening in 'Bartlett' pears treated after harvest with 1-MCP, *Postharvest Biol. Technol.*, 59, 1-9.

Wang, Z. and Dilley, D. R. (2000), Initial low oxygen stress controls superficial scald of apples, *Postharvest Biol. Technol.*, 18, 201-13.

Wang, Z., McCully, T. J. and Dilley, D. R. (1997), The effect of ultra low oxygen storage, initial oxygen stress and ethanol vapor treatments on controlling superficial scald of 'Granny Smith', 'Law Rome', 'Red Delicious' and 'Idared' apples, in E. J. Mitchman, (Ed.), *Apples and Pears*. Proc. 7th Int. Nat. CA Res. Conf., 2, pp. 105-111.

Watkins, C. B. (2010), Managing physiological processes in fruits and vegetables with inhibitors of ethylene biosynthesis and perception, *Acta Hortic.*, 880, 301-10.

Watkins, C. B. (2018), New technologies for storage of horticultural products - there is more to adoption than availability, *Acta Hortic.*, 1194, 1233-44.

Watkins, C. B. and Nock, J. F. (2012a), Controlled-atmosphere storage of 'Honeycrisp' apples, *HortScience*, 47(7), 886-92.

Watkins, C. B. and Nock, J. F. (2012b), Rapid 1-methylcyclopropene (1-MCP) treatment and delayed controlled atmosphere storage of apples, *Postharvest Biol. Technol.*, 69, 24-31.

Weber, A., Thewes, F. R., de Oliveira Anese, R., Both, V., Pavanello, E. P. and Brackmann, A. (2017), Dynamic controlled atmosphere (DCA): Interaction between DCA methods and 1-methylcyclopropene on 'Fuji Suprema' apple quality, *Food Chem.*, 235, 136-44.

Wilcox, M. H. (2001), Energy efficiency in fruit storage warehouses, *2001 Proceedings, Energy Efficiency in Fruit Storage Warehouses, WSU-TFREC Postharvest Information Network*, pp. 1-8, http://postharvest.tfrec.wsu.edu/PC2001T.pdf (accessed 20 December 2018).

Withnall, M. (2008), Alternative 'dynamically controlled atmosphere' technique for control of long-term storage disorders, *The Fruit Grower Magazine*, May 2008, Maidstone, UK, pp. 19-20.

Wright, A. H., DeLong, J. M., Franklin, J. L., Lada, R. R. and Prange, R. K. (2008), A new minimum fluorescence parameter, as generated using pulse frequency modulation, compared with pulse amplitude modulation: Fα versus Fo, *Photosyn. Research*, 97, 205-14.

Wright, H., DeLong, J., Harrison, P. A., Gunawardena, A. H. L. A. N. and Prange, R. (2010), The effect of temperature and other factors on chlorophyll a fluorescence and the lower oxygen limit in apples (*Malus domestica*), *Postharvest Biol. Technol.*, 55, 21-8.

Wright, A. H., DeLong, J. M., Gunawardena, A. H. L. A. N. and Prange, R. K. (2011), The interrelationship between the lower oxygen limit, chlorophyll fluorescence and the xanthophyll cycle in plants, *Photosynth. Res.*, 107, 223-35.

Wright, A. H., DeLong, J. M., Gunawardena, A. H. L. A. N. and Prange, R. K. (2012), Dynamic controlled atmosphere (DCA): Does fluorescence reflect physiology in storage? *Postharvest Biol. Technol.*, 64, 89-30.

Wright, A. H., DeLong, J. M., Arul, J. and Prange, R. K. (2015), The trend toward lower oxygen levels during apple (*Malus x domestica* Borkh) storage - a review, J. Hortic. Sci. Biotech., 90(1), 1-13.

Yang, S. F. (1985), Biosynthesis and action of ethylene, *HortScience*, 20, 41-5.

Yearsley, C. W., Lallu, N., Burmeister, D., Burdon, J. and Billing, D. (2003), Can dynamic controlled atmosphere storage be used for 'Hass' avocados? *N. Z. Avocado Grow. Assoc. Ann. Res. Report*, 3, 86-92.

Zanella, A. (2003), Control of apple superficial scald and ripening - a comparison between 1-methylcyclopropene and diphenylamine postharvest treatments, initial low oxygen stress and ultra low oxygen storage, *Postharvest Biol. Technol.*, 27, 69-78.

Zanella, A. and Rossi, O. (2015), Post-harvest retention of apple fruit firmness by 1-methylcyclopropene (1-MCP) treatment or dynamic CA storage with chlorophyll fluorescence (DCA-CF), *Eur. J. Hortic. Sci.*, 80(1), 11-17.

Zanella, A. and Stürz, S. (2013), Replacing DPA postharvest treatment by strategical application of novel storage technologies controls scald in 1/10th of EUs apples producing area, *Acta Hortic.*, 1012, 419-26.

Zanella, A. and Stürz, S. (2015), Optimizing postharvest life of horticultural products by means of dynamic CA: Fruit physiology controls atmosphere composition during storage, *Acta Hortic.*, 1071, 59–68.

Zanella, A., Cazzanelli, P., Panarese, A., Coser, M., Cecchinel, M. and Rossi, O. (2005), Fruit fluorescence response to low oxygen stress: Modern storage technologies compared to 1-MCP treatment of apple, *Acta Hortic.*, 682, 1535–42.

Zanella, A., Cazzanelli, P. and Rossi, O. (2008), Dynamic controlled atmosphere (DCA) storage by the means of chlorophyll fluorescence response for firmness retention in apple, *Acta Hortic.*, 796, 77–82.

Zerbini, P. E. and Grassi, M. (2010), Chlorophyll fluorescence and gas exchanges in 'Abbé Fétel' and 'Conference' pears stored in atmosphere dynamically controlled with the aid of fluorescence sensors, *Acta Hortic.*, 857, 469–74.

Chapter 3

Advances in modified atmosphere and active packaging of horticultural produce

Jeffrey S. Brandenburg, The JSB Group LLC, USA

1 Introduction

Modified atmosphere packaging (MAP) is the key packaging technology associated with the shelf life and quality extension of fresh produce, specifically fresh-cut produce. Fresh-cut produce is a submarket to the entire fresh produce market. This submarket is comprised of the following segments: packaged salads as well as value-added fruits and vegetables. As stated by the United Fresh Produce Association's Fresh Facts on Retail Year in Review 2018, the current market size in the US is in excess of $2.112 million.

This chapter will discuss the design elements, technology, materials and applications of MAP. Discussions will include the intrinsic relationship between the packaging and fresh produce and detail the critical importance of produce physiology when designing packaging systems. Specific polymers, films, structures and technologies will be reviewed and analyzed with respect to their individual properties and subsequent impact on MAP. An analysis of packaging types and formats will also be detailed. Additive technologies such as antimist, moisture, slip, ethylene control and antimicrobials and their impact on the functionality of MAP will be explored. The chapter will conclude

http://dx.doi.org/10.19103/AS.2019.0055.03

with some of the common reasons as to why MAP fails. MAP when combined with proper postharvest-handling procedures and temperature control management can have a positive impact on the quality and shelf life of fresh produce.

When designing MAP the convergence of three unique and separate sciences must take place. These are the sciences of:

- produce physiology;
- polymer engineering and;
- converting technology.

Effective MAP design can effectively take place only within the intersection of these three disciplines (Fig. 1).

In many package design scenarios, only two of the three disciplines are used. When this occurs, it is impossible to optimize the package for quality and shelf life. Surrounding these scientific disciplines is the impact and requirements of marketing, brand management and the desires and needs of the consumer. This often includes economics which cannot be understated as an influence. What makes fresh produce packaging unique is the impact of the fresh produce. Unlike almost all other packaging applications, produce MAP involves packaging of a living product. Therefore, a critical factor in the design of an effective package is a quantitative understanding of the physiological properties and requirements.

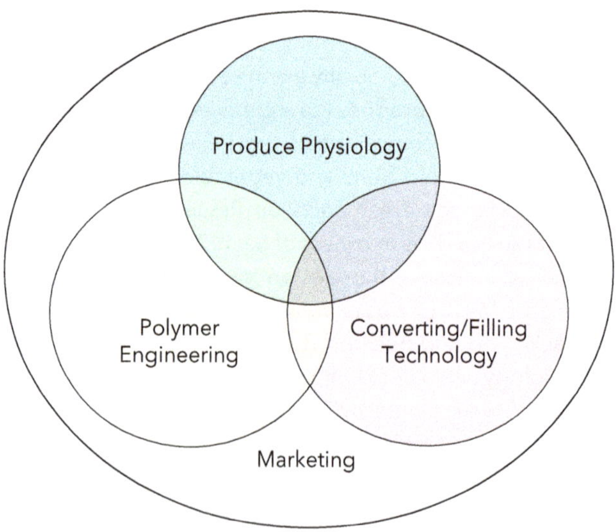

Figure 1 Sciences impacting modified atmosphere package design.

2 Modified atmospheres

Most extended shelf life packaging technologies involve the use of oxygen (O_2) and carbon dioxide (CO_2) in ratios that are different from those found in air. The partial pressures, diffusion properties and permeation rates through polymer films, microperforations and membranes of each gas are independent of the other gases present. This fact simplifies matters because gas interactions are irrelevant. Air is about 21% O_2, 0.04% CO_2, 0.9% Argon and the remainder nitrogen (N_2). If the internal atmosphere inside a package is different from air, then by scientific definition you have a modified atmosphere (MA).

Most of the reactions with food involving O_2 are a degradation process involving the oxidative breakdown of carbohydrates and other substrates. Many spoilage microorganisms such as pseudomonas, yeasts and molds require O_2 and will grow and cause off-odors in the presence of sufficient O_2. Oxygen is necessary for the normal respiratory metabolism of fresh fruits and vegetables, and normal atmospheric concentrations of O_2 encourage and facilitate senescence and degradation of quality. Therefore, many packaging strategies seek to exclude O_2 and thus slow these degradation processes. Only when the concentration of O_2 inside the package falls below at least 10% will the plant respiration start to slow.

Reducing O_2 concentrations below 8% slows the respiration rate in a sufficient manner for many fresh fruits and vegetables and therefore slows the rates at which they ripen, age and decay. Reducing O_2 concentration can, in products which are prone to this process, reduce oxidative browning reactions which can be of particular concern in precut leafy vegetables. Reduced O_2 can delay compositional changes such as fruit softening, pigment development, toughening of some vegetables (such as asparagus and broccoli) and development of flavor (Kader, 1986).

Ethylene is normally produced in various levels by many kinds of ripening fruit. In addition fruit varieties have a varied degree of sensitivity to ethylene. Elevated CO_2 (>~2%) can help reduce the damaging effects of ethylene by rendering plant tissues insensitive to ethylene (Herner, 1987; Kader et al., 1988). This may be one of the primary benefits of MAs for many commodities, even when ethylene concentrations in the packages are elevated. Elevated CO_2 can, like reduced O_2, also slow respiratory processes, thereby extending shelf life. Although the effects of elevated CO_2 on respiration are not as dramatic as those of low O_2, high CO_2 and low O_2 together can, in some cases, reduce respiration more than either gas alone (Kader et al., 1988).

CO_2 at concentrations greater than 10% has been shown to suppress the growth of a number of decay-causing fungi and bacteria. For example, 15–20% CO_2 is routinely applied around strawberries during shipment primarily to suppress growth of the mold *Botrytis cinerea*. It is important to note that these levels of CO_2 do not suppress some human pathogenic bacteria of potential concern on fresh produce. For example, *Clostridium botulinum* and *Listeria monocytogenes* are relatively resistant to the effects of CO_2 (Farber, 1991). There is some concern that elevated CO_2 could suppress spoilage microorganisms that would otherwise signal microbial growth and product spoilage while allowing potentially hazardous pathogens to continue to grow. For this reason, MAP should always work in conjunction with an excellent program of temperature management, sanitation and quality assurance. In addition, too much CO_2 can damage the plant tissues of certain fruits and vegetables that differ in their tolerance to CO_2.

3 Fruit and vegetable senescence

One of the keys to understanding and optimizing MAP is that plant tissues in fresh fruits and vegetables are still alive after harvest and even after minimal processing. To remain alive, they must derive energy, primarily through the process of respiration. Respiration involves the consumption of O_2 and the consequent production of energy, heat, CO_2 and moisture vapor. Different fruits and vegetables, and even different varieties of a given fruit or vegetable, will vary in their rates of respiration. Those that have high respiration rates (such as asparagus, strawberries and broccoli) tend to be the most perishable while those with low respiration rates (such as root vegetables, apples, onions and potatoes) tend to be the least perishable. In fresh-cut produce, the respiration rate is also strongly dependent on cut size. The smaller the cut size, greater exposed the cut surface area, the higher the respiration rate. The primary effects of MAP on produce are based on the slowing of plant respiration in low O_2 environments. When the respiration rate is reduced, fruits and vegetables become less perishable (Fig. 2).

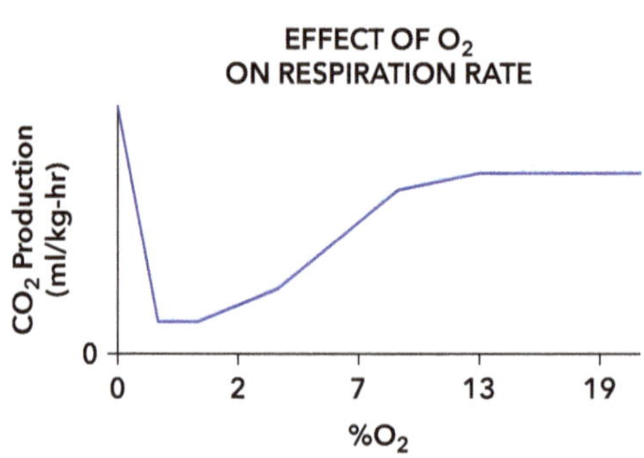

Figure 2 Effect of oxygen concentration on fruit or vegetable respiration rate.

Increases in storage and handling temperature can have the greatest impact on the respiration rate with increases that can more than double for every increase of 10°C. It is critical to note that the increase in the respiration rate with temperature (or cut size) is not a linear progression and sometimes can reach exponential rates. Given its importance, the best way to reduce the respiration rate is to reduce the temperature. All biological processes proceed more slowly at lower temperatures. Most fruits and vegetables will maintain their best quality at temperatures near 0°C. Keeping fresh produce at the lowest possible temperature without causing freezing or chilling injury is the surest way to maintain the quality and shelf life. As a supplement to good temperature control, MAP can further extend the quality and shelf life.

This suppression of respiration continues until O_2 reaches about 1-3% for most fruits or vegetables. If the concentration of O_2 drops below 1-3% (depending on the product and the temperature), anaerobic (fermentative) metabolism replaces normal aerobic metabolism and large amounts of CO_2, alcohols and aldehydes, which create off-flavors and off-odors, are produced. Under hypobaric (low pressure) conditions, extremely low O_2 partial pressures have been shown to be effective in decay control (Burg, 2004). Additional benefits of MAP include preservation of vitamins, particularly vitamin C (Zagory and Kader, 1989).

If CO_2 concentration gets too high, the production of undesirable volatiles and physiological injury can occur. As noted earlier, reduced O_2 and elevated CO_2 together can reduce respiration more than either alone (Zagory and Kader, 1989). In addition, elevated CO_2 can suppress plant tissue sensitivity to the effects of the ripening hormone ethylene. Brown (1922) reported that CO_2 concentrations of 10% or more can retard the sporulation and/or the growth of numerous fungal decay organisms, while O_2 typically has little effect on plant pathogen growth or survival at levels appropriate for use in MAP (Fig. 3).

Due to the slowing of oxidative metabolism associated with reduced O_2 and elevated CO_2, the ripening processes of many fruits and vegetables are retarded. Ethylene, a colorless, odorless and tasteless gas, is a plant hormone

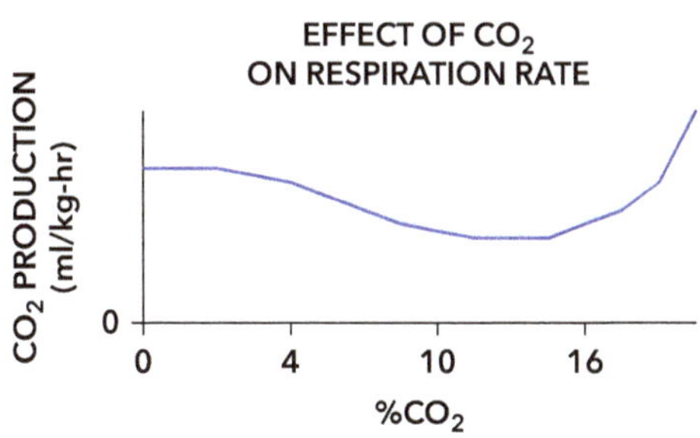

Figure 3 Effect of carbon dioxide concentration on fruit or vegetable respiration rate.

that has multiple physiological effects when present at concentrations as low as one part per million (ppm). Ethylene can induce rapid and irreversible ripening and softening of climacteric fruits such as avocados, tomatoes, kiwi, apples, stone fruit and mangoes. Ethylene may also cause yellowing of many vegetables such as spinach, broccoli, cucumbers and celery. Ethylene causes russet spot of iceberg lettuce, consisting of small, oval brown spots that develop on the lettuce midribs. Ethylene may also induce the formation of bitter-tasting isocoumarins in carrots, sprouting of potatoes and toughening of asparagus (Baldwin, 2004).

Elevated CO_2 can prevent or delay all of these effects by reducing the sensitivity of plant tissues to ethylene. In addition, low O_2 (below 2-4%) can reduce ethylene production by plant tissues (Table 1). Synthesis of some anthocyanin pigments is also retarded by reduced O_2, which slows color development of some fruits such as pomegranate. In order to take into account all of the above, it is critical that when measuring shelf life more than just a visual determination is made; O_2/CO_2 measurements as well as evaluations of organoleptic quality must be taken and analyzed.

It is also critically important to understand that there are conditions under which MAP will not work. First and foremost MAP is only optimally effective if there is consistent temperature management throughout the entire life cycle of the product. This includes processing as well as the entire distribution channel. Lack of temperature control will result in produce physiological variations which will impact the effectiveness of the packaging system. In addition, MAP will never improve the quality of the incoming raw material product. Under ideal circumstances the best that can be achieved is to maintain the existing

Table 1 Physiological effects of reduced O_2 and elevated CO_2 on fruits and vegetables

	General effects of	
Cause of deterioration	Reduced O_2	Elevated CO_2
Respiration rate	(>1%) –	(<15-20%) –
	(<1%) +	(>15-20%) +
Ethylene action	–	–
Chlorophyll degradation	–	–
Anthocyanin development	–	–
Carotenoid biosynthesis	–	–
Enzymatic browning	– (near 0%)	–
Off-flavors	+ (<1%)	+ (>15-20%)
Vitamin C loss	–	–
Fungal growth	– (<1%)	– (>10-15%)
Bacterial growth	– or 0	– or 0

– = decrease or inhibit, 0 = no effects, + = stimulate or increase.
Source: adapted from Kader et al. (1989).

quality level throughout the desired shelf life. In real-world applications, MAP will often maintain quality for the majority of the targeted shelf life, but due to parameter variations during distribution, quality will suffer at the very end of the desired shelf life. Since MAP will never improve incoming product quality, the need for optimal postharvest-handling procedures is paramount.

In addition, MAP cannot enhance the safety of fresh fruits and vegetables. Most of the human pathogens of concern are either not affected by MAP, or their growth may actually be enhanced by MAP. MAP will never improve the quality of incoming poor-quality products. At best, MAP can slow the processes of deterioration and so extend the quality and shelf life, but MAP cannot reverse those processes.

4 Respiration rate quantification

MAP depends upon the respiratory activity, as measured by respiration rate, of the enclosed product as a driving force for the atmosphere modification within the package. This in concert with the permeability of the packaging material are used to maintain atmospheres within desired limits. It is the continued depletion of O_2 and/or the release of CO_2 (and water vapor) by the product that enables the MA to become established within a sealed package. Factors that must be controlled or incorporated include film or technology permeability; film area; film thickness; temperature; type and weight of the product; and the associated respiration rate dependent upon variety, cut size and temperature of the product.

The dependence of MAP on the physiological properties of the product requires a quantitative understanding of the respiration of specific fresh produce. Produce respiration is determined through testing. The condition, age and temperature of the raw material product can significantly impact the outcome of the respiration test. There are a number of independent resources for product respiration calculations including University of California Davis (http://postharvest.ucdavis.edu/), the JSB Group LLC. (www.jsbgroup.com), the Q Fresh Lab (https://qfreshlab.com/) and Perfotec (http://perfotec.com).

The primary method for determining the required O_2 transmission rate (OTR) for a MAP is seen in the following equation. In addition to respiration rate (RR), parameters include package surface area, product weight, structure thickness and target final MA. A thorough understanding of the equation and how each parameter can affect the package OTR is critical for optimal package design. It is the interplay of these parameters that dictates that there is no one package that fits all applications. MAP must be designed on an individual application basis. Once the produce respiration is quantified and the balance of parameters are determined the target OTR can be calculated. Establishment of the target OTR allows the design process to move from the physiological research into the polymer science and packaging research. OTR for a target atmosphere is:

$$OTR = RR_{O2} \times t \times W/A \times (O_{2air} - O_{2pkg})$$

where

OTR = film O_2 permeability per mil
RR = respiration rate (O_2 consumption rate in ml/kg-h)
t = film thickness (mls)
W = product weight (kg)
A = film surface area (cm^2)
O_{2pkg} = desired O_2 concentration in the package (% O_2 target atmosphere)

Note: OTR in this example is calculated on a per mil basis (25 μm). Therefore, when a 2 mil thickness film is used, the OTR on a per mil basis must double to achieve the equivalent package OTR. Polymeric films change the transmission rate as thickness changes. However there is not always a linear relationship between the film thickness and transmission rate. Therefore transmission rates should be tested for each film thickness.

To achieve the target MA, the packaging films must be permeable to gases. Specifically they must have the required gas transmission properties to achieve the targeted OTR. The movement of gases across films depends on several physical factors that are related through Fick's Law as follows:

$$J_{gas} = \frac{A \times \Delta C_{gas}}{R}$$

where

J_{gas} = total flux of gas (cm^3/s)
A = surface area of the film (cm^2)
ΔC_{gas} = concentration gradient across the film
R = resistance of the film to gas diffusion (s/cm)

The gas flow across a film increases with increasing surface area and with increasing concentration gradient across the film. The gas flow across the film decreases with increasing film resistance to gas diffusion. Gases diffuse through polymeric films at different rates. CO_2 diffuses from two to five times faster than O_2. The ratio, within a polymer, film or structure of CO_2TR to OTR is termed the beta value. Polymeric films therefore have a beta value of 2-5:1, with an average of 3:1. The beta value of a MAP will have a direct bearing on the final atmosphere achieved within the package. For example, in polymer films it is possible to achieve a low O_2 level (2%) in combination with a mid-CO_2 level (7%). The impact of the beta value is dependent on the specific ideal MA for each individual product.

5 Polymer engineering

5.1 Polymers

There are a variety of polymers used in fresh produce MAP. A portion of these polymers is used in primarily flexible packaging structures, a portion is used in primarily rigid packaging structures and a portion is found in both applications. Each polymer has physical, chemical and transmission rate properties that are unique to that polymer (Table 2). The design of a packaging structure entails in matching the specific polymer properties to the requirements of the MAP application. A polymer gas transmission rate, specifically, OTR and CO_2TR are key attributes for the physiological portion of fresh-cut MA applications.

However, research, awareness and understanding of polymers' abilities to transmit, absorb and manage ethylene are gaining momentum. In addition, most polymers used in fresh produce packaging are water vapor barriers and as such residual water, humidity and water vapor were controlled through other technologies. Recently significant work is being carried out to explore, develop and commercialize polymers which have a significantly higher water transmission rate (WVTR). Polymers such as polystyrene and nylon are showing promise.

Table 2 List of the common polymers used in producing modified atmosphere packaging (MAP)

Polymer	Abbreviation	Characteristic	Typical OTR (cc/100 in²/mil/atm/day)	Application most commonly used
Low-density polyethylene	LDPE	General-purpose polymer	450–500	Both
Linear low-density polyethylene	LLDPE	Increased stiffness	480–500	Both
Linear medium-density polyethylene	MDPE	Increased stiffness, lower OTR, decreased clarity	300–350	Both
High-density polyethylene	HDPE	Relatively stiff, opaque	150	Rigid
Ultra low-density polyethylene	ULDPE	High OTR, decreased stiffness	900	Flexible
Plastomer metallocenes	-	Very high OTR, soft	1100	Flexible
Amorphous polyethyleneterapthalate	APET	Clear, rigid	5	Rigid
Poly vinyl chloride	PVC	Clear, rigid	10	Rigid
Polypropylene	PP	Decreased clarity	300	Both
Ethylene vinyl acetate	EVA	Sealability	600–900	Flexible
Polystyrene	PS	Stiffness	350	Rigid

Biopolymers are also receiving a lot of research and development attention. Polymers such as polylactic acid (PLA), polyhydroxyalkanoate (PHA), polynucleotides, polyamides, polysaccharides, polyoxoesters, polythioesters, polyanhydrides, polyisoprenoids and polyphenols are being evaluated as potential candidates for substitution of synthetic plastics in fresh produce applications. Cellulose polymers and cellophane are even being re-branded and experiencing a resurgence in use and applications. In addition to the above, many natural fiber-based materials (rice, sugarcane and wheat husks for example) are being evaluated for MAP. The challenge with the plant-based structures includes being able to quantify the transmission properties, determine property consistency and ensure that the products can be used in a direct food application.

One of the challenges in the biopolymer/sustainability area is combining the transmission requirements for produce with the desire for sustainability and compostability. Understanding the difference and requirements between industrial composting and home composting is essential in determining the functionality and applicability of bio-based polymers. Examining the entire lifecycle of the polymer and package is also critical in determining not only the best polymer for the application, but also the total cost of the technology.

5.2 Films and structures

MAP contains not only individual polymers but combinations of polymers, as well as other substrates such as paper, and mesh in the form of films and structures. These films and structures can consist of homopolymer monolayer films, blended monolayer films, coextruded films or combinations of both laminated, glued together. Determination of the structure gas transmission rate therefore requires knowledge of not only the individual polymers but also how they are combined (Fig. 4). A blend of polymers within one layer of film, referred to as a blend, will yield a transmission that is the weighted average of the transmission of the individual polymer components. Individual distinct layers of pure polymers, or blends, created within a single total film structure, are referred to as a coextrusion. These can be used as is, or combined with additional coextrusions or films, to create a lamination.

Since gases will move through each independent layer sequentially, the transmission of each independent layer of polymer must be determined and inserted into the following equation to calculate the overall structure transmission rate (OTR):

$$\mathrm{OTR} = \frac{1}{\frac{t_1}{\mathrm{OTR}_1} + \frac{t_2}{\mathrm{OTR}_2} + \frac{t_3}{\mathrm{OTR}_3}}$$

where

t = thickness of the individual layer

NB OTR = oxygen transmission rate of the individual layer at 1 mil

Based upon this equation the total structure transmission can never be greater than the lowest individual layer transmission. This limiting factor establishes a practical ceiling for polymer MAP. Depending on the type of structure and all of the converting, geometric and marketing requirements, this ceiling may be as low as 175 cc/100 in^2/mil/atm/day (2713 cc/m^2/mil/atm/day) or as high as 900–1000 cc/100 in^2/mil/atm/day (13 950–15 500 cc/m^2/mil/atm/day).

It is important to note that even though they are very thin layers, inks, adhesives and coatings, including antimist, can have a significant impact on the overall transmission rate. Therefore all layers of the laminations, not just the polymeric or paper layers, must be taken into account when calculating overall transmission rate.

The specific combination of polymers, paper, blends, coextrusions and laminations are governed by numerous parameters in addition to the physiological requirements mandated by the product being packaged. These additional functional requirements often act in an opposing manner. Therefore to effectively design a MAP it is necessary to have a full understanding of all of its desired features. Requirements impacting polymer and film choice can include stiffness, sealability, aesthetics, clarity, graphics, dimensions, economics, sustainability, runnability, packaging format, coefficient of friction (COF), antimicrobial additives and thickness. Examination of just one of the parameters demonstrates the results of these opposing parameters. As listed in Table 2, polymers that exhibit increased stiffness properties also exhibit lower gas transmission rate properties. Therefore if structure stiffness is of paramount importance, then the overall gas transmission rate of the package may be low, so low, in fact, that it does not meet the requirements of the packaged produce. We can see that this parameter alone has a significant impact on overall package design. Rarely, though, does one have the luxury in examining one parameter in isolation; therefore it is critical to have a thorough understanding of all applicable parameters and their relative importance.

5.3 Perforations

The increase in demand for higher respiring fresh produce, outside the traditional leafy green market, has necessitated an increase in MAP gas

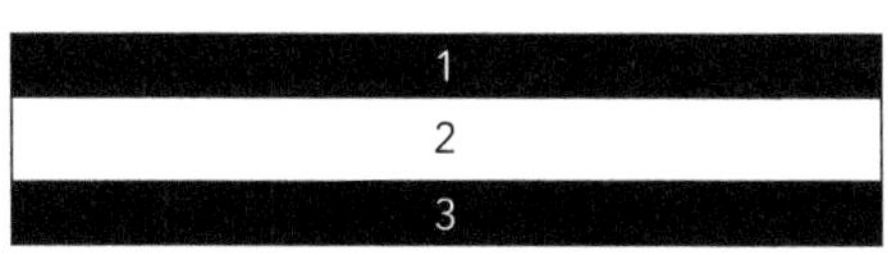

Figure 4 OTR calculations on multilayer coextruded films and structures.

transmission rates. A method for achieving high OTR packaging structures that is not limited by the upper end of polymer gas transmission rates and stiffness constraints is microperforation technology. This technology employs the science of placing micro holes in the packaging structure through the use of a laser. With microperforation technology, the gas transmission rate of the package is governed by the configuration of holes and their individual geometry and size (Fig. 5). The hole size and configuration can vary with the specific perforation method but all microperforations are not visible to the naked eye as they can range from 40 to 200 μm in diameter. The smaller the hole size, the greater number of holes can be placed in a structure while achieving the same overall transmission rate. There is a limit, however, as the diameter of the perforation should not be less than the thickness of the film. This is due to the laser energy that dissipates though the thickness of the film, which creates a cone shape hole where the diameter of the perforation on the inside surface of the film may be less than the diameter on the outside surface of the film. If the outside diameter and the film are too narrow, then you may actually not have a hole all the way through the film.

It is essential to have a complete understanding of the package geometry as well as produce geometry and 'bulk density' so that the holes are not blocked or obstructed in any way. This is critical to the success and control of gas transmissions. Microperforations also have transmission rate limitations that are at the lower end of the range. Typical OTRs for this packaging are from 250 cc/100 in^2/mil/atm/day and above. Gas transmission rates of microperforated structures are determined by the gas diffusion properties through the combined effect of the individual micro holes, their corresponding placement and in certain cases the OTR of the perforated structure. The number of holes decreases as the targeted OTR requirements are lower. Since microperforated structures cannot have less than one hole, the gas transmission rate through a single hole dictates the lower transmission rate level. MA microperforation packaging with only one hole can also be problematic. With two or more holes there is less risk from hole blockage as well as there is more uniform gas flow throughout the package

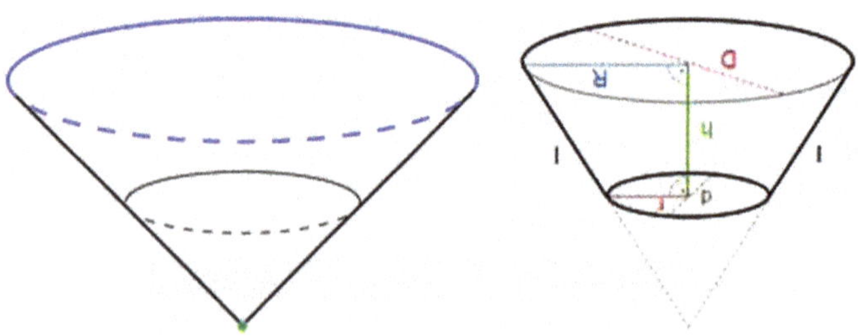

Figure 5 Microperforation hole geometry.

mitigating the possibility of package micro atmospheres. In order to avoid only one microperforated hole, alterations to the other key OTR control parameters, such as product weight, or package dimension, should be considered.

New micro perforation technologies, non-laser, are being actively researched, which if successful, could lower the effective achievable hole size. This would allow for a lower total package transmission rate as well as increase the number of holes per package creating a more uniform gas flow throughout the package and further mitigating microatmospheres. The majority of microperforations are incorporated into the package at the packaging supplier. The company 'Perfotec' has developed a unit which can be place directly on the filling line.

The diffusion rates of various gases through microperforations are very similar. In effect the diffusion rates of CO_2 and O_2 are virtually the same, a beta value of 1. Therefore, for a targeted 2% O_2, the CO_2 concentration will be 19%. The beta value difference between polymeric and microperforated structures create different final MAs. It is not possible with a microperforated film to achieve a MA consisting of low O_2 concentrations and low to moderate CO_2 concentrations. Conversely, if low O_2 concentrations in combination with high CO_2 concentrations are required then packaging comprising of engineered polymers is not suitable. This parameter needs to be accounted for in the packaging design process.

Macroperforations, with holes that are visible to the naked eye cannot consistently modify and control the gas concentrations within the package. Therefore attempting to create a MA with low O_2 is not feasible. This does not mean, however, that there is not a function and need for macroperforated structures. Since the gas transmission is so high a macroperforated structure will virtually never become anaerobic as a result of O_2 concentrations falling below 0%, even under temperature-abuse situations. If having a fresh-cut produce package, which does not become anaerobic under any circumstances, is the highest priority and atmospheric O_2 concentrations do not significantly impact shelf life, then macroperforation technology may be applicable. Historically this has been the technology of choice for mushroom packaging. Recent advances in microperforation technology have provided mushroom growers and processors alternatives. If one plans to pursue microperforations for mushroom packaging then it would be wise to make sure that all relevant government food safety requirements are met.

5.4 Breathable membranes

There have been a number of types of membrane technologies that have been developed over the years. Membranes generally are patches, or labels, placed over a die cut hole within the package. They are high transmission rate polymers which allow for the gas transmission of the package to be independent of the package make up. This allows them to be used in both flexible as well as rigid applications.

One of the most innovative to be commercialized is the Landec Apio Breatheway membrane. This technology not only allows the package designer to dictate the beta value of the membrane, allowing for specific final O_2/CO_2 design targets, but can adjust its transmission rate with temperature increases. This is a unique feature in MAP design which allows the package to increase its transmission rate during temperature abuse, when the produce is respiring at an increased level.

Breatheway membranes are a flexible and enabling technology that allows the decoupling of the packaging media from providing the permeability needed by the produce, while at the same time providing the atmospheres necessary for preserving produce (Clarke, 2011). The temperature switch function of side chain crystalline (SCC) polymer will maintain the atmospheres through moderate temperature fluctuations in the cold chain (Fig. 6).

5.5 Gas flushing

Passive MAP for fresh-cut products relies on the relationship between produce respiration and package transmission rate to alter the atmosphere within the package. This process generally takes a number of days to reach the target MA and equilibrium. For produce items which are prone to enzymatic browning reactions, 'pinking', which can be exacerbated by O_2 concentrations above 3%, this gradual decent may be too long. Gas flushing in MAP of fresh-cut products establishes an initial low alternative atmosphere within the package which can be beneficial in reducing enzymatic browning reactions 'pinking'. Gas flushing, however, is not a substitute for proper package design or leakers. Gas flushing is only optimized when it is employed in combination with proper package design and a leak-free package. Since the primary goal of gas flushing is to quickly reduce the initial O_2 level within the package, N_2 is the most effective and economic gas. However, an initial shot of CO_2 can be beneficial in creating an increased level of CO_2, which can inhibit the growth of spoilage organisms. Certain pad technologies in which CO_2 is released when liquid (purge) absorbs into the pad are useful in this application. As with gas flushing this technology is to be used in concert with properly designed MAP and not a substitute for proper package design. Having a CO_2 release within a sealed container without a proper design gas flow transmission rate can damage the integrity of the package.

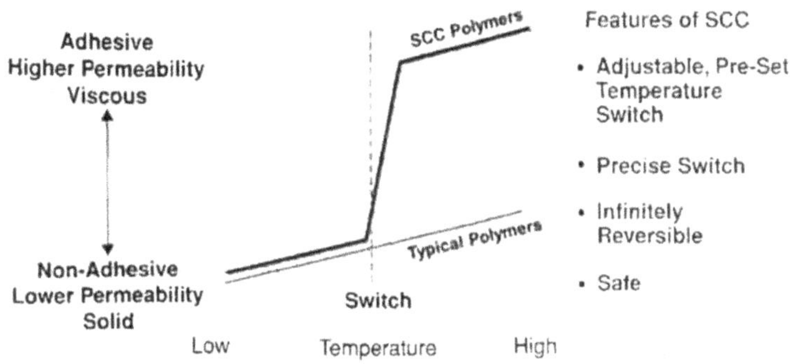

Figure 6 Temperature sensing capabilities with side chain crystalline (SCC) polymers.

6 Converting technology

6.1 Packaging format

Effective MAP design hinges on the proper balance of the physiological, polymer, converting and marketing requirements. Issues such as package configuration, package stiffness, graphics, filling method, economics, environmental impact as well as ancillary requirements such as cook-in, antimist, resealability and compostability all impact the design and makeup of the package, many of which are similar to those found in traditional packaging applications. The significant difference with MAP is the impact that these requirements have on the gas transmission rate properties of the package. The majority of these requirements fall under the science of converting technology. The science of converting technology combines the raw material polymers, films, adhesives, inks and additives in the proper sequence to create the desired package. Depending upon the format, packaging can be developed separately or in combination with automatic filling equipment. The result of these various converting requirements is the development of a wide variety of packaging formats. There is no right or wrong format. Rather it is important to choose the format with optimum suitability based on all the outlined requirements. Packaging formats can range from the very basic monolayer pre-formed side-weld bag to the very complex multilayer coextruded reverse print lamination and thermoformed multilayer tray with peelable lidding structure. Table 3 lists some of the more common packaging formats. It is important to note that this is not an all-inclusive list since there are many variations to these common formats which optimize specific desired properties to the specific produce packaged.

6.2 Flexible versus rigid packaging

As the names imply the fundamental difference between flexible versus rigid packaging relates to the stiffness and often thickness of the respective

Table 3 Common packaging formats

Package format	Suitable for MAP	Natural aspiration	Prone to leakers if properly manufactured	Common products packaged
Side-weld pre-made bag	Somewhat	Somewhat	Yes	Carrots
Pre-made pouch	Yes	No	No	
VFFS bag	Yes	No	Structure dependent	Leafy greens
Pre-made SUP	Yes	No	No	
VFFS SUP	Yes	No	No	
Thermoformed tray with attachable lid	No	Yes	N/A	In-store fresh-cut fruit
Clamshell tray	No	Yes	N/A	Berries
Thermoformed tray with sealable lid	Yes	No	Lidding dependent	
Tray with overwrap	Package dependent	Yes	Yes	Cut squash
Macroperforations	No	Yes	N/A	Mushrooms

packaging configuration. The relative stiffness of a given package is controlled by the choice of polymers and their respective stiffness properties as well as the thickness of the structure and geometry. In MAP, the effective surface area available to gas diffusion between rigid and flexible packaging is another difference. The polymers required to maintain the rigid form of a package generally exhibit low gas transmission rates. These low transmission rates in combination with the thickness required to maintain the package form effectively create a gas barrier structure. This dictates that when determining the effective 'breathable' surface area of the package, the rigid portion of the package generally cannot be included in the calculations. This means that the flexible lidding material sealed to the top of the tray must take on the entire burden of gas transmission.

When comparing a flexible package with a rigid tray type package, the gas transmission level of the effective surface area of the flexible package can have a significantly lower overall transmission rate. This is an application where the breathable membrane when incorporated into the rigid tray portion of the package can increase the overall breathability of the structure. When a rigid tray is used in combination with a rigid lid both components of the package are effective gas barriers. Therefore if a hermetic seal is created between lid and tray, effective gas transmission will not take place and depending upon the type and quantity of produce anaerobic conditions will rapidly develop.

Often a rigid tray and lid combination are not designed to create a complete seal. In this situation the package is not a designed MAP but rather a 'natural aspiration' package, meaning that depending on how effective the seal, the atmosphere inside could range from anaerobic to ambient. When designing the packaging for the produce market, it is important to decide which one of these packages is being designed. Both can be effective packaging formats depending upon the initial requirements and desired outcome. Self-life optimization can only be achieved through MAP technology.

Regardless of whether a flexible or rigid package is chosen, an optimal MAP requires the controlled and quantifiable transmission of gases through the package in concert with the physiological characteristics of the produce being packaged is necessary. If there is no control or quantification of gas transmission rates then optimal atmospheres cannot be guaranteed and the packaging system is thereby not controlled. There are a number of reasons for an out-of-control packaging system including improperly quantified produce physiological properties, an improperly specified package and out-of-specification raw materials. However the most common reason is due to a leaking package. If the package does not have a leak-free seal then gases will immediately begin to pass through the leak area. Depending upon the size of the leak the impact could range from missing the optimal target atmosphere to allowing the package to remain at ambient conditions. In either case optimal shelf life and quality will not be achievable.

One of the most important packaging parameters that must be considered is the selection of the sealant layer polymer and configuration. Choice of the correct sealing polymer and format is dependent upon operating parameters including package machine type, filling speed, package configuration, seal configuration, as well as product type and weight. Potential sealant layer polymers exhibit a wide range of seal characteristics. The common choices of sealant polyolefin sealant polymers include low-density polyethylene, ethyl vinyl acetate, ultra low-density polyethylene and plastomer metallocenes. Each polymer has its own sealing characteristics including ultimate seal strength, hot tack strength, seal initiation temperature, ability to seal through contamination and oxygen transmission rate. Careful consideration should be given in order to optimize the specific polymer characteristics to the specific requirements of the package necessary properties.

In certain circumstances a peelable seal is desired. This can be in both a bag or tray configuration however it is much more common in a tray configuration. There are a number of technologies that can be employed to achieve an optimal peelable seal. As Gorny and Brandenburg (2003) point out, peelable lidding stock technologies are generally complex; therefore, a dialogue with a packaging expert is recommended. The three most commonly available peelable technologies are controlled contamination, dissimilar resins and controlled delamination, with each having its own advantages and disadvantages.

The basic attributes of the three technologies are (Gorny and Brandenburg, 2003):

1 *Controlled contamination*: This technology utilizes a small amount of contaminant resin within the sealant layer to achieve a controlled weak seal or 'peel seal'. Common contaminant resins are various grades of polybutylene. This technology is used predominately for applications where the film is peelable to itself. Through engineered blending technology the film can be designed to be peelable within a specific temperature range. This results in peelable areas, as well as non-peelable areas within the same finished packaging material.
2 *Dissimilar resins*: This technology utilizes a dissimilar resin between the peelable lidstock and the bottom (thermoformed) tray stock. This technology is most commonly used in lidstock applications where a peelable lid is sealed to a rigid or flexible thermoformed bottom web or rigid container. The choice of peelable sealant layers is directly dependent upon the sealant layer of the bottom material. Whenever possible, the top sealant layers should be designed together so the optimum match is made and the desired peel seal is achieved.
3 *Controlled delamination*: This technology involves achieving an initial fusion seal between the sealant layers of the lidstock and the bottom web. The 'peel seal' is then achieved through the delamination of the coextruded sealant film. The delamination within the coextrusion occurs between the sealant layer and the next inner layer (often the core layer). It is possible to achieve a frosty or 'tamper evident' seal, as well as re-seal structure. This technology is often found in medical packaging.

It is critical to get the correct match between the peelable lidding material and the bottom web. There are a wide variety of polymers and constructions available, depending on the type of bottom web sealant layer, peel seal requirements, as well as the look and feel of the total structure. The companies familiar with peelable lidding materials should be consulted so that all of the necessary variables are taken into account. The need for direct dialogue with a packaging expert is even more critical when evaluating sustainable and compostable options for films and trays. Since many of the biopolymers and 'natural' products have very unique physical properties, special care should be taken during the design and evaluation phase to ensure a consistent peel seal system.

6.3 Packaging equipment

Packaging equipment used for MAP is as varied as the packaging types themselves. Packaging equipment can range from the basic hand-held

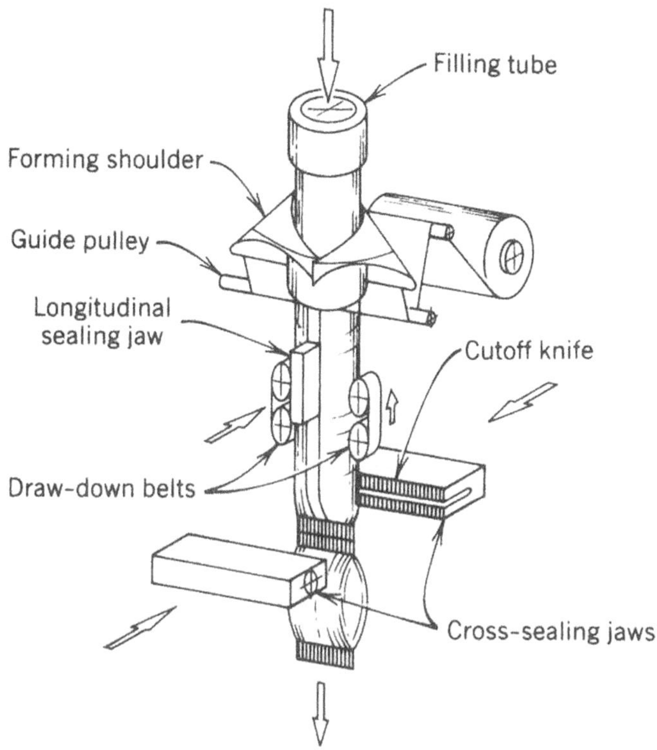

Figure 7 Vertical filling line. Source: adapted from Gorny and Brandenburg (2003).

impulse sealer to processor-controlled automatic form fill and seal lines. Within this plethora of options, however, are two fundamental categories: vertical and horizontal. Figures 7 and 8 are examples, respectively, of a vertical and horizontal filing line. Both styles of packaging equipment can be configured for either manual or automatic operation.

Although there are exceptions, generally vertical packaging machines are designed to run flexible style packaging and horizontal machines are designed to run rigid style packaging. Within the vertical configuration there are two seal configurations: fin and lap (Fig. 9). Determining which seal type will be used is a critical step in the package design since the seal type can impact polymer selection, graphics and material usage. The most significant difference between the two types is that the fin-seal configuration seals only the inside of the package whereas the lap seals the inside to the outside of the package. Therefore when a lap-seal configuration is used the polymers on both the inside and outside of the package must be sealable.

Traditionally using heat, from the outside to the inside of package, during sealing has been the sealing method of choice. This technology uses

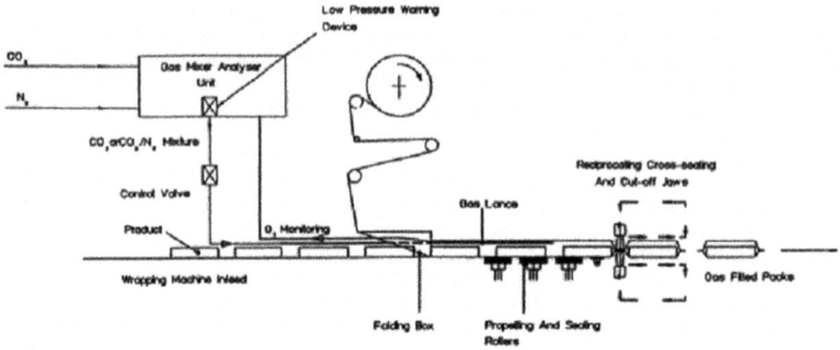

Figure 8 Horizontal form-fill-seal machine with box-motion sealing head. Source: adapted from Soltani et al. (2015).

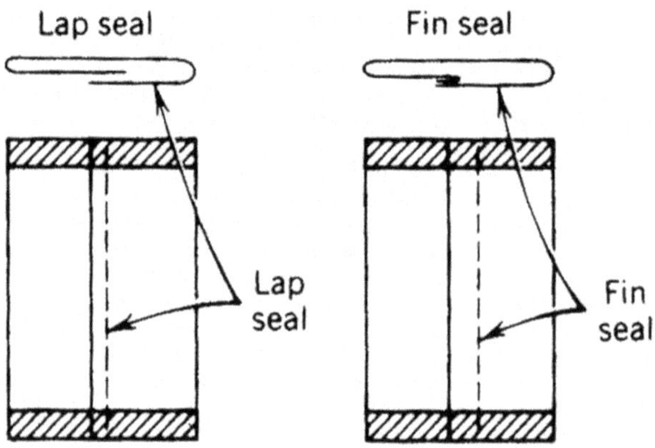

Figure 9 Lap and fin seals. Source: adapted from Brody and Marsh (1997).

a combination of heat, pressure and time (dwell) as the method to weld the sealant layer polymer together. During this welding process the sealant layer polymer transitions from solid to liquid and back to solid thus creating the seal. The hot tack of a seal is the strength of the seal when the polymer is still in the molten or liquid phase, whereas the ultimate seal strength, as the name implies, is the final re-solidified seal strength.

Recently ultrasonic sealing has been making inroads into fresh-cut produce packaging. With ultrasonic sealing, the heat required for melting is only generated inside the thermoplastic sealing layer (Herrmann Ultrasonics, 2017). For localized conversion of vibrations to friction heat, the anvil or sonotrode profiles are mostly linear and include radii or small plateaus. These profiles ensure focusing of the energy input and therefore shorten the sealing times between 100 and 200 ms. With ultrasonic sealing, heat is generated at the inside of the film, not by heat input from the outside as is the case with heat sealing.

7 Additive technologies

Additive technology, which includes slip and antiblock, antifog, UV inhibitors, antimicrobials, pigments, absorbers and absorbents, can greatly impact MAP design. Each has its own distinctive purpose and function.

7.1 Antifog

Antifog technology is often incorporated into produce packaging to prevent water condensation on the inside of the package, which would obscure a potential purchaser's view of the product (Gorny and Brandenburg, 2003). The functionality of antifog technologies is accomplished by coating the interior surface of the flexible packaging material with compounds that reduce water surface tension or reduce the ability of the water to adhere to the packaging material and thus cause the condensed water to run off the interior surface of the package. There are two categories of antifog technology: applied coatings and sealant layer incorporation with each yielding distinct advantages and disadvantages. Applied coatings are applied to the sealing surface of the finished package and historically have been superior to the incorporation method in that they provide superior optics and antifog properties. Within the category of applied antifog coatings there are two predominant formats: registered applied and flood coating. In the register-applied system the coating placement is controlled so that the antifog coating does not fall into the sealing area, thereby not impacting the sealing properties of the structure. To utilize this system specialized equipment is necessary to achieve registration (placement within the dimensions of the package) of the antifog coating to the package surface, as well as to achieve good adhesion without damaging the sealing surface. Not all converting companies have the equipment to properly apply and register these coatings. Flood coating does not apply the antifog in register but rather coats the entire sealing surface, thereby eliminating the need for the specialized register application equipment.

The main disadvantage with applied coatings is that the coatings are applied to the non-treated surface, and most applied coatings require a surface treatment to guarantee strong adhesion to the packaging structure. Therefore, antifog coating pick off (flaking) may occur. Coordination and good communication with both the antifog supplier and the converter is critical when planning to use this type of antifog coating.

Antifog sealant layer blends incorporates or blends an antifog compound into the sealant layer during the film manufacturing process. This system reduces the cost of adding antifog technology to produce packages since no specialized converting equipment is needed, and thus has advantages both for the converter and end user. The major disadvantage to antifog sealant layer blends is that the antifog is, by default, in the sealing area of the bag. Therefore weaker seals and potentially package leakers are a greater possibility. However, altering sealant layer formulations can often overcome this issue. The optics and performance of this type of antifog technology may be inferior to coated antifogs. Also, to properly function the incorporated antifog must 'bloom' (migrate and create a separate layer) to the surface and thus can cause a reduction in the gas transmission rates of the package.

7.2 Slip and antiblock

Slip and antiblock are key component of most packaging, especially flexible packaging. The addition of slip and antiblock compounds into the polymer film structures reduces the natural tackiness of certain polymers as well as increases the film's resistance to blocking. This is especially important if vertical form fill and seal (VFFS) equipment is used to form and seal the package. The ability of the packaging structure to slide over the metal forming collars and tubes is essential for proper structure formation. Like antifog additives, slip and antiblock can alter the gas transmission rate properties of the packaging structure. Care must be taken to properly quantify and minimize the effect upon gas transmission rates. Slip and/or the improper application of antiblock additives can also impact package forming and sealability, leading to an increased incidence of improperly sealed packages and leakers.

7.3 Antimicrobial films

Antimicrobial additives also bloom to the surface of the inside layer of the package. Therefore the issues of gas transmission rate reduction and sealing ability are equally present. Their method of operation is by contact of the antimicrobial component to the packaged product. Although the common place in MAP is for proteins, they are not widely employed in producing MAP. This is due to the relatively high amount of product surface area in relation to package surface area in fresh produce applications. This ratio imbalance prevents the majority of the product to come in contact with the microbial agent thus making these systems relatively ineffective. New technologies are currently under development with potentially exciting results.

7.4 Absorbing compounds

A wide variety of absorbing compounds are also being evaluated and in some cases already commercialized. Moisture, ethylene, O_2, CO_2 and odors are all compounds that absorbers are targeting. There are a number of challenges with absorbing compounds including US Food and Drug Administration (FDA) approval for direct food contact, capacity and consumer acceptance if a sachet is used. A full understanding of what these compounds can and cannot do is critical in deciding if they should be incorporated into the packaging structure.

In all additive applications, careful consideration must be given when considering the incorporation of any additive into a MA package. Although the addition may enhance or add certain desirable features or properties they will almost surely detract or negatively impact other properties or functions. The key property in fresh produce applications that is most likely to be negatively affected will be the overall gas transmission rate properties of the packaging system.

8 Applications

8.1 New produce items and geometries

As the fresh produce market has matured beyond the initial category of leafy green packaging, the technology of MAP has responded. New and novel types of vegetables, vegetable geometries, fruits, unique combinations of different types of fresh produce, ready meals and microwave and steam-in microwave applications are all examples of new applications and markets. Within the category of new produce items and geometries there has been tremendous growth in the past few years. New items such as cut Brussels sprouts, kale, power greens and cauliflower rice have been introduced, and are becoming common place. Spiralized geometries of root vegetables and summer squash are also becoming very popular in the marketplace. In each case, optimal MAP needs to be developed to insure both the optimal shelf life and quality. As with all new developments each new application challenges the science and technology of MAP and adds its own demands and requirements.

8.2 Microwave and steam-in

One of the fastest-growing segments of fresh-cut produce packaging is the use of microwaveable and steam-in microwaveable packages (Gorny and Brandenburg, 2003). There are a number of venting technologies available in the marketplace, including perforations, controlled delamination slits, valves and scoring. In considering which technology is best for your application it is important to keep the steam venting technology separate from the MA perforation technology. The venting utilizes holes but their design size and functionality are different from the MA perforation. To try and combine the two technologies will only result in both technologies becoming compromised.

In addition to the functional requirements generated by cooking product inside the package, there is an additional and often hidden requirement of migratory compliance. Polymers that are direct contact food compliant and act as 'functional barriers' at ambient temperatures may or may not remain so under elevated temperatures. In addition, as the internal package temperature rises, fewer polymers remain both functionally and regulatorally compliant. It is therefore critical to engage in a detailed dialogue with the packaging supplier and to make sure that all applicable regulations have been identified. The following sections of the US Code of Federal Regulations (CFRs) may be applicable when dealing with microwave and steam-in microwave applications:

- 21CFR § 177.1350
- 21CFR § 177.1390
- 21CFR § 177.1395
- 21CFR § 177.1520

It is important to assure that, before developing a microwave product, you conduct in-depth discussions with your packaging supplier to assure that the above-mentioned issues have been addressed and that you are in full regulatory compliance. Not all packaging materials are suitable for all microwave applications.

8.3 Varietal blends and novel produce combinations

The growth of varietal blends and novel combinations is a direct result of processors that continuously search and investigate ways to differentiate their product line. This need to differentiate has led to an explosion of vegetable and leafy green combinations, 'power blends' vegetable stew and soup mixes, as well as varieties and combinations of tropical fruits. These new opportunities place additional challenges and requirements on the technology of MAP. Specifically, each separate and unique raw material has its own physiological properties combined with its own individual optimal target atmosphere. Therefore not all fresh-cut fruit and vegetables are compatible. In other words what may be an optimal target for one product may be very detrimental to another. Compromises in atmosphere targets will most likely be required however it is critical to minimize the compromises. Instead, companies are quantifying the physiological properties of their individual raw materials and then grouping like products to create desirable blends and combinations without having to sacrifice quality and shelf life.

8.4 Ready meals

The combination of produce, protein and carbohydrate into a fresh ready meal format has increased greatly over the past few years. In addition, plant proteins are being added to kits so that the vegan consumer has more access to proteins. A technology, in which different types of MAP are used for each type of food category, is more wide spread in Europe than in other world markets and is rapidly gaining mainstream popularity. Hence these applications generally require multi-compartment trays to separate each food product with unique target atmospheres in each compartment. Since there are different and unique food groups within a single package, effective package design must cross over and combine, often competing requirements. Coordination between the packaging designer and food scientists from each of the participating food groups is essential.

One requirement that has come out of all of these new applications is the need to incorporate and coordinate the package design process into and with the new product development process. Leaving the package design until the end of the product development process prevents the optimization of both product and package. Since historically this has not been the norm, re-education and enhanced communication between the new product development group within the food company and packaging designers is required.

9 Conclusion and future trends

It is an exciting time within the MAP industry. The emerging technologies and opportunities will have a far-reaching impact on the marketplace. Issues such as sustainability in packaging and the impact that packaging has on current food safety issues are already providing both tremendous challenges and opportunities. The challenge will be how to incorporate all of the desired requirements into MAP without diluting its fundamental purpose. A package that tries to become all things to all applications becomes mediocre at best with respect to any one requirement.

10 References

Baldwin, A. 2004. Ethylene and postharvest commodities. *Hort. Science*. 39(7), 1-3.

Brandenburg, J. 2017. Packaging design: functions and materials. In: Gil, M. I. and Beaudry, R. (Eds), *Controlled and Modified Atmospheres for Fresh and Fresh-Cut Produce*. Elsevier.

Brandenburg, J. and Zagory, D. 2009. Modified and controlled atmosphere packaging technology and applications. In: Yahia, E. M. (Ed.), *Modified and Controlled Atmospheres for Transportation, Storage and Packaging of Horticultural Commodities*. CRC Press (Taylor & Francis).

Brody, A. L. and Marsh, K. S. (Eds). 1997. *The Wiley Encyclopedia of Packaging Technology* (2nd edn.).

Brown, W. 1922. On the germination and growth of fungi at various temperatures and in various concentrations of oxygen and carbon dioxide. *Ann. Bot.* 36, 257-83.

Burg, S. P. 2004. *Postharvest Physiology and Hypobaric Storage of Fresh Produce*. CAB International, Wallingford, UK.

Clarke, R. 2011. Modified atmosphere packaging for fresh-cut fruits and vegetables. In: Brody, A. L., Zhuang, H. and Han, J. H. (Eds), *Modified Atmosphere Packaging for Fresh-Cut Fruits and Vegetables*. Wiley-Blackwell.

Farber, J. M. 1991. Relative effect of CO2 on the growth of foodborne microorganisms. *J. Food Prot.* 54(1), 58–70. doi:10.4315/0362-028X-54.1.58.

Gorny, J. R. and Brandenburg, J. 2003. *Packaging Design for Fresh-Cut Produce*. International Fresh-cut Produce Association, Alexandria VA.

Hermann Ultrasonics. 2017. Available at: https://www.herrmannultrasonics.com/en-us/ultrasonic-sealing/.

Herner, R. C. 1987. High CO2 effects on plant organs. In: Weichmann, J. (Ed.), *Postharvest Physiology of Vegetables*. Marcel Dekker, New York, p. 239.

Kader, A. A.1986. Biochemical and physiological basis for effects of controlled and modified atmospheres on fruits and vegetables. *Food Technol.* 40, 99–100 & 102–4.

Kader, A. A., Zagory, D. and Kerbel, E. L. 1988. Modified atmosphere packaging of fruits and vegetables. *Crit. Rev. Food Sci. Nutr.* 28(1), 1–30. doi:10.1080/10408398909527506.

Kader, A. A., Zagory, D. and Kerbel, E. L. 1989. Modified atmosphere packaging of fruits and vegetables. *Crit. Rev. Food Sci. Nutr.* 28(1), 1–30.

Soltani, M., Alimardani, R., Mobli, H. and Mohtasebi, S. S. 2015. Modified atmosphere packaging: a progressive technology for shelf-life extension of fruits and vegetables. *Journal of Applied Packaging Research* 7(3), Article 2.

Zagory, D. and Kader, A. A. 1989. Quality maintenance in fresh fruits and vegetables by controlled atmospheres. In: Jen, J. J. (Ed.), *Quality Factors of Fruits and Vegetables: Chemistry and Technology*. American Chemical Society, Washington DC, pp. 174–88. doi:10.1021/bk-1989-0405.ch014.

Chapter 4

Advances in the use of barrier coatings and additives in the preservation of fresh horticultural produce

Elizabeth A. Baldwin, formerly USDA-ARS, USA; and Jeffrey K. Brecht, University of Florida, USA

1 Introduction

Edible coatings and surface treatments have been used to preserve fresh fruits and vegetables for centuries (Contreras-Medellin and Labuza, 1981; Hardenburg, 1967). Many reviews have been written on this topic over the years. A few examples include Flores-López et al. (2016), Romanazzi et al. (2017), Dhall (2013), Bai and Plotto (2012), Soliva-Fortuny et al. (2012), Valencia-Chamorro et al. (2011), Krochta (2002), Baldwin (1999, 2003), and Baldwin et al. (1995a). Horticultural food waste and losses can be from 25 to 50% of the total production in both developed and developing countries (Gustavsson et al., 2011; Muth et al., 2011; Parfitt et al., 2010). This technology offers ways to reduce these losses without use of plastic packaging, thus being more environmentally sustainable. Edible coatings are used to prevent water loss, delay ripening, prevent oxidation, reduce decay, improve appearance, and carry useful additives (Dea et al., 2012; Sanches-Silva et al., 2014; Valencia-Chamorro et al., 2011).

Major postharvest issues for maintaining quality of fruits and vegetables include appearance, since marketability depends on this factor, and water loss due to transpiration. Often, the process of washing fresh produce results in a partial removal or rearrangement of the natural waxy cuticle on the produce surface. This reduces the cuticle's efficacy in restricting water loss, necessitating

http://dx.doi.org/10.19103/AS.2019.0055.04

the application of hydrophobic coatings (Baldwin, 1994). Still another postharvest issue is decay. There are some reports that certain coatings reduce decay, but usually this is because they are carrying fungicides, anti-microbial essential oils, preservatives, or some other antimicrobial compound (Rohas-Graü et al., 2009; Valencia-Chamorro et al., 2011). Sometimes the coating polymer itself can reduce decay, as is the case for coatings containing chitosan (Romanazzi et al., 2017).

Finally, there is the problem of flavor life versus shelf life based on appearance during postharvest shipping, storage, and marketing. Coatings can affect produce appearance and metabolism depending on the coating permeability to oxygen (O_2), carbon dioxide (CO_2), ethylene, and aroma volatiles. Coatings improve appearance, driving initial sales, but internal quality involving flavor and texture often more strongly affects repeat sales (Baldwin et al., 2007). Some coatings help maintain flavor while other coatings can induce off-flavors depending on coating permeability to gases (Bai and Plotto, 2012; Baldwin et al., 2007).

Common materials used as edible coatings include lipids, waxes, resins, carbohydrates, and proteins. Coatings can contain other compounds that affect their performance and permeability such as plasticizers to impart flexibility (brittle coatings tend to crack and flake), surfactants to help adhere to produce surfaces, pH modifiers, and emulsifiers for emulsion coatings. These additives, used for the purpose of modifying the mechanical properties of the coating, can also change the water and gas barrier properties. For example, plasticizers generally increase water or gas permeability (Donhowe and Fennema, 1994) as they open the polymer structure, making it more elastic.

Lipids, waxes, and resins are most commonly used as coatings for whole fruits and vegetables (Hall, 2012). Some commercial coatings are emulsions, mixtures of more than one ingredient (hydrophilic and hydrophobic components) (Hagenmaier and Baker, 1996), or are applied as a bilayer of different materials (Arnon et al., 2014, 2015). Coatings are often applied on a packing line brush or sponge bed with coatings applied as sprays or foams directly on the produce or dripped onto rotating brushes or foam rollers that then coat the fruit moving over them. Dipping fruits and vegetables in coatings is also possible, although this technique is rarely used commercially (Hall, 2012).

Coatings, wraps, and surface treatments are also used to extend shelf life, improve appearance, reduce desiccation, and prevent microbial growth for fresh-cut products (Baldwin et al., 1995a; McHugh and Senesi, 2000; Romanazzi et al., 2017). For these products, the cut surface has damaged cellular material, making it difficult for wax and lipid coatings to adhere. Therefore, emulsions or aqueous coating materials like carbohydrate or protein-based films with additives, or additives alone are used (Dea et al., 2012).

2 Coatings for fruits and vegetables

To understand how edible coatings and surface treatments affect fresh produce quality, an understanding of postharvest physiology of fruits and vegetables is helpful. Fruits and vegetables are mostly composed of water and lose water to the atmosphere via transpiration. This process can be minimized by storage in high relative humidity (RH) (Mishra and Gamage, 2007; Wills and Golding, 2016), but will become an issue during marketing display in stores where the RH is relatively low. Water loss accelerates fruit and vegetable senescence and can cause drying or shriveling of the peel, reducing marketability. Hydrophobic edible coatings present a barrier to water and can help reduce water loss (Bai et al., 2002, 2003a; Hall, 2012).

Fresh produce, like all plant tissue, is metabolically active. Fruits and vegetables undergo respiration, consuming O_2 and producing CO_2 in the process. They also produce ethylene, a gaseous ripening and senescence hormone (Kader and Saltveit, 2003; Wills et al., 1981). Fresh produce respiration and ethylene production are both affected by temperature, being elevated at higher temperature and reduced at lower temperature (Kader and Saltveit, 2003). The optimum postharvest temperature for different fruits and vegetables is determined either by the tissue freezing point for temperate crops or the chilling threshold temperature for subtropical and tropical crops. Ideally, fresh fruits and vegetables should be cooled quickly and maintained at their lowest safe temperature after harvest.

Maintaining fresh produce postharvest constantly under appropriate storage and transport temperatures and humidity is known as the 'cold chain'. Breaks in the cold chain postharvest result in higher losses due to elevated temperature-induced increases in respiration, ethylene production, ripening, senescence, and decay (Yahia, 2010; Yahia et al., 2011). Additionally, it is difficult and usually impractical to re-cool produce after a break in the cold chain occurs. Fresh produce is usually shipped and stored at relatively low temperature (lowest tolerated by a particular type of produce), but ultimately, marketing of whole fruits and vegetables is usually done at ambient temperature. Design of coating or packaging materials must take into account the highest temperature and duration of exposure expected to be experienced by the produce, usually occurring during marketing display.

Fruits and vegetables differ greatly in their respiration rates, which reflect the diversity of metabolic rates exhibited by different detached plant organs. Design of coatings, therefore, must take into account the fruit or vegetable respiration rate, the fruit type, the harvest maturity, and the recommended storage temperature. The reason is that, similar to modified atmosphere packaging (Beaudry, 2000), coatings present a gas barrier that modifies the internal atmosphere of the coated produce based on the produce respiration.

The extent to which the atmosphere is modified depends on the permeability of the coating, the coated product respiration rate, and the storage temperature (Bai and Plotto, 2012).

Further complicating the situation, fruits differ in their ability to ripen after harvest (Bai and Plotto, 2012). Those fruits that ripen after harvest are often harvested prior to onset of ripening and subsequently ripen postharvest during transit, storage, and marketing. These fruits are termed 'climacteric' and their ripening is synchronized and accelerated by ethylene (Kader and Barrett, 2004; Mishra and Gamage, 2007). Those fruits that do not ripen after harvest, and must therefore be harvested ready to eat, are termed 'nonclimacteric'. Ethylene does not affect nonclimacteric fruit ripening, but it does accelerate the senescence of both nonclimacteric fruits and vegetables. Thus, ethylene can shorten their shelf life by inducing undesirable color, texture or flavor changes as well as general senescence (Baldwin, 2004).

Fresh-cut fruits and vegetables are more perishable than their intact counterparts, having the interior tissues exposed. This allows for more water loss and leaves the fresh-cut products more susceptible to decay organisms. Furthermore, fresh-cut fruits and vegetables are basically wounded tissue, for which the wound response triggers many metabolic events including increased respiration and ethylene production, oxidative browning, and production of secondary metabolites that can affect flavor (Baldwin and Bai, 2010; Brecht, 1995; Toivonen and DeEll, 2002). Fresh-cut products must always be refrigerated at <5°C, irrespective of chilling susceptibility and even during marketing display, because of their extreme perishability as well as to minimize microbial proliferation for food safety reasons. Coatings and aqueous surface treatments can minimize some of the undesirable quality effects associated with fresh-cut processing.

2.1 Coating purposes and properties

Coatings are used to extend shelf life of fresh produce, at least partly by replacing natural surface waxes on fruits and vegetables that may be lost during washing to remove dirt and chemical residues. Coatings may extend shelf life through reduction of water loss, improved appearance, retardation of ripening, and/or inhibition of decay. Secondly, by increasing glossiness, preventing water loss-induced shriveling, or inhibiting undesirable color changes, coatings increase marketability. Coatings are also used to reduce surface scarring and injury for tomatoes, cucumbers, eggplants, and bell peppers on packing lines using mineral oil coating as a lubricant (Baldwin, 1994; Hall, 2012). By reducing injury, this also reduces pathogen infection that would occur through points of injury. Coating permeability to gases plays an important role for retardation of ripening, while coating permeability to water vapor determines how effective the coating is at retarding water loss. The effect of coatings on fresh produce decay usually depends on antimicrobial additives (Avena-Bustillos and McHugh, 2012), but sometimes the film former itself can be antimicrobial (Romanazzi et al., 2017).

2.1.1 Improving appearance

Some coatings add gloss or shine, making coated products more attractive. Some coatings delay undesirable color changes and shriveling (due to water loss) and can enhance expected color (Bai and Plotto, 2012; Hall, 2013; Sun et al., 2017). Improving appearance of certain types of produce by using coatings that impart gloss or shine improves sales. This is especially true for apple and citrus, and to a lesser extent, papaya, mango, stone fruits, cucumber, pear, and other fruits (Bai and Plotto, 2012).

Shellac resin is used for shine as it is one of the glossiest of coating materials. It is a secretion of the *Laccifer lacca* insect found in India, is composed of aleuritic and shelloic acids and is permitted as an indirect food additive (21 CFR 175.300). The low gas permeability of shellac, however, can result in fermented off-flavor, especially in citrus (Hagenmaier, 2001, 2002; Hagenmaier and Baker, 1994; Hagenmaier and Shaw, 2002).

Carnauba wax is the exudate of palm tree leaves from the tree of life found in Brazil (21 CFR 184.1978, GRAS, coating agent, confections) (Baldwin et al., 2007); it is a better barrier to water vapor, and less of a barrier to gases, than is shellac. Carnauba wax is usually applied as an emulsion, and the smaller the lipid micelles, the shinier and more stable the coating. Therefore, carnauba wax is usually applied as a micro-emulsion, although it is still not considered as shiny as shellac. Shellac shine, however, diminishes after repeated changes in temperature (cold to warm temperature) resulting in condensation of water on the fruit surface. Temperature change occurs especially during international shipping, and turns shellac-coated fruit milky, termed 'white blush' and therefore, shellac is not used for exported fruit (Baldwin, 1994).

Commercially, shellac and carnauba wax are often used individually or combined for apples and citrus. Since shellac imparts more gloss, but carnauba is less permeable to water vapor, but more permeable to gases, combining these compounds optimizes shine, water loss control, and gas permeability. Corn zein also makes a shiny coating and has been tested on apples (Bai et al., 2003b), but also tends to whiten when in contact with water. Modifying the formulations with a plasticizer minimized the whitening reaction to water, but decreased the gloss.

Early season citrus fruit that have not experienced enough cold nights for the peel to turn orange receive ethylene treatments to degrade the green chlorophyll, resulting in a yellowish peel. The fruit then traditionally receive a dye treatment (citrus red #2) that turns the fruit orange before being coated with wax in a two-step treatment. Unfortunately, this dye has been labeled a carcinogen (Sun et al., 2015), thus new treatments were developed using natural pigments from annatto and paprika, which were incorporated into wax coatings to add orange color to early season oranges and tangerines in a one-step process (Sun et al., 2015, 2017). Papaya, mango, and other fruits are also sometimes coated with carnauba wax, giving the fruit an attractive sheen and reducing water loss-induced shriveling.

2.1.2 Reducing water loss

Hydrophilic coatings are mostly used to reduce water loss of produce. Available materials include lipids, waxes, and emulsions. Lipids such as white mineral oil (mix of paraffinic and naphthenic hydrocarbons) are used for lubrication through sorting and packing, but also improve appearance and prevent some water loss.

Waxes are more commonly used as fruit and vegetable coatings to prevent water loss, including beeswax, which is generally recognized as safe (GRAS), carnauba wax, polyethylene wax, and candelilla wax. Of these, polyethylene wax is a petroleum by-product and generally only used on fruit for which the peel is not consumed, like avocado, banana, coconut, citrus, mango, melon, pineapple, papaya, pumpkin, and certain nuts in shell. Paraffin wax is used on root crops like cassava, yams, rutabaga, and turnip. Waxes are used as storage, shipping, solvent or water waxes (Hall, 2012).

Resins are acidic substances secreted by certain trees and shrubs like shellac, copal, rosin, and amber, and are used in lacquers, varnishes, synthetic plastics, and pharmaceuticals (Hall, 2012). Of these, shellac and rosin can be used in fruit coatings. Carbohydrates and protein coatings do not present effective moisture barriers unless combined with lipids in an emulsion multi-component coating.

2.1.3 Decay prevention

Coatings can be used to reduce decay often induced by injury on packing lines, or by carrying antimicrobials. In some cases, however, they can promote decay by trapping spores (Yahia et al., 2011). The chitosan polymer is interesting because it has a dual mode of action on the pathogen and on the produce. It reduces growth of decay-causing fungi and food borne pathogens and induces plant defense responses depending on the polymer length (i.e., molecular weight). It is GRAS, and also acts as a semipermeable barrier to gases, thus can delay ripening of climacteric fruit (Romanazzi et al., 2017). It is (1-4) N-acetyl-D-glucosamine derived from chitin (B-1,4 linked 2-acetomido-2-deoxy-D-glucopyranosyl residues), which is the major organic skeletal substance of invertebrates, a cell wall constituent of fungi and green algae (Nisperos-Carriedo, 2004), and thus plentiful. Mechanisms of action include positive chitosan charges interacting with negatively charged phospholipids in fungal membranes, causing membrane permeability (Palma-Guerrero et al., 2010). Chitosan is also reported to increase phenylalanine ammonia lyase (PAL) (Cheng and Breen, 1991), accumulation of phytoalexin-acting phenols, as well as chitinase and β-1-3-glucanase, two plant resistance enzymes that can degrade fungal cell walls (Romanazzi et al., 2017; Van-Loon and Van-Strein, 1999). Chitosan coating with thyme oil reduced anthracnose decay during storage of mango fruit as effectively as the commercial fungicide prochloraz while inducing higher accumulation of phenolics during storage (Fig. 1)

Aloe vera gel is also reported to have anti-fungal and anti-bacterial properties due to anthraquinones, while also providing a barrier to gases and water vapor (Misir et al., 2014). The barrier to water vapor is surprising as the gel is mostly composed of polysaccharides. This material reduced decay of papaya, especially when combined with a papaya leaf extract. (Brishti et al., 2013). Aloe vera coating was as effective as chitosan coating in reducing anthracnose decay incidence on mango fruit during storage and there was some synergistic effect when the two coating materials were combined (Fig. 2).

Nanotechnology (nano = 1–100 nm) has offered new options for edible coatings (Flores-López et al., 2016) in the form of nano-emulsions, often used to carry antimicrobial essential oils such that they have greater effectiveness. Entrapment of compounds within nano-emulsions may enhance their stability and performance, especially oil-in-water nano-emulsions (Acevedo-Fani et al., 2017). Nano-emulsions showed enhanced stability and antimicrobial activity,

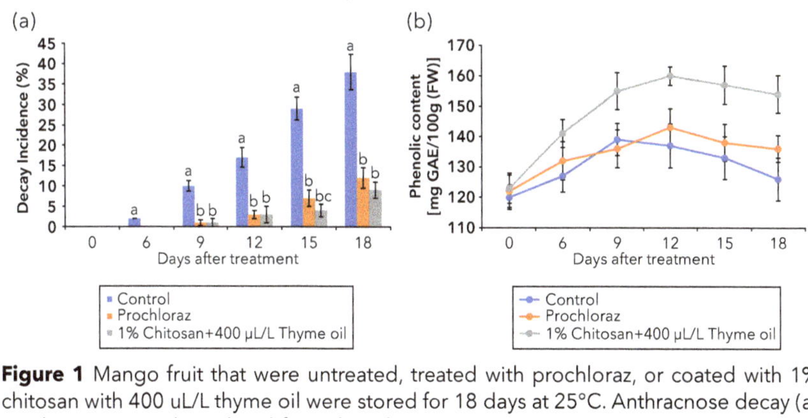

Figure 1 Mango fruit that were untreated, treated with prochloraz, or coated with 1% chitosan with 400 uL/L thyme oil were stored for 18 days at 25°C. Anthracnose decay (a) incidence was evaluated and fruit phenolic content (b) was measured every 3 days. (S. Shah et al., unpublished)

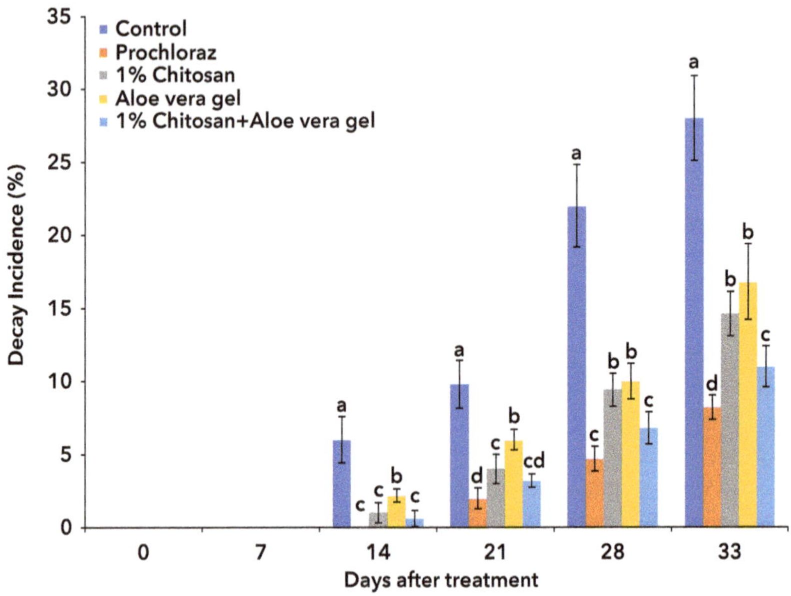

Figure 2 Mango fruit that were untreated, treated with prochloraz, or coated with 1% chitosan or aloe vera gel or a mixture of chitosan and aloe vera were stored for 28 days at 12°C followed by 5 days shelf life at 25°C. Anthracnose decay incidence was evaluated weekly during storage as well as after shelf life. (S. Shah et al., unpublished)

including inactivation of *Salmonella typhimurium* and *Escherichia coli* O157:H7 on grapes while reducing weight loss as well as loss of firmness and retaining phenolic compounds and antioxidant activity (Kim et al., 2014). Another nanotechnology is layer-by-layer deposition for nano-laminate coatings using opposite charged materials of nanometer dimension (Flores-López et al., 2016). For example, antimicrobial nano-emulsion coatings with sodium alginate and lemongrass essential oil were more effective in inactivation of *E. coli* on inoculated apple slices and psychrophilic bacteria during storage compared to conventional emulsions on fresh-cut 'Fuji' apples (Salvia-Trujillo et al., 2015).

2.1.4 Internal MA (semipermeable coatings)

For climacteric fruits, coatings presenting a semipermeable gas barrier can create a modified internal atmosphere of relatively low O_2 and high CO_2 within the fruit, which is similar to that obtained using controlled or modified atmosphere storage/packaging. Modification of the atmosphere slows down ethylene production, respiration, senescence and ripening, thus extending shelf life (Bai and Plotto, 2012; Baldwin et al., 1999; Dhall, 2013; Hagenmaier, 2002). For nonclimacteric fruits, the goals are reduction of water loss and undesirable color changes (Bai and Plotto, 2012). With 'Valencia' oranges, for example, use of shellac, which has relatively low gas permeability, resulted in fermented off-flavor due to increased ethanol and an altered aroma profile (Baldwin et al., 1995b). Comparing shellac and carnauba wax coatings on tangerine fruit, shellac coating resulted in significantly lower internal oxygen concentration and elevated ethanol, but the effect of the two coatings on water loss was similar (Fig. 3). For litchi fruit, coatings delayed the loss of red color and subsequent browning of the peel, due to polyphenyloxidase (PPO) activity, using chitosan (Jiang et al., 2005), carrageenan, hydroxypropylcellulose (HPMC), sucrose fatty acid esters (Semperfresh), pectin, and carnauba wax emulsion coatings (Jiang et al., 2005; Plotto et al., 2006).

2.2 Applications of coatings for whole and fresh-cut fruits and vegetables

Coatings are used commercially on apples and pears, stone fruits, grapefruit, oranges, limes and lemons, mangoes, papayas, avocados, passion fruit, litchi, pineapples, pomegranates, peppers, tomatoes, and cucumbers. Cellulose-derived polysaccharide-based coatings including methyl cellulose (MC), HPMC, carboxymethyl cellulose (CMC) with or without glycerol, oleic acid or stearic acid, and chitosan coatings were tested on mandarin fruit (Arnon et al., 2015). Then the CMC coating was tested as a bilayer with chitosan. The fatty acids did not improve water loss and enhanced fruit deterioration. The glycerol, often used in coatings as a plasticizer, had varying advantages for flavor and gloss at different amounts, but enhanced water loss. The bilayer CMC/chitosan coatings enhanced firmness and gloss while inhibiting weight loss and off-flavor, which

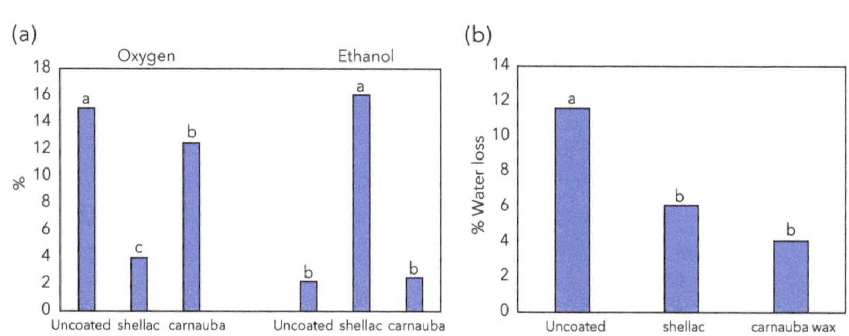

Figure 3 Tangerine fruit that were uncoated or coated with shellac or carnauba wax were stored for 7 days at 20°C. (a) Internal oxygen and ethanol concentrations were measured in the columella space within the fruit; (b) Water loss was measured as per cent of the initial fresh weight of the fruit. (E. A. Baldwin et al., unpublished)

is significant since mandarin citrus are especially susceptible to producing ethanol when coated, leading to fermented flavor.

A CMC/chitosan coating was applied to mandarins, oranges, and grapefruit as a bilayer coating and compared to commercial polyethylene wax (Arnon et al., 2014). The bilayer coating was comparable to the commercial wax for gloss, but was not effective in preventing weight loss, and both the commercial wax and the bilayer coating had minimal effect on modifying the atmosphere within the fruit. This resulted in only slight increases in ethanol and no flavor effects for oranges and grapefruits, but a gradual flavor decrease in acceptability for the mandarins, which are especially sensitive to off-flavor development (Arnon et al., 2014).

Coatings of chitosan alone, a trilayer of beeswax, chitosan and beeswax, and a composite coating of chitosan crosslinked with sodium tripolyphosphate were applied to strawberries (Velickova et al., 2013). The trilayer coating best delayed senescence and weight loss and modified respiration, but had a matte or slightly white appearance, while the chitosan and composite coatings were transparent. Therefore, the chitosan and composite coatings gave the best visual appearance as well as taste in a sensory test.

Composite coatings of galactomannans, collagen, and glycerol were applied to apples and mangoes, varying the amounts of the three ingredients (Lima et al., 2010). A mixture of 0.5% galactomannan, 1.5% collagen, and 1.5% glycerol reduced mango respiration and 0.5% galactomannan, 1.5% collagen, and 0% glycerol reduced respiration of apples, demonstrating that these coatings presented enough of a gas barrier to extend fruit shelf life.

2.3 Applications of coatings for fresh-cut fruits and vegetables

A composite of cassava starch, glycerol, carnauba wax, and stearic acid in an emulsion coating was applied to apple slices, varying the amounts of the ingredients; 1.5 g starch, 1.5 g glycerol, 0.2 g carnauba wax, and 0.8 g stearic acid per 100 g coating solution was optimal based on reduced respiration and water vapor resistance imparted by the coatings (Chiumarelli and Hubinger, 2012). Fresh-cut pineapples were dipped in ascorbic acid (5 g/L), citric acid (10 g/L) and compared to the same plus coated in cassava starch (20 g/L), and the same plus 10 g/L calcium lactate plus 20 g/L cassava starch (Bierhals et al., 2011). Treatments with the cassava starch, with or without calcium lactate, reduced respiration, weight loss, juice leakage, and maintained firmness, but exhibited browning, and less ascorbic acid content, and more microbial growth, resulting in 1 day less shelf life. All treatments, however, were acceptable by sensory tests. A nano multilayer system (five nano-layers) of k-carrageenan and the antimicrobial protein, lysozyme, was assembled on the surface of whole and fresh-cut pears (Medeiros et al., 2012). The coating prevented water (weight) loss on both whole and fresh-cut pears as well as reduced browning of fresh-cut pears. Despite such research on coatings for fresh-cut products, the only successful commercial application is coating of baby carrots to control the 'white blush' appearance of the product surface that is caused by water loss (Cisneros-Zevallos et al., 1997; Garcia and Barrett, 2002).

3 Additives for fruits and vegetables

The quality of fresh fruits and vegetables can deteriorate as a result of changes in appearance, texture, flavor, and nutritional value. These changes are the result of chemical, enzymatic, and physical processes as well as the effects of microbial activity. Additives refers to synthetic and natural chemical compounds that may be applied to whole and fresh-cut fruits and vegetables to extend their shelf life by limiting microbial survival and growth, as well as inhibiting undesirable physiological and biochemical processes such as softening, senescence-related tissue watersoaking, and oxidative browning reactions. Additives may also be used to maintain nutritional value or to nutritionally fortify fruits and vegetables. As mentioned previously in this chapter, coatings are applied to address some of the changes in fresh fruits and vegetables that result in loss of quality, especially the removal of natural surface waxes during washing and the resultant changes in appearance (i.e. glossiness) and water loss.

3.1 Additive properties

Fresh-cut fruits and vegetables are wounded as part of their preparation, which exposes the interior tissues to accelerated water loss and microbial colonization. In addition, wounding accelerates the rate of senescence of the tissues in general and initiates a number of specific physiological and biochemical processes related to the plant wound response (Baldwin and Bai, 2010; Brecht, 1995; Toivonen and DeEll, 2002). These include increased respiration, elevated ethylene and jasmonate production, synthesis of phenolic compounds and oxidative browning reactions, accelerated production of reactive oxygen species (ROS) resulting in membrane deterioration and loss of cellular compartmentation leading to watersoaking and tissue translucency, and other textural changes related to turgor loss and cell wall disassembly.

Like all postharvest treatments applied to fruits and vegetables, the specific additives applied to any product are chosen based on the most important shelf life limiting factors for each particular product. Sanitizer and antimicrobial compounds that are commonly applied to fresh horticultural crops to reduce decay and minimize food safety risks can be considered to be additives (Ayala-Zavala and Gonzalez-Aguilar, 2010; Ramos et al., 2013), but those are not the focus of this chapter. Gaseous treatments that are applied to fruits and vegetables, including ethylene to promote fruit ripening and citrus degreening (Brecht, 2019), reduced oxygen and/or elevated carbon dioxide in controlled or modified atmosphere storage (Yahia et al., 2019a) or modified atmosphere packaging (Caleb et al., 2013; Ghidelli and Perez-Gago, 2018) to extend storage life by restricting metabolic reactions, and fumigants such as methyl bromide for insect control (Yahia et al., 2019b) are also not covered in this chapter.

3.1.1 Anti-browning compounds

Enzymatic browning of plant tissues occurs as a consequence of damage to the tissues, allowing usually sequestered enzymes and substrates to come in contact with each other. Wounding also stimulates ethylene production and synthesis of the phenolic compounds that are the substrates for oxidative browning reactions via induction of PAL, which catalyzes the rate limiting step in biosynthesis of the phenolics that serve as substrates for the browning reactions. The discoloration is primarily the result of the action of a group of enzymes collectively called polyphenol oxidases (PPO) that catalyze hydroxylation of monophenols to o-diphenols and oxidation of o-diphenols to produce *o*-quinones, which although only slightly colored are highly reactive and can non-enzymatically react with other phenolic compounds, amino acids and amino groups, aromatic amines, and thiol compounds to produce polymers that are brown, reddish-brown, blue-gray, and black (Garcia and Barrett, 2002; Toivonen and Brummell, 2008). Peroxidase is another enzyme that increases in wounded and otherwise stressed plant tissues and, in the presence of hydrogen peroxide, can oxidize phenolics.

Control of browning reactions is most commonly accomplished using either physical or chemical approaches. The former approach involves using low temperature to reduce the intensity of the wound response and the browning reaction rates, as well as barriers (films, packaging) to restrict access of oxygen to the wounded tissue. Reducing wound ethylene production, either by directly inhibiting its synthesis or by inhibiting ethylene action, avoids the increased production of the phenolic substrates of the browning reactions that normally accompanies wounding and stress. Chemical additives that are used to control browning either remove the products or inhibit the PPO-mediated reactions and are commonly applied in combination.

Reducing agents such as ascorbic acid act as preferred substrates for oxidative reactions in general, and specifically convert the reactive *o*-quinones formed by PPO back to *o*-diphenols. Thiol-containing compounds such as cysteine and glutathione are another example of reducing agents that can be used to inhibit enzymatic browning, but these are usually used only as supplements to other reducing agents due to the unpleasant taste that may be imparted at high concentrations. A limitation for use of reducing agents to inhibit enzymatic browning is that the compounds are consumed in the process, thus allowing the browning reactions to continue once the supply of reducing agents has been exhausted.

The reactive site of the PPO enzyme contains copper, which can be complexed by chelating agents like ethylenediamine tetraacetic acid (EDTA), rendering the enzyme inactive. Other compounds like cyclodextrins can complex PPO substrates or products, mitigating its activity.

For most fruit and vegetable PPO enzymes, the optimum pH for activity is in the range of pH 6.0–6.5, with little activity below pH 4.5 (Whitaker, 1994). Therefore, acidulants like organic acids, most commonly citric acid, are a common component of anti-browning treatment cocktails. However, citric acid may not reduce browning of potato slices by reduction of pH, since the effectiveness of citric acid in reducing browning in fresh-cut potato could not be duplicated by simple pH adjustment of the immersion solution (Tsouvaltzis and Brecht, 2016). Ascorbic acid, in addition to being a reducing agent, also acts as a mild acidulant in reducing tissue browning.

3.1.2 Firming agents

Calcium increases or maintains fruit and vegetable firmness by cross-linking pectic polymers in the cell wall middle lamella (Tokala and Mahajan, 2018). The calcium ions bind to free carboxyl groups in adjacent polygalacturonic acid polymers, creating an 'egg box' structure that not only is structurally strong, but also is resistant to the action of cell wall degrading enzymes including polygalacturonase, pectin methylesterase (PME), beta-galactosidase, and cellulase (de Freitas and Nassur, 2017). Calcium also stabilizes plant cell membranes and reduces increases in membrane permeability by binding to membrane phospholipids and membrane-associated proteins (de Freitas and Nassur, 2017). Calcium delays membrane lipid degradation and changes in composition, favoring membrane restructuring processes, and inhibiting the enzyme lipoxygenase.

3.1.3 Antimicrobials and other additives

Sanitizers are chemicals used to reduce populations of microorganisms on the surfaces of whole and fresh-cut fruits and vegetables, and importantly reduce the possibility of cross-contamination during washing and other hydrohandling processes in order to reduce decay and incidence of foodborne illnesses. Chlorine in various forms is the most commonly used sanitizer compound; others include ozone, hydrogen peroxide, peroxyacetic acid, and trisodium phosphate. Very few synthetic chemical fungicides are applied to fresh fruits and vegetables after harvest for decay control due to environmental issues and human health concerns. The exceptions are fungicides applied to fruits for which the peel is not eaten, such as bananas and citrus fruits, or applied to vegetables like sweetpotatoes that are stored and cooked before consumption (Adaskaveg and Förster, 2009). The trend in recent years has been toward GRAS compounds that are exempt from residue tolerances on fresh fruits and vegetables (Palou et al., 2016). The most common antimicrobial agents applied to fresh fruits and vegetable are organic acids such as acetic, benzoic, citric, fumaric, lactic, malic, propionic, sorbic, succinic, and tartaric acid, among others. Potassium sorbate and sodium benzoate are the most widely used as antimicrobial food additives. Since low pH is integral to the effectiveness of these chemicals, the use of organic acids as preservatives is limited to fruits and vegetables with acidic pH. Potassium carbonate, sodium bicarbonate, sodium carbonate, sodium metabisulfite, and sodium salicylate have also been shown to be effective for control of postharvest diseases on certain crops (Palou et al., 2016).

3.2 Applications of additives for whole fruits and vegetables

Browning of fruits and vegetables after harvest is mainly a consequence of the wounding that occurs during harvest. In most cases, the resulting browning, such as the discoloration of the base of banana 'hands' where they were detached from the main bunch stem, is not considered to be important enough to affect marketability. Other crops such as lettuce, mushroom, and artichoke can develop browning where they were detached and in rare cases additives may be used to minimize the discoloration. However, other than for long international shipping times, for which controlled and modified atmospheres including carbon monoxide with lettuce may be used, rapid cooling and maintenance of low temperatures during postharvest handling and distribution are primarily relied on to control such discoloration in whole products (Thompson, 2015).

Scald (superficial scald, storage scald, senescent scald) is a storage disorder of apples and pears that causes brown to black areas to develop on the peel (Lurie and Watkins, 2012). The discoloration is the result of auto-oxidation of the volatile compound α-farnesene, synthesis of which is induced by ethylene. Subsequent production of trienols, possibly through free radical-mediated oxidation, causes tissue injury, which leads to phenolic oxidation via PPO and the scald symptoms. Pre-storage application of the antioxidants diphenylamine (DPA) and ethoxyquin (6-ethoxy-2,2,4-trimethyl-1,2-dihydroquinoline) are effective in inhibiting scald. However, these chemicals are not allowed in many countries, leading to the use of other treatment methods, primarily ultra-low oxygen and dynamic controlled atmosphere storage systems, and application of the ethylene action inhibitor, 1-methylcyclopropene (1-MCP; Lurie and Watkins, 2012).

Litchi is unusual in that loss of red color and enzymatic browning occur over the entire pericarp surface of the fruit during storage. Anthocyanins react with enzymatically produced *o*-quinones, changing the fruit color from red to brown and impairing marketability (Jiang, 2000). Fumigation with 1-2% sulfur dioxide (SO_2) coupled with application of 1 N hydrochloric acid (HCl) has traditionally been the preferred method to inhibit litchi pericarp discoloration, but use of sulfites on food products has been limited in recent years by government regulation. Therefore, application of acidulants, such as HCl or citric acid, along with ascorbic acid is now the main approach taken to inhibit litchi pericarp discoloration.

Bananas and mangoes exude latex when the fruit are detached, which can cause chemical injury of the epidermis leading to enzymatic browning. Latex discoloration can be prevented by holding the fruit stem-end down until latex flow ceases or by rinsing the freshly harvested fruit with copious water to remove the latex; if neither draining or water rinsing is practical, the fruit may be dipped in solutions of detergent, hydrated lime (calcium hydroxide), or alum (potassium aluminum sulfate) to remove the latex (Brecht, 2018; Thompson, 2015).

Calcium treatment of whole fresh fruits and vegetables has received much research attention over many years (de Freitas and Nassur, 2017; Martin-Diana et al., 2007; Valero and Serrano, 2010), showing delay of ripening and senescence and inhibition of respiration, cell wall breakdown, membrane deterioration, softening, and weight loss, as well as increased antioxidant capacity and resistance to decay and physiological disorders such as blossom end rot in tomato and melons, bitter pit in apples, cracking in cherry, and chilling injury of many crops (de Freitas and Nassur, 2017; Madani and Forney, 2016). However, there appears to be little commercial use of calcium treatments for whole fresh fruits and vegetables other than apples, possibly due to difficulty with achieving sufficient uniform penetration into the tissues of whole fruits and vegetables despite availability of vacuum and pressure infiltration techniques that improve calcium uptake over dipping treatments. Changes in food safety regulations have greatly reduced postharvest calcium usage as drenches are being phased out.

3.3 Applications of additives for fresh-cut fruits and vegetables

Vegetables such as eggplant, lettuce, mushroom, potato, and fruits such as apple, pear, avocado, banana, and peach can undergo substantial discoloration due to enzymatic browning when cut. Additives to control browning of fresh-cut fruits and vegetables are most often applied in combinations. Thus, combination treatments are commonly used that contain ascorbic acid, citric acid, and calcium chloride or calcium lactate, which combine the modes of action of antioxidant, acidulant, and membrane stabilizer. For example, while it was found that neither ascorbic acid or citric acid, alone or in combination, were effective in reducing browning of fresh-cut apple (Ponting et al., 1972) or potato (Mattila et al., 1993), with the addition of calcium chloride or calcium lactate, the resulting combination treatments were quite effective. The most commonly used commercial anti-browning treatment for fresh-cut products is NatureSeal®, which contains mainly calcium ascorbate.

Calcium ions can passively diffuse within the cell wall structure of plant cells, moving through the apoplast or intercellular spaces; thus, unlike whole fruits and vegetables, the interior tissues of fresh-cut fruits and vegetables are open to calcium infiltration, and the positively charged calcium ions are actually attracted by negatively charged carboxyl groups in the middle lamella and cell wall (Madani and Forney, 2016). Fresh-cut melons are prone to accelerated softening and development of watersoaking or translucency compared with the intact fruit. Fresh-cut apple and mango tissue can develop a watersoaked appearance and mealy texture. Fresh-cut watermelon and tomato leak copious amounts of the liquid contents of the cells. Calcium is commonly applied commercially to these products to help maintain the fresh texture and appearance (Martin-Diana et al., 2007).

While calcium chloride can cause bitter flavor at the concentrations required to effectively retain tissue integrity, other forms such as calcium lactate do not (Luna-Guzman and Barrett, 2000; Martin-Diana et al., 2007). Other additives for fresh-cut fruits and vegetables that have been examined by researchers are not currently used commercially (Ayala-Zavala and Gonzalez-Aguilar, 2010)

4 Combination of coatings and additives for fruits and vegetables

Various natural and chemical additives have been incorporated into edible coatings to modify their properties and particularly to address the specific shelf life limiting factors of the fresh fruits and vegetables being coated (Martin-Belloso et al., 2009; Palou, 2018; Parreidt et al., 2018; Quezado-Gallo, 2009; Rojas-Graü et al., 2009). Chief among the additives incorporated into coatings are antioxidants and antimicrobials. A great variety of antimicrobial agents have been incorporated into fruit and vegetable coatings, including acetic, benzoic, lactic and sorbic acids, trans-cinnamaldehyde, hypochlorous acid, and essential oils, while ascorbic acid, citric acid, cysteine, N-acetylcysteine, glutathione, and α-tocopherol are incorporated into coatings as antioxidants and anti-browning agents (Parreidt et al., 2018). For essential oils, the limiting factor is the strong odor and flavor that they can impart at effective concentrations, which is due to the presence of phenolic compounds such as abietane diterpenes, carnosol, and ursolic acid (Parreidt et al., 2018).

4.1 Additive and synergistic effects of coating and additive combinations

Incorporating antioxidants and antimicrobials into coatings tends to improve their efficacy due to the possibility for gradual release of the additives, which can result in maintenance of a critical, effective concentration of the additive over a longer period compared with direct application to the fruit or vegetable (Parreidt et al., 2018).

4.2 Applications of coating and additive combinations for whole fruits and vegetables.

Coatings used commercially on whole fruits and vegetables are proprietary products; thus, the components of the coatings are not generally known. However, research has demonstrated the potential usefulness of various coating additives in improving the effect of various coatings to improve fruit and vegetable quality and shelf life. For example, carnauba wax with oleic acid emulsifier, polyethylene glycol humectant, sodium alginate thickener, and sodium dodecyl sulfate surfactant was applied to eggplant fruit and was found to extend their shelf life from 3 days to 12 days under ambient conditions of 18–21°C and RH of 50–55% (Singh et al., 2016).

Cassava starch coatings, with and without the antimicrobial potassium sorbate, were tested on strawberries (Garcia et al., 2010). It was concluded that 3% starch plus 0.05% potassium sorbate was optimal for reducing respiration, increasing water vapor resistance, and maintaining appearance and sensory acceptance. Potassium sorbate has been found to be quite effective for control of citrus fruit decay when incorporated into coatings (Youssef et al., 2012). Several commercial decay control products utilizing potassium sorbate are available for postharvest treatment of fresh fruits in combination with coatings, especially citrus and stone fruits.

Coatings can also be designed to support and carry antagonistic yeasts or bacteria that either outcompete pathogenic organisms or produce substances toxic to other organisms. This was demonstrated on citrus fruits for *Candida guilliermondii* and *Debaryomyces* sp. incorporated into cellulase-based fruit coating in comparison with coating containing the commercial fungicide imazalil (Portjewijd et al., 1995). Integration of calcium chloride into chitosan coating, which has anti-fungal properties itself, with the biocontrol yeast, *Cryptococcus laurentii*, improved the control of postharvest blue mold decay of pears over that of either the chitosan or yeast alone (Yu et al., 2012).

4.3 Applications of coating and additive combinations for fresh-cut fruit and vegetables

The exposure of interior tissues that occurs during fresh-cut processing has led to the idea that edible coatings with moisture and oxygen barrier properties, containing various additives, could replace the epidermis, offsetting many, if not all, of the negative consequences of cutting the fruits and vegetables (Baldwin et al., 1995a; Corbo et al., 2015; Oms-Oliu et al. 2008). This may include films based on fruit and vegetable purees, pomaces, and extracts (Otoni et al., 2017). An example is the 'apple wrap' (McHugh and Senesi, 2000), an edible coating made from 70% apple puree/27% vegetable oil/1.5% ascorbic acid/1.5% citric acid that was reported to reduce moisture loss and browning in fresh-cut apples. An alginate (2%), pectin (2%) or gellan based (0.5%) coating with calcium chloride (9%) for cross-liking with or without N-acetylcysteine (0.75%) and glutathione (0.75%), and with glycerol (0.5–0.6%) as plasticizer (w/v), emulsified with sunflower oil was tested on fresh-cut 'Fuji' apples (Rojas-Graü et al., 2007). The polysaccharide coatings increased water vapor resistance and reduced ethylene production, while the additives reduced microbial growth and browning. Coatings with alginate or pectin with additives maintained sensory quality for 2 weeks at 14°C. The main potential benefit of edible coatings for fresh-cut produce is stabilization of the product, thereby extending product shelf life. More specifically, coatings have the potential to reduce moisture loss, restrict oxygen entrance and create an internal modified atmosphere, lower respiration, retard ethylene production, seal in flavor volatiles, and carry additives that retard discoloration and microbial growth.

A novel approach to increase fresh-cut fruit firmness, which was demonstrated with strawberries, is infiltration of the enzyme PME plus calcium into the cut fruit pieces (Carnelossi et al., 2018). PME cleaves the methyl groups from galacturonic acid polymers (pectin) in plant cell walls. Infiltration of PME into fresh-cut products thus creates free carboxyl groups that can be crosslinked by the calcium ions, resulting in enhanced tissue firmness.

5 Conclusion and future trends

Much research has been conducted on the use of coatings and additives to overcome the shelf life-limiting factors that accompany deterioration of whole and fresh-cut fruits and vegetables after harvest and processing. Successful commercial application of these research findings will be closely linked with the creation of formulations and application methods that not only extend the product shelf life in terms of appearance and texture, but also preserve the important sensory properties of the whole and fresh-cut fruits and vegetables. Edible coatings that reproduce or enhance the properties of natural fruit coatings through their physical properties and by the action of various additives incorporated into the coatings hold great promise for improving the shelf life, quality, and safety of fresh horticultural crops.

6 References

Acevedo-Fani, A., Soliva-Fortuny, R. and Martín-Belloso, O. (2017), 'Nanoemulsions as edible coatings', *Curr. Opin. Food Sci.*, 15, 43–9.

Adaskaveg, J. E. and Förster, H. (2009), 'New developments in postharvest fungicide registrations for edible horticultural crops in the United States', in D. Prusky and M. L. Gullino (Eds), *Postharvest Pathology in the 21st Century*, Vol. 2. Springer, pp. 107–17.

Arnon, H., Zaitsev, Y., Porat, R. and Poverenov, E. (2014), 'Effects of carboxymethyl cellulose and chitosan bilayer edible coating on postharvest quality of citrus fruit', *Postharv. Biol. Technol.*, 87, 21–6.

Arnon, H., Granit, R., Porat, R. and Poverenov, E. (2015), 'Development of polysaccharides-based edible coatings for citrus fruits: A layer-by-layer approach', *Food Chem.*, 166, 465–72.

Avena-Bustillos, R. J. and McHugh, T. H. (2012), 'Role of edible film and coating additives', in E. A. Baldwin, R. Hagenmaier and J. Bai (Eds), *Edible Coatings and Films to Improve Food Quality*, 2nd edn. CRC Press, pp. 157–84.

Ayala-Zavala, J. F. and Gonzalez-Aguilar, G. A. (2010), 'Use of additives to preserve the quality of fresh-cut fruits and vegetables', in O. Martin-Belloso and R. Soliva-Fortuny (Eds), *Advances in Fresh-Cut Fruits and Vegetables Processing*. CRC Press, pp. 232–54.

Bai, J., Hagenmaier, R. D. and Baldwin, E. A. (2002), 'Volatile response of four apple varieties with different coatings during marketing at room temperature', *J. Agric. Food Chem.*, 50, 7660–8.

Bai, J., Hagenmaier, R. D. and Baldwin, E. A. (2003a), 'Coating selection for "Delicious" and other apples', *Postharv. Biol. Technol.*, 28, 381–90.

Bai, J., Alleyne, V., Hagenmaier, R. D., Mattheis, J. P. and Baldwin, E. A. (2003b), 'Formulation of zein coatings for apples (*Malus domestica* Borkh)', *Postharv. Biol. Technol.*, 28, 259–68.

Bai, J. and Plotto, A. (2012), 'Coatings for fresh fruits and vegetables', in E. A. Baldwin, R. Hagenmaier and J. Bai (Eds), *Edible Coatings and Films to Improve Food Quality*, 2nd edn. CRC Press, pp. 185–242.

Baldwin, E. A. (1994), 'Edible coatings for fresh fruits and vegetables: Past, present, and future', in J. Krochta, E. A. Baldwin and M. O. Nisperos-Carriedo (Eds), *Edible Coatings and Films to Improve Food Quality*. Technomic Publishing, pp. 25–64.

Baldwin, E. A. (1999), 'Surface treatment and edible coating in food preservation', in M. S. Rahman (Ed.), *Handbook of Food Preservation*, 2nd edn. CRC Press, pp. 477–508.

Baldwin, E. A. (2003), 'Coatings and other supplemental treatments to maintain vegetable quality', in J. A. Bartz and J. K. Brecht (Eds), *Postharvest Physiology of Vegetables*, 2nd edn. Marcel Dekker Inc., pp. 413–35.

Baldwin, E. A. (2004), 'Edible coatings', in S. Ben-Yehoshua (Ed.), *Environmentally Friendly Technologies for Agricultural Produce Quality*. CRC Taylor & Francis, pp. 301–14.

Baldwin, E. A. and Bai, J. (2010), 'Physiology of fresh-cut fruits and vegetables', in O. Martin-Belloso and R. Soliva-Fortuny (Eds), *Advances in Fresh-Cut Fruits and Vegetables Processing*. CRC Press, pp. 87–113.

Baldwin, E. A., Burns, J. K., Kazokas, W., Brecht, J. K., Hagenmaier, R. D., Bender, R. J. and Pesis, E. (1999), 'Effect of two edible coatings with different permeability characteristics on mango (*Mangifera indica* L.) ripening during storage', *Postharv. Biol. Technol.* 17, 215–26.

Baldwin, E. A., Nisperos-Carriedo, M. O. and Baker, R. A. (1995a), 'Use of edible coatings to preserve quality of lightly (and slightly) processed products', *Crit. Rev. Food Sci. and Nutr.*, 35, 509–24.

Baldwin, E. A., Nisperos, M. O., Shaw, P. E. and Burns, J. K. (1995b), 'Effect of coatings and prolonged storage conditions on fresh orange flavor volatiles, degrees brix, and ascorbic acid levels', *J. Agric. Food Chem.*, 43, 1321–31.

Baldwin, E. A., Plotto, A. and Goodner, K. (2007), 'Shelf-life versus flavour-life for fruits and vegetables: How to evaluate this complex trait', *Stewart Postharv. Rev.*, 1(3), 1–10.

Beaudry, R. M. (2000), 'Responses of horticultural commodities to low oxygen: Limits to the expanded use of modified atmosphere packaging', *HortTechnology*, 10, 491–500.

Bierhals, V. S., Chiumarelli, M., Hubinger, M. (2011), 'Effect of cassava starch coating on quality and shelf life of fresh-cut pineapple (Ananas comosus L. Merril cv "Perola")', *J. Food Sci.*, 766, E62–72.

Brecht, J. K. (1995), 'Physiology of lightly processed fruits and vegetables', *HortScience*, 30, 18–22.

Brecht, J. K. (2018), 'Mango', in S. Pareek and S. Tonetto de Freitas (Eds), *Postharvest Physiological Disorders in Fruits and Vegetables*. CRC Press, pp. 443–66.

Brecht, J. K. (2019), 'Ethylene technology', in E. M. Yahia (Ed.), *Postharvest Technology of Perishable Horticultural Commodities*. Elsevier, pp. 481–97.

Brishti, F. H., Misir, J. and Sarker, A. (2013), 'Effect of biopreservatives on storage life of papaya (*Carica papaya* L.)', *Intl. J. Food Studies*, 2, 126–36.

Caleb, O. J., Mahajan, P. V., Al-Said, F. A. and Opara, U. L. (2013), 'Modified atmosphere packaging technology of fresh and fresh-cut produce and the microbial consequences – a review', *Food Bioprocess Technol.*, 6, 303–29.

Carnelossi, M., Brecht, J. K., Huber, D., Carvalho, L. M. and Sargent, S. A. (2018), 'Vacuum infusion of pectin methylesterase and calcium maintains firmness of the fresh-cut strawberry', *Actas Portuguesas de Horticultura*, 29, 132–40.

Cheng, G. W. and Breen, P. J. (1991), 'Activity of phenylalanine ammonialyase (PAL) and concentrations of anthocyanins and phenolics in developing strawberry fruit', *J. Amer. Soc. Hort. Sci.*, 116, 865–9.

Chiumarelli, M. and Hubinger, M. D. (2012), 'Stability, solubility, mechanical and barrier properties of cassava starch – carnauba wax edible coatings to preserve fresh-cut apples', *Food Hydrocoll.*, 28, 59–67.

Cisneros-Zevallos, L., Saltveit, M. E. and Krochta, J. M. (1997), 'Hygroscopic coatings control surface white discoloration of peeled (minimally processed) carrots during storage', *J. Food Sci.*, 62, 363–8.

Contreras-Medellin, R. Labuza, T. P. (1981), 'Prediction of moisture protection requirements for foods', *Cereal Foods World*, 26, 335–7.

Corbo, M. R., Campaniello, D., Speranza, B., Bevilacqua, A. and Sinigaglia, M. (2015), 'Non-conventional tools to preserve and prolong the quality of minimally-processed fruits and vegetables', *Coatings*, 5, 931–61.

de Freitas, S. T. and Nassur, R. C. M. R. (2017), 'Calcium treatments', in S. Pareek (Ed.), *Novel Postharvest Treatments of Fresh Produce*. CRC Press, pp. 51–78.

Dea, S., Ghidelli, C., Perez-Gago, M. B. and Plotto, A. (2012), 'Coatings for minimally processed fruits and vegetables', in E. A. Baldwin, R. Hagenmaier and J. Bai (Eds), *Edible Coatings and Films to Improve Food Quality*, 2nd edn. CRC Press, pp. 243–89.

Dhall, R. K. (2013), 'Advances in edible coatings for fresh fruits and vegetables: A review', *Crit. Rev. Food Sci. & Nutr.*, 53, 435–50.

Donhowe, I. G. and Fennema, O. (1994), 'Edible films and coatings: Characteristics, formation, definitions and testing methods', in J. M. Krochta, E. A. Baldwin and M. Nisperos-Carriedo (Eds), *Edible Coatings and Films to Improve Food Quality*. Technomic Publishing Co. Inc., pp. 1–24.

Flores-López, M. L., Cerqueira, M. A., de Rodríguez, D. J. and Vicente, A. A. (2016), 'Perspectives on utilization of edible coatings and nano-laminate coatings for extension of postharvest storage of fruits and vegetables', *Food Eng. Rev.*, 8, 292–305.

Garcia, E. and Barrett, D. M. (2002), 'Preservative treatments for fresh-cut fruits and vegetables', in O. Lamikanra (Ed.), *Fresh-Cut Fruits and Vegetables: Science, Technology and Market*. CRC Press, pp. 267–304.

Garcia, L. C., Pereira, L. M., de Luca Sarantopoulos, C. L. G. and Hubinger, M. D. (2010), 'Selection of an edible starch coating for minimally processed strawberry', *Food Bioprocess Technol.*, 3, 834–42.

Ghidelli, C. and Perez-Gago, M. B. (2018), 'Recent advances in modified atmosphere packaging and edible coatings to maintain quality of fresh-cut fruits and vegetables', *Crit. Rev. Food Sci. Nutr.*, 58, 662–79.

Gustavsson, J., Cederberg, C., Sonesson, U., van Otterdijk, R. and Meybeck, A. (2011), *Global Food Losses and Food Waste*. Food and Agriculture Organization of the United Nations, Rome, 29 pp.

Hagenmaier, R. D. (2001), 'Ethanol content of 'Murcott' tangerines harvested at different times and treated with coatings with different O_2 permeability', *Proc. Fla. State Hort. Soc.*, 114, 170–3.

Hagenmaier, R. D. (2002), 'The flavor of mandarin hybrids with different coatings', *Postharv. Biol. Technol.*, 24, 79–87.

Hagenmaier, R. D. and Baker, R. A. (1994), 'Internal gases, ethanol content and gloss of citrus fruit coated with polyethylene wax, carnauba wax, shellac or resin at different application levels', *Proc. Fla. State Hort. Soc.*, 107, 261–5.

Hagenmaier, R. D. and Baker, R. A. (1996), 'Edible coatings from candelilla wax microemulsions', *J. Food Sci.*, 61, 562–5.

Hagenmaier, R. D. and Shaw, P. E. (2002), 'Changes in volatile components of stored tangerines and other specialty citrus fruits with different coatings', *J. Food Sci.*, 67, 1742–5.

Hall, D. J. (2012), 'Edible coatings from lipids, waxes, and resins', in E. A. Baldwin, R. Hagenmaier and J. Bai (Eds), *Edible Coatings and Films to Improve Food Quality*, 2nd edn. CRC Press, pp. 79–101.

Hall, D. J. (2013), 'The color-add process as applied in Florida', *Proc. Fla. State Hort. Soc.*, 126, 220–4.

Hardenburg, R. E. (1967), 'Wax and related coatings for horticultural products; A bibliography', U.S. Dept. of Agriculture, Agricultural Research Service, Washington DC, 15 pp.

Jiang, Y. (2000), 'Role of anthocyanins, polyphenol oxidase and phenols in lychee pericarp browning', *J. Sci. Food Agric.*, 80, 305–10.

Jiang, Y., Li, J. and Jiang, W. (2005), 'Effects of chitosan coating on shelf life of cold-stored litchi fruit at ambient temperature', *LWT Food Sci. Technol.*, 38, 757–61.

Kader, A. A. and Barrett, D. M. (2004), 'Classification, composition of fruits and postharvest maintenance of quality', in D. M. Barrett, L. Somogyi, H. Ramaswamy (Eds), *Processing Fruits: Science and Technology*, 2nd edn. Technomic Publishing Company, pp. 3-22.

Kader, A. A. and Saltveit, M. E. (2003), 'Respiration and gas exchange', in J. A. Bartz and J. K. Brecht (Eds), *Postharvest Physiology and Pathology of Vegetables*, 2nd edn. Marcel Dekker Inc., pp. 7–29.

Kim, I. H., Oh, Y. A., Lee, H., Song, K. B. and Min, S. C. (2014), 'Grape berry coatings of lemongrass oil-incorporating nanoemulsion', *LWT-Food Sci Technol.*, 58, 1–10.

Krochta, J. M. (2002), 'Proteins as raw materials for films and coatings: Definitions, current status and opportunities', in A. Gennadios (Ed), *Protein-based Films and Coatings*. CRC Press, pp. 1–41.

Lima, A. L., Cerqueira, M. A., Souza, B. W. S., Santos, E. C. M., Teixeira, J. A., Moreira, R. A. and Vicente, A. A. (2010), 'New edible coatings composed of galactomannans and collagen blends to improve the postharvest quality of fruits – Influence on fruits gas transfer rate', *J. Food Eng.*, 97, 101–9.

Luna-Guzman, I. and Barrett, D. (2000), 'Comparison of calcium chloride and calcium lactate effectiveness in maintaining shelf stability and quality of fresh-cut cantaloupes', *Postharv. Biol. Technol.*, 19, 61–72.

Lurie, S. and Watkins, C. B. (2012), 'Superficial scald, its etiology and control', *Postharv. Biol. Technol.*, 65, 44–60.

Madani, B. and Forney, C. F. (2016), 'Recent research on calcium and postharvest behavior', in R. B. H. Wills and J. Golding (Eds), *Advances in Postharvest Fruit and Vegetable Technology*. CRC Press, pp. 19–33.

Martin-Belloso, O., Rojas-Grau, M. A. and Soliva-Fortuny, R. (2009), 'Delivery of flavor and active ingredients using edible films and coatings', in M. E. Embuscado and K. C. Huber (Eds), *Edible Films and Coatings for Food Applications*. Springer, pp. 295–313.

Martin-Diana, A. B., Ricoa, D., Frias, J. M., Barat, J. M., Henehan, G. T. M. and Barry-Ryan, C. (2007), 'Calcium for extending the shelf life of fresh whole and minimally processed fruits and vegetables: A review', *Trends Food Sci. Technol.*, 18, 210–18.

Mattila, M., Ahvenainen, R. and Hurme, E. (1993), 'Prevention of browning of pre-peeled potato', in J. De Baerdemaker, B. McKenna, M. Janssens, A. Thompson, F. Artes Calero, E. Hohn and Z. Somogyi (Eds), *Proceedings COST 94 Workshop: Systems and Operations for Post-Harvest Quality*. Commission of the European Communities, pp. 225-34.

McHugh, T. H. and Senesi, E. (2000), 'Apple wraps: A novel method to improve the quality and extend the shelf life of fresh-cut apples', *J. Food Sci.*, 65, 480-5.

Medeiros, B. G. S., Pinheiro, A. C., Teixeira, J. A., Vicente, A. A. and Carneiro-da-Cunha, M. G. (2012), 'Polysaccharide/protein nanomultilayer coatings: Construction, characterization and evaluation of their effect on 'Rocha' pear (*Pyrus communis* L.) shelf life', *Food Bioprocess Technol.*, 5, 2435–45.

Mishra, V. K. and Gamage, T. V. (2007), 'Postharvest physiology of fruit and vegetables', in M. S. Rahman (Ed.), *Handbook of Food Preservation*, 2nd edn. CRC Press, pp. 19–48.

Misir, J., Brishti, F. H. and Hoque, M. M. (2014), 'Aloe vera gel as a novel edible coating for fresh fruits: A review'. *Amer. J. Food Sci. Technol.*, 2(3), 93–7.

Muth, M. K., Karns, S. A., Nielsen, S. J., Buzby, J. C. and Wells, H. F. (2011), 'Consumer-level food loss estimates and their use in the ERS loss-adjusted food availability data', *USDA Econ. Res. Serv. Tech. Bull.* No. 1927, USDA, ERS, Washington, DC.
Nisperos-Carriedo, M. O. (2004), 'Edible coatings and films based on polysaccharides', in J. M. Krochta, E. A. Baldwin and M. O. Nisperos-Carriedo (Eds), *Edible Coatings and Films to Improve Food Quality*. Technomic Publishing Company, pp. 305–35.
Oms-Oliu, G., Soliva-Fortuny, R. and Martín-Belloso, O. (2008), 'Edible coatings with antibrowning agents to maintain sensory quality and antioxidant properties of fresh-cut pears', *Postharv. Biol. Technol.*, 50, 87–94.
Otoni, C. G., Avena-Bustillos, R. J., Azeredo, H. M. C., Lorevice, M. V., Moura, M. R., Mattoso, L. H. C. and McHugh, T. H. (2017), 'Recent advances on edible films based on fruits and vegetables – A review', *Comp. Rev. Food Sci. Food Saf.*, 16, 1151–69.
Palma-Guerrero J., Lopez-Jimenez J. A., Pérez-Berná, A. J., Huang, I. C., Jansson, H. B., Salinas, J., Villalaín, J., Read, N. D. and Lopez-Llorca L. V. (2010), 'Membrane fluidity determines sensitivity of filamentous fungi to chitosan', *Molec. Microbiol.*, 75, 1021–32.
Palou, L. (2018), 'Postharvest treatments with GRAS salts to control fresh fruit decay', *Horticulturae*, 4, 46–60.
Palou, L., Ali, A., Fallik, E. and Romanazzi, G. (2016), 'GRAS, plant- and animal-derived compounds as alternatives to conventional fungicides for the control of postharvest diseases of fresh horticultural produce', *Postharv. Biol. Technol.*, 122, 41–52.
Parfitt, J., Barthel, M. and Macnaughton, S. (2010), 'Food waste within food supply chains: Quantification and potential for change to 2050', *Philos. Trans. Royal Soc. B*, 365, 3065–81.
Parreidt, T. S., Muller, K. and Schmid, M. (2018), 'Alginate-based edible films and coatings for food packaging', *Foods*, 7(10), 170–207.
Plotto, A., Narciso, J., Baldwin, E. A. and Rattanapanone, N. (2006), 'Edible coatings and other surface treatments to maintain color of lychee fruit in storage', *Proc. Fla. State. Hort. Soc.*, 119, 323–331.
Ponting, J. D., Jackson, R. and Waters, G. (1972), Refrigerated apple slices: Preservative effects of ascorbic acid, calcium and sulfites. *J. Food Sci.*, 37, 434–6.
Portjewijd, R., Nisperos, M. O., Burns, J. K., Parish, M. and Baldwin, E. A. (1995), 'Cellulose-based coatings as carriers for *Candida guillermondii* and *Debaryomyces* sp. in reducing decay of oranges', *HortScience*, 30, 1417–21.
Quezado-Gallo, J.-A. (2009), 'Delivery of food additives and antimicrobials using edible films and coatings', in M. E. Embuscado and K. C. Huber (Eds), *Edible Films and Coatings for Food Applications*. Springer, pp. 315–33.
Ramos, B., Miller, F. A., Brandao, T. R. S., Teixeira, P. and Silva, C. L. M. (2013), 'Fresh fruits and vegetables – An overview on applied methodologies to improve its quality and safety', *Postharv. Biol. Technol.*, 20, 1–15.
Rojas-Graü, M. A., Tapia, M. S., Rodriguez, F. J., Carmoa, A. J. and Martin-Belloso, O. (2007), 'Alginate and gellan-based edible coatings as carriers of antibrowning agents applied on fresh-cut Fuji apples', *Food Hydrocolloids*, 21, 118–27.
Rojas-Graü, M. A., Soliva-Fortuny, R. and Martín-Belloso, O. (2009), 'Edible coatings to incorporate active ingredients to fresh-cut fruits: A review', *Trends Food Sci. & Technol.*, 20, 438–47.
Romanazzi, G., Feliziani, E., Baños, S. B. and Sivakumar, D. (2017), 'Shelf life extension of fresh fruit and vegetables by chitosan treatment'. *Crit. Rev. Food Sci. Nutr.*, 57(3), 579–601.

Salvia-Trujillo, L., Rojas-Graü, M. A., Soliva-Fortuny, R. and Martín-Belloso, O. (2015), 'Use of antimicrobial nanoemulsions as edible coatings: Impact on safety and quality attributes of fresh-cut Fuji apples', *Postharv. Biol. Technol.*, 105, 8–16.

Sanches-Silva, A., Costa, D., Albuquerque, T. G., Buonocore, G. G., Ramos, F., Castilho, M. C., Machado, A. V. and Costa, H. S. (2014), 'Trends in the use of natural antioxidants in active food packaging: A review', *Food Additives & Contaminants: Part A*, 31, 374–95.

Singh, S., Khemariya, P., Rai, A., Rai, A. C., Koley, T. K. and Singh, B. (2016), 'Carnauba wax-based edible coating enhances shelf-life and retain quality of eggplant (Solanum melongena) fruits', *LWT*, 74, 420–6.

Soliva-Fortuny, R., Rojas-Graü, M. A. and Martín-Belloso, O. (2012), 'Polysaccharide coatings', in E. A. Baldwin, R. Hagenmaier and J. Bai (Eds), *Edible Coatings and Films to Improve Food Quality*, 2nd edn. CRC Press, pp. 103–36.

Sun, X., Baldwin, E. A., Ritenour, M., Plotto, A. and Bai, J. (2015), 'Evaluation of natural colorants and their application on citrus fruit as alternatives to Citrus Red No. 2', *HortScience*, 50, 1353–7.

Sun, X., Baldwin, E. A., Ritenour, M. Hagenmaier, R. and Bai, J. (2017), 'Formulating a natural colorant containing wax for a one-step color-add application for fresh citrus', *HortScience*, 52, 408–12.

Thompson, A. K. (2015), *Fruit and Vegetables: Harvesting, Handling and Storage*, 3rd edn. Wiley, 981 pp.

Toivonen, P. M. A. and Brummell, D. A. (2008), 'Biochemical bases of appearance and texture changes in fresh-cut fruit and vegetables', *Postharv. Biol. Technol.*, 48, 1–14.

Toivonen, P. M. A. and DeEll, J. R. (2002), Physiology of fresh-cut fruits and vegetables', in O. Lamikanra (Ed.), *Fresh-Cut Fruits and Vegetables: Science, Technology and Market*. CRC Press, pp. 91–124.

Tokala, V. Y. and Mahajan, B. V. C. (2018), 'Calcium: An indispensable element affecting postharvest life of fruits and vegetables', in K. Barman, S. Sharma and M. W. Siddiqui (Eds), *Emerging Postharvest Treatment of Friuits and Vegetables*. CRC Press, pp. 59–83.

Tsouvaltzis, P. and Brecht, J. K. (2016), 'Inhibition of enzymatic browning of fresh-cut potato by immersion in citric acid is not solely due to pH reduction of the solution', *J. Food Proc. Pres.*, doi:10.1111/jfpp.12829.

Valencia-Chamorro, S. A., Paulo, L, Del Río, M. A. and Pérez-Gago, M. B. (2011), 'Antimicrobial edible films and coatings for fresh and minimally processed fruits and vegetables: A review', *Crit. Rev. Food Sci. & Nutr.*, 51, 872–900.

Valero, D. and Serrano, M. (2010), 'Calcium treatments', in *Postharvest Biology and Technology for Preserving Fruit Quality*. CRC Press, pp. 109–24.

Van-Loon, L. C. and Van-Strein, E. A. (1999), 'The families of pathogenesis-related proteins, their activities and comparative analysis of PR-1 type proteins', *Physiol. Molec. Plant Pathol.*, 55, 85–97.

Velickova, E., Winkelhausen, E., Kuzmonova, S., Alves, V. D. and Moldao-Marins, M. (2013), 'Impact of chitosan-beeswax edible coatings on the quality of fresh strawberries (*Fragaria ananassa* cv Camerosa) under commercial storage conditions', *LWT-Food Sci Technol.*, 52, 80–92.

Whitaker, J. R. (1994), *Principles of Enzymology for the Food Sciences*, 2nd edn. Marcel Dekker, 648 pp.

Wills, R. H. and Golding, J. (2016), *Postharvest: An Introduction to the Physiology and Handling of Fruit and Vegetables*, 6th edn. CABI.

Wills, R. H. H., Lee, T. H., Graham, D., McGlasson, W. B. and Hall, E. G. (1981), *Postharvest: An Introduction to the Physiology and Handling of Fruit and Vegetables*, 1st edn. CABI.

Yahia, E. M. (2010), 'Cold chain development and challenges in the developing countries', *Acta Horticulturae*, 877, 127–32.

Yahia, E. M., Ornelas-Paz, J. J. and Elansari, A. (2011), 'Postharvest technologies to maintain the quality of tropical and subtropical fruits', in *Postharvest Biology and Technology of Tropical and Subtropical Fruits*. Woodhead Publishing Series in Food Science, Technology and Nutrition, pp. 142–93.

Yahia, E. M., Fadanelli, L. Matte, P. and Brecht, J. K. (2019a), 'Controlled atmosphere storage', in E. M. Yahia (Ed.), *Postharvest Technology of Perishable Horticultural Commodities*. Elsevier, pp. 439–79.

Yahia, E. M., Neven, L. G. and Jones, R. W. (2019b), 'Postharvest insects and their control', in E. M. Yahia (Ed.), *Postharvest Technology of Perishable Horticultural Commodities*. Elsevier, pp. 529–62.

Youssef, K., Ligorio, A., Nigro, F. and Ippolito, A. (2012), 'Activity of salts incorporated in wax in controlling postharvest diseases of citrus fruit', *Postharv. Biol. Technol.*, 65, 39–43.

Yu, T., Yu, C., Chen, F., Sheng, K., Zhou, T., Zunun, M., Abudu, O., Yang, S. and Zheng, X. (2012), 'Integrated control of blue mold in pear fruit by combined application of chitosan, a biocontrol yeast and calcium chloride', *Postharv. Biol. Technol.*, 69, 49–53.

Part 2

Safety management and disinfection techniques

Chapter 5

Post-harvest risk management of biological hazards encountered in horticultural produce

Keith Warriner and Mahdiyeh Hasani, University of Guelph, Canada

1 Introduction

The consumption of fresh fruits and vegetables in the daily diet has been promoted by health agencies around the globe. Such initiatives have been successful with a rapid expansion of the fresh produce sector that is only second to the bottled water market. To meet consumer demand for fresh produce, the industry has become more globalized and centralized with complex distribution networks. With the changes in fresh produce production has come an increase in foodborne illness outbreaks caused by a diverse range of enteric pathogens that include viral, bacterial and protozoal (Table 1). The rise in foodborne illness outbreaks can be attributed to increased volumes of product consumed and better surveillance tools such as next generation sequencing that facilitates source attribution (Warriner et al., 2009). In addition, the rise in pre-packed or processed fruit and vegetables has also contributed to the rise of foodborne illness cases (Warriner et al., 2009). Certainly, the increase in foodborne illness cases mirrors that of the introduction of processed fresh produce most notably with bagged salad (Warriner et al., 2009). Indeed, a tour

http://dx.doi.org/10.19103/AS.2019.0055.06

Table 1 Examples of foodborne illness outbreaks linked to fresh produce in North America, Europe and New Zealand from 2010–2018

Year	Product	Pathogen	# Cases
Fruit			
2018	Pre-cut melon	*Salmonella adelaide*	77
2018	Frozen raspberries	*Norovirus*	724
2017	Papayas	*Salmonella kiambu, Thompson, Agona, Gaminara*	173
2016	Frozen strawberries	Hepatitis A	143
2014	Frozen strawberries	Norovirus	11, 000
2014	Caramel apples	*Listeria monocytogenes*	35
2013	Frozen berries	Hepatitis A	1442
2012	Mangoes	*Salmonella* Braenderup	25
2012	Cantaloupe	*Salmonella enterica* - Typhimurium and Newport	261
2012	Mango	*Salmonella enterica* - Braenderup	127
2011	Cantaloupe	*Listeria monocytogenes*	147
2011	Cantaloupe	*Salmonella enterica* - Panama	20
2011	Papaya	*Salmonella agona*	106
Leafy greens			
2018	Romaine lettuce	*E. coli* O157:H7	72
2018	Romaine lettuce	*E. coli* O157:H7	210
2017	Romaine lettuce	*E. coli* O157:H7	45
2015	Packaged salads	*Listeria monocytogenes*	55
2015	Packaged salads	*E. coli* O157:H7	13
2013	Watercress	*E. coli* O157:H7	19
2013	Shredded lettuce	*E. coli* O157:H7	30
2013	Ready-to-eat salad	*E. coli* O157:H7	29
2013	Lettuce/cilantro	*Cyclospora*	631
2012	Organic spinach/ spring mix blend	*E. coli* O157:H7	33
2012	Romaine lettuce	*E. coli* O157:H7	24
2012	Lettuce	*Cryptosporidium*	300
2011	Romaine lettuce	*E. coli* O157:H7	58
2010	Romaine lettuce	*E. coli* O145	26
Herbs			
2018	Cilantro	*Cyclospora*	56
2015	Cilantro	*Cyclospora*	90
2014	Cilantro	*Cyclospora*	25
2013	Cilantro	*Cyclospora*	38

Year	Product	Pathogen	# Cases
Vine crops			
2018	Cucumbers	*Salmonella* Infantis	56
2015	Cucumbers	*Salmonella* Poona	907
2015	Tomato	*Salmonella* Newport	115
2014	Cucumbers	*Salmonella Agona*	147
2014	Cucumbers	*Salmonella enterica* - Newport	275
2013	Cucumbers	*Salmonella* Saintpaul	73
2012	Cucumbers	*Salmonella enterica* - Saintpaul	84
Prepared vegetables			
2018	Frozen sweetcorn, peas, carrots, broccoli	*Listeria monocytogenes*	45
2018	Vegetable trays	*Cyclospora*	250
2016	Frozen vegetables	*Listeria monocytogenes*	9

Source: adapted from CDC (2017).

around any supermarket will illustrate the rows of processed fresh produce available compared to intact ones (Anon., 2019). Moreover, the current trend of in-store prepared fruit and salads is becoming more common with retail outlets being considered mini-processing plants.

The specific source of foodborne pathogens implicated in outbreaks linked to fresh produce remains unclear. Yet, it can be hypothesized that contamination acquired in the field can be disseminated during post-harvest processing with cutting/shredding operations along with post-harvest washing (Murray et al., 2017). The field environment represents an open system with contamination being derived from organic nutrient amendments, irrigation water, animals and via transfer from workers (Jongman and Korsten, 2018). Although the greenhouse environment outwardly appears enclosed, human pathogens can enter via irrigation water, foot-traffic, workers, containers and tools. Reducing the risk of crop contamination in the production phase is attempted by Good Agricultural Practice (GAP) and initiatives such as the Leafy Greens Agreement, along with provisions in the Final Produce Rule of Food Safety Modernization Act (USFDA, 2019). The following chapter will focus on the potential sources and dissemination of human pathogens during post-harvest along with risk management options.

2 Foodborne illness outbreaks linked to fresh produce

Fresh produce represents a significant food safety risk across the globe principally because of the open nature of production coupled with a lack of a kill step (Gutierrez-Rodriguez and Adhikari, 2018). The actual burden of foodborne illness is difficult to estimate given the short shelf life of the product, which has either been consumed or discarded by the time the outbreak is detected (Arendt et al., 2013). Moreover, the production, processing and distribution networks for fresh produce make traceability back to the origins challenging. Yet, despite the lack of accurate outbreak data, a common theme is that fresh produce is responsible for the majority of foodborne illness outbreaks within North America and Europe although meat remains the main cause of foodborne illness cases (Olaimat and Holley, 2012). Within the US, fresh produce was responsible for 80% of the outbreaks reported between 1996 and 2014 (Li et al., 2018b). Within Europe, over 150 outbreaks have been linked to fresh produce between 2004 and 2012 (Callejon et al., 2015). In Australia, the outbreaks linked to fresh produce account for 7% of the known outbreaks reported between 2001 and 2009 (Astridge et al., 2011). From a review of produce-related foodborne illness outbreaks, 998 with known aetiologies were identified that included 45 723 cases. With regards to different countries, New Zealand had the highest number of outbreaks per million person years and it was 0.76. This compared to 0.25, 0.26, 0.13, 0.12 and 0.05 for Australia, US, EU, Canada and Japan, respectively (Li et al., 2018b).

Leafy greens, tomatoes, melons, soft fruit and herbs account for 88% of fresh-produce-related foodborne illness outbreaks (Li et al., 2018b). The reasons for these products being implicated in foodborne illness outbreaks can be partly explained by volumes consumed and, to a degree, consumer handling. However, there exists pathogen–produce pairing that would suggest additional factors are involved (Anderson et al., 2011). For example, from reviewing outbreaks linked to leafy greens, *Escherichia coli* O157:H7 and other Shiga toxin-producing *E. coli* are frequently implicated (Table 1). In contrast, outbreaks linked to melons, papayas, mangos and tomatoes typically implicate *Salmonella*. With other pathogens, there are produce associations that have a tendency to have a broader range of vehicles (Table 1). For example, *Listeria monocytogenes* has a tendency to be associated with a broader range of produce types ranging from leafy greens to fruit. Enteric viruses such as Norovirus and Hepatitis A are commonly linked to soft fruit with both being implicated in large-scale outbreaks involving contaminated frozen berries. However, a broader range of produce types are implicated in foodborne illness cases if contamination is introduced at the time of food service given that both pathogens are shed and then transferred via handling (Perez-Rodriguez et al., 2019).

Enteric protozoans, such as enteric viruses, have the gastrointestinal tract as a primary habitat and consequently are associated with manure along with sewage (Koutsoumanis et al., 2018; Murray et al., 2017). As with enteric viruses, protozoans have been recovered on a broad range of produce types but have a specific association with leafy greens (Dixon, 2016). For example, there is an interesting association of *Cyclospora* with cilantro outbreaks (Table 1).

The underlying basis for produce-pathogen pairing remains to be identified, but it is thought to include geographical regions and adaption of pathogens to persist on plants (Warriner and Namvar, 2010). With regards to the former, the centralization of production would make certain crops more prone to contamination depending on the pathogens present in the environment. For example, the majority of leafy greens in the US are produced in Arizona and California where it is common to encounter cattle feedlots that are considered a source of *E. coli* O157:H7. In a similar manner, regions that produce tomatoes have been shown to have an environmental burden of *Salmonella* (Pagadala et al., 2015). In the case of cilantro, which is primarily grown in Mexico, there is an endemic population of *Cyclospora* in the region thereby increasing the risk of contamination by the pathogen (Bonilla and Vielma, 2018; Marques et al., 2017). Therefore, in risk assessment terms, the underlying reasons for pathogen-produce pairing could be simply the probability of produce being contacted by the pathogen. However, there is also evidence that pathogens have adapted to persist on produce that would suggest evolutionary adaption (Schleker et al., 2015; Crozier et al., 2016). A description of pathogens associated with produce along with potential sources is provided in Table 2.

3 Sources of human pathogens in post-harvest operations

A large knowledge gap exists with regards to the original source of pathogens implicated in produce-related foodborne illness outbreaks. It is likely that the contamination is introduced at the pre-harvest stage via irrigation water, flooding events, organic amendments and ingress of wild animals in most cases (Olaimat and Holley, 2012). Therefore, post-harvest contamination events are largely more due to dissemination of pathogens between batches as opposed to introduction at the facility level. However, pathogens, such as *L. monocytogenes*, can become endemic within processing plants, thereby acting as a continuous source of contamination. In a similar manner, pathogens can be passed on by food handlers shedding pathogens such as norovirus at food service. Therefore, in some aspects, the potential for pathogens to become associated with produce is related to the point of entry within the chain.

3.1 Harvest

Contamination can be spread during harvesting via contact surfaces that includes the harvesters in addition to tools such as knives (McEvoy et al., 2009). It is also possible for produce to be contaminated by handling if the farm worker is shedding enteric pathogens and not following good hygienic

Table 2 Characteristics and sources of foodborne pathogens linked to fresh produce

Pathogen	Symptoms	Infectious dose and sources
Bacterial		
Bacillus cereus	Emetic syndrome: 0.5-6 h incubation period. Nausea and vomiting eventually diarrhoea Diarrhoegenic syndrome: 12-24 h incubation period. Abdominal pain, watery diarrhoea	10^6–10^9 cells Grains, nuts, soil and manure
Campylobacter jejuni	2-5 day incubation period. Explosive diarrhoea that may be bloody, weakness, dizziness and fever. Secondary condition: Guillian Barre syndrome (paralysis).	500 cells. Commonly encountered in poultry, cattle, pigs and asymptomatic carriers. Manure and contaminated water.
Clostridium botulinum	18-36 h incubation period. Progressive paralysis initially in the face, then progresses down the body. Can be fatal or long-term paralysis.	Ingestion of 20 ng toxin/kg body weight. Soil, river sediments, gastrointestinal tract of fish and root crops.
Listeria monocytogenes	48 h–4 week incubation period. Flu-like symptoms progressing to nausea, vomiting, diarrhoea and fever. Can lead to paralysis, meningitis, septicaemia and in pregnant women - stillbirth.	$10^3 10^8$ cells. Decaying plant material, water, soil and manure. Commonly endemic within produce-processing facilities.
Shiga toxin-producing *Escherichia coli*	2-8 days incubation period. Nausea, vomiting, diarrhoea, abdominal pain. Bloody urine that can lead to haemolytic urinary syndrome (HUS), paralysis and septicaemia.	10-100 cells. Manure of ruminants. Raw sewage, contaminated soil and water.
Salmonella	24-72 h incubation period. Nausea, vomiting, diarrhoea, abdominal pain and dehydration. Secondary symptoms includes arthritis (Reiter's syndrome)	1000–10^6 cells. Manure derived from poultry, pigs and cattle. Insects, rodents and sewage. Contaminated soil and water.
Shigella	Same as for Shiga toxin-producing *E. coli*. Dysentery	10–100 cells. Person-to-person. Raw sewage.
Enteric protozoan		
Cyclospora	1-6 weeks incubation period with relapse. Bloating, stomach cramps, fatigue and loss of appetite.	Unknown but low (20 oocysts). Raw sewage, contaminated soil and water. Not person-to-person transmission.

Pathogen	Symptoms	Infectious dose and sources
Cryptosporidium	1-2 week incubation period. Fever, stomach cramps, watery diarrhoea, vomiting and weight loss.	Infectious dose unknown, but low. Person-to-person transmission, manure, sewage, contaminated soil and water.
Giardia	9-15 day incubation period. Abdominal pain, flatulence, explosive diarrhoea, vomiting and weight loss.	Infectious dose unknown but low. Person-to-person transmission, manure, sewage, contaminated soil and water.
Viruses		
Adenovirus	2-4 day incubation period. Sore throat, bronchitis, fever, diarrhoea possible leading to meningitis.	Unknown but considered low (20 virons). Person-to-person transmission. Sewage and contaminated water.
Hepatitis A	10-50 days incubation period. Fatigue, fever, dark urine, nausea, vomiting and jaundice.	Unknown. Person-to-person transmission. Sewage and contaminated water.
Norovirus	12-48 h incubation period. Dry mouth, nausea, vomiting, diarrhoea and abdominal pain.	Unknown but considered low. Person-to-person transmission. Sewage and contaminated water.

practice (Monaghan et al., 2017). When bins containing produce are transferred to cans also result in spread of pathogens, especially, if not sanitized between uses. A high risk part of the process, in terms of spreading pathogens, occurs during the vacuum cooling step used to reduce respiration rates and microbial growth (Li et al., 2008). Removing field heat from crops can also be undertaken using ice or via hydrocooling. If the ice or water is contaminated, then this is a way of directly introducing contamination or spread between batches. Although vacuum cooling negates the need for water or ice, it has been shown to draw microbes from the sub-surface into the cut edges and stomata. Once internalized, pathogens would be protected from washing and be transferred through the chain (Li et al., 2008).

3.2 Transportation and storage

The transporter trucks used to carry fresh produce from the field to the packing facility or processing plant have also been considered as a point of introducing contamination. This was illustrated in a *Listeria monocytogenes* outbreak implicating contaminated cantaloupes (Buchanan et al., 2017). Among other infractions, it was found that the trailer used to transport the fruit had previously been used to carry animals. Although the trailer was not directly linked to the source of *L. monocytogenes*, it can be envisaged that the risk existed.

Fresh produce is commonly held within storage rooms before processing or re-packing. The storage room temperature and humidity are controlled to reduce respiration and dehydration. The psychrotrophic nature of *L. monocytogenes* enables the pathogen to become established on condenser coils and units thereby representing a continuous source of contamination with distribution throughout the storage room being facilitated by air flow (Ndraha et al., 2018). The air flow also results in organic debris to become trapped on meshes and fans that can sustain the formation of biofilms that can be long lived and difficult to remove.

3.3 Processing

Fresh produce contacts several surfaces such as tables, conveyors and bins during processing. Produce such as peaches and cantaloupes are scrubbed by brushes thereby disseminating contamination between batches (Chardon et al., 2016). The spread of contamination is not limited to food contact surfaces, but also the vicinity around the processing line such as floors, drains and frames (Buchholz et al., 2012, 2014; Perez Rodriguez et al., 2011). Processing facilities typically have standing water with plant materials that are ideal conditions for supporting biofilms that can harbour pathogens. Biofilms were recovered within a cantaloupe-processing facility that processed fruit contaminated with *L. monocytogenes*. *Listeria* biofilms were also recovered from the facility that processed pre-prepared salads that resulted in 55 cases of foodborne illness (CDC, 2016). Cold conditions within facilities do not support the growth of toxigenic *E. coli* or *Salmonella* but enable *L. monocytogenes* to become established and it can persist over extended periods (Smith et al., 2018).

The most significant cross-contamination point in produce processing is the wash step (Barrera et al., 2012). There are different formats of washing produce from batch tanks through to sequential multiple wash tanks with air agitation (Murray et al., 2017). Previously, washing was considered to be a decontamination step whereby field-acquired contamination could be removed. However, the post-harvest wash process has limited efficiency to remove pathogens on produce and therefore represents a potential cross-contamination step rather than help in decontamination (Fu et al., 2018). Deficiencies in the wash process have been highlighted in several high-profile foodborne illness outbreaks (Murray et al., 2017). The outbreak of *L. monocytogenes* linked to contamination cantaloupes previously mentioned is a prime example where the sanitizer was removed from the water to obtain organic certification. The net result was that *Listeria* on the cantaloupe was deposited in the wash water which then spread to other fruit during processing. A further example was the 2006 landmark outbreak of *E. coli* O157:H7 linked to spinach (Sharapov et al., 2016). The original source of *E. coli* O157:H7 was considered to be wild pigs and irrigation water, but was thought to be disseminated to further batches via the wash process due to the lack of attention given to maintaining the free-chlorine concentrations within the tank. Post-harvest washing cannot be ensured to remove contamination, but rather there is a risk of disseminating pathogens if the process is not controlled.

3.4 Food service and domestic kitchens

The final step in the fresh produce chain is serving preparation prior to the serving. In many instances, the produce is pre-prepared (for example, bagged leafy greens) and hence has been through post-harvest washing. However, some do select to wash bagged salads that can lead to contamination being introduced from the kitchen environment with dish sponges and cloths being of prime concerns (Ovca et al., 2016). Consequently, lettuce should not be re-washed unless carried out using running water.

As with processing, many of the outbreaks linked to fresh produce at the food service level can be traced to downstream sources (Choi et al., 2016). This was thought to be the case in an outbreak of *Salmonella* typhimurium linked to tomatoes served in restaurants that caused over 180 confirmed cases (Behravesh et al., 2012). An outbreak centred around a Sudbury (Canada) restaurant that received a batch of onions contaminated with *E. coli* O157:H7. The actual outbreak lasted 10 days with the majority of the 280 cases being linked to onions cross-contaminated via a dicer that was inadequately sanitized between uses. However, enteric viruses such as norovirus and hepatitis A are more associated with food service centre outbreaks compared to bacterial pathogens. The challenge with norovirus is its high infectivity as an enteric viruses, whereby only 20 virons would be sufficient to cause illness and consequently hard to control (Chatziprodromidou et al., 2018). Therefore, foods consumed raw, such as fresh produce, are especially at risk from the pathogen.

4 Risk assessment

Quantitative microbial risk assessment (QMRA) is aimed at collecting available data to generate the degree of risk to public health related to exposure to pathogens. The general approach for undertaking risk assessments is to identify the hazard and then calculate the probability of consumers being exposed to the selected hazard at levels that could cause illness. The same approach can be taken to identify the points in the production, processing and distribution chain where pathogens could be introduced, transferred, allowed to grow or be reduced. Yet, as with all risk assessments, the probabilities generated are only as good as the data used to construct models. In this respect there remains many knowledge gaps in how pathogens behave within the fresh produce chain and the methods used to study pathogen:plant interactions make comparing studies challenging. There is also the fact that contamination of fresh produce by pathogens is sporadic and typically involves unpredictable breakdown of control points within the chain (Allende et al., 2018). However, QMRA still provides a valuable tool in identifying critical points in the chain where pathogens can be introduced and then assessing their fate.

QMRA has been performed in assessing the pre-harvest risks of introducing contamination in leafy greens via irrigation water (Allende et al., 2018). The risk assessment identified that the source of water (canals vs. ground) in combination with level of pathogens carried an increased probability of transferring contamination via processing (Allende et al., 2018). In a risk assessment model of post-harvest leafy greens processing, the probability of fresh produce carrying pathogens was related to the free-chlorine concentration in wash tanks and the frequency by which the water is changed or recharged (Mokhtari et al., 2018). The conclusion was further supported by the importance of sanitizers in minimizing the risk derived from carriage of *Listeria* on lettuce (Guzel et al., 2017). In general, the risk assessments performed identify the need to minimize contamination in pre-harvest operations and implement post-harvest operations that help in reducing carriage of pathogens along with minimizing microbial growth (Mogren et al., 2018).

Another model predicted the critical points that contributed to the microbial loading and growth within the tomato supply chain (Zoellner et al., 2018). The model was based on using field data along with published data to identify which steps added contamination to tomatoes and which removed microbes, permitted growth or facilitated transfer. From the model it was suggested that holding of tomatoes at retail was the key stage that defined the total aerobic count and coliforms on tomato fruit. Extended holding time coupled with temperature was responsible for the increased risk factors (Zoellner et al., 2018; Roy and Melotto, 2019). The authors acknowledged that the model would need to be verified if it were to represent the risks derived from human pathogens.

Collectively, the risk assessments to date have highlighted that control of contamination at the pre-harvest stage is the most effective strategy along with ensuring that the facility is sanitary and that there is implementation of an effective decontamination step.

5 Risk management of post-harvest fresh produce

Risk management is part of the risk analysis that proposes interventions, procedures and practices that can be applied to reduce hazards. In pre-harvest operations these are described in guidelines issued by organizations such as the Leafy Greens Marketing Agreement (LGMA) among others

(Hoagland et al., 2018; Gutierrez-Rodriguez and Adhikari, 2018; Sangshin et al., 2012). In post-harvest operations there remains a strong reliance on the Good Manufacturing Practice (GMP) and washes although research on decontamination technologies and risk prevention remains an active area (Mogren et al., 2018).

5.1 Harvest

There is evidence that the presence of standing water, after a flood or precipitation event, increases the levels of pathogens such as *L. monocytogenes* (Weller et al., 2015). Therefore, by avoiding parts of the field with standing water and leaving a gap of a 7-day period from a watering (natural or via irrigation) should decrease the likelihood of transferring contamination through to processing/market. Although it is possible that pathogen movement can occur through soil via watering events, it is also possible that the increased prevalence could be attributed to revival of dormant bacteria.

At harvest there has been a move to make workers wear protective clothing (gloves, hairnets) to reduce the risk of transferring pathogens from infected field workers. This coupled with food safety training represents a further initiative (Monaghan et al., 2017). However, it has also been demonstrated under laboratory conditions that there is an exchange of *Salmonella* between tomatoes and gloves, but this is independent of the degree of soils on the gloves (Brar and Danyluk, 2013). The cloths used to wipe produce off dirt (for example, on tomatoes) were also found to transfer *Salmonella,* especially, if the cloths are damp (Sreedharan et al., 2014).

Decontamination of cutting surfaces such as mechanical harvesting blades and knives has been evaluated using a range of sanitizers (Tapp et al., 2013). The sanitizers included sodium metasilicate, quaternary ammonium salts and hypochlorite with log reductions of *E. coli* O157:H7 ranging from 1.16 to 3.82 log cfu being reported (Tapp et al., 2013). Bacteriophages against *E. coli* O157:H7 supported a 4.45 log cfu reduction of the pathogen inoculated on the blades of harvesting equipment (Patel et al., 2011). The design of knives by reducing welding surfaces has been demonstrated to decrease the attachment of *E. coli* O157:H7 and aid in sanitation while using chlorine in combination with ultrasound (Zhou et al., 2012).

The bins used to transport produce to market, packing stations or processing facilities have been identified as a further point of cross-contamination (Rapusas and Rolle, 2009). The issue is of lesser importance for single-use packaging based on paperboard, but more significant for reusable plastic or wooden crates and bins. It has been reported that *Salmonella* and *Listeria* can survive on both plastic and wooden surfaces over extended time periods, thereby, necessitating effective sanitation between uses (Li et al., 2018a).

The decontamination of produce during vacuum cooling has been assessed as a point for decreasing the carriage of pathogens before entering distribution, packing or processing. For example, ozone introduced into vacuum cooling units at a rate of 1.5 g/kg produce reduced *E. coli* O157:H7 on spinach leaves by 3 log cfu (Yesil et al., 2017). The same approach was undertaken, but an initial spray of a quaternary ammonia-based sanitizer following the vacuum cooling and ozone gas treatment reduced *E. coli* O157:H7 by 3.9 log cfu (Pyatkovskyy et al., 2017). Although effective, applying ozone prior to the vacuum cooling step is limited to short contact times and can also result in significant corrosion of stainless steel.

5.2 Storage

The pathogen loading on harvested fruit and vegetables can be reduced through introduction of anti microbial gases into the storage room headspace. For example, ozone applied at 0.5–0.8 mg/L into a storage room of apples decreased *Listeria* counts by 1 log cfu over 30 days but had a detrimental effect on fruit quality (Sheng et al., 2018). A more effective ozone treatment for apples is that based on a forced-air reactor whereby the ozone is introduced into a stream of air through an apple column (Murray et al., 2018b). Here, ozone introduced into the apple column at a concentration of 23 mg/L supported a 2.12–3.07 log cfu reduction of *L. monocytogenes* with a 20-min treatment with no detrimental effect on fruit quality (Murray et al., 2018b).

Chlorine dioxide gas is highly oxidative and has been evaluated for decontamination of a diverse range of produce in storage (Sun et al., 2019). Although effective at pathogen reduction, the negative effects on product quality and the potential for chlorite left behind as residue have restricted commercial application (Sun et al., 2019).

5.3 Processing

Processing represents a key point where contamination can become distributed across batches and potential endemic strains of pathogens such as *L. monocytogenes* become established. Common contact surfaces such as conveyors, belts, brushes and slicing/shredding areas represent high-risk points for cross-contamination from one batch to the next. Sanitary design of equipment, whereby minimizing accumulation of organic matter on lines along with ease of access during sanitation, are key requirements. Indeed, the benefits of sanitation can be lost early into processing due to the deposition of contamination on the line and formation of biofilms due inability to adequately sanitize equipment. In this regard, interest in antimicrobial coatings on surfaces that inactivate microbes and/or prevent attachment has increased (Bastarrachea et al., 2015; Huang et al., 2017; Kugel et al., 2011). Equipment is available whereby the surface topography has been modelled similar to surfaces encountered in nature that prevent microbial adherence (Bastarrachea et al., 2015). For example, surfaces shaped like shark skin or plant leaves have been shown to reduce microbial adherence and, thereby, initiation of biofilm formation (Bastarrachea et al., 2015). Other approaches have been to impregnate antimicrobials into the polymer base of conveyors and belts. Tested antimicrobial agents have included quaternary ammonium salts, Triclosan, and copper and silver ions (Bastarrachea et al., 2015). The main limitation of the slow-release antimicrobials is that once depleted, the agents need to be replaced, which is often economically unfeasible. This limitation is overcome to some extent by the deposition of shelf-healing films that are composed of *N*-halamines that can be regenerated by reacting with hypochlorite in the course of sanitation (Bastarrachea and Goddard, 2016). In laboratory-based studies, *N*-halamine films have been demonstrated to inactivate *L. monocytogenes* and prevent biofilm formation (Bastarrachea and Goddard, 2013). The main practical challenge in its implementation is the negative impact on antimicrobial activity by an organic material coupled with instability following repeated microbial generations (Bastarrachea and Goddard, 2016; Qiao et al., 2018). An alternative, or addition to, is in-process sanitation, whereby the sanitizer flows constantly across the contact surface (for example, belt, slicer) and, thereby, prevents accumulation. A limitation of in-process sanitation is that the sanitizer residues can be transferred to the product being processed. Consequently, sanitizers such as ozone that do not leave disinfection by-products are preferred (Smolinski et al., 2018). A further approach is the positioning of UV-C lamps close to the surface, thereby, enabling continuous disinfection although safety hazards such as shattered glass and exposure to workers exist.

One of the most critical control points in fresh produce processing is the wash step. As previously described, the wash process was considered a decontamination step but in commercial operations has represented a key cross-contamination point. Washing cannot effectively remove contamination due to the internalization of pathogens or association with biofilms and, therefore, the wash processes are validated in terms of preventing cross-contamination rather than decontaminating the produce (Gombas et al., 2017). Although there have been many studies performed in relation to evaluating different sanitizers for produce washing, these have been primarily devoted to the decontamination efficacy of the produce rather than the prevention of cross-contamination (Beuchat et al., 2004). In contrast, relatively few reports on the efficacy of different sanitizers to prevent contamination have been published (Banach et al., 2015). To date the most widely applied sanitizer in the fresh produce sector is hypochlorite due to low cost and historical use. However, a disadvantage in using hypochlorite is the reactivity/sequestering of organic and inorganic constituents, thereby, reducing its antimicrobial activity (Murray et al., 2018a). It follows that depletion of free chlorine correlates with cross-contamination events, whereby, pathogens are transferred during the wash process (Fu et al., 2018). Therefore, attempts to control the cross-contamination of produce by ensuring a minimum level of free-chlorine has been considered as a monitoring step and constituting a part of an HACCP-based system. A further point of consideration is the level of free-chlorine required to be maintained to inactivate pathogens, which lies between 2 and 20 mg/L (Gombas et al., 2017). However, even measuring free-chlorine within a tank is challenging given that hypochlorite is constantly changing and hence the value measured is dependent on the location of the sensor placed in the tank. As a supplement or alternative, attempts have been made to look at different water parameters that correlate with antimicrobial (free-chlorine) capacity (Shen et al., 2016). The parameters are, for example, turbidity, total organic content (TOC), chemical oxygen demand (COD), biological oxygen demand (BOD) and oxidation reduction potential (ORP) (Shen et al., 2013; Zhou et al., 2014, 2015; Barrera et al., 2012; Chen and Hung, 2016). Of these, only COD has shown close correlation with free-chlorine concentrations although the most rapid test to determine the oxygen demand takes more than two hours and cannot be performed in-line (Murray et al., 2017, 2018a). It has been proposed to measure the COD indirectly by taking turbidity and conductivity readings and then using an algorithm to calculate the chlorine demand of wash tanks in order to determine the amount of chlorine to be added (Zhou et al., 2014). Such feed-forward control would be more reliable compared to the current feed-back based on ORP measurements; however, this has yet to be applied commercially.

A further approach to reduce cross-contamination within wash tanks is to apply more stable sanitizers relative to hypochlorite. For example, chloroamines (termed stabilized chlorine) are more stable than hypochlorite and also stable in the presence of organics, but only have a faction of the antimicrobial activity (Teng et al., 2018; Fu et al., 2018). However, other sanitizers with greater stability in the presence of organics, such as peroxyacetic acid and chlorine dioxide, are seen reducing the transfer of pathogens during post-harvest washing (Murray et al., 2018a).

Post-harvest contact with water is inevitable with most produce requiring field-acquired debris, such as soil and insects, to be removed, in addition to be provided with a suitable means of transportation without causing damage. Consequently, moving towards a water-free processing system to reduce the potential of cross-contamination is challenging. Yet, using mechanical brushes to remove soil from produce can be employed to decrease carriage of debris into wash tanks as well as reducing the water usage as part of a water-management system. In terms of decontamination, there are post-wash technologies that can be applied to inactivate pathogens present on produce (Table 3).

More attention on water-free, post-harvest decontamination methods includes the application of irradiation such as x-rays and e-beams (Pinela and Ferreira, 2017). Irradiation can be considered to be the closest to a Golden Bullet in that pathogens can be inactivated without changing the quality of the produce. Currently low-dose irradiation can be applied for insect disinfection as a means of pest control, ripening prevention and spoilage microflora reduction (Jeong and Jeong, 2018). The inactivation of pathogens on fresh produce using irradiation has been reported although it has not been widely applied despite being approved for use within the US for treatments employing doses up to 1 kGy (Gomes et al., 2009).

Electron beams (e-beams) is a type of irradiation that uses lower doses, but they are not dependent on the decay of radioisotopes (Pillai and Shayanfar, 2018). In this regard, it has been demonstrated that a >3 log cfu reduction of *E. coli* O157:H7 on spinach can be achieved, provided the depth of produce is kept low due to the limited penetration of the ions (Moosekian et al., 2014) (Table 3). The underlying limitation of radiation is the softening and discolouration of the produce using irradiation doses >1 kGy. A further limitation is that enteric viruses and protozoans are relatively resistant to irradiation compared to bacteria and are thereby likely to survive the irradiation process (Olaimat and Holley, 2012). However, the antimicrobial efficacy of irradiation can be enhanced by combining it with modified atmosphere packaging (MAP). For example, by using 100% oxygen MAP, it is possible to achieve a >5 log cfu reduction of *E. coli* O157:H7 on spinach using 0.7 kGy irradiation treatment (Fan et al., 2008). A combination of irradiation and ozone enhanced the inactivation of *E. coli* and *Listeria* on cauliflower florets compared with separate treatments (Boumail et al., 2016).

A more acceptable form of irradiation is UV-C that inactivates microbes by disrupting their DNA, thereby, resulting in lethal mutations (Fan et al., 2017). The main challenge with UV-C-based decontamination methods is the direct contact of the UV photons with the target. Even smooth surfaces of fresh produce have areas of shadowing caused by microscopic ridges and other surface structures (Fan et al., 2017). Shadowing is more pronounced with rough surfaces such as the cantaloupes or stomata on leafy greens. To address the issues of shadowing, one strategy is to apply high-intensity light to increase the probability of lethal strikes against the target. However, this approach can lead to overdosing samples with UV, resulting in activation of polyphenol oxidase that negatively affects the colour and flavour along with the weakening of plant cells and causing surface dehydration.

Table 3 Non-aqueous interventions for decontaminating post-harvest fresh produce

Intervention	Produce type	Microbes/log reduction	Source
Advanced oxidative process	Apples	H_2O_2 6% v/v, 2 g/h ozone, 54 mJ/cm^2. 3 log cfu reduction of *L. monocytogenes.*	Murray et al. (2018b)
Advanced oxidative process	Iceberg lettuce	H_2O_2 1.5%, 37.8 mJ/cm^2. 4.12 log reduction *Salmonella*.	Hadjok et al. (2008a)
Chlorine dioxide gas	Cantaloupe	5 ppm for 5.5 min. 3-5 log reduction of *E. coli*, *Salmonella* and *L. monocytogenes.*	Mahmoud et al. (2008)
Chlorine dioxide gas	Cantaloupes, strawberries and tomatoes	10 ppm for 3 min. 5 log reduction of *Salmonella*. 3 log reduction *E. coli* O157:H7 and *Listeria.*	Trinetta et al. (2013)
Chlorine dioxide gas	Apples	12 mg/L for 10 min. 8 log reduction *E. coli* O157:H7	Du et al. (2003)
Chlorine dioxide gas	Spinach	0.13 mg/L for 20 min. 2.45 log reduction E. coli O157:H7 and *Salmonella*	Park and Kang (2015)
Chlorine dioxide gas	Tomato	0.85 mg/L for 20 min. 3.95 log reduction.	Netramai et al. (2016)
Gas plasma	Tomato	300 s air plasma. 3.14 log reduction *E. coli*	Ziuzina et al. (2014)
	Lettuce	300 s air plasma. 3.3 log reduction *E. coli*	Ziuzina et al. (2015)
Ozone gas	Apples	23 mg/L ozone for 20 min. 3.07 log reduction of *L. monocytogenes*	Murray et al. (2018b)
	Spinach	935 mg/L 30 min. 5 log reduction *E. coli* O157:H7	Vurma et al. (2009)
	Cantaloupe	3 mg/L 5 min. 1 log reduction *E. coli* O157:H7	Rodgers et al. (2004)
	Tomato	10 mg/L 20 min. 7 log reduction *Salmonella*	Das et al. (2006)
Irradiation	Spinach	0.7 kGy. 5 log reduction *Salmonella* and *Listeria*	Gomes et al. (2008)
	Strawberries	28.7 kGy. 6 log reduction of norovirus and Tulane virus	DiCaprio et al. (2016)
	Cantaloupe	1.5 kGy. 3.6 log reduction *Salmonella*.	Palekar et al. (2015)
	Lettuce	1 kGy. Log reduction of 5 log of *E. coli* O157:H7	Jeong et al. (2010)

Intervention	Produce type	Microbes/log reduction	Source
	Cucumber	0.38 kGy. Log reduction of 1 log *Salmonella*.	Joshi et al. (2018)
Pulsed light	Spinach	40 mJ/cm^2. 2.2 log reduction *E. coli* O157:H7	Aguero et al. (2016)
	Watermelon	12 J/cm^2. 5 log reduction *E. coli* O157:H7	Ramos-Villarroel et al. (2015)
UV-C	Strawberries, blueberries and raspberries	A dose of 212 mJ/cm^2 supported 2-3 log reduction of hepatitis A, 1 log reduction of *E. coli, Salmonella* and *Listeria*	Butot et al. (2018)
	Lettuce	81.6 mJ/cm^2. 0.31 log reduction of *E. coli* O157:H7.	Kim et al. (2013)
	Cantaloupe	11.9 mJ/cm^2. 1 log reduction *E. coli* O157:H7	Adhikari et al. (2015)
	Tomato	6.0 mJ/cm^2. 2.8 log reduction of *Salmonella*	Mukhopadhyay et al. (2014)
UV-C and chlorine dioxide	Tomatoes	10 mg/L chlorine dioxide in presence of UV-C for 15 min. >5 log reduction *E. coli* O157:H7, *Salmonella* and *Listeria*.	Park et al. (2018)
UV-A: phenolic acids	Spinach	*E. coli* O157 and *Listeria*. 2 log cfu reduction	Cossu et al. (2018)

Source: adapted from Murray et al. (2017).

A further approach is to construct reactors such as drums that continuously turn, thereby, presenting a different surface towards the UV-C lamps. However, more common is to have multiple arrays of lamps combined with reflective surfaces that result in photon scattering. Regardless of the method applied to expose the shaded areas to UV photons, the treatments are ultimately limited by pathogens being located in sub-surfaces such as cut edges and stomata (Fan et al., 2017).

LED-UV lights are an alternative to UV-lamps with the advantages of being more compact units with less heat generation and making available a more diverse range of wavelengths that have a higher penetration than 254 nm (Green et al., 2018). Studies performed with LED-UV have shown that wavelengths of 259 and 268 nm have higher antimicrobial activity compared with that of 254 nm. Moreover, a combination of wavelengths of 259 and 289 nm has synergistic activity compared to monochromatic light (Green et al., 2018). Yet, inactivation of *Listeria* on apples still leads to tailing within inactivation curves due to shadowing effects.

The availability of an extended range of wavelengths enables the application of photosensitizing agents to inactivate pathogens on the surface of the produce. By using wavelengths longer than 254 nm, it is possible to use higher powered light sources and activate photoactive antimicrobials on the surface of the produce that overcomes the shadowing effect. In one example, chlorophyllin was used as the photosensitizing agent and deposited on the cantaloupe that was subsequently illuminated at 405 nm (Josewin et al., 2018). A treatment of 1200 J/cm^2 at 405 nm reduced *L. monocytogenes* by 3 log cfu although there was no significant difference between LED alone being used or when combined with photosensitizer suggesting that the photons from the light source were the main mode of inactivation. Also, the combination of UV-A and phenolic acids exhibit a synergistic antimicrobial effect against *E. coli* and *Listeria* with a 2 log cfu reduction being reported without changes to the product quality (Cossu et al., 2018)

A produce-decontamination method based on AOP has been adapted into a continuous system (Murray et al., 2018b; Xie et al., 2008; Hadjok et al., 2008). AOP is based on the UV-C-mediated degradation of hydrogen peroxide generating antimicrobial free radicals that inactivate relevant pathogens on the surface and sub-surface. The process decontaminated a diverse range of vegetable and fruit types, in addition to mushrooms, with a 3-5 log cfu reduction of *Salmonella*, *E. coli* O157:H7 and *L. monocytogenes,* with no negative effects on quality (Murray et al., 2015, 2018b; Hadjok et al., 2008). The process has been scaled up and units are commercially available for fresh produce decontamination.

A further potential post-harvest decontamination technology that has yet to be commercialized for produce decontamination is gas plasma (Pignata et al., 2017; Ekezie et al., 2017). Gas plasma is referred to as the fourth state of matter that essentially consists of ionized gas produced by placing gas (air, oxygen, helium or a combination) within a high voltage field. The generated plasma contains a mixture of radicals, free electrons and UV photons, similar in some respects to AOP. Previously, gas plasma had to be performed under high pressure with high voltages and the ionized gas transferred to a treatment chamber. With advances in the area it is possible to produce plasmas under atmospheric conditions, although the process is restricted to treating small batch samples. One promising treatment is the atmospheric plasma jet that generates a continuous plasma stream by exposing an oxygen-rich air flow to microwave or radiation sources (Chen et al., 2019). The generated ionized gas stream can then be directed towards the surface to be treated. In one example, *E. coli* O157:H7 on lettuce was reduced by 1.5 log cfu after a 10-min treatment that compares to 2.7 log cfu reduction of *Salmonella* (reviewed in Pignata et al., 2017). Although marginally effective, the challenge is in scale-up along with extended treatment times and these are obvious limitations of gas plasma treatment (Chen et al., 2019).

From a comparison of different non-wash-based interventions the most effective are those based on irradiation, gas phase treatments and AOP (Table 3). Although log reductions are an important metric, the negative impacts on produce quality are typically not described in detail. This is especially relevant with gas phase treatment and irradiation whereby tissue softening along with discolouration can occur.

6 Case study: validation of post-harvest interventions using Response Surface Methodology

In the course of developing post-harvest decontamination methods there is a need to optimize operating parameters to ensure pathogen reduction without adversely affecting the sensory quality of the produce. Therefore, a balance needs to be made between the process variables to identify the optimal settings. Previously, this would involve laborious testing of a matrix of different treatment combinations to identify optimal conditions for supporting inactivation of target microbes without negatively affecting quality. To aid validation trials there is an increased use of multivariate statistic methods such as Response Surface Methodology (RSM). RSM is a statistical technique for designing, modelling and analysis of problems where a response, or responses, of interest are affected by several different parameters. The main advantage of this method is the small number of experiments compared with the conventional approach of applying all the possible combinations (Bezerra et al., 2008; Takashina et al., 2018). RSM has been successfully used for optimizing the different food processes such as extraction, drying, blanching, enzymatic hydrolysis and clarification, production of microbial metabolites and formulation (Bezerra et al., 2008).

An example of the application of RSM in post-harvest interventions is in the validation of an AOP-based process for decontaminating lemons (Hasani et al., 2019). A central composite rotational design (CCRD) with three independent variables of the AOP process and RSM is employed to obtain a target of a 2 log cfu reduction in *Aspergillus* spore levels without causing unacceptable colour changes of the lemons. *A. niger* spores were selected as the target microbe due to their inherent stress tolerance that generated quantitative log reduction data and spoilage potential.

The independent variables were intensity of UV-C (0.6–3.6 mW/cm^2), dose (36–80 mJ/cm^2) and concentration of H_2O_2 (1–6% v/v). Each of the process parameters was evaluated at five different levels of combining factorial points (-1, +1), axial points (-α, +α) and central point (0). A total of 20 runs, with six replications in the centre points, were performed in duplicate. To find the optimum parameters of the AOP system, a second-order polynomial equation was used:

$$Y = a_0 + \sum_{i=1}^{3} a_i x_i + \sum_{i=1}^{3} a_{ii} x^2 + \sum_{i=1}^{2} \sum_{j=i+1}^{3} a_{ij} x_i x_j + \varepsilon$$

where *Y* is the predicted response; x_i and x_j are the factors that influence the predicted response *Y*; a_i, a_{ii} and a_{ij} are the coefficient of linear, quadratic and interaction terms, respectively, and ε is a random error.

The fit and significance of the developed regression models were appraised through the analysis of variance, which consists mainly of the correlation coefficients, the F-value and P-value. A predictive model was developed for each response and the optimum levels of the selected variables were obtained by solving the regression equation by analysing the response surface contours and surface plots. Finally, in order to validate the accuracy of the predictive models, additional experiments were tested in the range of the optimized conditions.

By using the RSM approach, an interaction between hydrogen peroxide concentration and UV-C intensity was found to have the most significant contribution towards the inactivation of *A. niger* spores. Also, the quadratic regression models identified hydrogen peroxide concentration as the key parameter for both antimicrobial efficacy and discolouration of lemons (Figures 1 and 2). A synergy was observed between UV-C intensity and hydrogen peroxide concentration for the inactivation of *A. niger* spores (Figure 1). The results confirm that the intensity of UV-C directly affects the extent of free radical generation from the degradation of hydrogen peroxide rather than direct UV-mediated inactivation of *A. niger*.

To arrive at the optimum conditions, the process parameters were kept within the studied range and the log count reduction value of *Aspergillus* spores

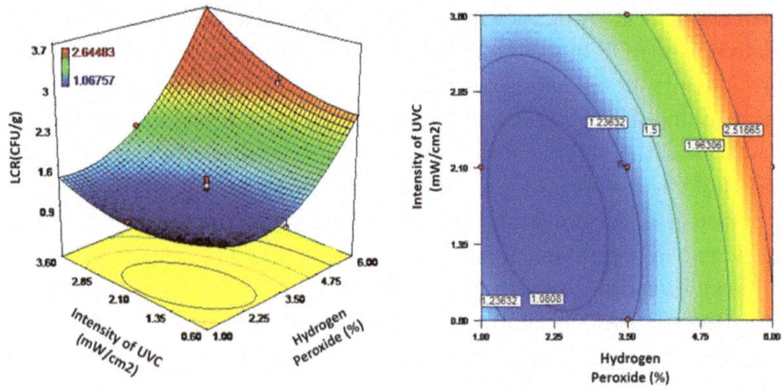

Figure 1 3-D response surfaces and counter plots to illustrate the effect of hydrogen peroxide concentration and intensity of UV-C light on the log count reduction of *Aspergillus niger* inoculated onto lemons and then treated with an advanced oxidative process. Source: taken from Hasani et al. (2019).

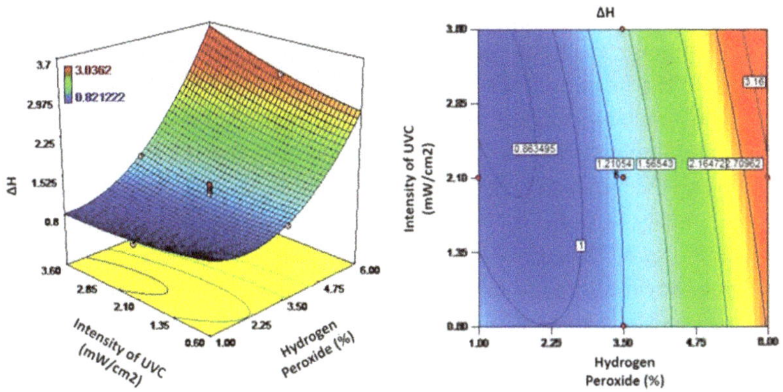

Figure 2 3-D response surfaces and counter plots to illustrate the effect of hydrogen peroxide concentration and intensity of UV-C light on the change in delta hue (Δ*H*; colour) of lemons treated with an advanced oxidative process. Source: taken from Hasani et al. (2019).

as a response was selected to be in the range of 1.5–2.0 log cfu and minimal Δ*H* (Hue) and Δ*E* (colour perception) values for the colour of lemons. The software predicted that the optimal treatment (desirability value (*d*) approaching 1.0) for decontaminating lemons while retaining colour was to apply a hydrogen peroxide concentration of 1.88% v/v with a UV-C intensity of 3.26 mW/cm^2 and 20 s treatment time. It was predicted that the treatment would result in a log cfu reduction of *Aspergillus* spores (Y1) of 1.56 with a Δ*H* of 0.74 (Y2) and Δ*E* of 1.26 (Y3). In verification trials the treatment supported a 1.54 ± 0.04 log cfu decrease in *Aspergillus* spores with Δ*H* and Δ*E* changes of 0.73 ± 0.07 and 1.20 ± 0.29, respectively. These values confirmed that the model was suitable and could be applied to ensure consistent and predictable AOP treatment. Also, the results showed that the optimized AOP treatment did not result in changes in the shelf-life of lemons compared to non-treated controls.

The case study illustrates how CCRD and RSM can be applied in optimizing treatment processes. Although the current modelling was for AOP, the same approach can be applied to a diverse range of intervention methods.

7 Future trends and conclusion

Fresh produce remains a significant food safety risk despite the introduction of Good Agricultural Practices and stringent regulations such as those within the Final Fresh Produce Rule under FSMA. Pre-harvest research is primarily focussed on assessing pathogen dissemination and survival in the environment along with validating risk assessment models and diagnostics for screening inputs into crop production such as irrigation water. Advances in post-harvest operations have focussed on developing intervention methods and improved traceability of produce back to the field. There have been attempts to introduce traceability systems into the fresh produce sector although all have failed due to the need for growers, packers, processors and retailers to adopt the system. The more recent technological advances and the increasingly prominent role of retailers in controlling the food safety narrative have made a universal traceability system more of a reality. Specifically, Blockchain coupled with continuous monitoring via The Internet of Things (IoT) could aid trace-back in addition to identifying when breakdown within the chain (for example, low chlorine concentration in tanks) occurs (Mahajan et al., 2017). An attempt to introduce Blockchain in the fresh produce sector is being undertaken in North America, but ultimately this maybe limited due to the complexity of the distribution network in the industry, which is composed of layers of contractors and sub-contractors.

An active area of research deals with post-harvest decontamination methods than can support a defined log cfu reduction of relevant pathogens and can be monitored as part of a risk prevention control strategy. This has involved attempts to improve post-harvest wash processes with new sanitizers or at least metrics (for example, turbidity of water) to define the efficacy of the process. Alternatives to post-harvest washing (irradiation, AOP) are typically more defined and have predictable performance. A common challenge with both water and aqueous-free interventions is, however, a standardized method for validating and selection of appropriate surrogates that can be applied within processing facilities. The selection of surrogates is a subject of on-going research and it is likely that there will be no single indicator identified given the different modes of inactivation by interventions. This along with other aspects of post-harvest risk management is much needed by the industry, given the on-going food safety issues faced by the fresh produce sector.

8 Where to look for further information

Leafy Greens Marketing Agreement Alliance. Formed in 2007, the LGMA is an industry-government association to develop food safety standards and provide guidance to farmers on Good Agricultural Practice, available at: https://lgma.ca.gov.

UC Davis Center for Fresh Produce Safety. The Center of Fresh Produce Safety (CPS) was established in the wake of the *E. coli* O157:H7 outbreak linked to spinach. The CPS in an industry-government-academic network that funds food safety research linked to fresh produce, available at: https://www.centerforproducesafety.org/.

Fresh Produce Safety Center Australia and New Zealand. The centre was established in 2014 by a collaboration between Produce Marketing Association and the University of Sydney. The network provides food safety outreach, auditing and supports fresh-produce-related research, available at: https://fpsc-anz.com.

9 References

Adhikari, A., Syamaladevi, R. M., Killinger, K. and Sablani, S. S. 2015. Ultraviolet-C light inactivation of *Escherichia coli* O157:H7 and *Listeria monocytogenes* on organic fruit surfaces. *International Journal of Food Microbiology* 210, 136–42. doi:10.1016/j.ijfoodmicro.2015.06.018.

Aguero, M. V., Jagus, R. J., Martin-Belloso, O. and Soliva-Fortuny, R. 2016. Surface decontamination of spinach by intense pulsed light treatments: impact on quality attributes. *Postharvest Biology and Technology* 121, 118–25. doi:10.1016/j.postharvbio.2016.07.018.

Allende, A., Truchado, P., Lindqvist, R. and Jacxsens, L. 2018. Quantitative microbial exposure modelling as a tool to evaluate the impact of contamination level of surface irrigation water and seasonality on fecal hygiene indicator *E. coli* in leafy green production. *Food Microbiology* 75, 82–9. doi:10.1016/j.fm.2018.01.016.

Anderson, M., Jaykus, L. A., Beaulieu, S. and Dennis, S. 2011. Pathogen-produce pair attribution risk ranking tool to prioritize fresh produce commodity and pathogen combinations for further evaluation (P(3)ARRT). *Food Control* 22(12), 1865–72. doi:10.1016/j.foodcont.2011.04.028.

Anon. 2019. Fresh trends: tracking the four trends driving growth across the fresh section. Available at: https://www.nielsen.com/us/en/insights/news/2019/fresh-trends-tracking-the-four-trends-driving-growth-across-fresh.html (accessed July 2019).

Arendt, S., Rajagopal, L., Strohbehn, C., Stokes, N., Meyer, J. and Mandernach, S. 2013. Reporting of foodborne Illness by US consumers and healthcare professionals. *International Journal of Environmental Research and Public Health* 10(8), 3684–714. doi:10.3390/ijerph10083684.

Astridge, K., McPherson, M., Kirk, M., Knope, K., Gregory, J., Lardamanidis, K. and Bell, K. 2011. Foodborne disease outbreaks in Australia 2001–2009. *Food Australia* 63, 44–50.

Banach, J. L., Sampers, I., Van Haute, S. and Van Der Fels-Klerx, H. J. 2015. Effect of disinfectants on preventing the cross-contamination of pathogens in fresh produce washing water. *International Journal of Environmental Research and Public Health* 12(8), 8658–77. doi:10.3390/ijerph120808658.

Barrera, M. J., Blenkinsop, R. and Warriner, K. 2012. The effect of different processing parameters on the efficacy of commercial post-harvest washing of minimally processed spinach and shredded lettuce. *Food Control* 25(2), 745–51. doi:10.1016/j.foodcont.2011.12.013.

Bastarrachea, L. J. and Goddard, J. M. 2013. Development of antimicrobial stainless steel via surface modification with N-halamines: characterization of surface chemistry and N-halamine chlorination. *Journal of Applied Polymer Science* 127(1), 821–31. doi:10.1002/app.37806.

Bastarrachea, L. J. and Goddard, J. M. 2016. Self-healing antimicrobial polymer coating with efficacy in the presence of organic matter. *Applied Surface Science* 378, 479–88. doi:10.1016/j.apsusc.2016.03.198.

Bastarrachea, L. J., Denis-Rohr, A. and Goddard, J. M. 2015. Antimicrobial food equipment coatings: applications and challenges. *Annual Review of Food Science and Technology* 6(6), 97–118. doi:10.1146/annurev-food-022814-015453.

Behravesh, C. B., Blaney, D., Medus, C., Bidol, S. A., Phan, Q., Soliva, S., Daly, E. R., Smith, K., Miller, B., Taylor, T., Nguyen, T., Perry, C., Hill, T. A., Fogg, N., Kleiza, A., Moorhead, D., Al-Khaldi, S., Braden, C. and Lynch, M. F. 2012. Multistate outbreak of *Salmonella* serotype Typhimurium infections associated with consumption of restaurant tomatoes, USA, 2006: hypothesis generation through case exposures in multiple restaurant clusters. *Epidemiology and Infection* 140(11), 2053–61. doi:10.1017/S0950268811002895.

Beuchat, L. R., Adler, B. B. and Lang, M. M. 2004. Efficacy of chlorine and a peroxyacetic acid sanitizer in killing *Listeria monocytogenes* on iceberg and romaine lettuce using simulated commercial processing conditions. *Journal of Food Protection* 67(6), 1238–42. doi:10.4315/0362-028X-67.6.1238.

Bezerra, M. A., Santelli, R. E., Oliveira, E. P., Villar, L. S. and Escaleira, L. A. 2008. Response surface methodology (RSM) as a tool for optimization in analytical chemistry. *Talanta* 76(5), 965–77. doi:10.1016/j.talanta.2008.05.019.

Bonilla, L. C. and Vielma, J. R. 2018. *Cyclosporiasis*: distribution, prevalence and control. *Investigacion Clinica* 59, 67–93.

Boumail, A., Salmieri, S. and Lacroix, M. 2016. Combined effect of antimicrobial coatings, gamma radiation and negative air ionization with ozone on *Listeria innocua*, *Escherichia coli* and mesophilic bacteria on ready-to-eat cauliflower florets. *Postharvest Biology and Technology* 118, 134–40. doi:10.1016/j.postharvbio.2016.04.005.

Brar, P. K. and Danyluk, M. D. 2013. *Salmonella* Transfer potential during hand harvesting of tomatoes under laboratory conditions. *Journal of Food Protection* 76(8), 1342–9. doi:10.4315/0362-028X.JFP-13-048.

Buchanan, R. L., Gorris, L. G. M., Hayman, M. M., Jackson, T. C. and Whiting, R. C. 2017. A review of *Listeria monocytogenes*: an update on outbreaks, virulence, dose-response, ecology, and risk assessments. *Food Control* 75, 1–13. doi:10.1016/j.foodcont.2016.12.016.

Buchholz, A. L., Davidson, G. R., Marks, B. P., Todd, E. C. D. and Ryser, E. T. 2012. Quantitative Transfer of *Escherichia coli* O157:H7 to equipment during small-scale production of fresh-cut leafy greens. *Journal of Food Protection* 75(7), 1184–97. doi:10.4315/0362-028X.JFP-11-489.

Buchholz, A. L., Davidson, G. R., Marks, B. P., Todd, E. C. D. and Ryser, E. T. 2014. Tracking an *Escherichia coli* O157:H7-contaminated batch of leafy greens through a pilot-scale fresh-cut processing line. *Journal of Food Protection* 77(9), 1487–94. doi:10.4315/0362-028X.JFP-14-058.

Butot, S., Cantergiani, F., Moser, M., Jean, J., Lima, A., Michot, L., Putallaz, T., Stroheker, T. and Zuber, S. 2018. UV-C inactivation of foodborne bacterial and viral pathogens and surrogates on fresh and frozen berries. *International Journal of Food Microbiology* 275, 8–16. doi:10.1016/j.ijfoodmicro.2018.03.016.

Callejon, R. M., Rodriguez-Naranjo, M. I., Ubeda, C., Hornedo-Ortega, R., Garcia-Parrilla, M. C. and Troncoso, A. M. 2015. Reported foodborne outbreaks due to fresh produce in the United States and European Union: trends and causes. *Foodborne Pathogens and Disease* 12(1), 32–8. doi:10.1089/fpd.2014.1821.

Chardon, J., Swart, A., Evers, E. and Franz, E. 2016. Public Health relevance of cross-contamination in the fresh-cut vegetablei ndustry. *Journal of Food Protection* 79(1), 30-6. doi:10.4315/0362-028X.JFP-15-272.

Chatziprodromidou, I. P., Bellou, M., Vantarakis, G. and Vantarakis, A. 2018. Viral outbreaks linked to fresh produce consumption: a systematic review. *Journal of Applied Microbiology* 124(4), 932–42. doi:10.1111/jam.13747.

Chen, X. and Hung, Y. C. 2016. Predicting chlorine demand of fresh and fresh-cut produce based on produce wash water properties. *Postharvest Biology and Technology* 120, 10–5. doi:10.1016/j.postharvbio.2016.05.007.

Chen, B. Y., Zhu, C. P., Fei, J. T., Jiang, Y. F., Yin, C., Su, W., He, X., Li, Y., Chen, Q., Ren, Q. G. and Chen, Y. W. 2019. Reaction kinetics of phenols and p-nitrophenols in flowing aerated aqueous solutions generated by a discharge plasma jet. *Journal of Hazardous Materials* 363, 55–63. doi:10.1016/j.jhazmat.2018.09.051.

Choi, J., Norwood, H., Seo, S., Sirsat, S. A. and Neal, J. 2016. Evaluation of food safety related behaviors of retail and food service employees while handling fresh and fresh-cut leafy greens. *Food Control* 67, 199–208. doi:10.1016/j.foodcont.2016.02.044.

Cossu, A., Huang, K., Cossu, M., Tikekar, R. V. and Nitin, N. 2018. Fog, phenolic acids and UV-A light irradiation: a new antimicrobial treatment for decontamination of fresh produce. *Food Microbiology* 76, 204–8. doi:10.1016/j.fm.2018.05.013.

Crozier, L., Hedley, P. E., Morris, J., Wagstaff, C., Andrews, S. C., Toth, I., Jackson, R. W. and Holden, N. J. 2016. Whole-transcriptome analysis of verocytotoxigenic *Escherichia coli* O157:H7 (Sakai) suggests plant-species-specific metabolic responses on exposure to spinach and lettuce extracts. *Frontiers in Microbiology* 7, 1088. doi:10.3389/fmicb.2016.01088.

Das, E., Gurakan, G. C. and Bayindirli, A. 2006. Effect of controlled atmosphere storage, modified atmosphere packaging and gaseous ozone treatment on the survival of *Salmonella* Enteritidis on cherry tomatoes. *Food Microbiology* 23(5), 430–8. doi:10.1016/j.fm.2005.08.002.

Dicaprio, E., Phantkankum, N., Culbertson, D., Ma, Y. M., Hughes, J. H., Kingsley, D., Uribe, R. M. and Li, J. R. 2016. Inactivation of human Norovirus and Tulane virus in simple media and fresh whole strawberries by ionizing radiation. *International Journal of Food Microbiology* 232, 43–51. doi:10.1016/j.ijfoodmicro.2016.05.013.

Dixon, B. R. 2016. Parasitic illnesses associated with the consumption of fresh produce - an emerging issue in developed countries. *Current Opinion in Food Science* 8, 104–9. doi:10.1016/j.cofs.2016.04.009.

Du, J., Han, Y. and Linton, R. H. 2003. Efficacy of chlorine dioxide gas in reducing *Escherichia coli* O157 : H7 on apple surfaces. *Food Microbiology* 20(5), 583-91. doi:10.1016/S0740-0020(02)00129-6.

Ekezie, F. G. C., Sun, D. W. and Cheng, J. H. 2017. A review on recent advances in cold plasma technology for the food industry: current applications and future trends. *Trends in Food Science and Technology* 69, 46–58. doi:10.1016/j.tifs.2017.08.007.

Fan, X., Niemira, B. A. and Prakash, A. 2008. Irradiation of fresh fruits and vegetables. *Food Technology* 62, 36–41.

Fan, X. T., Huang, R. Z. and Chen, H. Q. 2017. Application of ultraviolet C technology for surface decontamination of fresh produce. *Trends in Food Science and Technology* 70, 9–19. doi:10.1016/j.tifs.2017.10.004.

Fu, T. J., Li, Y. C., Awad, D., Zhou, T. Y. and Liu, L. 2018. Factors affecting the performance and monitoring of a chlorine wash in preventing *Escherichia coli* O157:H7 cross-contamination during postharvest washing of cut lettuce. *Food Control* 94, 212–21. doi:10.1016/j.foodcont.2018.06.035.

Gombas, D., Luo, Y., Brennan, J., Shergill, G., Petran, R., Walsh, R., Hau, H., Khurana, K., Zomorodi, B., Rosen, J., Varley, R. and Deng, K. 2017. Guidelines to validate control of cross-contamination during washing of fresh-cut leafy vegetables. *Journal of Food Protection* 80(2), 312–30. doi:10.4315/0362-028X.JFP-16-258.

Gomes, C., Moreira, R. G., Castell-Perez, M. E., Kim, J., Da Silva, P. and Castillo, A. 2008. E-beam irradiation of bagged, ready-to-eat spinach leaves (*Spinacea oleracea*): an engineering approach. *Journal of Food Science* 73(2), E95–E102. doi:10.1111/j.1750-3841.2007.00629.x.

Gomes, C., Da Silva, P., Moreira, R. G., Castell-Perez, E., Ellis, E. A. and Pendleton, M. 2009. Understanding *E. coli* internalization in lettuce leaves for optimization of irradiation treatment. *International Journal of Food Microbiology* 135(3), 238–47. doi:10.1016/j.ijfoodmicro.2009.08.026.

Green, A., Popovic, V., Pierscianowski, J., Biancaniello, M., Warriner, K. and Koutchma, T. 2018. Inactivation of *Escherichia coli*, *Listeria* and *Salmonella* by single and multiple wavelength ultraviolet-light emitting diodes. *Innovative Food Science and Emerging Technologies* 47, 353–61. doi:10.1016/j.ifset.2018.03.019.

Gutierrez-Rodriguez, E. and Adhikari, A. 2018. Preharvest farming practices impacting fresh produce safety. *Microbiology Spectrum* 6(2), 20. doi:10.1128/microbiolspec.PFS-0022-2018.

Guzel, M., Moreira, R. G., Omac, B. and Castell-Perez, M. E. 2017. Quantifying the effectiveness of washing treatments on the microbial quality of fresh-cut romaine lettuce and cantaloupe. *LWT-Food Science and Technology* 86, 270–6. doi:10.1016/j.lwt.2017.08.008.

Hadjok, C., Mittal, G. S. and Warriner, K. 2008. Inactivation of human pathogens and spoilage bacteria on the surface and internalized within fresh produce by using a combination of ultraviolet light and hydrogen peroxide. *Journal of Applied Microbiology* 104(4), 1014–24. doi:10.1111/j.1365-2672.2007.03624.x.

Hasani, M., Chudyk, J., Murray, K., Limm, L.-T., Lubitz, D.. and Warriner, K. 2019. Inactivation of *Salmonella*, *Listeria monocytogenes*, *Aspergillus* and *Penicillium* on lemons using advanced oxidation process optimized through response surface methodology. *Innovative Food Science and Emerging Technologies* 54, 182–91. doi:10.1016/j.ifset.2019.04.010.

Hoagland, L., Ximenes, E., Ku, S. and Ladisch, M. 2018. Foodborne pathogens in horticultural production systems: ecology and mitigation. *Scientia Horticulturae* 236, 192–206. doi:10.1016/j.scienta.2018.03.040.

Huang, S. Q., Guild, B., Neethirajan, S., Therrien, P., Lim, L. T. and Warriner, K. 2017. Antimicrobial coatings for controlling *Listeria monocytogenes* based on polylactide modified with titanium dioxide and illuminated with UV-A. *Food Control* 73, 421–5. doi:10.1016/j.foodcont.2016.08.030.

Jeong, M. A. and Jeong, R. D. 2018. Applications of ionizing radiation for the control of postharvest diseases in fresh produce: recent advances. *Plant Pathology* 67(1), 18–29. doi:10.1111/ppa.12739.

Jeong, S., Marks, B. P., Ryser, E. T. and Moosekian, S. R. 2010. Inactivation of *Escherichia coli* O157:H7 on lettuce, using low-energy X-ray irradiation. *Journal of Food Protection* 73(3), 547–51. doi:10.4315/0362-028X-73.3.547.

Jongman, M. and Korsten, L. 2018. Irrigation water quality and microbial safety of leafy greens in different vegetable production systems: a review. *Food Reviews International* 34(4), 308–28. doi:10.1080/87559129.2017.1289385.

Josewin, S. W., Kim, M. J. and Yuk, H. G. 2018. Inactivation of *Listeria monocytogenes* and *Salmonella* spp. on cantaloupe rinds by blue light emitting diodes (LEDs). *Food Microbiology* 76, 219–25. doi:10.1016/j.fm.2018.05.012.

Joshi, B., Moreira, R. G., Omac, B. and Castell-Perez, M. E. 2018. A process to decontaminate sliced fresh cucumber (*Cucumis sativus*) using electron beam irradiation. *LWT-Food Science and Technology* 91, 95–101. doi:10.1016/j.lwt.2018.01.034.

Kim, Y. H., Jeong, S. G., Back, K. H., Park, K. H., Chung, M. S. and Kang, D. H. 2013. Effect of various conditions on inactivation of *Escherichia coli* O157:H7, *Salmonella* Typhimurium, and Listeria monocytogenes in fresh-cut lettuce using ultraviolet radiation. *International Journal of Food Microbiology* 166(3), 349-55. doi:10.1016/j.ijfoodmicro.2013.08.010.

Koutsoumanis, K., Allende, A., Alvarez-Ordonez, A., Bolton, D., Bover-Cid, S., Chemaly, M., Davies, R., De Cesare, A., Herman, L., Hilbert, F., Lindqvist, R., Nauta, M., Peixe, L., Ru, G., Simmons, M., Skandamis, P., Suffredini, E., Caccio, S., Chalmers, R., Deplazes, P., Devleesschauwer, B., Innes, E., Romig, T., Van Der Giessen, J., Hempen, M., Van Der Stede, Y., Robertson, L. and Biohaz, E. P. B. H. 2018. Public health risks associated with food-borne parasites. *EFSA Journal* 16, 113-9.

Kugel, A., Stafslien, S. and Chisholm, B. J. 2011. Antimicrobial coatings produced by 'tethering' biocides to the coating matrix: a comprehensive review. *Progress in Organic Coatings* 72(3), 222–52. doi:10.1016/j.porgcoat.2011.07.004.

Li, H. P., Tajkarimi, M. and Osburn, B. I. 2008. Impact of vacuum cooling on *Escherichia coli* O157: H7 infiltration into lettuce tissue. *Applied and Environmental Microbiology* 74(10), 3138–42. doi:10.1128/AEM.02811-07.

Li, K. W., Khouryieh, H., Jones, L., Etienne, X. and Shen, C. L. 2018a. Assessing farmers market produce vendors' handling of containers and evaluation of the survival of *Salmonella* and *Listeria monocytogenes* on plastic, pressed-card, and wood container surfaces at refrigerated and room temperature. *Food Control* 94, 116–22. doi:10.1016/j.foodcont.2018.06.036.

Li, M., Baker, C. A., Danyluk, M. D., Belanger, P., Boelaert, F., Cressey, P., Gheorghe, M., Polkinghorne, B., Toyofuku, H. and Havelaar, A. H. 2018b. Identification of biological hazards in produce consumed in industrialized countries: a review. *Journal of Food Protection* 81(7), 1171–86. doi:10.4315/0362-028X.JFP-17-465.

Mahajan, P. V., Caleb, O. J., Gil, M. I., Izumi, H., Colelli, G., Watkins, C. B. and Zude, M. 2017. Quality and safety of fresh horticultural commodities: recent advances and future perspectives. *Food Packaging and Shelf Life* 14, 2–11. doi:10.1016/j.fpsl.2017.08.001.

Mahmoud, B. S. M., Vaidya, N. A., Corvalan, C. M. and Linton, R. H. 2008. Inactivation kinetics of inoculated *Escherichia coli* O157: H7, *Listeria monocytogenes* and *Salmonella* Poona on whole cantaloupe by chlorine dioxide gas. *Food Microbiology* 25(7), 857–65. doi:10.1016/j.fm.2008.05.009.

Marques, D. F. P., Alexander, C. L., Chalmers, R. M., Elson, R., Freedman, J., Hawkins, G., Lo, J., Robinson, G., Russell, K., Smith-Palmer, A., Kirkbride, H., Chiodini, P. and Godbole, G. 2017. Cyclosporiasis in travellers returning to the United Kingdom from Mexico in summer 2017: lessons from the recent past to inform the future. *Euro Surveillance:*

Bulletin Europeen sur les Maladies Transmissibles = European Communicable Disease Bulletin 22(32), 2-5. doi:10.2807/1560-7917.ES.2017.22.32.30592.

McEvoy, J. L., Luo, Y., Conway, W., Zhou, B. and Feng, H. 2009. Potential of *Escherichia coli* O157:H7 to grow on field-cored lettuce as impacted by postharvest storage time and temperature. *International Journal of Food Microbiology* 128(3), 506-9. doi:10.1016/j.ijfoodmicro.2008.08.008.

Mogren, L., Windstam, S., Boqvist, S., Vagsholm, I., Soderqvist, K., Rosberg, A. K., Linden, J., Mulaosmanovic, E., Karlsson, M., Uhlig, E., Hakansson, Å and Alsanius, B. 2018. The Hurdle Approach-A Holistic concept for controlling food safety risks associated with pathogenic bacterial contamination of leafy green vegetables. A review. *Frontiers in Microbiology* 9, 1965. doi:10.3389/fmicb.2018.01965.

Mokhtari, A., Oryang, D., Chen, Y. H., Pouillot, R. and Van Doren, J. 2018. A mathematical model for pathogen cross-contamination dynamics during the postharvest processing of leafy greens. *Risk Analysis: an Official Publication of the Society for Risk Analysis* 38(8), 1718-37. doi:10.1111/risa.12960.

Monaghan, J. M., Augustin, J. C., Bassett, J., Betts, R., Pourkomailian, B. and Zwietering, M. H. 2017. Risk assessment or assessment of risk? Developing an evidence-based approach for primary producers of leafy vegetables to assess and manage microbial risks. *Journal of Food Protection* 80, 725-33. doi:10.4315/0362-028X.JFP-16-237.

Moosekian, S. R., Jeong, S. and Ryser, E. T. 2014. Inactivation of sanitizer-injured *Escherichia coli* O157:H7 on baby spinach using X-ray irradiation. *Food Control* 36(1), 243-7. doi:10.1016/j.foodcont.2013.08.024.

Mukhopadhyay, S., Ukuku, D. O., Juneja, V. and Fan, X. 2014. Effects of UV-C treatment on inactivation of Salmonella enterica and *Escherichia coli* O157:H7 on grape tomato surface and stem scars, microbial loads, and quality. *Food Control* 44, 110-7. doi:10.1016/j.foodcont.2014.03.027.

Murray, K., Wu, F., Aktar, R., Namvar, A. and Warriner, K. 2015. Comparative study on the efficacy of bacteriophages, sanitizers, and UV light treatments to control *Listeria monocytogenes* on sliced mushrooms (*Agaricus bisporus*). *Journal of Food Protection* 78(6), 1147-53. doi:10.4315/0362-028X.JFP-14-389.

Murray, K., Wu, F., Shi, J., Jun Xue, S. and Warriner, K. 2017. Challenges in the microbiological food safety of fresh produce: limitations of post-harvest washing and the need for alternative interventions. *Food Quality and Safety* 1(4), 289-301. doi:10.1093/fqsafe/fyx027.

Murray, K., Aldossari, H., Wu, F. and Warriner, K. 2018a. Dynamic changes in free-chlorine levels within a commercial post-harvest wash and prevention of cross-contamination between shredded lettuce batches. *Food Control* 85, 127-34. doi:10.1016/j.foodcont.2017.09.029.

Murray, K., Moyer, P., Wu, F., Goyette, J. B. and Warriner, K. 2018b. Inactivation of *Listeria monocytogenes* on and within apples destined for caramel apple production by using sequential forced air ozone gas followed by a continuous advanced oxidative process treatment. *Journal of Food Protection* 81(3), 357-64. doi:10.4315/0362-028X.JFP-17-306.

Ndraha, N., Hsiao, H. I., Vlajic, J., Yang, M. F. and Lin, H. V. 2018. Time-temperature abuse in the food cold chain: review of issues, challenges, and recommendations. *Food Control* 89, 12-21. doi:10.1016/j.foodcont.2018.01.027.

Netramai, S., Kijchavengkul, T., Sakulchuthathip, V. and Rubino, M. 2016. Antimicrobial efficacy of gaseous chlorine dioxide against *Salmonella enterica* Typhimurium on grape tomato (*Lycopersicon esculentum*). *International Journal of Food Science and Technology* 51(10), 2225-32. doi:10.1111/ijfs.13209.

Olaimat, A. N. and Holley, R. A. 2012. Factors influencing the microbial safety of fresh produce: a review. *Food Microbiology* 32(1), 1-19. doi:10.1016/j.fm.2012.04.016.

Ovca, A., Jevsnik, M., Jereb, G. and Raspor, P. 2016. Effect of educational intervention on young people, targeting microbiological hazards in domestic kitchens. *Food Policy* 61, 156-62. doi:10.1016/j.foodpol.2016.03.004.

Pagadala, S., Marine, S. C., Micallef, S. A., Wang, F., Pahl, D. M., Melendez, M. V., Kline, W. L., Oni, R. A., Walsh, C. S., Everts, K. L. and Buchanan, R. L. 2015. Assessment of region, farming system, irrigation source and sampling time as food safety risk factors for tomatoes. *International Journal of Food Microbiology* 196, 98-108. doi:10.1016/j.ijfoodmicro.2014.12.005.

Palekar, M. P., Taylor, T. M., Maxim, J. E. and Castillo, A. 2015. Reduction of *Salmonella enterica* serotype Poona and background microbiota on fresh-cut cantaloupe by electron beam irradiation. *International Journal of Food Microbiology* 202, 66-72. doi:10.1016/j.ijfoodmicro.2015.02.001.

Park, S. H. and Kang, D. H. 2015. Combination treatment of chlorine dioxide gas and aerosolized sanitizer for inactivating foodborne pathogens on spinach leaves and tomatoes. *International Journal of Food Microbiology* 207, 103-8. doi:10.1016/j.ijfoodmicro.2015.04.044.

Park, S. H., Kang, J. W. and Kang, D. H. 2018. Inactivation of foodborne pathogens on fresh produce by combined treatment with UV-C radiation and chlorine dioxide gas, and mechanisms of synergistic inactivation. *Food Control* 92, 331-40. doi:10.1016/j.foodcont.2018.04.059.

Patel, J., Sharma, M., Millner, P., Calaway, T. and Singh, M. 2011. Inactivation of *Escherichia coli* O157:H7 attached to spinach harvester blade using bacteriophage. *Foodborne Pathogens and Disease* 8(4), 541-6. doi:10.1089/fpd.2010.0734.

Perez Rodriguez, F. P., Campos, D., Ryser, E. T., Buchholz, A. L., Posada-Izquierdo, G. D., Marks, B. P., Zurera, G. and Todd, E. 2011. A mathematical risk model for *Escherichia coli* O157:H7 cross-contamination of lettuce during processing. *Food Microbiology* 28(4), 694-701. doi:10.1016/j.fm.2010.06.008.

Perez-Rodriguez, F., Kwon, J., Bolivar, A., Sauer, K., Ryu, D. and Todd, E. 2019. Probabilistic risk model of Norovirus transmission during handling and preparation of fresh produce in school foodservice operations. *International Journal of Food Microbiology* 290, 159-69. doi:10.1016/j.ijfoodmicro.2018.09.029.

Pignata, C., D'Angelo, D., Fea, E. and Gilli, G. 2017. A review on microbiological decontamination of fresh produce with nonthermal plasma. *Journal of Applied Microbiology* 122(6), 1438-55. doi:10.1111/jam.13412.

Pillai, S. D. and Shayanfar, S. 2018. Electron beam processing of fresh produce - a critical review. *Radiation Physics and Chemistry* 143, 85-8. doi:10.1016/j.radphyschem.2017.09.008.

Pinela, J. and Ferreira, I. C. 2017. Nonthermal physical technologies to decontaminate and extend the shelf-life of fruits and vegetables: trends aiming at quality and safety. *Critical Reviews in Food Science and Nutrition* 57(10), 2095-111. doi:10.1080/10408398.2015.1046547.

Pyatkovskyy, T., Shynkaryk, M., Yousef, A. and Sastry, S. K. 2017. Fresh produce sanitization by combination of gaseous ozone and liquid sanitizer. *Journal of Food Engineering* 210, 19-26. doi:10.1016/j.jfoodeng.2017.03.031.

Qiao, M. Y., Liu, Q. S., Yong, Y., Pardo, Y. D., Worobo, R., Liu, Z., Jiang, S. and Ma, M. L. 2018. Scalable and rechargeable antimicrobial coating for food safety applications. *Journal of Agricultural and Food Chemistry* 66(43), 11441–50. doi:10.1021/acs.jafc.8b03864.

Ramos-Villarroel, A. Y., Martin-Belloso, O. and Soliva-Fortuny, R. 2015. Combined effects of malic acid dip and pulsed light treatments on the inactivation of *Listeria innocua* and *Escherichia coli* on fresh-cut produce. *Food Control* 52, 112–8. doi:10.1016/j.foodcont.2014.12.020.

Rapusas, R. S. and Rolle, R. S. 2009. Management of reusable plastic crates in fresh produce supply chains: a technical guide. Available at: http://www.fao.org/3/i0930e/i0930e00.pdf (accessed July 2019).

Rodgers, S. L., Cash, J. N., Siddiq, M. and Ryser, E. T. 2004. A comparison of different chemical sanitizers for inactivating *Escherichia coli* O157 : H7 and *Listeria monocytogenes* in solution and on apples, lettuce, strawberries, and cantaloupe. *Journal of Food Protection* 67(4), 721–31. doi:10.4315/0362-028X-67.4.721.

Roy, D. and Melotto, M. 2019. Stomatal response and human pathogen persistence in leafy greens under preharvest and postharvest environmental conditions. *Postharvest Biology and Technology* 148, 76–82. doi:10.1016/j.postharvbio.2018.10.013.

Park, S., Szonyi, B., Gautam, R., Nightingale, K., Anciso, J. and Ivanek, R. 2012. Risk factors for microbial contamination in fruits and vegetables at the preharvest level: a systematic review. *Journal of Food Protection* 75(11), 2055–81. doi:10.4315/0362-028X.JFP-12-160.

Schleker, S., Kshirsagar, M. and Klein-Seetharaman, J. 2015. Comparing human-*Salmonella* with plant-Salmonella protein-protein interaction predictions. *Frontiers in Microbiology* 6, 45. doi:10.3389/fmicb.2015.00045.

Sharapov, U. M., Wendel, A. M., Davis, J. P., Keene, W. E., Farrar, J., Sodha, S., Hyytia-Trees, E., Leeper, M., Gerner-Smidt, P., Griffin, P. M., Braden, C. and OUTBREAK INVESTIGATION TEAM. 2016. Multistate outbreak of *Escherichia coli* O157:H7 infections associated with consumption of fresh spinach: United States, 2006. *Journal of Food Protection* 79(12), 2024–30. doi:10.4315/0362-028X.JFP-15-556.

Shen, C. L., Luo, Y. G., Nou, X. W., Wang, Q. and Millner, P. 2013. Dynamic Effects of free chlorine concentration, organic load, and exposure time on the inactivation of *Salmonella*, *Escherichia coli* O157:H7, and non-O157 Shiga Toxin-Producing *E. coli*. *Journal of Food Protection* 76(3), 386–93. doi:10.4315/0362-028X.JFP-12-320.

Shen, C. L., Norris, P., Williams, O., Hagan, S. and Li, K. W. 2016. Generation of chlorine by-products in simulated wash water. *Food Chemistry* 190, 97–102. doi:10.1016/j.foodchem.2015.04.146.

Sheng, L. N., Hanrahan, I., Sun, X. F., Taylor, M. H., Mendoza, M. and Zhu, M. J. 2018. Survival of *Listeria innocua* on Fuji apples under commercial cold storage with or without low dose continuous ozone gaseous. *Food Microbiology* 76, 21–8. doi:10.1016/j.fm.2018.04.006.

Smith, A., Moorhouse, E., Monaghan, J., Taylor, C. and Singleton, I. 2018. Sources and survival of *Listeria monocytogenes* on fresh, leafy produce. *Journal of Applied Microbiology* 125(4), 930–42. doi:10.1111/jam.14025.

Smolinski, H. S., Wang, S. Y., Ren, L., Chen, Y. H., Kowalcyk, B., Thomas, E., Van Doren, J. V. and Ryser, E. T. 2018. Transfer and redistribution of *Salmonella* Typhimurium LT2 and *Escherichia coli* O157:H7 during pilot-scale processing of baby spinach, cilantro, and romaine lettuce. *Journal of Food Protection* 81(6), 953–62. doi:10.4315/0362-028X.JFP-17-420.

Sreedharan, A., Schneider, K. R. and Danyluk, M. D. 2014. *Salmonella* transfer potential onto tomatoes during laboratory-simulated in-field debris removal. *Journal of Food Protection* 77(7), 1062–8. doi:10.4315/0362-028X.JFP-13-509.

Sun, X. X., Baldwin, E. and Bai, J. H. 2019. Applications of gaseous chlorine dioxide on postharvest handling and storage of fruits and vegetables – a review. *Food Control* 95, 18–26. doi:10.1016/j.foodcont.2018.07.044.

Takashina, T. A., Leifeld, V., Zelinski, D. W., Mafra, M. R. and Igarashi-Mafra, L. 2018. Application of response surface methodology for coffee effluent treatment by ozone and combined ozone/UV. *Ozone: Science and Engineering* 40(4), 293–304. doi:10.1080/01919512.2017.1417112.

Tapp, W. N., Gragg, S. E., Brooks, J. C., Miller, M. F. and Brashears, M. M. 2013. Reduction of *Escherichia coli* O157:H7 and *Salmonella* after application of various sanitizing treatments to harvesting knives. *Journal of Food Protection* 76(2), 200–4. doi:10.4315/0362-028X.JFP-12-345.

Teng, Z., Luo, Y. G., Alborzi, S., Zhou, B., Chen, L., Zhang, J. L., Zhang, B. C., Millner, P. and Wang, Q. 2018. Investigation on chlorine-based sanitization under stabilized conditions in the presence of organic load. *International Journal of Food Microbiology* 266, 150–7. doi:10.1016/j.ijfoodmicro.2017.11.027.

Trinetta, V., Linton, R. H. and Morgan, M. T. 2013. The application of high-concentration short-time chlorine dioxide treatment for selected specialty crops including Roma tomatoes (*Lycopersicon esculentum*), cantaloupes (*Cucumis melo* ssp melo var. cantaloupensis) and strawberries (*Fragaria x ananassa*). *Food Microbiology* 34(2), 296–302. doi:10.1016/j.fm.2012.12.010.

US Center of Disease Control and Prevention (CDC). 2016. Annual summaries of foodborne illness outbreaks. Available at: https://www.cdc.gov/fdoss/pdf/2016_FoodBorneOutbreaks_508.pdf (accessed July 2019).

US Center of Disease Control and Prevention (CDC). 2017. List of selected multistate foodborne outbreak investigations. Available at: https://www.cdc.gov/foodsafety/outbreaks/multistate-outbreaks/outbreaks-list.html (accessed July 2019).

USFDA. 2019. FSMA final rule on produce safety. Available at: https://www.fda.gov/food/food-safety-modernization-act-fsma/fsma-final-rule-produce-safety.

Vurma, M., Pandit, R. B., Sastry, S. K. and Yousef, A. E. 2009. Inactivation of *Escherichia coli* O157:H7 and natural microbiota on spinach leaves using gaseous ozone during vacuum cooling and simulated transportation. *Journal of Food Protection* 72(7), 1538–46. doi:10.4315/0362-028X-72.7.1538.

Warriner, K. and Namvar, A. 2010. The tricks learnt by human enteric pathogens from phytopathogens to persist within the plant environment. *Current Opinion in Biotechnology* 21(2), 131–6. doi:10.1016/j.copbio.2010.01.004.

Warriner, K., Huber, A., Namvar, A., Fan, W. and Dunfield, K. 2009. Recent advances in the microbial safety of fresh fruits and vegetables. In: Taylor, S. L. (Ed.), *Advances in Food and Nutrition Research* (vol. 57). Elsevier Academic Press Inc, San Diego, CA.

Weller, D., Wiedmann, M. and Strawn, L. K. 2015. Spatial and temporal factors associated with an increased prevalence of *Listeria monocytogenes* in spinach fields in New York State. *Applied and Environmental Microbiology* 81(17), 6059–69. doi:10.1128/AEM.01286-15.

Xie, Y., Hajdok, C., Mittal, G. S. and Warriner, K. 2008. Inactivation of MS2 F (+) coliphage on lettuce by a combination of UV light and hydrogen peroxide. *Journal of Food Protection* 71(5), 903–7. doi:10.4315/0362-028X-71.5.903.

Yesil, M., Kasler, D. R., Huang, E. and Yousef, A. E. 2017. Efficacy of gaseous ozone application during vacuum cooling against *Escherichia coli* O157:H7 on spinach leaves as influenced by bacterium population size. *Journal of Food Protection* 80(7), 1066–71. doi:10.4315/0362-028X.JFP-16-284.

Zhou, B., Lu, Y. G., Millner, P. and Feng, H. 2012. Sanitation and design of lettuce coring knives for minimizing *Escherichia coli* O157:H7 contamination. *Journal of Food Protection* 75(3), 563–6. doi:10.4315/0362-028X.JFP-11-218.

Zhou, B., Luo, Y. G., Nou, X. W. and Millner, P. 2014. Development of an algorithm for feed-forward chlorine dosing of lettuce wash operations and correlation of chlorine profile with *Escherichia coli* O157:H7 inactivation. *Journal of Food Protection* 77(4), 558–66. doi:10.4315/0362-028X.JFP-13-352.

Zhou, B., Luo, Y., Nou, X., Lyu, S. and Wang, Q. 2015. Inactivation dynamics of *Salmonella enterica*, *Listeria monocytogenes*, and *Escherichia coli* O157:H7 in wash water during simulated chlorine depletion and replenishment processes. *Food Microbiology* 50, 88–96. doi:10.1016/j.fm.2015.03.004.

Ziuzina, D., Petil, S., Cullen, P. J., Keener, K. M. and Bourke, P. 2014. Atmospheric cold plasma inactivation of *Escherichia coli*, *Salmonella enterica* serovar Typhimurium and *Listeria monocytogenes* inoculated on fresh produce. *Food Microbiology* 42, 109–16. doi:10.1016/j.fm.2014.02.007.

Ziuzina, D., Han, L., Cullen, P. J. and Bourke, P. 2015. Cold plasma inactivation of internalised bacteria and biofilms for *Salmonella enterica* serovar Typhimurium, *Listeria monocytogenes* and *Escherichia coli*. *International Journal of Food Microbiology* 210, 53–61. doi:10.1016/j.ijfoodmicro.2015.05.019.

Zoellner, C., Al-Mamun, M. A., Grohn, Y., Jackson, P. and Worobo, R. 2018. Postharvest supply chain with microbial travelers: a Farm-to-Retail microbial simulation and visualization framework. *Applied and Environmental Microbiology* 84(17), 13–18. doi:10.1128/AEM.00813-18.

Chapter 6

Advances in understanding pathogens contaminating horticultural produce

P. Truchado, A. Allende and M. I. Gil, CEBAS-CSIC, Spain

1 Introduction

There is an overall trend in most of the developed countries that confirms the rise in the number of outbreaks associated with fresh products. Many comprehensive reviews have highlighted the most probable reasons that might explain this trend but, as happens in many other areas, the explanation is probably multifactorial. Major factors that contribute are global consumer demand, increasing worldwide trade and the development of new molecular tools that allow the identification of the causative agents and the outbreak traceability. Despite the great advances in the application of genetic methods, there is still substantial knowledge gaps concerning the major risk factors for microbial contamination of fresh produce. Fresh fruits and vegetables (FFV) are produced in very different environments using a multitude of agricultural practices as well as using diverse processing and storage facilities, and distribution channels. Therefore, the FFV supply chain is complex, encompassing several crucial steps. Microbial contamination can occur during any of the steps through the farm-to-consumer including production, harvest, fresh-cut processing, wholesale

http://dx.doi.org/10.19103/AS.2019.0055.07

storage, transportation or retailing and handling at home. The contamination can originate from multiple sources.

Grower associations and competent authorities on food safety have increasingly emphasized the need for the implementation of good agricultural practices (GAP), good manufacturing practices and good hygiene practices to reduce microbial risks along the supply chain. This has been translated into the publication of new regulations (FSMA Final Rule on Produce Safety) and new guidances (EC, 2017; FDA, 2018a) with the aim to reduce foodborne outbreaks linked to FFV. However, despite all these efforts, it is clear that more research is needed to understand the ability of enteric pathogens to colonize, survive and grow on FFV. A clear example is the three outbreaks linked to consumption of romaine lettuce in the United States in 2018, which have created social alarm and the breakdown of the market. Although there are numerous research fields that need to be addressed to understand better the behaviour of pathogens in plant tissues and the major pathways they use to colonize and invade, efforts are now being directed to understand how interactions between the natural microbiota and the pathogens affect their survival and growth. This is a new field of study for which many hopes are placed to develop new intervention strategies that allow the selection of specific microbial groups that can compete directly with pathogenic bacteria and prevent their proliferation. During the last 20 years, science has made great progress in the identification of potential risk factors and understanding the behaviour of enteric pathogens in the plant tissue. However, current trends predict a future in which the implication of FFV in outbreaks will continue. Efforts that are being made by the industry and research organisms on prevention strategies will hopefully bear fruit to reduce current microbiological risks.

2 Microbial contamination of horticultural produce: surveillance, source attribution and microbial risk assessment of foodborne pathogens

The implication of foodborne illness in FFV was shown in 2018 with three different outbreaks linked to romaine lettuce. The leafy-greens growers' associations and associated authorities are implementing drastic measures to curb food poisoning. The Centers for Disease Control and Prevention (CDC) advised restaurants and retailers to not sell or serve any romaine lettuce harvested from the Central Coastal growing regions of northern and central California for more than a month (CDC, 2018). This generated social alarm as well as the collapse of the markets.

In general, more and more developed countries are being affected by an intensification of outbreaks associated with fresh products. However, it is difficult to determine the reasons for this amplification, although several explanations have been suggested. These include the dynamic genetic content of relevant microorganisms, changes in agricultural production systems due to the introduction of new technologies, climate change and increasing worldwide trade in food products and raw material as a global consumer demand (Morse, 1995; Franz et al., 2014). In the last 10 years, there has been a revolution in the development of advanced tools that allow the identification of outbreaks of human illness as well as the link from these outbreaks back to the food source responsible for the illness. Scientific advances on new information technologies and laboratory techniques have improved the ability to identify outbreaks, which improved the protection of the public from outbreaks, but at the same time, the number of outbreaks may have increased since we are now able to identify problems that were previously invisible to us (FDA, 2018b). In general, researchers and authorities do not believe that foodborne outbreaks are increasing because the food is safer than ever before but nowadays, the identification of outbreaks, when they occur, is easier and faster than ever before.

Franz et al. (2014) described the differences between the use of conventional techniques and advanced molecular tools. Briefly, traditional microbiology techniques require several steps including enrichment and culture on selective media as well as serotyping to detect and characterize pathogens. However, advanced molecular techniques based on DNA sequencing or proteins enable faster and more accurate identification of pathogens. The use of polymerase chain reaction for the detection of specific virulent serotypes known to cause human disease has been a great improvement for diagnosis. However, it is necessary to identify marker genes associated with pathogen stress survival, growth and/or virulence, and for this, genomics data becomes essential. Genomics data can also be used to correlate disease incidence with the distribution and diversity of specific genomic sequences within a geographic area, to anticipate changes in pathogenicity and pathogen behaviour in food chains among others.

Advanced genetic methods that allow the identification of outbreaks of human illness include the use of whole-genome sequencing (WGS) and next-generation sequencing (NGS) tools, which have been proven to be very useful in foodborne outbreak investigations, particularly in the identification and evaluation of hazards (Ronholm et al., 2016). Rapid developments in WGS and NGS have led to an increase in information about foodborne microbes, improving our understanding of how they survive in foods, how they cause disease and why some strains are more virulent than others (Haddad et al., 2018). Sequence data obtained using WGS can be processed to yield further details such as serotype, virulence and antibiotic resistance profiles, as well as genetic variations such as single-nucleotide polymorphisms (Njage et al., 2019).

Public health authorities and regulatory bodies are using WGS on a regular basis to characterize clinical isolates of selected foodborne pathogens and to support epidemiological investigations, which allow comparison of different isolates after different periods (Rantsiou et al., 2018). The application of genomics data can be illustrated with one of the latest *Listeria monocytogenes* outbreak investigations that happen in Europe. A multi-country outbreak of *L. monocytogenes* causing 53 cases and 10 deaths (2015–18 period) was linked to frozen vegetables. In this case, experts used WGS to identify the food source, which initially was thought to be limited to frozen corn. Based on the investigations, the same strains of *L. monocytogenes* were detected in several types of frozen vegetables produced by the same Hungarian company in 2016, 2017 and 2018, indicating the persistence of the bacteria in the processing plant (EFSA, 2018). Taking one step forward, the FDA is routinely sequencing all *L. monocytogenes* and has established the GenomeTrakr Network to facilitate sharing the genomic and other isolate-related information between laboratories (Rantsiou et al., 2018). This network encompasses several laboratories all over the world with the aim of collecting the largest number of sequences to create the largest database (Allard et al., 2016).

Most of the literature highlights that the most significant advantage of WGS for the pathogenic microorganism is that it allows typing at a much greater resolution than traditional molecular typing methods (Franz et al., 2016). However, as stated by Rantsiou et al. (2018), routine use or complete replacement of this typing technique by WGS will need adequate attention to overcome existing challenges and limitations such as cost, human resources and laboratory and bioinformatics infrastructure among others. What is clear is that the use of WGS is increasing the number of outbreaks detected: many of them would probably pass unnoticed using the conventional molecular typing methods (Franz et al., 2016).

The use of the omics data not only is changing outbreak surveillance, but it also has great potential for microbial risk assessment (MRA). MRA provides an objective, transparent, evidence-based assessment of the health risk of exposure pathways to support risk management decisions (De Keuckelaere et al., 2015). MRA consists of four steps: (1) hazard identification; (2) exposure assessment; (3) hazard characterization; and (4) risk characterization (Codex, 1999). Hazard identification is usually based on available data sources that provide relevant information to identify and evaluate hazards including the Rapid Alert System for Food and Feed (RASFF) portal (RASFF, 2016), the World Health Organization (WHO) databases, reports from European scientific agencies such as the European Food Safety Authority and the European Centre for Disease Prevention and Control as well as the data from EU MS national monitoring programmes (Allende et al., 2018a). The second step, the exposure assessment, aims to evaluate qualitatively and/or quantitatively the likely intake of microbial agents via food, as well as exposure from other sources if relevant (den Besten et al., 2018; Codex, 1999). In the case of a quantitative or probabilistic exposure assessment, large amounts of data (e.g. prevalence/concentration of pathogenic microorganisms, chain information such as steps in production of foods, time/temperature conditions, quantitative information on factors contributing to increase of pathogens, reduction, multiplication or growth potentials) as well as food consumption patterns are needed (Allende et al., 2018a). Genomics represents a promising tool to be used in MRA helping to understand the dynamics of pathogens in a complex food ecosystem as well as predicting pathogen behaviour variability. Omics technologies, including genome-wide sequencing tools, genome-wide transcript and protein analysis and assessment of metabolic profile of microorganisms, can be used to gain more insight in the microbial ecology of a processing environment and how it relates to the food ecology (den Besten et al., 2018). WGS provides a greater potential input for improved MRA because data are not restricted to a specific choice of target genes.

As already highlighted by several authors, the potential of omics technologies to be used in surveillance and outbreak investigation as well as risk assessment are clear. However, there are still several challenges that should be overcome before WGS and NGS can be used on a routine basis, including the harmonization of the methods (Franz et al., 2016).

3 Main sources for pathogen contamination: primary production, processing, distribution and cold chain

FFV are produced, transported, processed, stored and marketed under a wide range of environments that are potential sources of foodborne pathogens (FAO/WHO, 2008). They are grown indoors and outdoors, harvested and either field-packed or transported to the packinghouse or the processing plant for processing and shipping through the cold chain before reaching the final consumer. In this supply chain of fresh produce from 'farm to fork', sources of produce contamination can vary because of the combination of growing environment, land-use interactions and climate (Alegbeleye et al., 2018). Produce contamination also depends on the type of product as there are some crops that are grown close to the ground such as leafy greens (lettuce, spinach), edible herbs (basil, cilantro), strawberries and tree nuts harvested on the ground of higher risk of faecal contamination by wild animals (Jay-Russell, 2013). The risk also depends upon the way the crop is prepared and eaten (with or without skin, cooked or raw), and upon the history of its pathogen contamination. Thus, Monaghan (2006) classified them into four categories according to the risk the fresh produce pose. In category 1, the crops included are those eaten raw with no protective skin that includes leafy vegetables and fresh herbs. In category 2, the crops included are those that can be eaten raw and either have no protective skin or have some risk or history of pathogen contamination (broccoli, cabbage and carrots). In category 3, the crops included are those that can be eaten raw and either have a protective skin or grow clear of the ground and have no history of pathogen contamination including tree fruit, pineapple and green beans. In category 4, the crops included are always cooked before eaten such as asparagus, Brussels sprouts and eggplant. In line with this classification, crops have also been divided into different categories based on their potential microbiological risks in recent publications from grower associations and competent authorities (EFSA, 2013; EC, 2017). In agreement with Monaghan (2006), the starting point from 'farm to fork' is that there is no guarantee of pathogen-free produce in any of the different steps that the fresh produce goes through. Combinations of peculiar environmental risk factors influence the frequency and transmission of foodborne pathogens and affect the risk of produce contamination (Strawn et al., 2013). Therefore, to minimize the risk factors, it is important first to identify the potential sources of contamination and also understand the establishment of pathogens in produce.

The major potential sources of microbial foodborne hazards that may lead to a contaminated product have been identified, and several classifications have been suggested from specific sources to more general ones (FAO/WHO, 2008; EC, 2017). In this section, the main sources for pathogen contamination are discussed as they have been recently identified and grouped by the European Commission (EC, 2017) (Fig. 1).

3.1 Environmental factors, including animal reservoirs

Available scientific studies have identified environmental factors including animal reservoirs as the main sources of foodborne contamination and persistence on plants (Suslow et al., 2003; Matthews, 2014; Liu et al., 2018). However, the specific influence and interactions among the production environment and crop management practices are not sufficiently understood for microbial risk prevention. More knowledge is needed on these topics to assist growers to enforce hygiene requirement programmes in the production and handling of FFV.

For environmental factors, one of the key areas of concern is animal rearing and animal reservoirs, mostly because domesticated animal and wildlife activity on the crop is particularly risky for the introduction of pathogens. Foodborne pathogens shed by animals can be spread to plants directly through faecal deposition or indirectly via faecal contamination of agriculture water (AW), soil, compost, farm equipment and workers' boots and clothing (Jay-Russell, 2013). Animal feeding operations in concentrated areas – Concentrated Animal Feeding Operations (CAFOs) – are common on farms and ranches. More

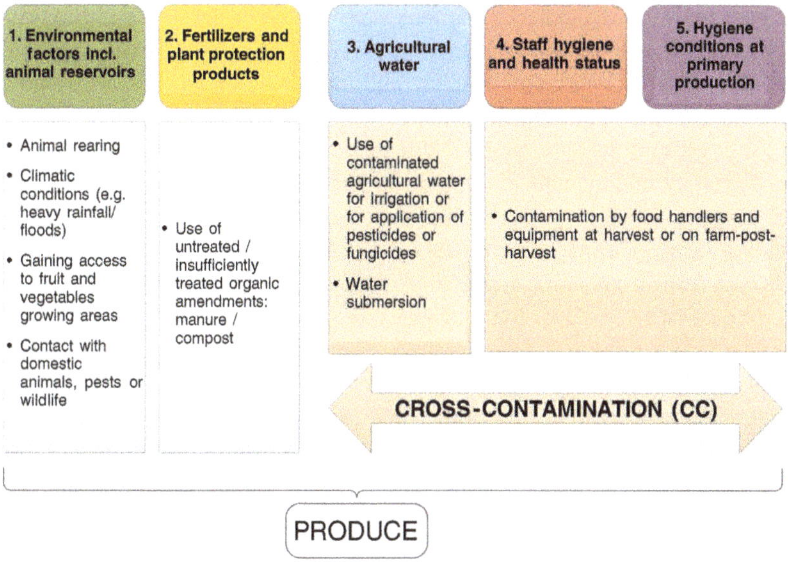

Figure 1 Main risk factors for microbiological pathogens in FFV identified by EFSA. Source: Adapted from EC (2017).

rigorous risk assessments will be in place whenever leafy greens are grown in the general area of a CAFO, including changes for surface water use and buffer zones (FDA, 2018a). These updates to minimum distances and the adoption of more rigorous water metrics reflect an abundance of caution while relying upon the best available scientific studies to date. Through collaborative efforts between the leafy greens industry and stakeholders, new guidance documents provide further considerations for production near CAFOs. FSMA rules (FDA, 2018a) address concerns about the feasibility of compliance for farms that rely on grazing animals (such as livestock) or working animals for various purposes. The same standards for these animals as for the intrusion of wild animals (such as deer or feral swine) are established. Farmers must visually examine the growing area and the produce to be harvested, and take measures reasonably necessary to assist later during harvest such as placing flags outlining the affected area. Farms are not required to exclude animals from outdoor growing areas, destroy animal habitat or clear borders around growing or drainage areas (FDA, 2018a). However, the extent of potential illness from contaminated produce is more likely to be broad in locations where large-scale animal production is immediately adjacent to edible crop production (Suslow et al., 2003). Another type of situation that occurs is wild birds and rodents contaminated with verocytotoxin-producing *Escherichia coli* (VTEC) in the proximity to farms (Nielsen et al., 2004; Jay-Russell, 2013) or sources of water for irrigation including surface water (open canals, impounded water such as ponds, reservoirs) (Uyttendaele et al., 2015). Irrigation canals and ponds may also serve as wild animal habitat, and therefore should be considered as potential sources of pathogen transfer.

The geographical characteristics of the growing site including the topology may influence the magnitude and frequency of transfer of pathogenic microorganisms from environmental sources to the crop (FAO/WHO, 2008). When growing in farms located in valleys, the run-off of water or dust circulation and drift from higher elevations and distant sources of contaminated sites must be considered. Over the past years, the effect of environmental weather conditions (e.g. temperature, humidity, sunlight/radiation and pollution) on the survival of foodborne pathogens has been examined (Tang, 2009). Changes in expected weather patterns, particularly extreme climatic events, have been highlighted as possibly having an important role as reported in the most recent outbreak related to fresh produce on romaine lettuce in 2018 (American Vegetable Grower, 2018). A direct relationship between foodborne outbreaks and weather, particularly extreme weather events, has been associated. Recently, the FDA investigated the last multistate outbreak of *E. coli* O157:H7 infections linked to romaine lettuce in the United States. WGS of the *E. coli* O157:H7 strains tested to date from patients in this current outbreak were similar to strains of *E. coli* O157:H7 associated with a previous outbreak from the fall of 2017. Romaine lettuce is again the suspected vehicle for both the US and Canadian outbreaks in the fall of 2018. Evidence suggests a combination of unusual weather events and plant diseases that may have been contributing factors to pathogens, such as *E. coli* entering the leaves. Despite several other leafy greens growing in the area during that time, only romaine was infected. Since romaine is more vascular than other types of leafy greens, frost can blister the skin and make it more susceptible than other crops (American Vegetable Grower, 2018). This is in agreement with previous studies that highlight that changes in weather (e.g. temperature and RH) may affect the growth and physiological conditions of leafy vegetables and thus their susceptibility to pathogen contamination (FAO/WHO, 2008). Arizona and California Leafy Green Marketing Agreements have made recommendations including an additional environmental assessment following weather events such as flooding, frost or high winds. Dry periods can also cause dust storms that settle dust particles on leafy vegetables, which may harbour foodborne pathogens (FAO/WHO, 2008).

3.2 Fertilizers and plant protection products

In general, agricultural inputs have a very diverse composition including organic fertilizers (e.g. livestock manure, slurries and sewage sludge), and also inorganic fertilizers (chemical fertilizers), which are not as critical as organic fertilizers from the microbiological point of view (EC, 2017). Organic fertilizers, if they are not used correctly, can be a source of both microbiological (e.g. *Salmonella* spp., VTEC, Norovirus) and chemical contamination (e.g. heavy metals) (Ingham et al., 2004; Islam et al., 2004a,b, 2005). Any biological soil amendment of animal origin (BSAAO) has the potential to be a route for on-farm contamination of produce (FDA, 2018a). The use of unsafe agricultural practices, particularly involving an untreated BSAAO, could lead to foodborne illness from the consumption of contaminated produce. Organic amendments are added to the soil to improve the physical and chemical characteristics of plant growth, but they have to be properly treated to avoid the transfer of foodborne pathogens to the crop. Significant populations of *E. coli* and related pathogens of faecal origin are limited to soil contaminated with faecal material and held to be transitory (Suslow et al., 2003). However, in some environments, pathogens may exist as a soil resident associated with decomposed organic matter as has been reported for *L. monocytogenes* (Dowe et al., 1997; FAO/WHO, 2008). Animal waste production and animal production operations are geographically concentrated and generate a large volume of waste that must be composted before farm application. Among the growing sector of organic producers, the practice of using manure as a fertilizer is common (Suslow et al., 2003). GAP recommendations have been revised to include treatment and application of animal waste (EC, 2017; FDA, 2018a).

3.3 Water for primary production and associated operations at harvest and postharvest

There are several risk factors linked to the use of AW of poor microbial quality; irrigation water represents one of the major sources of microbial contamination of fresh produce. Some of these factors are source of water, type of irrigation (drip, sprinkler irrigation etc.), application of the water (e.g. whether the edible portion of the produce is in direct contact with irrigation water) and use of water disinfection treatments by the grower and the timing of irrigation (EC, 2017). Irrigation water can originate from multiple sources such as municipal supplies, rainwater, groundwater, surface water (open canals, impounded water such as ponds, reservoirs and lakes) and reclaimed plants (Uyttendaele et al., 2015). As Liu et al. (2018) mentioned, developing general standards for water used for irrigation is complicated because the source of water for irrigation can vary greatly with region, season and type of crop as well as the microbial quality of the water, which varies in a dynamic way. Precipitation, the distance to domesticated animal raising facilities and the number of wild or domestic animals in the proximity of the water can also result in substantial changes to the microbial quality of the water (Liu et al., 2018).

The source of water used to irrigate produce is usually only minimally treated or untreated in many cases (Steele and Odumeru, 2004; Alegbeleye et al., 2018). Water disinfection treatments can be used to improve the microbiological safety of agricultural water. As mentioned in our previous study (López-Gálvez et al., 2017), growers should ascertain the minimum effective disinfectant doses able to achieve the desired microbiological goals, particularly recommended after extreme weather events such as flooding. The consequences of the treatment of irrigation water with chlorine dioxide (ClO_2) used for sprinkler irrigation of baby spinach in two commercial agricultural fields were assessed, and low concentrations of ClO_2 were effective in reducing *E. coli* present in the water (López-Gálvez et al., 2018b).

The type of irrigation system also has significant impacts on the risk of microbiological contamination. Among all the irrigation systems, sprinkler irrigation poses the highest risk of microbiological contamination because of the direct contact between edible portions of certain plants (leafy vegetables provide a larger surface area for contact and possible microbial attachment) and irrigation water (Berger et al., 2010; Kisluk and Yaron, 2012). On the contrary, surface furrow and drip irrigation systems minimize the contact with the edible portion of the product (Alegbeleye et al., 2018). Contaminated irrigation water poses the most significant risk when crops are irrigated close to the harvest time because harvesting produce containing viable pathogens is more likely. The FSMA Final Rule on Produce Safety indicates a 0.5 log cfu/g decline each day between the last irrigation event and harvest (Allende et al., 2018b; FDA, 2018a).

One of the postharvest activities in which water comes into contact with produce in a variety of ways is during ice cooling and hydrocooling. When cooling produce using ice or hydrocooling, the water must be potable as the water can be a potential source of pathogenic contamination (Coetzer, 2006). Hydrocooling systems require daily cleaning and water needs to be changed as required to maintain quality. Water that is reused to cool continuous loads of produce is a potential source of cross-contamination as it can contaminate new incoming produce.

3.4 Hygiene conditions during fresh produce activities at farm and processing levels and health status of workers

In many of the operations that occur in the field including harvest and field packing as well as in the processing plant, direct contact between fresh produce and workers (handlers) occurs. This contact can often be the way harmful pathogens are transmitted to the produce and the more the produce is handled, the higher the likelihood of introducing foodborne pathogens. Handling of produce is mostly unavoidable, and personal hygiene is therefore extremely important (Coetzer, 2006). Measures must be in place to eliminate or diminish the possibility of contamination of produce. All workers must have training in basic hygiene and they should be informed about all the possible risks related to the contamination of the product (EC, 2017). Some produce-related outbreaks and contamination events have identified personnel health and hygiene practices on the farm and produce handling and storage practices as likely contributing factors to these events (FDA, 2018a). Furthermore, a study of food-related outbreaks found more than 800 instances where personnel practices were instrumental in or contributory to the outbreak (Greig et al., 2007).

Poor personal hygiene practices and bad work habits and conditions that can introduce pathogens include (Coetzer, 2006):

1 Not washing hands after using toilet facilities,
2 Not washing hands properly,
3 Inadequate hand-washing and toilet facilities,
4 Employees with contagious diseases, open wounds or infections.

It is particularly critical that workers with any illness and/or diarrhoea must be excluded from working to reduce potential contamination of produce and must notify the supervisor. Injuries including cuts, burns, boils and skin eruptions should be reported to the supervisor because employees are the most important link in preventing foodborne illness. Good personal hygiene and cleanliness, including proper and frequent hand-washing, is the best way to prevent foodborne illness. The main intent of any working apparel policy is to protect the edible produce from all possible sources of contamination including clothing. To prevent potential contamination, suitable clean-washed and well-maintained work clothing must be provided to workers while on the job. Work clothing provided for the producer should be worn alone or over any piece of worker clothing that may come into contact with food products. In an attempt to minimize risk from contamination of protective clothing outside of the site, all protective clothing, including head covering, must not leave the site for any reason, except for laundering by a contractor (Monaghan, 2006). It is not acceptable for personnel to launder their protective clothing at home. The preferred option for personal clothing is to use removable clothing.

FFV should be stored and transported under good hygienic conditions. Storage facilities and vehicles for transporting the harvested FFV should be constructed in a manner that will minimize damage to FFV and prevent access by pests such as insects, rodents and birds (EC, 2017). All containers and equipment that come into contact with the produce must be washable and must be kept clean and sanitized. These include harvest containers, harvesting machinery, knives, packaging materials, brushes, pallets, and trailers in which produce is transported (Coetzer, 2006). Good practice recommendations regarding harvesting and packaging operations at the field or farm level have been described (EC, 2017). Specific good practice recommendations for tomatoes, berries, melons and watermelons in relation with some operations such as loading/transport/storage and cleaning/disinfection operations to minimize the risks of pathogen contamination have also been prepared (EC, 2017).

Regarding contact surfaces, the microbiological analysis of food contact and non-food contact surfaces in fresh-cut processing facilities after routine daily sanitization showed that among 23 genera detected, some included soil bacteria, plant-related bacteria, coliforms and opportunistic plant- or human-pathogenic bacteria (Liu et al., 2013).

3.5 Fresh-cut processing

Various steps are included in the preparation of fresh-cut products, each of which is a specific unit operation. Each unit operation must be performed properly to ensure that the finished product quality and safety are satisfactory (Gil and Selma, 2006). In the fresh-cut preparation, the unit operations that may be included are raw material receiving and storage, preliminary washing and sorting, peeling, size reduction/cutting, defect sorting, washing/cooling, dewatering, weighing and packaging, storage and transportation. FFV may be contaminated with pathogens when they enter the processing chain (Nguyen-the and Carlin, 1994). During handling, contamination may occur via unhygienic personnel or inadequately cleaned equipment. Cutters and slicers can be potent sources of contamination because they usually provide inaccessible sites and increase the surface area for produce contamination and growth (Gil and Selma, 2006). During this step, the risk of pathogen contamination such as *L. monocytogenes* and *Salmonella* increases. Velani and Roberts (1991) found that substantial contamination by *L. monocytogenes* occurred during chopping, mixing and packaging.

Washing is a very critical part of any produce preparation process especially if raw, processed fresh produce is sold as 'ready-to-eat' (Allende et al., 2008; Gil et al., 2009). Fresh-cut produce can become contaminated with foodborne pathogens in contact with contaminated wash water. Several investigations of fresh produce pointed out that improper wash water management has been one of the most important contributing factors to large disease outbreaks (CDC, 2011). Despite the general idea of most of the processors that disinfectants are used to reduce the microbial population on the produce, its main effect is maintaining the microbial quality of the process water (Gil et al., 2009). Apart from eliminating soil residues and plant debris, washing with tap water has a limited effect on fresh produce microbiota because of the high aerobic bacterial count of the raw material, with process wash water quickly becoming contaminated. Washing has been identified as a potential pathway for the dispersion of microorganisms during the production of FFV and fresh-cut produce and if pathogens are present, they are introduced to the end product via cross-contamination (Allende et al., 2008; López-Gálvez et al., 2009; Holvoet et al., 2012). A low amount of contaminated product present in a batch can affect the safety of the entire lot that passes through the washing tank (Gombas et al., 2017).

4 Routes of pathogens contaminating FFV: attachment and internalization

There are recurring plant-pathogen combinations associated with produce-related outbreaks and recalls that suggest that in some degree foodborne pathogens may adapt to the plant environment (Jay-Russell, 2013). When a foodborne pathogen is introduced onto the plant surface, the epiphytic microbiota face a hostile environment and die-off within hours to days (López-Gálvez et al., 2018a). However, there are numerous ways and mechanisms through which pathogens may gain access to produce, thereby compromising produce microbiological safety (Alegbeleye et al., 2018). The main entry points for pathogens to humans via produce is represented in Fig. 2 (Alegbeleye et al., 2018).

Increasing numbers of outbreaks involving environmental contamination of produce has resulted in questions about the attachment and interactions of pathogens on and in plants (Hirneisen et al., 2012). There is growing evidence showing the potential for internalization of foodborne pathogens in the field. This is a very relevant topic because internalization of bacteria into plants represents an added risk because bacteria located inside the plant tissue can

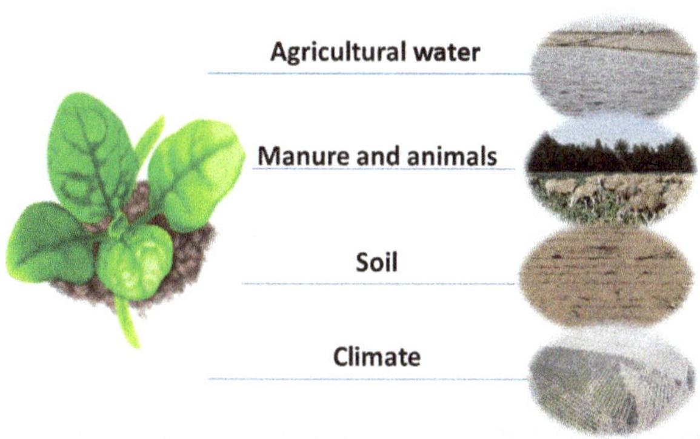

Figure 2 Main entry points for pathogens in FFV.

be protected from external stress factors (e.g. solar radiation, wind, rain) as well as from disinfection treatments used by the industry during washing of FFV.

One of the topics that has received much debate over the last years has been the potential for uptake of the foodborne pathogen through the roots into the crop by internalization. Hirneisen et al. (2012) defined internalization as the uptake of human enteric pathogens through the roots into the intercellular spaces between plant cells and in the plant vasculature tissues, xylem and phloem. Many researchers have also focused on the methods of bacterial internalization, including contaminated seeds, roots, stomata or damage to the epidermis (Wright et al., 2017).

Various factors have been shown to affect the ability of human pathogens to internalize including growth substrate (soil vs. hydroponic solution). In the review by Hirneisen et al. (2012) several studies have been discussed. Some of them had shown bacterial internalization into plant tissues through the roots when plants were grown in inoculated hydroponic solution in contrast with those grown in inoculated soil, in which little or no internalization was observed. If roots are damaged, there are more internalized bacterial populations in plants grown in contaminated soil versus hydroponic solution (Franz et al., 2007). When lettuce plants were grown in soil with more root damage than in hydroponic media with less root damage, larger numbers of *E. coli* O157:H7 cells were internalized in leaves (Franz et al., 2007). However, this effect was not observed in spinach plants as mechanical damage of root tissues did not increase the likelihood of internalization through uptake (Hora et al., 2005). Therefore, plant species and cultivar seem to influence internalization. Jablasone et al. (2005) suggested that the antimicrobial constituents of some root vegetables may be responsible for limited bacterial growth and internalization of human-pathogenic bacteria in some plant species and cultivars than in others. Plant developmental stage also plays a role in possible internalization of the bacterial pathogen.

Several studies have focused on the impact of the maturity stage. Leaf age has been shown to influence the growth and survival of *Salmonella* and *E. coli* O157:H7, with younger plants being more susceptible than more mature ones (Brandl and Amundson, 2008). However, in other studies, human enteric bacteria uptake in older plants is more pronounced than in younger plants (Bernstein et al., 2007). Several mechanisms of bacterial internalization ('active' vs. 'passive') in plant roots have been hypothesized, but inconclusive results have been published. However, it has now become clear that enteric pathogens are adapted to survive for prolonged periods outside of their animal hosts, attaching tenaciously to plants via biofilm formation and internalization in plant tissues (Hoagland et al., 2018).

After harvest, enteric pathogens have been shown to enter and attach to plant leaf stomata where the cells protected them from washing and chemical sanitizing agents used by the fresh-cut produce industry (Kroupitski et al., 2009). These authors suggested that pathogens are attracted to nutrients produced *de novo* by photosynthetically active cells. In our research group, we investigated the impact of three postharvest handling conditions such as (1) illumination, (2) negative temperature differential and (3) relative humidity (RH) before washing on the internalization of pathogens during washing (Gómez-López et al., 2013). Results showed that internalization of *Salmonella* into baby spinach leaves always occurs but can be minimized under specific postharvest handling conditions such as low RH. Exposure of leaves to low-RH conditions before washing reduced the tissue water content and the fragility and decreased internalization of *Salmonella* compared with internalization when leaves were exposed to high RH (Gómez-López et al., 2013).

5 Factors affecting the survival-inactivation of pathogens in horticultural products

Foodborne outbreaks associated with FFV have been linked to a wide variety of microorganisms. A few years ago, EFSA elaborated a risk ranking of food/pathogen combinations to identify the top five combinations linked to foodborne outbreaks (EFSA, 2013). The top-ranking food/pathogen combination was *Salmonella* spp. and leafy greens eaten raw followed by (in equal rank) *Salmonella* spp. and bulb and stem vegetables, *Salmonella* spp. and tomatoes, *Salmonella* spp. and melons, and pathogenic *E. coli* and fresh pods, legumes or grain. Therefore, although *L. monocytogenes* has been implicated in several outbreaks, enteric pathogens, that are capable of colonizing animal intestine, are currently of the greatest concern for causing illness in fresh produce (Hoagland et al., 2018).

For many years, it was assumed that the leaf surface was a hostile environment for enteric pathogens, defining them as transient inhabitants. However, it is now assumed that enteric pathogens show a relative fitness in the phyllosphere which allows them to survive in the environment, including soil, water, organic fertilizers and crops (Lindow and Brandl, 2003; Brandl et al., 2013; Hoagland et al., 2018). Brandl (2013) suggested that plants might be valuable alternative hosts for enteric pathogens by providing a refuge after excretion from the animal intestinal tract. The hypothesis was that the ability of enteric pathogens such as *Salmonella* to colonize edible plants might be an effective survival strategy as it provides a direct route from its excretion in the environment back to its numerous herbivorous and omnivorous hosts.

Recent findings have shown that the survival of bacteria on FFV appears to be affected by many different factors including external ones such as agricultural practices and weather conditions (daily cycles of temperature, solar radiation and RH), and internal ones such as phyllosphere composition, plant species and leaf age (Rastogi et al., 2012; Wei et al., 2016; Truchado et al., 2019). Among these factors, the interaction between the phyllosphere and the environmental factors seems to be the most relevant.

5.1 Environmental factors

Factors such as temperature, humidity (both relative and absolute) and sunlight (ultraviolet (UV) light) exposure can inactivate or promote the survival of foodborne pathogens (Truchado et al., 2019). Increased temperature can increase the rate of microbial growth in soil and on the leaves while increased UV may be responsible for the decrease of potential pathogens (FDA/WHO, 2008). Most of the studies focused on the effects of weather factors on pathogen survival on different FFV use controlled environment chambers allowing the control of weather variables while avoiding other factors that may affect the survival of microbial pathogens.

Under a model approach system, several authors have demonstrated that the survival of human pathogens on plant surfaces is affected by RH. Devleesschauwer et al. (2017) reported that tomatoes that were incubated at higher humidity the week before harvest had lower postharvest *Salmonella* proliferation than tomatoes incubated at lower RH. López-Gálvez et al. (2018a) also observed that *Salmonella* Typhimurium survived better on baby lettuce plants at high RH while Brandl and Mandrell (2002) observed a decline of *Salmonella* Typhimurium on cilantro leaves under low RH.

As confirmed by Truchado et al. (2017), solar radiation is a very important stress factor for bacterial colonization and survival in the phyllosphere. Jacobs and Sundin (2001) evaluated the culturable bacterial population in peanuts demonstrating that solar UV-B radiation is an important environmental stress, especially for non-pigmented bacteria. Although little information is still available, studies performed in commercial growing fields have also demonstrated that solar radiation is an important abiotic factor affecting microbial decay in leafy greens during primary production (Whitman et al., 2004; Castro-Ibañez et al., 2015). However, a better understanding of the impact of different climatic conditions that occur during primary production, including solar radiation, temperature and RH, on the survival of foodborne pathogens is still needed to determine the risk and if possible to reduce or prevent microbial growth during production (Liu et al., 2013).

5.2 Pathogens-microbiota interactions

FFV harbour a large and diverse variety of microorganisms (e.g. bacteria, virus and protozoa) which include epiphytes and endophytes bacteria, plant pathogenic bacteria as well as human pathogens (Bulgarelli et al., 2013; Müller et al., 2016). Bacteria are the most common residents on the surface of FFV and the aerial parts of plants, the so-called phyllosphere (Vorholt, 2012). The composition of the phyllosphere has been estimated in about 10^8 cells/g of plant material (Hirano and Upper, 2000). The phyllosphere microbiota include a high diversity of bacteria of about 10^6 of taxes (Truchado et al., 2018). The phyla *Proteobacteria,* particularly *Alphaproteobacteria*, *Betaproteobacteria* and *Gammaproteobacteria*, as well as *Firmicutes*, *Bacteroidetes* and *Actinobacteria* are the most abundant bacteria species found within the phyllosphere of different products such as sprouts, fresh herbs and fruits (Bulgarelli et al., 2013; Leff and Fierer, 2013; Jackson et al., 2015). At the genus level, *Pseudomonas*, *Bacillus*, *Massilia*, *Arthrobacter*, *Flavobacterium*, *Acinetobacter*, *Bacillus* and *Pantoea* have been identified as part of the phyllosphere microbiota across different types of crops (Rastogi et al., 2012; Jackson et al., 2013; Leff and Fierer, 2013; Erlacher et al., 2014; Dees et al., 2015).

Although the literature on the topic is scarce, some studies have evaluated the interactions between the native epiphytic microbiota and foodborne pathogenic bacteria in the plant tissue. In some cases, the natural phyllosphere represents an added stress in the adaptation of the enteric pathogens. For example, Matos and Garland (2005) observed that *Pseudomonas fluorescens* isolated from the native microflora of sprouts reduced the growth of *Salmonella* on alfalfa sprouts (Matos and Garland, 2005). Similarly, Cooley et al. (2003) isolated several strains from Arabidopsis, including *Wausteria paucula* and *Enterobacter asburiae*, and evaluated their effects on the survival and growth of co-inoculated *E. coli* O157:H7 in the phyllosphere of growth-chamber-grown lettuces (Cooley et al., 2006). The obtained results showed that *Wausteria paucula* increased the growth of *E. coli* O157:H7 on the lettuce while *E. asburiae* inhibited it. The presence of *E. coli* O157:H7 also altered the diversity of the epiphytic spinach microbiota after harvest being affected by time and temperature of storage, which influenced in the establishment and fate of *E. coli* O157:H7 (Lopez Velasco et al., 2010). Later, Lopez-Velasco et al. (2012) demonstrated that strains isolated from the bacterial community of spinach, which belonging to the *Firmicutes* phylum and the *Enterobacteriaceae* family such as *Bacillus pumillus* and *Erwinia persicina* respectively, when co-cultured with *E. coli* O157:H7 inhibited growth in both *in vitro* and on detached spinach leaves. This finding highlights the importance of the composition of the phyllosphere bacterial community in the potential colonization and growth of enteric bacteria on the tissue, particularly on *E. coli* O157:H7 adaptation. Poza-Carrion et al. (2013) showed that the survival of viable populations of *S. enterica* on the leaves of edible plants varied depending on the presence of common native bacteria. They observed that native epiphytic bacteria such as *Pseudomonas fluorescens* and *Erwinia herbicola* (*Pantoea agglomerans*) increased the level of *Salmonella* inoculated on plants when compared with control plants. Additionally, *Salmonella enterica* and *E. coli* O157:H7 survival on the surface of tomatoes and lettuce have been shown to be affected by the composition of the bacterial community, which could also be associated with the crop genotype (Barak et al., 2011; Quilliam et al., 2012). All these studies suggested that the native microbiota composition of FFV seems to impact the survival and proliferation of enteric pathogens that may contaminate the products.

Poza-Carrion et al. (2013) suggested that environmental factors that affect the plant tissue, such as desiccation on the leaf, could influence the survival of enteric pathogenic bacteria due to a change in the epiphytic bacteria composition. In general, the composition, the structure and the proportions of the microbiota are in constant change due to intrinsic and extrinsic factors. Among all the extrinsic factors, plant development, agricultural practices and environmental factors (e.g. weather conditions and geographical location) have been shown to affect the plant phyllosphere (Zwielehner et al., 2008; Hunter et al., 2010; Jackson and Denney, 2011; Rastogi et al., 2012; Izhaki et al., 2013; Jense et al., 2013; Asakura et al., 2016; Redford et al., 2010).

Agricultural practices also play a role in the changes of the phyllosphere. Williams et al. (2013) observed that this type of irrigation system affected the bacterial community of lettuce, while water source did not show significant impact on the bacterial composition of the tomato fruit surface (Telias et al., 2011; Ottesen et al., 2016). Wei et al. (2016) reported differences in the composition of phyllosphere microbiota of strawberry plants, particularly the bacterial abundance, between strawberry plants cultivated in open field or under protection. A similar trend was obtained in the phyllosphere microbiota composition of red-pigmented baby leaf lettuce when cultivated uncovered and covered during the last 3 weeks before harvest (Truchado et al., 2017). Williams and Marco (2014) also described differences in the bacterial community composition between laboratory-acclimatized lettuces and lettuce plants collected from the field. The cultivation system can also cause changes that affect the bacterial populations on fresh produce. Koo et al. (2016) reported differences in the bacterial community composition of Tatsoi between two hydroponic cultivation systems, including the ebb-and-flow system and the trickle feed system modulated. A recent study has demonstrated that the family *Enterobacteriaceae* was reduced after irrigation with agricultural water constantly treated with a low residual concentration of ClO_2 (0.25 ppm) during the growing cycle of baby spinach (Truchado et al., 2018). This is remarkable because a large proportion of foodborne and pathogenic bacteria associated with fresh produce belong to these families. Therefore, more studies that evaluate if the changes of plant-associated bacterial communities can reduce the presence of foodborne pathogenic bacteria are needed to guarantee the safety of horticultural produce.

Concerning the environmental conditions, several studies have demonstrated that season, solar radiation and ambient temperature affected the phyllosphere-associated bacteria on fresh produce (Jackson and Denney, 2011; Rastogi et al., 2012; Dees et al., 2015). A recent study has demonstrated that global warming impacts the diversity and composition of the bacterial leaf microbiota of the herbaceous plant *Galium album* (Aydogan et al., 2018). These authors observed a reduction on the beneficial bacteria and an enhancement of potentially pathogenic bacteria in the phyllosphere of plants. They suggested that this aspect of the ecosystem can be a potential risk for pathogen transmission in agro-ecosystems soon (Aydogan et al., 2018). Overall, all these studies showed that the microbial ecology of fresh produce could be modified due to agricultural factors and climatic factors. However, the role of plant microbiota on the survival of foodborne pathogens and the population dynamics affecting the prevalence of pathogenic microorganisms is still an unresolved issue that will need more research.

In general, the composition of the phyllosphere-associated bacteria is evaluated using conventional culture-based methods. However, it has been reported that cultivation-dependent assays only represent between 1% and 10% of the true bacterial community (Pace, 1997; Rastogi et al., 2010). Recently, the use of molecular approaches and bioinformatics tools provide the opportunity to make significant progress in this area of research. For example, the use of high-throughput DNA sequencing allowed the analysis of the phyllosphere community on romaine lettuce inoculated or not with *E. coli* O157:H7, showing that the presence of pathogenic bacteria altered the microbial communities of the crop (Williams et al., 2013).

6 Conclusions

This chapter describes information on the safety of FFV that highlights the reasons for increasing trends of foodborne illnesses associated with fresh produce. Reflections have been provided regarding the most recent outbreaks linked to leafy greens including outbreaks in the United States and Europe. The research related to the source attribution and the potential risk factors associated with the production, processing and storage of FFV has also been described.

The main part of this chapter includes a comprehensive revision of the various sources and pathways through which bacteria contaminate fresh produce. Recently, several growers' associations and competent authorities in the United States and Europe have put their efforts into the development and publication of guidelines and new regulations, where the major risk factors associated with the fresh produce supply chain are revised, and intervention strategies are suggested to mitigate these factors. This chapter revises the most recent publications and summarizes the top five risk factors linked to FFV including environmental factors, water, organic fertilizers, hygiene conditions during production and processing.

The third part of the chapter refers to the routes of contamination that pathogens use to colonize and survive in FFV. The literature on this topic is relatively scarce, but several authors have identified the major paths used by pathogens for attachment and internalization. A summary of the most relevant findings are presented in this chapter.

The last part of the chapter focuses on the factors affecting the survival-inactivation of pathogens in horticultural products. Pathogen die-off rate on contaminated produce is affected by the characteristics of fresh produce, agricultural practices, climatic conditions and geographical locations. It has been reported that persistence of pathogenic and indicator microorganisms in leafy greens is predominantly limited by water availability, strain source and localization within the plant. However, there is still a need for studies able to isolate the effect of water, weather, strain, produce commodity and location on the pathogen die-off rate.

7 Future trends

The development of new advanced laboratory methods based on molecular techniques represents a great step forward for the outbreak traceability as they represent tools that increase food safety and reduce consumer concerns. However, there are still many knowledge gaps regarding the risk factors associated with FFV and the routes of contamination. Bacterial internalization has been defined as a bacterial strategy to survive in hostile environments. However, in most of the cases, studies focused on the ability of pathogens to become internalized in the plant tissue have generated controversial results. The evaluation of potential pathways used by enteric pathogens to colonize the roots and to reach the edible parts of the plants have been performed under well-controlled conditions which are far from representing real conditions in the field, and results are difficult to extrapolate. However, pilot studies under industrial conditions have demonstrated the ability of enteric pathogens to penetrate the plant tissue via stomata. Therefore, more research is needed to investigate the strategies to avoid internalization during production and processing, particularly during washing of fresh produce.

The ability of foodborne pathogens to colonize and survive in the plant tissue seems to be affected by many different factors including agricultural practices, weather conditions and the phyllosphere composition. All these factors seem to be interconnected because external factors such as climatic conditions and agricultural practices affect both the fate of pathogens and also the phyllosphere, but the changes in the natural microbiota also affect the ability of pathogens to survive. This is why the surface of plant tissue has been defined as a dynamic environment, in constant change, mostly generated by the interactions between all these factors. Understanding the significance of these interactions is important to gain insight into the survival of pathogens to develop new and more efficient prevention and intervention strategies that allow reduction of foodborne outbreaks associated with FFV.

8 Where to look for further information

Center for Produce Safety: https://www.centerforproducesafety.org/
Leafy Greens Marketing Association: https://lgma.ca.gov/food-safety-practices/
HUPLANTcontrol: https://huplantcontrol.igzev.de/

9 References

Alegbeleye, O. O., Singleton, I. and Sant'Ana, A. S. 2018. Sources and contamination routes of microbial pathogens to fresh produce during field cultivation: a review. *Food Microbiol.* 73, 177–208. doi:10.1016/j.fm.2018.01.003.

Allard, M. W., Strain, E., Melka, D., Bunning, K., Musser, S. M., Brown, E. W. and Timme, R. 2016. The practical value of food pathogen traceability through building a whole genome sequencing network and database. *J. Clin. Microbiol.* 54(8), 1975–83. doi:10.1128/JCM.00081-16.

Allende, A., Selma, M. V., López-Gálvez, F., Villaescusa, R. and Gil, M. I. 2008. Impact of wash water quality on sensory and microbial quality, including *Escherichia coli* cross-contamination of fresh-cut escarole. *J. Food Prot.* 71(12), 2514–8. doi:10.4315/0362-028X-71.12.2514.

Allende, A., Truchado, P., Lindqvist, R. and Jacxsens, L. 2018a. Quantitative microbial exposure modelling as a tool to evaluate the impact of contamination level of surface irrigation water and seasonality on fecal hygiene indicator *E. coli* in leafy green production. *Food Microbiol.* 75, 82–9. doi:10.1016/j.fm.2018.01.016.

Allende, A., Datta, A. R., Smith, W. A., Adonis, R., MacKay, A. and Adell, A. D. 2018b. Implications of new legislation (US FSMA) and guidelines (EC) on the establishment of management systems for agricultural water. *Food Microbiol.* 75, 119–25.

American Vegetable Grower. 2018. FDA says leafy green industry must improve safety. Miller, C. (Ed.). Available at: https://www.growingproduce.com/vegetables/fda-says-leafy-green-industry-must-improve-safety/.

Asakura, H., Tachibana, M., Taguchi, M., Hiroi, T., Kurazono, H., Makino, S. I., Kasuga, F. and Igimi, S. 2016. Seasonal and growth-dependent dynamics of bacterial community in radish sprouts. *J. Food Saf.* 36(3), 392–401. doi:10.1111/jfs.12256.

Aydogan, E. L., Moser, G., Müller, C., Kämpfer, P. and Glaeser, S. P. 2018. Long-term warming shifts the composition of bacterial communities in the phyllosphere of *Galium album* in a permanent grassland field-experiment. *Front. Microbiol.* 9, 144. doi:10.3389/fmicb.2018.00144.

Barak, J. D., Kramer, L. C. and Hao, L. Y. 2011. Colonization of tomato plants by *Salmonella enterica* is cultivar dependent, and type 1 trichomes are preferred colonization sites. *Appl. Environ. Microbiol.* 77(2), 498–504. doi:10.1128/AEM.01661-10.

Berger, C. N., Sodha, S. V., Shaw, R. K., Griffin, P. M., Pink, D., Hand, P. and Frankel, G. 2010. Fresh fruit and vegetables as vehicles for the transmission of human pathogens. *Environ. Microbiol.* 12(9), 2385–97. doi:10.1111/j.1462-2920.2010.02297.x.

Bernstein, N., Sela, S. and Neder-Lavoni, S. 2007. Assessment of contamination potential of lettuce by *Salmonella enterica* serovar Newport added to the plant growing substrate. *J. Food Prot.* 70(7), 1717–22. doi:10.4315/0362-028X-70.7.1717.

Brandl, M. T. and Amundson, R. 2008. Leaf age as a risk factor in contamination of lettuce with *Escherichia coli* O157:H7 and *Salmonella enterica*. *Appl. Environ. Microbiol.* 74(8), 2298–306. doi:10.1128/AEM.02459-07.

Brandl, M. T. and Mandrell, R. E. 2002. Fitness of *Salmonella enterica* serovar Thompson in the cilantro phyllosphere. *Appl. Environ. Microbiol.* 68(7), 3614–21. doi:10.1128/AEM.68.7.3614-3621.2002.

Brandl, M. T., Cox, C. E. and Teplitski, M. 2013. *Salmonella* interactions with plants and their associated microbiota. *Phytopathology* 103(4), 316–25. doi:10.1094/PHYTO-11-12-0295-RVW.

Bulgarelli, D., Schlaeppi, K., Spaepen, S., Ver Loren van Themaat, E. and Schulze-Lefert, P. 2013. Structure and functions of the bacterial microbiota of plants. *Annu. Rev. Plant Biol.* 64, 807–38. doi:10.1146/annurev-arplant-050312-120106.

Castro-Ibañez, I., Gil, M. I., Tudela, J. A. and Allende, A. 2015. Microbial safety considerations of flooding in primary production of leafy greens: a case study. *Food Res. Int.* 68, 62–9. doi:10.1016/j.foodres.2014.05.065.

CDC, Centers for Disease Control and Prevention. 2011. Multistate outbreak of listeriosis linked to whole cantaloupes from Jensen Farms, Colorado. Available at: www.cdc.gov/listeria/outbreaks/cantaloupes-jensen-farms/index.html (accessed on 9 April 2019).

CDC, Centers for Disease Control and Prevention. 2018. Outbreak of *E. coli* infections linked to romaine lettuce. Available at: https://www.cdc.gov/ecoli/2018/o157h7-11-18/index.html (accessed December 2018).

Codex (Codex Alimentarius Commission). 1999. Principles and guidelines for the conduct of Microbiological Risk Assessment. CAC/GL-30. Food and Agriculture Organization of the United Nations, Rome, pp. 1–6.

Coetzer, E. 2006. Microbiological risk in produce from the field to packing. In: Hoboken, N. J. (Ed.), *In Microbial Hazard Identification in Fresh Fruit and Vegetables*. John Wiley & Sons, Inc., Hoboken, NJ, pp. 73–93.

Cooley, M. B., Miller, W. G. and Mandrell, R. E. 2003. Colonization of *Arabidopsis thaliana* with *Salmonella enterica* and enterohemorrhagic *Escherichia coli* O157:H7 and competition by *Enterobacter asburiae*. *Appl. Environ. Microbiol.* 69(8), 4915–26. doi:10.1128/AEM.69.8.4915-4926.2003.

Cooley, M. B., Chao, D. and Mandrell, R. E. 2006. *Escherichia coli* O157:H7 survival and growth on lettuce is altered by the presence of epiphytic bacteria. *J. Food Prot.* 69, 2329–35.

De Keuckelaere, A., Jacxsens, L., Amoah, Ph., Medema, G. J., McClure, P., Jaykus, L. A. and Uyttendaele, M. 2015. Zero risk does not exist: lessons learnt from microbiological risk assessment related to use of water and safety of fresh produce. *Comp. Rev. Food Sci. Food Saf.* 14(4), 387–410. doi:10.1111/1541-4337.12140.

Dees, M. W., Lysøe, E., Nordskog, B. and Brurberg, M. B. 2015. Bacterial communities associated with surfaces of leafy greens: shift in composition and decrease in richness over time. *Appl. Environ. Microbiol.* 81(4), 1530–9. doi:10.1128/AEM.03470-14.

den Besten, H. M. W., Amézquita, A., Bover-Cid, S., Dagnas, S., Ellouze, M., Guillou, S., Nychas, G., O'Mahony, C., Pérez-Rodriguez, F. and Membré, J. M. 2018. Next generation of microbiological risk assessment: potential of omics data for exposure assessment. *Int. J. Food Microbiol.* 287, 18–27. doi:10.1016/j.ijfoodmicro.2017.10.006.

Devleesschauwer, B., Marvasi, M., Giurcanu, M. C., Hochmuth, G. J., Speybroeck, N., Havelaar, A. H. and Teplitski, M. 2017. High relative humidity pre-harvest reduces post-harvest proliferation of *Salmonella* in tomatoes. *Food Microbiol.* 66, 55–63. doi:10.1016/j.fm.2017.04.003.

Dowe, M. J., Jackson, E. D., Mori, J. G. and Bell, C. R. 1997. *Listeria monocytogenes* survival in soil and incidence in agricultural soils. *J. Food Prot.* 60(10), 1201–7. doi:10.4315/0362-028X-60.10.1201.

EC, European Commission. 2017. European Commission Notice No. 2017/C 163/01 guidance document on addressing microbiological risks in fresh fruit and vegetables at primary production through good hygiene. Available at: http://eur-lex.europa.eu/legalcontent/EN/TXT/PDF/?uri¼CELEX:52017XC0523(03)&from¼EN (accessed October 2018).

EFSA, European Food Safety Agency. 2013. EFSA Panel on Biological Hazards (BIOHAZ). scientific opinion on the risk posed by pathogens in food of non-animal origin. Part 1 (outbreak data analysis and risk ranking of food/pathogen combinations). *EFSA J.* 11(3025). Available at: www.efsa.europa.eu/efsajournal.

EFSA, European Food Safety Agency. 2018. Urgent scientific and technical assistance to provide recommendations for sampling and testing in the processing plants of frozen vegetables aiming at detecting *Listeria monocytogenes*. *EFSA Journal* 15, 1445E-86E. doi:10.2903/sp.efsa.2018.EN-1445.

Erlacher, A., Cardinale, M., Grosch, R., Grube, M. and Berg, G. 2014. The impact of the pathogen *Rhizoctonia solani* and its beneficial counterpart *Bacillus amyloliquefaciens* on the indigenous lettuce microbiome. *Front. Microbiol.* 5, 175. doi:10.3389/fmicb.2014.00175.

FAO/WHO, Food and Agriculture Organization of the United Nations/World Health Organization. 2008. Microbiological hazards in fresh leafy vegetables and herbs. Meeting report. Microbial risk assessment series No. 14. Rome, pp. 151.

FDA. 2018a. FSMA Final Rule on Produce Safety. Standards for the growing, harvesting, packing, and holding of produce for human consumption. Available at: https://www.fda.gov/food/guidanceregulation/fsma/ucm334114.htm.

FDA. 2018b. Statement from FDA Commissioner Scott Gottlieb, M.D., on developments in the romaine outbreak investigation, recent outbreaks and the use of modern tools to advance food safety. Available at: https://www.fda.gov/NewsEvents/Newsroom/PressAnnouncements/ucm612187.htm (accessed December 2018).

Franz, E., Visser, A. A., Van Diepeningen, A. D., Klerks, M. M., Termorshuizen, A. J. and van Bruggen, A. H. 2007. Quantification of contamination of lettuce by GFP-expressing *Escherichia coli* O157:H7 and *Salmonella enterica* serovar Typhimurium. *Food Microbiol.* 24(1), 106–12. doi:10.1016/j.fm.2006.03.002.

Franz, E., Delaquis, P., Morabito, S., Beutin, L., Gobius, K., Rasko, D. A., Bono, J., French, N., Osek, J., Lindstedt, B. A., et al. 2014. Exploiting the explosion of information associated with whole genome sequencing to tackle Shiga toxin-producing *Escherichia coli* (STEC) in global food production systems. *Int. J. Food Microbiol.* 187, 57–72. doi:10.1016/j.ijfoodmicro.2014.07.002.

Franz, E., Mughini Gras, L. M. and Dallman, T. 2016. Significance of whole genome sequencing for surveillance, source attribution and microbial risk assessment of foodborne pathogens. *Curr. Opin. Food Sci.* 8, 74–9. doi:10.1016/j.cofs.2016.04.004.

Gil, M. I. and Selma, M. V. 2006. Overview of hazards in fresh-cut produce production: control and management of food safety hazards. In: Hoboken, N. J. (Ed.), *In Microbial Hazard Identification in Fresh Fruit and Vegetables*. John Wiley & Sons, Inc., Hoboken, NJ, pp. 155–20. Chapter 6.

Gil, M. I., Selma, M. V., López-Gálvez, F. and Allende, A. 2009. Fresh-cut product sanitation and wash water disinfection: problems and solutions. *Int. J. Food Microbiol.* 134(1–2), 37–45. doi:10.1016/j.ijfoodmicro.2009.05.021.

Gombas, D., Luo, Y., Brennan, J., Shergill, G., Petran, R., Walsh, R., Hau, H., Khurana, K., Zomorodi, B., Rosen, J., et al. 2017. Guidelines to validate control of cross-contamination during washing of fresh-cut leafy vegetables. *J. Food Prot.* 80(2), 312–30. doi:10.4315/0362-028X.JFP-16-258.

Gómez-López, V. M., Marín, A., Allende, A., Beuchat, L. R. and Gil, M. I. 2013. Postharvest handling conditions affect internalization of *Salmonella* in baby spinach during washing. *J. Food Protect.* 76, 1145–51.

Greig, J. D., Todd, E. C. D., Bartleson, C. A. and Michaels, B. S. 2007. Outbreaks where food workers have been implicated in the spread of foodborne disease. Part 1. Description of the problem, methods, and agents involved. *J. Food Prot.* 70(7), 1752–61. doi:10.4315/0362-028X-70.7.1752.

Haddad, N., Johnson, N., Kathariou, S., Métris, A., Phister, T., Pielaat, A., Tassou, C., Wells-Bennik, M. H. J. and Zwietering, M. H. J. 2018. Next generation microbiological risk assessment–potential of omics data for hazard characterization. *Int. J. Food Microbiol.* 287, 29–39.

Hirano, S. S. and Upper, C. D. 2000. Bacteria in the leaf ecosystem with emphasis on *Pseudomonas syringae* a pathogen, ice nucleus, and epiphyte. *Microbiol. Mol. Biol. Re.* 64(3), 624–53. doi:10.1128/MMBR.64.3.624-653.2000.

Hirneisen, K. A., Sharma, M. and Kniel, K. E. 2012. Human enteric pathogen internalization by root uptake into food crops. *Foodborne Pathog. Dis.* 9(5), 396–405. doi:10.1089/fpd.2011.1044.

Hoagland, L., Ximenes, E., Ku, S. and Ladisch, M. 2018. Foodborne pathogens in horticultural production systems: ecology and mitigation. *Sci. Hortic.* 236, 192–206. doi:10.1016/j.scienta.2018.03.040.

Holvoet, K., Jacxsens, L., Sampers, I. and Uyttendaele, M. 2012. Insight into the prevalence and distribution of microbial contamination to evaluate water management in the fresh produce processing industry. *J. Food Prot.* 75(4), 671–81. doi:10.4315/0362-028X.JFP-11-175.

Hora, R., Warriner, K., Shelp, B. J. and Griffiths, M. W. 2005. Internalization of *Escherichia coli* O157:H7 following biological and mechanical disruption of growing spinach plants. *J. Food Prot.* 68(12), 2506–9. doi:10.4315/0362-028X-68.12.2506.

Hunter, P. J., Hand, P., Pink, D., Whipps, J. M. and Bending, G. D. 2010. Both leaf properties and microbe-microbe interactions influence within-species variation in bacterial population diversity and structure in the lettuce (*Lactuca species*) phyllosphere. *Appl. Environ. Microbiol.* 76(24), 8117–25. doi:10.1128/AEM.01321-10.

Ingham, S. C., Losinski, J. A., Andrews, M. P., Breuer, J. E., Breuer, J. R., Wood, T. M. and Wright, T. H. 2004. *Escherichia coli* contamination of vegetables grown in soils fertilized with non-composted bovine manure: garden-scale studies. *App. Environ. Microbiol.* 70(11), 6420–7. doi:10.1128/AEM.70.11.6420-6427.2004.

Islam, M., Doyle, M. P., Phatak, S. C., Millner, P. and Jiang, X. 2004a. Persistence of enterohemorrhagic *Escherichia coli* O157:H7 in soil and on leaf lettuce and parsley grown in fields treated with contaminated manure composts or irrigation water. *J. Food Prot.* 67(7), 1365–70. doi:10.4315/0362-028X-67.7.1365.

Islam, M., Morgan, J., Doyle, M. P., Phatak, S. C., Millner, P. and Jiang, X. 2004b. Persistence of *Salmonella enterica* serovar Typhimurium on lettuce and parsley in soils on which they were grown in fields treated with contaminated manure composts irrigation water. *Foodborne Path. Dis.* 1, 27–35.

Islam, M., Doyle, M. P., Phatak, S. C., Millner, P. and Jiang, X. 2005. Survival of *Escherichia coli* O157:H7 in soil and on carrots and onions grown in fields treated with contaminated manure composts or irrigation water. *Food Microbiol.* 22(1), 63–70. doi:10.1016/j.fm.2004.04.007.

Izhaki, I., Fridman, S., Gerchman, Y. and Halpern, M. 2013. Variability of bacterial community composition on leaves between and within plant species. *Curr. Microbiol.* 66(3), 227–35. doi:10.1007/s00284-012-0261-x.

Jablasone, J., Warriner, K. and Griffiths, M. 2005. Interactions of *Escherichia coli* O157:H7, *Salmonella Typhimurium* and *Listeria monocytogenes* plants cultivated in a gnotobiotic system. *Int. J. Food Microbiol.* 99(1), 7–18. doi:10.1016/j.ijfoodmicro.2004.06.011.

Jackson, C. R. and Denney, W. C. 2011. Annual and seasonal variation in the phyllosphere bacterial community associated with leaves of the southern magnolia (*Magnolia grandiflora*). *Microb. Ecol.* 61(1), 113–22. doi:10.1007/s00248-010-9742-2.

Jackson, C. R., Randolph, K. C., Osborn, S. L. and Tyler, H. L. 2013. Culture dependent and independent analysis of bacterial communities associated with commercial salad leaf vegetables. *BMC Microbiol.* 13, 274. doi:10.1186/1471-2180-13-274.

Jackson, C., Stone, B. and Tyler, H. 2015. Emerging perspectives on the natural microbiome of fresh produce vegetables. *Agriculture* 5(2), 170–87. doi:10.3390/agriculture5020170.

Jacobs, J. L. and Sundin, G. W. 2001. Effect of solar UV-B radiation on a phyllosphere bacterial community. *Appl. Environ. Microbiol.* 67(12), 5488–96. doi:10.1128/AEM.67.12.5488-5496.2001.

Jay-Russell, M. T. 2013. What is the risk from wild animals in food-borne pathogen contamination of plants? *CAB Reviews* 8(40). Available at: http://www.cabi.org/cabreviews. doi:10.1079/PAVSNNR20138040.

Jensen, B., Knudsen, I. M. B., Andersen, B., Nielsen, K. F., Thranes, U., Jensen, D. F. and Larsen, J. 2013. Characterization of microbial communities and fungal metabolites on field grown strawberries from organic and conventional production. *Int. J. Food Microbiol.* 160(3), 313–22. doi:10.1016/j.ijfoodmicro.2012.11.005.

Kisluk, G. and Yaron, S. 2012. Presence and persistence of *Salmonella enterica* serotype Typhimurium in the phyllosphere and rhizosphere of spray-irrigated parsley. *Appl. Environ. Microbiol.* 78(11), 4030–6. doi:10.1128/AEM.00087-12.

Koo, O. K., Kim, H., Kim, H. J., Baker, C. A. and Ricke, S. C. 2016. Bacterial community analysis of Tatsoi cultivated by hydroponics. *J. Environ. Sci. Health* 51(7), 490–6. doi:10.1080/03601234.2016.1159462.

Kroupitski, Y., Golberg, D., Belausov, E., Pinto, R., Swartzberg, D., Granot, D. and Sela, S. 2009. Internalization of *Salmonella enterica* in leaves is induced by light and involves chemotaxis and penetration through open stomata. *Appl. Environ. Microbiol.* 75(19), 6076–86. doi:10.1128/AEM.01084-09.

Leff, J. W. and Fierer, N. 2013. Bacterial communities associated with the surfaces of fresh fruits and vegetables. *PLOS ONE* 8(3), e59310. doi:10.1371/journal.pone.0059310.

Lindow, S. E. and Brandl, M. T. 2003. Microbiology of the phyllosphere. *Appl. Environ. Microbiol.* 69(4), 1875–83. doi:10.1128/AEM.69.4.1875-1883.2003.

Liu, N. T., Lefcourt, A. M., Nou, X., Shelton, D. R., Zhang, G. and Lo, Y. M. 2013. Native microflora in fresh-cut produce processing plants and their potentials for biofilm formation. *J. Food Prot.* 76(5), 827–32. doi:10.4315/0362-028X.JFP-12-433.

Liu, H., Whitehouse, C. A. and Li, B. 2018. Presence and persistence of *Salmonella* in water: the impact on microbial quality of water and food safety. *Front. Public. Health* 6(159), 159. doi:10.3389/fpubh.2018.00159.

López-Gálvez, F., Allende, A., Selma, M. V. and Gil, M. I. 2009. Prevention of *Escherichia coli* cross-contamination by different commercial sanitizers during washing of fresh-cut lettuce. *Int. J. Food Microbiol.* 133(1–2), 167–71. doi:10.1016/j.ijfoodmicro.2009.05.017.

López-Gálvez, F., Sampers, I., Gil, M. I. and Allende, A. 2017. Modelling of *E. coli* inactivation by chlorine dioxide in irrigation water. *Agric. Water Manag.* 192, 98–102. doi:10.1016/j.agwat.2017.07.001.

López-Gálvez, F., Gil, M. I. and Allende, A. 2018a. Impact of relative humidity, inoculum carrier and size, and native microbiota on *Salmonella* ser. Typhimurium survival in baby lettuce. *Food Microbiol.* 70, 155–61. doi:10.1016/j.fm.2017.09.014.

López-Gálvez, F., Gil, M. I., Meireles, A., Truchado, P. and Allende, A. 2018b. Demonstration tests of irrigation water disinfection with chlorine dioxide in open field cultivation of baby spinach. *J. Sci. Food Agric.* 98(8), 2973–80. doi:10.1002/jsfa.8794.

Lopez-Velasco, G., Davis, M., Boyer, R. R., Williams, R. C. and Ponder, M. A. 2010. Alterations of the phylloepiphytic bacterial community associated with interactions of *Escherichia coli* O157:H7 during storage of packaged spinach at refrigeration temperatures. *Food Microbiol.* 27(4), 476–86. doi:10.1016/j.fm.2009.12.010.

Lopez-Velasco, G., Tydings, H. A., Boyer, R. R., Falkinham, J. O. and Ponder, M. A. 2012. Characterization of interactions between *Escherichia coli* O157:H7 with epiphytic bacteria *in vitro* and on spinach leaf surfaces. *Int. J. Food Microbiol.* 153(3), 351–7. doi:10.1016/j.ijfoodmicro.2011.11.026.

Matos, A. and Garland, J. L. 2005. Effects of community versus single strain inoculants on the biocontrol of *Salmonella* and microbial community dynamics in alfalfa sprouts. *J. Food Prot.* 68(1), 40–8. doi:10.4315/0362-028X-68.1.40.

Matthews, K. 2014. *The Produce Contamination Problem*. Academic Press.

Monaghan, J. M. 2006. United Kingdom and European approach to fresh produce food safety and security. *HortTechnology* 16, 559–62. doi:10.21273/HORTTECH.16.4.0559.

Morse, S. S. 1995. Factors in the emergence of infectious-diseases. *Emerg. Infect. Dis.* 1(1), 7–15. doi:10.3201/eid0101.950102.

Müller, D. B., Schubert, O. T., Röst, H., Aebersold, R. and Vorholt, J. A. 2016. Systems-level proteomics of two ubiquitous leaf commensals reveals complementary adaptive traits for phyllosphere colonization. *Mol. Cell Proteomics* 15(10), 3256–69. doi:10.1074/mcp.M116.058164.

Nguyen-the, C. and Carlin, F. 1994 The microbiology of minimally processed fresh fruits and vegetables. *Crit. Rev. Food Sci. Nut.* 34(4), 371–401. doi:10.1080/10408399409527668.

Nielsen, E. M., Skov, M. N., Madsen, J. J., Lodal, J., Jespersen, J. B. and Baggesen, D. L. 2004. Verocytotoxin-producing *Escherichia coli* in wild birds and rodents in close proximity to farms. *Appl. Environ. Microbiol.* 70(11), 6944–7. doi:10.1128/AEM.70.11.6944-6947.2004.

Njage, P. M. K., Leekitcharoenphon, P. and Hald, T. 2019. Improving hazard characterization in microbial risk assessment using next generation sequencing data and machine learning: predicting clinical outcomes in Shiga toxigenic *Escherichia coli*. *Int. J. Food Microbiol.* 292, 72–82. doi:10.1016/j.ijfoodmicro.2018.11.016.

Ottesen, A., Telias, A., White, J. R., Newell, M. J., Pahl, D., Brown, E. W., Musser, S.. and Walsh, C. 2016. Bacteria of tomatoes managed with well water and pond water: impact of agricultural water sources on carposphere microbiota. *Int. J. Environ. Agric. Res.* 2, 7–14.

Pace, N. R. 1997. A molecular view of microbial diversity and the biosphere. *Science* 276(5313), 734–40. doi:10.1126/science.276.5313.734.

Poza-Carrion, C., Suslow, T. V. and Lindow, S. E. 2013. Resident bacteria on leaves enhance survival of immigrant cells of *Salmonella enterica*. *Phytopathology* 103(4), 341–51. doi:10.1094/PHYTO-09-12-0221-FI.

Quilliam, R. S., Williams, A. P. and Jones, D. L. 2012. Lettuce cultivar mediates both phyllosphere and rhizosphere activity of *Escherichia coli* O157:H7. *PLOS ONE* 7(3), e33842.

Rantsiou, K., Kathariou, S., Winkler, A., Skandamis, P., Saint-Cyr, M. J., Rouzeau-Szynalski, K. and Amézquita, A. 2018. Next generation microbiological risk assessment: opportunities of whole genome sequencing (WGS) for foodborne pathogen surveillance, source tracking and risk assessment. *Int. J. Food Microbiol.* 287, 3–9. doi:10.1016/j.ijfoodmicro.2017.11.007.

RASFF. 2016. Rapid alert system for food and feed. Directorate-General for Health and Food Safety and European Commission, Brussels, Belgium. Available at: https://ec.europa.eu/food/sites/food/files/safety/docs/rasff_annual_report_2016.pdf.

Rastogi, G., Tech, J. J., Coaker, G. L. and Leveau, J. H. 2010. A PCR-based toolbox for the culture-independent quantification of total bacterial abundances in plant environments. *J. Microbiol. Methods* 83(2), 127–32. doi:10.1016/j.mimet.2010.08.006.

Rastogi, G., Sbodio, A., Tech, J. J., Suslow, T. V., Coaker, G. L. and Leveau, J. H. J. 2012. Leaf microbiota in an agroecosystem: spatiotemporal variation in bacterial community composition on field-grown lettuce. *ISME J.* 6(10), 1812–22. doi:10.1038/ismej.2012.32.

Redford, A. J., Bowers, R. M., Knight, R., Linhart, Y. and Fierer, N. 2010. The ecology of the phyllosphere: geographic and phylogenetic variability in the distribution of bacteria on tree leaves. *Environ. Microbiol.* 12(11), 2885–93. doi:10.1111/j.1462-2920.2010.02258.x.

Ronholm, J., Nasheri, N., Petronella, N. and Pagotto, F. 2016. Navigating microbiological food safety in the era of whole-genome sequencing. *Clin. Microbiol. Rev.* 29(4), 837–57. doi:10.1128/CMR.00056-16.

Steele, M. and Odumeru, J. 2004. Irrigation water as source of foodborne pathogens on fruit and vegetables. *J. Food Prot.* 67, 2839–49.

Strawn, L. K., Fortes, E. D., Bihn, E. A., Nightingale, K. K., Grohn, Y. T., Worobo, R. W., Wiedmann, M. and Bergholz, P. W. 2013. Landscape and meteorological factors affecting prevalence of three food-borne pathogens in fruit and vegetable farms. *Appl. Environ. Microbiol.* 79(2), 588–600. doi:10.1128/AEM.02491-12.

Suslow, T. V., Oria, M. P., Beuchat, L. R., Garrett, E. H., Parish, M. E., Harris, L. J., Farber, J. N. and Busta, F. F. 2003. Production practices as risk factors in microbial food safety of fresh and fresh-cut produce. *Compr. Rev. Food Sci. Food Saf.* 2(s1), 38–77. doi:10.1111/j.1541-4337.2003.tb00030.x.

Tang, J. W. 2009. The effect of environmental parameters on the survival of airborne infectious agents. *J. R. Soc. Interface* 6, S737–46. doi:10.1098/rsif.2009.0227.focus.

Telias, A., White, J. R., Pahl, D. M., Ottesen, A. R. and Walsh, C. S. 2011. Bacterial community diversity and variation in spray water sources and the tomato fruit surface. *BMC Microbiol.* 11, 81. doi:10.1186/1471-2180-11-81.

Truchado, P., Gil, M. I., Reboleiro, P., Rodelas, B. and Allende, A. 2017. Impact of solar radiation exposure on phyllosphere bacterial community of red-pigmented baby leaf lettuce. *Food Microbiol.* 66, 77–85. doi:10.1016/j.fm.2017.03.018.

Truchado, P., Gil, M. I., Suslow, T. and Allende, A. 2018. Impact of chlorine dioxide disinfection of irrigation water on the epiphytic bacterial community of baby spinach and underlying soil. *PLOS ONE* 13(7), e0199291. doi:10.1371/journal.pone.0199291.

Truchado, P., Gil, M. I., Moreno-Candel, M. and Allende, A. 2019. Impact of weather conditions, leaf age and irrigation water disinfection on the major epiphytic bacterial genera of baby spinach grown in an open field. *Food Microbiol.* 78, 46–52. doi:10.1016/j.fm.2018.09.015.

Uyttendaele, M., Jaykus, L. A., Amoah, P., Chiodini, A., Cunliffe, D., Jacxsens, L., Holvoet, K., Korsten, L., Lau, M., McClure, P., et al. 2015. Microbial hazards in irrigation water: standards, norms, and testing to manage use of water in fresh produce primary production. *Compr. Rev. Food Sci. Food Saf.* 14(4), 336–56. doi:10.1111/1541-4337.12133.

Velani, S. and Roberts, D. 1991. *Listeria monocytogenes* and other *Listeria* spp. in prepacked mixed salads and individual salad ingredients, PHLS microbiol. *Digestion* 8, 21–2.

Vorholt, J. A. 2012. Microbial life in the phyllosphere. *Nat. Rev. Microbiol.* 10(12), 828–40. doi:10.1038/nrmicro2910.

Wei, F., Hu, X. and Xu, X. 2016. Dispersal of *Bacillus subtilis* and its effect on strawberry phyllosphere microbiota under open field and protection conditions. *Sci. Rep.* 6, 22611. doi:10.1038/srep22611.

Whitman, R. L., Nevers, M. B., Korinek, G. C. and Byappanahalli, M. N. 2004. Solar and temporal effects on *Escherichia coli* concentration at a Lake Michigan swimming beach. *Appl. Environ. Microbiol.* 70(7), 4276–85. doi:10.1128/AEM.70.7.4276-4285.2004.

Williams, T. R. and Marco, M. L. 2014. Phyllosphere microbiota composition and microbial community transplantation on lettuce plants grown indoors. *mBio* 5(4), e01564-14. doi:10.1128/mBio.01564-14.

Williams, T. R., Moyne, A. L., Harris, L. J. and Marco, M. L. 2013. Season, irrigation, leaf age, and *Escherichia coli* inoculation influence the bacterial diversity in the lettuce phyllosphere. *PLOS ONE* 8(7), e68642. doi:10.1371/journal.pone.0068642.

Wright, K. M., Crozier, L., Marshall, J., Merget, B., Holmes, A. and Holden, N. J. 2017. Differences in internalization and growth of *Escherichia coli* O157:H7 within the apoplast of edible plants, spinach and lettuce, compared with the model species *Nicotiana benthamiana*. *Microb. Biotechnol.* 10(3), 555–69. doi:10.1111/1751-7915.12596.

Zwielehner, J., Handschur, M., Michaelsen, A., Irez, S., Demel, M., Denner, E. B. and Haslberger, A. G. 2008. DGGE and real-time PCR analysis of lactic acid bacteria in bacterial communities of the phyllosphere of lettuce. *Mol. Nutr. Food Res.* 52(5), 614–23. doi:10.1002/mnfr.200700158.

Chapter 7

Advances in postharvest sanitizing regimes for horticultural produce

Joshua B. Gurtler, Xuetong Fan, Tony Jin and Brendan A. Niemira, USDA-ARS, USA

1 Introduction

There is increased demand for and greater sales of fresh and fresh-cut fruits and vegetables (Heaton and Jones, 2008; Olaimat and Holley, 2012; Baselice et al., 2017). The demand is driven from a desire for healthier eating as well as quick and convenient food, as a result of demanding lifestyles, increased numbers of spouses working outside the home, as well as an increased number of one- or two-person homes. These factors decrease the amount of food preparation done at home, which lead to the purchase of fresh-cut produce. However, fresh produce has been known to be contaminated with many foodborne pathogens, which have resulted in numerous foodborne illness outbreaks (Van Haute et al., 2015; Olmez and Kretzchmar, 2009). The processing, cutting, slicing, chopping, and preparation of fresh produce has caused contamination leading to foodborne illnesses, which affect (in total) over 48 million people per year in the United States with a cost of over $50 billion (Bermudez-Aguirre and Barbosa-Canovas, 2013). Overall, a large

http://dx.doi.org/10.19103/AS.2019.0055.09
Published by Burleigh Dodds Science Publishing Limited, 2020.

number of the illnesses stem from the consumption of contaminated fresh produce (Callejon et al., 2015; Slayton et al., 2013). Outbreaks have impacted consumer confidence in eating certain types of fresh produce, especially raw sprouts and leafy greens.

Of all types of fresh produce involved in foodborne illnesses internationally, leafy greens have been implicated in numerous outbreaks and have been classified as the fresh produce commodity most likely to be involved in outbreaks (FAO/WHO, 2008; CDC et al., 2013). Fresh-cut lettuce, as an example of leafy greens, has often been associated with foodborne illness outbreaks (Doyle and Erickson, 2008) including 110 outbreaks and 2343 illnesses between 2010 and 2017 (NORS, 2019). More broadly, leafy greens include commodities such as kale, chard, arugula, escarole, cabbage, endive, spinach, iceberg lettuce, spring mix, baby leaf lettuce, Romaine lettuce, butter lettuce, green leaf lettuce, and red leaf lettuce (Gombas et al., 2017). *Salmonella* and leafy greens have also been determined to be the pathogen/food combination with the greatest chance of producing human illness in the European Union (EU) from 2007 to 2011 (Da Silva Felicio et al., 2015). One reason for the high risk associated with lettuces and other leafy greens may be that they are more difficult to decontaminate with sanitizing washes than have been other types of fresh produce (Prado-Silva et al., 2015). Pathogen internalization in fresh produce tissues may preclude sanitizers from decontaminating microorganisms by access to harborage points such as biofilms, crevices, cracks, stomata, and cut surfaces (Gómez-López et al., 2013; Whipps et al., 2008). Further, Prado-Silva et al. (2015) revealed that sanitizing treatments were less effective on leafy vegetables than on other fresh produce. Nevertheless, sanitizers present in wash waters have been useful in preventing cross-contamination via water to uncontaminated fresh produce (López-Gálvez, 2009). Washing produce may not only reduce pathogens but can also reduce the background aerobic mesophilic microflora on the surface of fresh produce by up to 2–3 logs (Beuchat et al., 2004; Gil et al., 2009; Selma et al., 2008). Nevertheless, after storage, bacterial populations were the same regardless of whether produce was washed with unamended water or with a sanitizing solution (Allende et al., 2008).

Contaminated irrigation water is one common method used to spread contamination to fresh produce (Holvoet et al., 2012). Foodborne illness outbreaks have traced the source of contamination to pathogen-laden irrigation water (Nygard et al., 2008; Greene et al., 2008; Behravesh et al., 2011; Söderström et al., 2008; CFERT, 2007; Butot et al., 2009; De Giusti et al., 2010; Predmore and Li, 2011). Irrigation water can be remediated by various processes or chemical sanitizers including ultraviolet light, membrane filtration, NaOCl, ClO_2, H_2O_2 ozonation, and peracetic acid (Van Haute et al., 2015). Because produce is frequently consumed raw or minimally processed, there is a need for a decontamination step in the production continuum to reduce numbers of spoilage aerobic mesophilic bacteria (which can range as high as 10^3–10^6 CFU/g) as well as inactivate pathogenic bacteria such as *Salmonella, E. coli* O157:H7, or *Listeria monocytogenes* on the product, but especially in the wash waters (Vanderkinderen, 2009). Van Haute et al. (2015) list the resistance ranking of foodborne pathogens to chemical sanitizers in increasing order as follows: Gram-negative bacteria < Gram-positive bacteria < viruses < bacterial spores and protozoal parasites. Based on continuing foodborne illness outbreaks associated with the consumption of fresh produce, advanced research into the development of novel sanitizers is needed (WHO, 2009). To be commercially useful, these must achieve the desired antimicrobial reductions, while avoiding negative effects on the organoleptic properties of the foods (Rico et al., 2007).

Many factors must be considered in the decontamination of produce. These include sanitizer (type, concentration, and contact time), water quality (organic matter, turbidity, temperature, and pH), time interval between contamination and washing, types of produce including texture characteristics and surface area, microorganisms targeted (specific foodborne pathogens, fungi, spoilage organisms or background aerobic mesophilic microbiota), and method of sanitizer application (dipping, immersing, spraying, sequential washing, and post-washing potable water rinsing) (Gil et al., 2009).

Produce washing is primarily conducted to remove gross sediment and debris, including soil, stems, leaves, insects, and other plant residues, at which time cross-contamination of fresh produce can occur by microbial pathogens transported or disseminated in the water (Luo, 2007; Shen, 2014; Luo et al., 2011; Zagory, 1999). Also, one piece of contaminated produce could potentially transfer pathogens to wash waters and contaminate other produce being washed (Zagory, 1999). Hence, sanitizers are often used to inactivate pathogenic or spoilage microorganisms on the produce and to kill bacteria, fungi, and even viruses in the wash water. The US Environmental Protection Agency (EPA) classifies fresh produce sanitizers as pesticides, which must demonstrate efficacy in a range of conditions before being approved for use on fresh produce and sold to the public (USEPA, 2018). Nevertheless, the ability of sanitizers used in fresh produce wash waters to inactivate pathogenic bacteria on fruits and vegetables is usually limited to a reduction of <2.0 log (López-Gálvez, 2010b). A series of one to three recirculating or uncirculating, single-use or multiple-use washes are often used. The final wash is often used to remove any antimicrobial agents applied in previous wash(es) and may be recirculated into an earlier wash containing antimicrobial sanitizer(s). Because of increasing costs for water, as well as environmental concerns, processors are desirous of minimizing the amount of water used in washing produce (Ongeng et al., 2006; Olmez and Kretzschmar, 2009). Although the sanitizing agents used in an aqueous sanitizing step may be effective in reducing numbers of pathogenic bacteria, this is not their primary purpose (Gombas et al., 2017). A major goal in washing produce with aqueous sanitizers is to prevent cross-contamination from the water to the produce being washed, and does not ensure pathogen-free produce (Goodburn, 2013; Olaimat and Holley, 2012; Luo, 2017; Buchholz et al., 2012; Davidson et al., 2013; López-Gálvez et al., 2009; Al-Nabulsi et al., 2014; Suslow, 2001). Nevertheless, efforts have been made to reduce the levels of sanitizers applied with the goal of lowering the risk of generating hazardous by-products, responding to environmental concerns, and eliminating potential effects on product quality, not to mention cost reduction (López-Gálvez et al., 2009; Food and Drug Administration, Center for Food Safety and Applied Nutrition, 2008; Suslow, 2001). Although only occasional pieces of produce may be contaminated with pathogenic bacteria, contamination can disseminate throughout a fresh produce water flume or dump tank and subsequently contaminate other produce pieces (Doyle and Erickson, 2008). Other processing steps (e.g. harvesting, trimming, cutting, shipping, handling, vacuum-cooling, and packaging) also can serve as steps where produce may become contaminated either prior to or following pre-washing, washing, rinsing, or dewatering (López-Gálvez et al., 2010b; Suslow, 2001).

Process water disinfection is a process that takes places *in-situ* in flumes or dump tanks, which is distinguished from water reprocessing, which involves the inactivation of pathogens in water outside the wash tank (Banach et al., 2015). Gombas et al. (2017) have identified a number of factors and considerations that need to be taken into account when selecting an antimicrobial sanitizer to be used in washing of fresh produce, including (1) a broad spectrum of effectiveness against an array of foodborne microbial pathogens, (2) regulatory approval, (3) USDA and/or Organic Materials Review Institute (OMRI) certified organic listing approval (if marketing the fresh produce as USDA certified organic), (4) minimal environmental impact, and ease of processing by wastewater treatment, (5) stability during storage or application, (6) no adverse effect on organoleptic properties of the commodity, (7) the degree of corrosion potential on processing equipment, (8) the suitability of its application in light of worker safety issues and OSHA compliance, (9) if the produce must undergo a potable water rinse following the antimicrobial rinse (Artes and Allende, 2005), (10) if the pH should be adjusted, and (11) the organic load tolerance of the sanitizer (Li et al., 2001; Gombas et al., 2017). Goodburn and Wallace (2013) cited other factors that would affect the efficacy of sanitizing fresh produce, such as (1) the time and temperature of the washing treatment, (2) the type of sanitizer, (3) the initial pathogen population on the commodity, and (4) whether the pathogen is internalized in the fresh produce or not.

A significant issue with chemical sanitizers in wash water is the buildup of organic debris or organic load from dirt, filth, sugars, starch, or even the latex and other exudates secreted from the tissues of fresh fruits and vegetables (Zhang et al., 2009). The organic load is known to inhibit the antimicrobial action of some sanitizers including, especially, chlorine, added in the form of sodium hypochlorite (NaOCl or commercial bleach) and, to a lesser extent, peracetic acid ($C_2H_4O_3$) and chlorine dioxide (ClO_2) by neutralizing their viable antimicrobial molecules (Gombas et al., 2017; Zhang et al., 2009). In addition to these sanitizers, acidic, neutral, or alkaline electrolytic oxidizing water (AcEOW, NEOW, or AlEOW, respectively), acidified sodium chlorite (ASC), which is a combination of sodium chlorite and any food grade acid, such as citric acid or phosphoric acid, and ozone (O_3) (Ruiz-Cruz et al., 2007; US FDA, 2014) can be used. This chapter will primarily cover these six chemical antimicrobial sanitizing treatments currently used in wash waters for fresh and fresh-cut fruits and vegetables. Processes and sanitizing agents that will not be examined include H_2O_2, organic acids, ultrasonication, and ultraviolet light, which each have major limitations that preclude their use as commercial wash water sanitizers in the processing of fresh produce (Banach, 2015).

2 Chlorine in the form of sodium hypochlorite (NaOCl) yielding hypochlorous acid (HOCl)

Chlorine sanitizers (specifically, NaOCl) are the most commonly used water disinfectants for preventing the cross-contamination of bacterial pathogens to fresh produce (Fatica and Schneider, 2009; Gombas et al., 2017) and also the least expensive of all the common sanitizers (Garret et al., 2003). In one survey, 76% of industry respondents reported using hypochlorite for fresh produce washes (Seymor, 1999). NaOCl is relatively inexpensive, and is converted into hypochlorous acid (HOCl, a form of free chlorine) in wash waters, and the concentration may be simply assessed by measurement of free chlorine. HOCl is formed from NaOCl by the following chemical reaction:

$$NaOCl + H_2O \leftrightarrow HOCl + NaOH$$

Measuring and adjusting pH levels in wash waters is a critically important factor, because higher pH values lead to lower concentrations of HOCl and, hence, free chlorine in aqueous systems. At pH 4.5, 100% of chlorine is present as HOCl, at pH 7.0, 75%, at pH 8.0, 23%, and at pH 9.0, only 4% of chlorine is in the form of HOCl (Cords et al., 2005). Because one of the decomposition compounds associated with NaOCl is NaOH, the pH in NaOCl-treated waters often rises to values of 9.5 or higher. For this reason, commercial washing operations often adjust the pH of NaOCl wash waters, attempting to take advantage of the greater disinfection efficacy by lowering the pH to between 6.5 and 7.0, sometimes with citric acid (which is a relatively inexpensive, commercially available organic acid). A recent study demonstrated that citric acid reacted with chlorine and produced high amounts of a chlorine by-product (Fan and Sokorai, 2015), suggesting that citric acid should be replaced with inorganic acid to avoid the accumulation of possible human carcinogens in wash waters and fresh produce. The pH must also be kept above 4.0 as pH values lower than that can lead to the release of toxic chlorine gas (Fatica and Schneider, 2009). At high pH values, HOCl dissociates into the hypochlorite ion (OCl-), while at lower pH values it can dissociate into chlorine gas (Cl_2) (Banach et al., 2015).

Several mechanisms of action have been proposed for the biocidal activity of NaOCl, including (1) acting on cellular proteins to inhibit active transport, substrate dehydrogenation, and respiration; (2) oxidizing the purine and pyrimidine bases of DNA; (3) affecting cell membrane activity; (4) damaging sulfhydryl enzymes and glucose transport enzymes; (5) disrupting of protein synthesis; (6) formation of N-chlorine toxic by-products via cytosine; and (7) the creation of chromosomal aberrations (Cords et al., 2005; Vanderkinderen, 2009). NaOCl is versatile in that it can be used for sanitizing fresh produce wash waters, inactivating bacteria and fungi on the surfaces of fresh produce and additionally may also be used to sanitize hard surfaces within produce washing facilities as well as packinghouses, processing equipment, crates and storage bins, and restaurants.

There is a current trend toward avoiding 'hyperchlorination' of fresh produce wash waters, targeting the minimum effective concentrations of free chlorine for economic reasons as well as to prevent the formation of carcinogenic chlorine decomposition by-products (Gomez-Lopez et al., 2014). Some produce companies utilize automated, digital, real-time monitoring systems to assess free chlorine concentrations in wash water, which is generally used at a level of ≤200 ppm free chlorine and washes are applied for ≤2 min (Beuchat, 1998) as NaOCl use is mandated by the US Code of Federal Regulations to have free chlorine concentrations <200 ppm in the United States. However, in other countries such as France, the limit for free chlorine is set at 80 ppm and must be followed by a potable water rinse (DGCCRF, 2009). Fresh produce wash with chlorinated water is outlawed in Belgium, Denmark, Germany, Switzerland, and the Netherlands (Artes and Allende, 2005; Rico et al., 2007; Artes et al., 2009). It has been proposed that 1–5 ppm free chlorine is needed to inactivate bacterial pathogens in the wash water, whereas 20–200 ppm or more are necessary to inactivate the same pathogens on fresh produce (Banach et al., 2015). One study determined that in waters used to wash artificially inoculated spinach, a concentration of 7 ppm free chlorine was needed to keep the water free from the inoculated *E. coli* O157:H7 (Gomez-Lopez et al., 2014). Sreedharan et al. (2017) tested the minimum concentration of free chlorine, in the form of HOCl, necessary to prevent cross-contamination of *Salmonella* to tomatoes in a model flume system. The authors determined that under chemical oxygen demand (COD) of 650 mg/L and 1000 mg/L (a quantification of the amount of oxygen depleted by reactions in a measured solution), 100 ppm of chlorine was needed to prevent cross-contamination, while reducing *Salmonella* populations on tomatoes by >4.5 log CFU/tomato. The COD in tomato flume water can increase to 800 mg/L (Zhou et al., 2014), but one study tested 1000 mg/L to simulate a worst-case scenario (Sreedharan et al., 2017). In addition, 75 ppm of chlorine did not prevent cross-contamination of *Salmonella* to tomatoes in the presence of organic loading.

Florida Tomato Good Agricultural Practices require the use of at least 150 ppm free chlorine from HOCl at a pH level of 6.5–7.5 (or a comparable sanitizer) in order to prevent cross-contamination in tomato flume water (Sreedharan et al., 2017; Chang et al., 2010). Gereffi et al. (2015) found that without HOCl treatment in flume water, tomatoes became contaminated within 15 s; however, in the presence of 500 mg/L COD, 25 ppm chlorine in the form of HOCl prevented cross-contamination to tomatoes. Studies have also been performed in which leafy greens (lettuce or spinach) were washed (with no additional organic load added) and the minimum free chlorine concentration required to prevent cross-contamination of inoculated *E. coli* O157:H7 was 10 ppm HOCl when the pH was held between 6.5 and 7.0 (Luo et al., 2011, 2012). Shen (2014) found no cross-contamination when washing lettuce inoculated with *E. coli* O157:H7 at HOCl concentrations of between 9.4 and 40.8 ppm; however, cross-contamination occurred when the free chlorine concentration dropped to ≤4.6 ppm. Gomez-Lopez et al. (2014) washed spinach for 1 h and found that 5 ppm free chlorine inactivated *E. coli* O157:H7 during the entire period.

Automated systems are available to not only monitor the concentration of free chlorine, but to also add additional NaOCl to the wash water in order to keep the chlorine levels at their target concentrations. The oxidation-reduction potential (ORP) is sometimes monitored as it can be correlated with the HOCl in the solution and the amount of free chlorine where 650–700 mV corresponds to ca. 3–5 ppm of free chlorine, while readings are generally targeted at 850 mV or higher (Luo et al., 2012; Chaidez et al., 2012; Van Haute et al., 2015; Suslow, 2004; Barrera et al., 2012). Because free chlorine from NaOCl is so readily inactivated by organic load (Gombas et al., 2017), it is imperative that wash waters be continuously monitored for free chlorine concentrations to prevent cross-contamination events from one or more pieces of contaminated produce. Water temperatures must be closely monitored as waters ≤15°C cooler than the fresh produce promote absorption and infiltration of waters into porous areas of the fresh produce pieces by temperature-generated pressure differentials, hence promoting internal contamination if waters are not completely free of pathogens or spoilage organisms (Fatica and Schneider, 2009). For this reason, it has been recommended that wash waters be held 10°C higher than the produce that is being washed (Parish et al., 2003).

HOCl has been shown to inactivate 3 log CFU/mL of *E. coli* in water by the addition of only 0.01 ppm free chlorine at 20°C in 5 min, >5.0 log CFU/mL of *E. coli* at 6 ppm free chlorine and pH 6.0 at 25°C for 15 s, and >5.0 log CFU/mL of *E. coli* at 12.5 ppm free chlorine and pH 7.7 at 25°C for 15 s. Although vegetative cells of bacteria are most susceptible to the action of HOCl, spores may also be inactivated, although typically with higher HOCl concentrations and longer contact times. Up to 4 log CFU/mL of *Clostridium botulinum* type A spores were shown to be inactivated by the addition of 4.5 ppm HOCl (by measurement of free chlorine) at pH 6.5, a water temperature of 25°C and exposure for 10.5 min (Cords, 2005). Nevertheless, washing of fresh produce generally takes place at refrigeration temperatures (ca. 4°C), in order to preserve product quality and prevent the growth of spoilage organisms (Castro-Ibanez et al., 2017). Exposure of the spores of the spoilage bacterium *Alicyclobacillus acidoterrestris* to 200 ppm of free chlorine from NaOCl for 10 min resulted in a 2.2 log CFU/mL reduction of the spores (Orr and Beuchat, 2000).

Aside from decontamination of water and preventing cross-contamination of fruits and vegetables, varying levels of pathogen inactivation have been reported for NaOCl on fresh produce, including a 0.7 log reduction of *L. monocytogenes* on lettuce after a 1 min wash at 100 ppm free chlorine (Hellstrom et al., 2006); a 1.2 log reduction of *L. monocytogenes* on lettuce after a 30 s wash at 20 ppm free chlorine (Li et al., 2002); a 1.30 log reduction of aerobic bacteria on lettuce after a 30 s wash in 200 ppm free chlorine (Simmons et al., 2006); a 0.0 log reduction of aerobic bacteria on diced tomatoes after a 30 s wash in 200 ppm free chlorine (Simmons et al., 2006); a 1.1 log reduction of *E. coli* O157:H7 on lettuce after a 1 min wash in 20 ppm free chlorine (Li et al., 2001); and a ca. 1 log reduction of *E. coli* O157:H7 and *Salmonella*, respectively, on lettuce after a 1 min wash in 200 ppm free chlorine (Koseki et al., 2003). Beuchat and Ryu (1997) demonstrated that treating alfalfa sprouts with 500 ppm chlorine for 2 min inactivated only 2.0 log of *Salmonella,* while treating with 2000 ppm reduced populations to less than 1 log CFU/mL. NaOCl reduced aerobic mesophilic microbiota on the surfaces of fresh and fresh-cut produce (Vandekinderen et al., 2009). Examples include 1.7 log reductions in carrots (Klaiber et al., 2004) and ca. 1.7 log inactivation on lettuce (Akbas and Ölmez, 2007). Treating shredded cabbage or lettuce for 10 min with 200 ppm chlorine only reduced populations of *L. monocytogenes* by 1.2 and 1.7 log CFU, respectively (Zhang and Farber, 1996). Li et al. (2001), however, reported that that there was no difference in the survival of *E. coli* O157:H7 when cut lettuce pieces were immersed for 90 s in plain water vs. a 90 s treatment in 20 ppm chlorine at either 20 or 50°C. Beuchat (1999) reported that there was no difference in the reduction of *E. coli* O157:H7 between spraying inoculated lettuce with 200 ppm chlorine and treating with deionized water.

The additive T128 (composed of propylene glycol and phosphoric acid) has been researched to enhance the biocidal activity of HOCl by stabilizing the antimicrobial in the water and preventing its breakdown (Lemons and Taylor Fresh Food Inc., 2009; Luo et al., 2012) and has also been used to stabilize HOCl concentrations in wash water by the fresh produce industry. T128 was demonstrated as stabilizing chlorine concentrations in water amended with up to 2% soil. In a semi commercial pilot scale set of experiments, T128 was shown to stabilize chlorine concentrations during the washing of spinach leaves; however, in chlorinated water without T128, chlorine levels dropped and resulted in cross-contamination of *E. coli* O157:H7 onto spinach (Lou et al., 2012). Luo et al. (2011) demonstrated that T128 plus low concentrations of HOCl provided a synergistic effect such that no *E. coli* O157:H7 were detected in T128-amended wash water with free chlorine concentrations as low at 0.05 mg/L in the treatment; nevertheless, in the absence of T128 the pathogen survived in waters with <0.4 mg/L free chlorine.

The drawbacks to using NaOCl include effects on the skin and the respiratory tract of workers (Abadias et al., 2008), formation of chloramines and trihalomethanes, which are known carcinogens (Richardson et al., 1998), and the corrosion of metal tanks and other metal implements used during the washing process (Beucaht, 1998). Efficacy of NaOCl is also affected by the type of produce washed, with greater reduction of pathogens when washed with NaOCl on lettuce than on cabbage (Nguyen and Carlin, 1994). Furthermore, NaOCl (or other sanitizers, for that matter) do not completely eliminate pathogens on produce, which could lead to foodborne illness outbreaks (Ruiz-Cruz et al., 2007; Felkey et al., 2006; Parish et al., 2003). Additionally, HOCl is also known to be rendered inactive by exposure to metals, light, and air.

3 Chlorine dioxide

Chlorine dioxide (ClO_2) gas may be produced either by reacting sodium chlorite with an acid or by treating sodium chlorite with gaseous chlorine (Olmez and Kretzschmar, 2009). Chlorine dioxide is permitted for use in the US food industry, if used in the aqueous form, for sanitization of wash waters at concentrations ≤3 ppm (Chlorine dioxide) and has been reported as retaining its antimicrobial properties at concentrations as low as 0.1 ppm (Artes et al., 2009; Joshi et al., 2013). However, ClO_2 washing of fresh produce must be followed by a potable water rinse. Furthermore, for sanitizing hard equipment, ClO_2 may be used at concentrations as high as 200 ppm. Gaseous ClO_2 has not been approved by the FDA for use on fresh produce prior to customer distribution, unless dissolved in water first, in which case, it is then in the aqueous form. ClO_2 has a distinct action and mechanism from that of NaOCl and HOCl in a number of ways. First, it is a gas, yet remains soluble in water (Gordon and Rosenblatt, 2005). It creates fewer carcinogenic compounds than was previously mentioned being generated by NaOCl (viz., chloramines and trihalomethanes) (López-Gálvez et al., 2010a; Gomez-Lopez, 2012), although it does generate chlorates and chlorites. Other advantages include having a higher oxidizing capacity than HOCl (up to 2.5 times greater) (Beucaht et al., 2004; Allende, 2009; Richardson et al., 2000; Hua and Reckhow, 2007; Gomez-Lopez et al., 2009; Bernarde et al., 1967) and a wide viable pH range of 6.0–10.0, which allows it to be used at alkaline conditions and, hence, prevent damage to metal flumes, dump tanks, pipes, and so forth (Fatica and Schneider, 2009), and being more effective than HOCl at lower concentrations due to the higher oxidizing capacity. An important characteristic that sets it apart from HOCl is its lower ability to react with organic molecules and organic precursors, and, therefore, its greater stability over a longer period of time (Winniczuk and Parish, 1997). Because of this characteristic, ClO_2 is more suitable for the destruction of biofilms than is the use of NaOCl (Joshi et al., 2013). The inactivation mechanism of ClO_2 involves damage to the proteins, cell membrane, RNA, and amino acids affecting cellular transport, permeability, and enzymatic reactions (Fatica and Schneider, 2009). ClO_2 is also more biocidal against viruses than is ozone or HOCl (Artes, 2009).

The antibacterial mechanism of aqueous ClO_2 is due to the fact that it has one free electron, which reacts with electron-rich molecules within organic compounds (Lin et al., 2005). These reactions oxidize the enzymes and membrane proteins in bacterial cells resulting in leakage of cellular contents (Huang et al., 1997; Jeng and Woodworth, 1990). Further, it disrupts the electron transport system of bacteria as well as attacking sulfhydryl groups (Young and Setlow, 2003; Berg et al., 1986). Joshi et al. (2013) reported that ClO_2 increases its antimicrobial efficacy with increasing pH, but its effectiveness decreases as temperature decreases (Joshi et al., 2013).

Being a gas, it is technically capable of being applied aqueously as well as in gaseous form. Nevertheless, when applied as a gas (typically to food contact surfaces and for decontamination of food processing facilities), high humidity levels are needed as ClO_2 loses its antimicrobial efficacy as conditions become more dry. Nevertheless, gaseous ClO_2 might be more easily able to penetrate and inactivate pathogens in hard-to-reach places, such as the stem scars of tomatoes and cucumbers or the calyx of the apple (Sadiq and Rodriguez, 2004; Reina et al., 1995). With regard to protozoal microorganisms, gaseous ClO_2 has proven effective against *Cryptosporidium parvum* but not against *Cyclospora cayetanensis* (Ortega et al., 2008). Gaseous ClO_2 is also known to be effective against foodborne pathogens on produce such as basil, bell peppers, apples, and tomatoes (Han et al., 2000; Du et al., 2002; Mahovic et al., 2009; Ortega et al., 2008; Wang et al., 2019).

Although less effective than gaseous ClO_2, the aqueous form is effective against *E. coli* O157:H7, *Bacillus subtilis,* as well as *Yersinia enterocolitica, Staphylococcus aureus, Salmonella, Pseudomonas aeruginosa,* and *L. monocytogenes* in solution or on produce such as blueberries, cucumbers, green peppers, lettuce, tomatoes, apples, and cabbage (Zhang and Farber, 1996; Young and Setlow, 2003; Gomez-Lopez et al., 2009; Joshi et al., 2013). One report noted that ClO_2 may have an additional benefit of being able to remove undesirable tastes or flavors from foods (Joshi et al., 2013).

One drawback of the use of ClO_2 over NaOCl is that the FDA mandates that fresh produce washed with ClO_2 must be rinsed with potable water prior to distribution to customers. Another drawback is that in some studies the use of gaseous ClO_2 has resulted in the bleaching of some fresh produce (Gomez-Lopez, 2012). On the other hand, Du et al. (2009) reported that prolonged washing of fresh lotus root at high concentrations of ClO_2 caused darkening. Some studies have also demonstrated that injured produce may contain harborages that safely entrap pathogenic microorganisms and prevent inactivation by gaseous ClO_2 (Han et al., 2000; Du et al., 2002). ClO_2 is also extremely volatile, sensitive to light, decomposes at temperatures >30°C, and is explosive under pressure; thus, it must be generated at point-of-use (Fatica and Schneider, 2009).

López-Gálvez et al. (2010b) reported that following a cross-contamination event to fresh-cut lettuce in wash waters, subsequent treatment of the waters with either 3 ppm of ClO_2 or 100 ppm of free chlorine via NaOCl failed to disinfect fresh-cut lettuce, purportedly due to the uptake of *E. coli* into the stomata of lettuce as visualized with electron microscopy. It has also been shown that *E. coli* is able to strongly attach to the surface of leafy greens (Shaw, 2008). The subsequent treatments did not eliminate *E. coli* from the lettuce pieces, but prevented cross-contamination from contaminated to fresh uncontaminated lettuce added to the wash waters.

López-Gálvez (2009) compared the efficacy of ClO_2 (3 ppm) to NaOCl (100 ppm) for the reduction of mesophiles, psychrophiles, *Pseudomonas* spp. *Enterobacteriaceae,* lactic acid bacteria, yeasts, and molds on fresh-cut lettuce and found that each treatment performed equally well, except that ClO_2-treated lettuce resulted in higher levels of yeast after 10 days of storage.

4 Acidified sodium chlorite

Acidified sodium chlorite (ASC) is a combination of sodium chlorite and any food grade acid, such as citric acid or phosphoric acid, approved for use on fresh produce with a pH of 2.3–2.9 and a concentration of 500–1200 ppm sodium chlorite, and (like ClO_2) must be followed by a potable water rinse (US FDA, 2018). ASC applied at the above concentration range has shown efficacy in inactivating *Salmonella, E. coli* O157:H7, *A. acidoterrestris* spores, as well as yeast and mold on products such as strawberries, peppers, honeydew melons, cantaloupes, asparagus, apples, and shredded carrots (Fatica and Schneider, 2009). ASC was more effective against microorganisms than either HOCl or ClO_2 (Gonzalez et al., 2004). The mechanism of ASC has to do with the oxidative intermediate products produced (such as ClO_2) and the acidic conditions, which may cause damage to produce at concentrations ≥1000 ppm; hence, lower concentrations applied for longer periods of time may be more effective (Fatica and Schneider, 2009). In a meta-analysis of 40 studies of fresh produce decontamination, and 15 sanitizers, ASC, slightly AEW, and ClO_2 were the three most effective at inactivating *E. coli* O157:H7, *Salmonella,* and *L. monocytogenes* (Prado-Silva et al., 2015). In a forest plot of the hypothetical overall mean log reduction of *E. coli* O157:H7, *Salmonella,* and *L. monocytogenes* on fresh produce by sanitizers at concentrations of 0.001 g/100 mL, ACS was the most effective with 3.86, 4.40, and 2.47 log reductions, respectively.

Taormina and Beuchat (1999) reported that treating alfalfa seeds with 500 ppm ASC reduced *E. coli* O157:H7 by greater than 2 log/g of seeds. When Tan et al. (2015) tested acidified electrolyzed water (5 min), cetylpyridinium chloride (1% for 3 min), ozonated water (2 ppm for 5 min), and sodium dichloroisocyanurate (150 ppm for 10 min) for efficacy in reducing *Salmonella* as well as background microbiota on peeled turnips (Jicama or Mexican turnip, *Pachyrhizus erosus*), ASC 1200 ppm for 3 min proved the most effective, reducing aerobic mesophilic bacteria by 2.09 logs, yeast and mold by 2.13 logs, and *Salmonella* by 3.91 logs (Tan et al., 2015). Phua et al. (2014) evaluated the efficacy of sanitizers to inactivate *Salmonella* and *E. coli* O157:H7 on mung bean sprouts and found that 1200 ppm ASC (pH 2.3) inactivated ca. 1.8 log CFU/g of *E. coli* O157:H7, which was greater than that inactivated by AEW (75 ppm free chlorine, pH 2.8), 2% cetylpyridinium chloride, 10% trisodium phosphate, or 2 ppm ozonated water. However, with respect to *Salmonella,* ASC only inactivated greater populations than were reduced via ozonated water or trisodium phosphate.

Inatsu et al. (2010) treated tomatoes, dip-inoculated with *E. coli* O157:H7 in ASC solution made with 1.0 g/L of either phytic acid or lactic acid and 0.5 g/L of sodium chlorite and found that this combination was capable of inactivating 3.5 log CFU/g of *E. coli.* Nei et al. (2009) found that treating spinach leaves with either ASC (15 mg/L sodium chlorite and 200 mg/L citric acid) or chlorine (100 mg/L) reduced bacterial populations between 1.1 and 1.9 log CFU/g, although there was no significant difference between the reduction elicited by the two sanitizers. Liao (2009) extended the standard 15 min treatment time for alfalfa seeds with sanitizer to 45 min and determined that 800 ppm of ACS reduced

Salmonella populations by 3.9 log, which was 1.2 log higher than the standard treatment of 20000 ppm of calcium hypochlorite. Furthermore, Liao found that seed germination was not affected by the ASC treatment, although it was reduced by 20% following the calcium hypochlorite treatment. Allende et al. (2009) washed fresh-cut cilantro in either 0.2 g/L of free chlorine or various concentrations of ASC for 1 min and reported reductions of 1.0–1.3 log CFU/g of background mesophilic microbiota or *E. coli* O157:H7, while 1g/L of ASC reduced populations by ca. 3.5 log CFU/g.

Fan et al. (2009) treated whole cantaloupes with 1000 ppm ASC for 10 min, which reduced populations of total mesophilic aerobic bacteria, and yeast and mold by 1.1 and 0.5 CFU/g, respectively. When the sanitizer at the above concentration was applied to cantaloupe rinds inoculated with *Salmonella* Poona, 1.7 log CFU/cm was inactivated. Stopforth et al. (2008) washed fresh mixed greens in 1200 ppm of ASC (pH 2.5) for 60 s, which reduced populations of *E. coli* O157:H7, *Salmonella,* or *L. monocytogenes* by 3.0–3.8 log CFU/g.

5 Peroxyacetic acid (peracetic acid)

One alternative to chlorine that has been used in wash waters by the fresh produce industry is peroxyacids, which are quaternary compounds created chemically by combining hydrogen peroxide with one or more organic carboxylic acids–the most commonly used in the fresh produce industry being peroxyacetic acid (PAA), or peracetic acid. Peracetic (CH_3COOOH), or PAA, which is the peroxide of acetic acid (Banach et al., 2015) is created by the reaction between hydrogen peroxide and acetic acid (CH_3COOH), creating a new compound–a peracid or peroxyacid. Unlike chlorine, PAA is non corrosive to aluminum and stainless steel (Beuchat, 2000). PAA is marketed in aqueous form as a combination of PAA, acetic acid, water, and H_2O_2 (Vandekinderen et al., 2009). PAA has been approved for use on fresh fruits and vegetables in the United States at concentrations ≤80 ppm. Because PAA degrades into only acetic acid, oxygen, and water, it is considered an organic treatment, when applied on fresh and fresh-cut fruits and vegetables (Koivunen and Heinonen-Tanski, 2005). Although PAA is still susceptible to inactivation by organic materials, it is much less subject to organic-mediated neutralization when compared with HOCl, which makes PAA especially effective against biofilms (Beuchat, 2000; Hilgren et al., 2007; Koivunen and Heinonen-Tanski, 2005). However, PAA is susceptible to neutralization by sulfurous compounds (such as are in cabbage, broccoli, cauliflower, and onions) as well as antioxidants such as vitamins E and A (Gombas et al., 2017).

Another advantage of PAA is that it maintains its biocidal efficacy over a broader pH range (viz., 3.0–7.5) (Sapers et al., 1999), although it loses its efficacy at higher pH values (Joshi et al., 2013). Further, PAA may have a synergistic activity at higher temperatures (Joshi et al., 2013). PAA has been used extensively for more than a century as a sanitizing solution and, more recently, PAA and other peroxyacids have been studied for use by the food industry, and cited in numerous publications as an adjuvant for flume or dump tank water treatment or as a general sanitizer against various human pathogens (Gurtler et al., 2012, 2014; Martin and Maris, 2012). Bacteria and viruses are generally more susceptible to the sanitizing effects of PAA, while spores of bacteria and protozoal cysts are more recalcitrant (Kitis, 2004). Mohan et al. (2009) found that peroxyacetic and peroxyoctanoic acids were highly effective against *Bacillus atropheus* spores (black-pigmented bacteria and used in biomedicine as an indicator strain for heat and chemical decontamination), especially at increasing temperatures (i.e., 50–70°C), based on research they performed in determining the concentration of free radicals produced by either formula. Hilgren et al. (2007) also found that PAA was more effective than chlorine in inactivating *Bacillus anthracis* spores in the presence of organic food matter than was chlorine, and its efficacy increased with increasing temperatures of 10–30°C, where 5 log reductions were generally effected between 0.2% and 2% concentrations (i.e. 2000–20,000 ppm). PAA is effective against *Salmonella, E. coli* O157:H7, *L. monocytogenes,* and spoilage microorganisms on commodities such as apples, tomatoes, honeydew melons, fresh carrots, lettuce, leek, cantaloupes, and shredded cabbage (Fatica and Schneider, 2009; Hellstrom et al., 2006; Vandekinderen et al., 2009). PAA has been reported to have greater biocidal efficacy at higher pH values when combined with lactic acid (Banach et al., 2015; Grace Ho et al., 2011). It is also known to have increased efficacy against microorganisms when supplemented with either octanoic acid (caprylic acid) or increased concentrations of H_2O_2 (Alasri et al., 1993; Hilgren and Salverda, 2000).

PAA is more bactericidal than hydrogen peroxide alone, is more effective at lower concentrations, as well as at lower temperatures, and has greater stability in the presence of organic matter than hydrogen peroxide alone (Cords et al., 2005). PAA was first shown to be an excellent fungicide and bacteriocide in 1902 (Freer and Novy, 1902) and was approved for use as a sanitizing solution on food contact surfaces in processing plants in the United States in 1990. Release of hydroxyl radicals by PAA leads to biocidal effects, although greater sporicidal efficacy may be attained with the addition of lower chlorine or bromine concentrations. Nevertheless, PAA alone has shown sporicidal activity at concentrations as low as 0.05% and 0.5% (500 and 5000 ppm). PAA has also been shown to be more effective against biofilms than some other common sanitizers (Fatemi and Frank, 1999). The mechanism of inactivation by PAA is based on its production of reactive oxygen species, which are known to damage cell membranes, cell walls, cell lipids, DNA, as well as enzymes and proteins following penetration of the cell membrane by the acetate fatty acid end of the molecule (Gombas et al., 2017; Hilgren et al., 2007; Small et al., 2007; Koivunen and Heinonen-Tanski, 2005). One study found reductions of 0.5–1.5 log CFU/g of either mesophilic bacteria or coliforms on lettuce after spraying or dipping treatments with PAA (Allende et al., 2008). PAA in combination with octanoic (caprylic) acid has been shown to increase the fungicidal activity in comparison to PAA alone in fresh produce wash waters (Hilgren and Salverda, 2000). Octanoic acid has been added to PAA to increase efficacy against yeasts and mold in vegetable processing (Alasri, 1993). Various manufacturers and distributers of PAA add different concentrations of H_2O_2 to their formula to increase its biocidal activity. Drawbacks to using peracetic acid include higher cost in comparison to NaOCl (Banach et al., 2015) as well as its instability at concentrations higher than 15% (Kitis, 2004). Gombas et al. (2017) report that PAA was EPA registered, and that 30–80 ppm PAA was capable of inactivating 3 log CFU/mL of *L. monocytogenes, Salmonella,* or *E. coli* after a 90 s wash.

Zhang et al. (2009) evaluated PAA at 30 ppm and a mixed peracid solution (of PAA + peroxyoctanoic acid) at 10, 20, and 30 ppm as well as NaOCl at 30 and 50 ppm for the reduction of *E. coli* O157:H7 in processing water with and without a 10% liquefied lettuce juice organic load. When no organic load was introduced, the last six treatments, respectively, reduced *E. coli* O157:H7 in processing water by 1.83, 1.73, 1.50, 1.83, 1.34, and 1.83 log CFU/mL. However, when a 10% organic load was included in the wash, PAA at 10 and 20 ppm and chlorine at 30 ppm resulted in *E. coli* O157:H7 populations that were not significantly different from water rinses alone.

Peracetic acid is unstable and storage containers must provide proper venting to allow off-gassing. To further stabilize PAA and prevent accelerated decomposition, HEDP (1-Hydroxyethylidene-1,1-diphosphonic acid) is added to PAA in the United States at levels not to exceed 4.8 ppm in wash water when used for treating fresh produce. If used at a concentration ≤80 ppm, PAA does not require a potable water rinse following the sanitizing rinse (Gombas et al., 2017).

Based on the cost, pH of the aqueous solution, and increasing the organic load of wash waters, López-Gálvez (2009) concluded that NaOCl at the minimum effective dose still represents the best option for ensuring disinfection of bacterial pathogens from fresh produce wash waters.

6 Electrolyzed oxidizing water

Electrolyzed oxidizing water (EOW) was first discovered in the late 1800s in Great Britain and was used in the medical field through the early 1900s (Gunarathna et al., 2014). More recently (since the 1980s), EOW has been used by the Japanese for the antibacterial treatment of food contact surfaces in food processing plants, seafood, fresh produce, meat, eggs, and poultry, and has been approved for use in the food industry by the FDA, USDA, and EPA

(Gunarathna et al., 2014). EOW may be used in the form of acidic, neutral, or alkaline (AcEOW, NEOW, or AlEOW). EOW is formed by the electrolysis of a solution containing KCl-$MgCl_2$ or a solution of NaCl alone (Hricova, 2008).

Vanderkinderen et al. (2009) reported that there are two general types of generators capable of making EOW. The first is a membrane-free generator in which tap water or a dilute NaCl solution is electrolyzed to produce EOW. The second type are membrane generators which produce the ionized AlEOW product at the anode and then redirects that charged chemical species either *toward* the cathode to produce NEOW, or *away from* the cathode to produce a mixture of AlEOW and AcEOW. Gunarathna et al. (2014) reported that AlEOW has a pH of 10.0–11.5 and an ORP of −800mV to −900mV with a voltage of −700mV to +200mV and a final end product of NaOH. AcEOW, on the other hand, has a pH of 2.5–3.5, an ORP of +1000mV to +1200mV, and a voltage of +200mV to +820mV and a final end product of HOCl and dilute HCl. Commercially available systems allow the users to (1) select the brine-flow rate, while the machine adjusts voltages and/or amperages, (2) select the amperages and/or voltages, while the machine adjusts the brine-flow rate, and (3) select a preset chlorine concentration and the machine changes the brine-flow rate and amperages and/or voltages accordingly (Gunarathna et al., 2014).

When NaCl is electrolyzed, it produces a number of antimicrobial reactive species or radicals, including O_3, HOCL, OCl^-, Cl_2, and H_2O_2 (Abadias et al., 2008; Ruiz-Cruz et al., 2007; Gomez-Lopez et al., 2007; Liao et al., 2007). Because of the additional reactive species produced, EOW has greater biocidal effects on microorganisms than does NaOCl (Feng et al., 2004; Li et al., 2004; Bergmann et al., 2008; Jeong et al., 2007; Fang et al., 2006; Martinez-Huitle and Brillas, 2008). EOW has a free chlorine concentration of ca. 11.0–86.0 ppm (Gunarathna et al., 2014), or up to 90 ppm (Joshi et al., 2013). AcEOW has a pH of 2.0–4.0, which leads to greater HOCl efficacy, in addition to a high redox potential of greater than 1000 mV, which can damage the bacterial membrane. Although the biocidal mechanism against microorganisms is not fully understood, at a minimum, the presence of O_3 and HOCL make it antimicrobial. The high or low pH could lead to membrane damage, inhibition of carbohydrate metabolism, glucose oxidation and oxygen uptake, disruption of protein synthesis, damage to key enzymes and nucleic acids, and oxidative phosphorylation (Gunarathna et al., 2014). EOW has been shown to be effective at inactivating various microbes such as *E. coli, Salmonella, Pseudomonas aeruginosa,* and *Pseudomonas solanacearum, Citrobacter freundii, Flavobacter, Proteus vulgaris, Alcaligenes faecalis, Aeromonas liquefaciens, Campylobacter jejuni, Enterococcus faecalis, Listeria monocytogenes, Staphylococcus aureus, Bacillus cereus, Enterobacter aerogenes, Erwinia carotovora, Xanthomonas* spp., *Vibrio parahaemolyticus, Burkholderia glumae, Cladosporium, Aspergillus, Botrytis, Colletotrichum,*

Curvularia, Fusarium, Helminthosporium, Magnaporthe, Monillnia, Penicillium, Phytophthora, and Tilletia indica (Gunarathna et al., 2014).

Benefits of NEOW over AlEOW are that the more neutral pH (ca. 6–9) is less acidic than AcEOW and thus less corrosive on metal equipment as well as less skin irritating to personnel applying the product (Abadias et al., 2008). AlEOW, on the other hand, seems to have the least antimicrobial efficacy, in comparison with AcEOW and NEOW, largely due to the fact that at higher pH values, less HOCl is available as free chlorine. General benefits of using EOW include low operating costs, simple components (water and salt), no effect on the organoleptic properties of the food, and no need for transportation, handling, or storage of toxic chemicals (Gunarathna et al., 2014). Nevertheless, Rico et al. (2008) reported that treating lettuce with NEW negatively impacted the color and firmness of lettuce. However, others report that the treatment does not affect the color or appearance of fresh produce, following washing (Issa-Zachariah et al., 2011). Drawbacks to using EOW include the need for costly equipment, corrosiveness to metal, and neutralization by organic materials.

AcEOW is also more antibacterial at lower free chlorine concentrations than NaOCl, as similar log inactivation on fresh produce was achieved with only 20 ppm free chlorine of AcEOW versus 150 ppm of free chlorine required for NaOCl (Kim et al., 2000). In another study, comparable antibacterial results in the reduction of aerobic mesophilic background microbiota were obtained with 50 ppm free chlorine of NEOW as were attained with 120 ppm free chlorine from NaOCl (Abadias et al., 2008). Ongeng et al. (2006) reported that the reduction of aerobic mesophilic background microbiota in fresh vegetables by NEOW was negligible, compared with unamended water; however, significantly greater reductions were achieved by the use of AcEOW. AcEOW has been adopted for use as a sanitizing product by the food industry in Japan, including washing of fresh-cut vegetables (Rahman, 2016; Izumi, 1999). AcEOW is known to be used for decontaminating alfalfa seeds, apples, bell pepper, carrots, celery, cucumbers, lettuce, peaches, potatoes, salmon, spinach, strawberries, tomatoes, tilapia, and yellow-fin tuna (Gunarathna et al., 2014; Joshi et al., 2013). EOW has been shown to inactivate pathogens and spoilage microorganisms in various classes such as aerobic bacteria, coliforms, mold, and yeast (Artes et al., 2009), *E. coli* O157:H7, *Salmonella, and L. monocytogenes* (Park et al., 2001), as well as reducing microbial counts on commodities such as lettuce, cucumbers, spinach, potatoes, bell peppers, alfalfa seeds and sprouts, and carrots (Park et al., 2001; Rico et al., 2008; Kim et al., 2003).

AcEOW may be more antimicrobial than HOCl even at lower concentrations of free chlorine. For example, similar inactivation levels were achieved when using 150 ppm HOCl (pH 9.6) as when using AcEOW at 20 ppm free chlorine and pH 6.1. Paolo et al. (2005) reported that adding 0.6% acetic acid to EOW resulted in greater log reductions of *L. monocytogenes* on lettuce than EOW alone.

7 Ozone (O_3)

Ozone (O_3) is a highly reactive, unstable, and volatile gas with a short half-life that ranks second in the oxidizing capacity of reactive oxygen species, the most reactive being the hydroxyl radical (OH^-) (Banach et al., 2015). O_3 is combustible at higher concentrations, and therefore must be created on-site by application of corona discharge, due to its toxic and explosive properties. Electrical discharges split the O_2 molecule into two singlet oxygen molecules, which, when recombined with available O_2, subsequently forms O_3. The resulting O_3 molecule is then free to react with microorganisms, and is known to be bactericidal (Guzel-Seydim et al., 2004). O_3 decomposes rather rapidly and during the decomposition process and can lead to formation of the hydroxyl (OH^-), superoxide ($\cdot O_2^-$), and hydroperoxyl ($\cdot HO^2$) radicals, which are highly oxidizing and bactericidal (Joshi et al., 2013). Ozone may be applied in gaseous form or infused into water, and fresh produce may be washed in the resulting solution which is an effective antimicrobial against microorganisms even at low concentrations for relatively short contact times (Graham, 1997; Aguayo et al., 2006). Ozone has been approved by the FDA for treatment of fresh produce and reducing bacterial populations on radish sprouts, bean sprouts, and carrots (Singla et al., 2011; Bermudez-Aguirre and Barbosa-Canovas, 2013). However, in gaseous form, it has also been reported to negatively affect the color of commodities such as carrots and peaches (Joshi et al., 2013). Ozone has been shown to be effective against *Staphylococcus aureus, Salmonella* Typhimurium, *Y. enterocolitica, L. monocytogenes, L. mesenteroides, P. fluorescens, Rhizopus stolonifer, C. parvum,* and Giardia (Kim et al., 1999b; Restaino, 1995; Sarig et al., 1996; Finch and Fairbairn, 1991; Peeters et al., 1989; Van Haute et al., 2015). Fan et al. (2012) tested the application of in-package ozonation (1000 ppm concentration) of *Salmonella-* and *E. coli* O157:H7-inoculated tomatoes and found reductions in the range of 2–3 log after 2–3 min treatments. Sheng et al. (2018) found that continuous low doses (50.0–87.0 ppb) of gaseous ozone application in controlled atmosphere storage reduced *L. innocua* by ca. 5.0 log CFU/apple during 30 weeks of storage.

O_3 readily reacts with proteins, DNA and RNA, amino acids, amines, and carbon-to-carbon double bonds, while being less reactive with fatty acids and carbohydrates (Banach et al., 2015). Ozone is most effective at a pH of 6.0 (although it has been shown to inactivate most microorganisms at pH ranges of 6–9) and at lower temperatures, whereas it decomposes more rapidly at higher pH values and higher temperatures (Kim et al., 1999a). Ozone is a more efficient oxidizer and has greater antimicrobial efficacy in a shorter amount of time than does chlorine (Fatica and Schneider, 2009). Ozone has also been shown to oxidize ethylene, hence slowing down senescence and increasing the shelf life of fresh produce (Skog and Chu, 2001). Additionally, the use of gaseous ozone during product storage is known to reduce odors as well as populations of spoilage microorganisms (Joshi et al., 2013).

The drawbacks to O_3 include exposure to high voltages in generating the gas, potentially exposing workers to O_3 based on off-gassing, its explosive nature at concentrations ≥30%, its short half-life and being only slightly soluble in water (Banach, 2015). It has been reported that concentrations greater than 0.2 ppm can cause respiratory tract damage (Artes et al., 2009) and maximum exposure has been set at 0.1 ppm for up to 8 h or 0.3 ppm for up to 15 min (Fatica and Schneider, 2009). Concentrations >1.0 ppm are known to corrode stainless steel (Fatica and Schneider, 2009). However, O_3 has greater solubility in water as the temperature decreases (Bablon et al., 1991). Infusion of O_3 into water is also increased with mixing or turbulence (Artes et al., 2009). Because the pH needs to be lowered to ca. 6.0 for optimum efficacy, an acid such as citric acid must be used to acidify wash waters. Ozone has also been known to damage the organoleptic properties of fresh fruits and vegetables at higher concentrations for extended contact times. For example, ozone treatment during storage of produce has been shown to reduce anthocyanin concentrations on blackberries and strawberries (Barth et al., 1995; Perez et al., 1999). High reactivity with organic compounds in wash waters requires real-time monitoring and adjusting O_3 levels with automated systems.

Other means of decontaminating fresh produce from foodborne bacterial pathogens have demonstrated promising results, although these technologies have not been widely applied commercially due to cost, consumer acceptance issues, and technical feasibility. Huang et al. (2006) reported 2.5–4.3 log CFU inactivation of *E. coli* and *Salmonella* on apples by combining 20–40 ppm ClO_2 with ultrasound technology at 170 kHz. Ionizing radiation eliminated pathogens from fresh produce without negatively affecting nutritional or organoleptic properties (Fan and Sokorai, 2008; Niemira, 2008; Niemira et al., 2003, 2018; Bari et al., 2004; Mintier and Folley, 2006; Dhokane et al., 2006; Lacombe et al., 2017). In August 2008, the FDA approved the use of ionizing radiation for inactivating populations of foodborne pathogens on spinach and iceberg lettuce. Application of high-intensity light, monochromatic light, and UV light have shown promise (Kingsley et al., 2018a,b; Lacombe et al., 2016; Milly et al., 2007). Innovative means of minimizing turbidity and shadowing effects in process wash waters have been investigated (Milly et al., 2007). Cold plasma is an emerging antimicrobial process that has been applied to a variety of commodities including fresh and fresh-cut fruits and vegetables with lots of potential benefits (Hertrich et al., 2017; Min et al., 2017, 2018; Niemira, 2018).

8 Conclusion

The continuous product recalls and outbreaks of foodborne illnesses resulting from pathogen-contaminated fresh produce calls for ongoing and advanced studies into the efficacy of commercial and novel sanitizers to reduce bacterial populations on fresh produce. A major hurdle will be to find effective sanitizers that can be used at especially low concentrations to minimize excessive water use and environmental issues, comply with regulations, as well as reduce costs, all of which can enhance their utilization by the commercial fresh produce industry. Other relevant factors include whether the sanitizer can be classified as 'organic', and if it affects the sensory properties of fruits and vegetables. In addition, consideration of whether a final potable water rinse is required following treatment, and how tolerant the sanitizer is with respect to the presence of an organic load is important. Further research needs to be conducted into developing new compounds or combinations of compounds, which can meet the aforementioned criteria. Collaborative efforts among horticulturalists, microbiologists, food scientists, and chemists will be needed to advance this field and find novel solutions to prevent or lessen the number of foodborne outbreaks, as well as product recalls, associated with fresh produce.

9 Acknowledgements

The authors would like to thank Brianna Boleratz and Sharon Driedger for their assistance in formatting this manuscript. The U.S. Department of Agriculture is an equal opportunity provider and employer.

10 Where to look for further information

A number of pertinent publications may be useful for the reader interested in investigating the topic of fresh produce sanitizers further. Some of these include the following:

Gombas, D., Luo, Y., Brennan, J., Shergill, G., Petran, R., Walsh, R., Hau, H., Khurana, K., Zomorodi, B., Rosen, J., Varley, R. and Deng, K. 2017. Guidelines to validate control of cross-contamination during washing of fresh-cut leafy vegetables. *J. Food Prot.* 80(2), 312–30. doi:10.4315/0362-028X.JFP-16-258.

Prado-Silva, L., Cadavez, V., Gonzales-Barron, U., Rezende, A. C. B. and Sant'Ana, A. S. 2015. Meta-analysis of the effects of sanitizing treatments on *Salmonella*, *Escherichia coli* O157:H7 and *Listeria monocytogenes* inactivation in fresh produce. *Appl. Environ. Microbiol.* 81(23), 8008–21. doi:10.1128/AEM.02216-15.

van Haute, S., Sampers, I., Jacxsens, L. and Uyttendaele, M. 2015. Selection criteria for water disinfection techniques in agricultural practices. *Crit. Rev. Food Sci. Nutr.* 55, 1529–51.

11 References

Abadias, M., Usall, J., Oliveira, M., Alegre, I. and Viñas, I. 2008. Efficacy of neutral electrolyzed water (NEW) for reducing microbial contamination on minimally processed vegetables. *Int. J. Food Microbiol.* 123(1-2), 151-8. doi:10.1016/j.ijfoodmicro.2007.12.008.

Aguayo, E., Escalona, V. and Artés, F. 2006. Effect of the cyclic exposure to ozone gas on phytochemical, sensorial and microbial quality in whole and sliced tomatoes. *Postharvest Biol. Technol.* 39, 166–77.

Akbas, M. Y. and Ölmez, H. 2007. Effectiveness of organic acid, ozonated water and chlorine dippings on microbial reduction and storage quality of fresh-cut iceberg lettuce. *J. Sci. Food Agric.* 87(14), 2609-16. doi:10.1002/jsfa.3016.

Alasri, A., Valverde, M., Roques, C., Michel, G., Cabassud, C. and Aptel, P. 1993. Sporocidal properties of peracetic-acid and hydrogen peroxide, alone and in combination, in comparison with chlorine and formaldehyde for ultrafiltration membrane disinfection. *Can. J. Microbiol.* 39(1), 52–60. doi:10.1139/m93-008.

Allende, A., Selma, M. V., López-Gálvez, F., Villaescusa, R. and Gil, M. I. 2008. Role of commercial sanitizers and washing systems on epiphytic microorganisms and sensory quality of fresh-cut escarole and lettuce. *Postharvest Bio. Technol.* 49(1), 155–63. doi:10.1016/j.postharvbio.2007.12.010.

Allende, A., McEvoy, J., Tao, Y. and Luo, Y. 2009. Antimicrobial effect of acidified sodium chlorite, sodium chlorite, sodium hypochlorite, and citric acid on *Escherichia coli* O157:H7 and natural microflora of fresh-cut cilantro. *Food Control* 20(3), 230–4. doi:10.1016/j.foodcont.2008.05.009.

Al-Nabulsi, A. A., Osaili, T. M., Obaidat, H. M., Shaker, R. R., Awaisheh, S. S. and Holley, R. A. 2014. Inactivation of stressed *Escherichia coli* O157:H7 cells on the surfaces of rocket salad leaves by chlorine and peroxyacetic acid. *J. Food Prot.* 77(1), 32–9. doi:10.4315/0362-028X.JFP-13-019.

Artés, F. and Allende, A. 2005. Processing lines and alternative preservation techniques to prolong the shelf-life of minimally fresh processed leafy vegetables. *Eur. J. Hortic. Sci.* 70, 231–45.

Artés, F., Gómez, P., Aquayo, E., Escalona, V. and Artés-Hernández, F. 2009. Sustainable sanitation techniques for keeping quality and safety of fresh-cut plant commodities. *Postharvest Biol. Technol.* 51(3), 287–96. doi:10.1016/j.postharvbio.2008.10.003.

Bablon, G., Belamy, W., Billen, G., Bourbigot, M. and Daniel, F. 1991. Practical applications of ozone: principles and case studies. In: Langlais, G., Reckhow, D. A. and Brink, D. R. (Eds), *Ozone in Water Treatment: Application and Engineering*. Lewis Publisher, Inc., Chelsea, MI, pp. 133–316.

Banach, J. L., Sampers, I., Van Haute, S. and van der Fels-Klerx, H. J. 2015. Effect of disinfectants on preventing the cross-contamination of pathogens in fresh produce washing water. *Int. J. Environ. Res. Public Health* 12(8), 8658–77. doi:10.3390/ijerph120808658.

Bari, M. L., Al-Haq, M. I., Kawasaki, T., Nakauma, M., Todoriki, S., Kawamoto, S. and Isshiki, K. 2004. Irradiation to kill *Escherichia coli* O157:H7 and *Salmonella* on ready-to-eat radish and mung bean sprouts. *J. Food Prot.* 67(10), 2263-8. doi:10.4315/0362-028X-67.10.2263.

Barrera, M. J., Blenkinsop, R. and Warriner, K. 2012. The effect of different processing parameters on the efficacy of commercial post-harvest washing of minimally processed spinach and shredded lettuce. *Food Control* 25(2), 745–51. doi:10.1016/j.foodcont.2011.12.013.

Barth, M. M., Zhou, C., Mercier, J. and Payne, F. A. 1995. Ozone storage effects on anthocyanin content and fungal growth in blackberries. *J. Food Sci.* 60(6), 1286–8. doi:10.1111/j.1365-2621.1995.tb04575.x.

Baselice, A., Colantuoni, F., Lass, D. A., Nardone, G. and Stasi, A. 2017. Trends in EU consumers' attitude towards fresh-cut fruit and vegetables. *Food Qual. Pref.* 59, 87–96. doi:10.1016/j.foodqual.2017.01.008.

Behravesh, C. B., Mody, R. K., Jungk, J., Gaul, L., Redd, J. T., Chen, S., Cosgrove, S., Hedican, E., Sweat, D., Chavez-Hauser, L., Snow, S. L., Hanson, H., Nguyen, T. A., Sodha, S. V., Boore, A. L., Russo, E., Mikoleit, M., Theobald, L., Gerner-Smidt, P., Hoekstra, R. M., Angulo, F. J., Swerdlow, D. L., Tauxe, R. V., Griffin, P. M. and Williams, I. T. 2011. 2008 outbreak of *Salmonella* Saint Paul infections associated with raw produce. *N. Engl. J. Med.* 364(10), 918–27. doi:10.1056/NEJMoa1005741.

Benarde, M. A., Snow, W. B., Olivieri, V. P. and Davidson, B. 1967. Kinetics and mechanism of bacterial disinfection by chlorine dioxide. *Appl. Microbiol.* 15, 257–65.

Berg, J. D., Roberts, P. V. and Matin, A. 1986. Effect of chlorine dioxide on selected membrane functions of *Escherichia coli*. *J. Appl. Bacteriol.* 60(3), 213–20. doi:10.1111/j.1365-2672.1986.tb01075.x.

Bergmann, H., Koparal, A. T., Koparal, A. S. and Ehrig, F. 2008. The influence of products and by-products obtained by drinking water electrolysis on microorganisms. *Microchem. J.* 89(2), 98–107. doi:10.1016/j.microc.2007.12.007.

Bermúdez-Aguirre, D. and Barbosa-Cánovas, G. V. 2013. Disinfection of selected vegetables under nonthermal treatments: chlorine, acid citric, ultraviolet light and ozone. *Food Control* 29(1), 82–90. doi:10.1016/j.foodcont.2012.05.073.

Beuchat, L. R. 1998. Surface decontamination of fruits and vegetables eaten raw: a review. WHO/FSF/FOS/98.2. Food Safety Unit. World Health Organization. Available at: http://apps.who.int/iris/handle/10665/64435 (accessed on 17 July 2018).

Beuchat, L. R. 1999. Survival of Enterohemorrhagic *Escherichia coli* O157:H7 in bovine feces applied to lettuce and the effectiveness of chlorinated water as a disinfectant. *J. Food Prot.* 62(8), 845–9. doi:10.4315/0362-028X-62.8.845.

Beuchat, L. R. 2000. Use of sanitizers in raw fruit and vegetable processing. In: Alzamora, S. M., Tapia, M. S. and Lopez-Malo, A. (Eds), *Minimally Processed Fruits and Vegetables: Fundamental Aspects and Applications*. Aspen Publishers, Inc., Gaithersburg, MD, pp. 63–78.

Beuchat, L. R., Adler, B. B. and Lang, M. M. 2004. Efficacy of chlorine and a peroxyacetic acid sanitizer in killing *Listeria monocytogenes* on iceberg and romaine lettuce using simulated commercial processing conditions. *J. Food Prot.* 67(6), 1238–42. doi:10.4315/0362-028X-67.6.1238.

Beuchat, L. R., Ryu, J-H. 1997. Produce handling and processing practices. *Emerg. Infect. Dis.* 3(4), 459–65.

Buchholz, A. L., Davidson, G. R., Marks, B. P., Todd, E. C. D. and Ryser, E. T. 2012. Quantitative transfer of *Escherichia coli* O157:H7 to equipment during small-scale production of fresh-cut leafy greens. *J. Food Prot.* 75(7), 1184–97. doi:10.4315/0362-028X.JFP-11-489.

Butot, S., Putallaz, T., Amoroso, R. and Sanchez, G. 2009. Inactivation of enteric viruses in minimally processed berries and herbs. *Appl. Environ. Microbiol.* 75(12), 4155–61. doi:10.1128/AEM.00182-09.

California Food Emergency Response Team. 2007. Investigation of an *Escherichia coli* 0157:H7 outbreak associated with dole pre-packaged spinach. Available at: http://www.marlerclark.com/2006_Spinach_Report_Final_01.pdf (accessed on 27 December 2012).

Callejón, R. M., Rodríguez-Naranjo, M. I., Úbeda, C., Hornedo-Ortega, R., García-Parrilla, M. C. and Troncoso, A. M. 2015. Reported foodborne outbreaks due to fresh produce in the United States and European Union: trends and causes. *Foodborne Pathog. Dis.* 12(1), 32-8. doi:10.1089/fpd.2014.1821.

Castro-Ibáñez, I., Gil, M. I. and Allende, A. 2017. Ready-to-eat vegetables: current problems and potential solutions to reduce microbial risk in the production chain. *LWT Food Sci. Technol.* 85, 284-92. doi:10.1016/j.lwt.2016.11.073.

Centers for Disease Control and Prevention, Gould, L. H., Walsh, K. A., Vieira, A. R., Herman, K., Williams, I. T., Hall, A. J. and Cole, D.. 2013. Surveillance for foodborne disease outbreaks - United States, 1998-2008. *Morb. Mortal. Wkly. Rep. Surveill. Summ.* 62(2), 1-34.

Chaidez, C., Castro-del Campo, N., Heredia, J. B., Contreras-Angulo, L., González-Aguilar, G. and Ayala-Zavala, J. F. 2012. Chlorine. In: Gómez-López, V. M. (Ed.), *Decontamination of Fresh and Minimally Processed Produce* (1st edn.). Wiley-Blackwell, Oxford, UK, pp. 121-33.

Chang, A., Balaguero, A., Goodrich-Schneider, R. M. and Schneider, K. R. 2010. Food safety on the farm: good agricultural practices and good handling practices-traceback. University of Florida-IFAS Extension document FSHN10-04. Available at: http://edis.ifas.ufl.edu (accessed on 11 October 2016).

Cords, B. R., Burnett, S. L., Hilgren, J., Finley, M. and Magnuson, J. 2005. Sanitizers: halogens, surface-active agents, and peroxides. In: Davidson, P. M., Sofos, J. N. and Branen, A. L. (Eds), *Antimicrobials in Food* (3rd edn.). CRC Press, Boca Raton, FL, pp. 507-72.

Da Silva Felício, M. T., Hald, T., Liebana, E., Allende, A., Hugas, M., Nguyen-The, C., Johannessen, G. S., Niskanen, T., Uyttendaele, M. and McLauchlin, J. 2015. Risk ranking of pathogens in ready-to-eat unprocessed foods of non-animal origin (FoNAO) in the EU: initial evaluation using outbreak data (2007-2011). *Int. J. Food Microbiol.* 195, 9-19. doi:10.1016/j.ijfoodmicro.2014.11.005.

Davidson, G. R., Buchholz, A. L. and Ryser, E. T. 2013. Efficacy of commercial produce sanitizers against nontoxigenic *Escherichia coli* O157:H7 during processing of iceberg lettuce in a pilot-scale leafy green processing line. *J. Food Prot.* 76(11), 1838-45. doi:10.4315/0362-028X.JFP-13-111.

De Giusti, M., Aurigemma, C., Marinelli, L., Tufi, D., De Medici, D., Di Pasquale, S., De Vito, C. and Boccia, A. 2010. The evaluation of the microbial safety of fresh ready-to-eat vegetables produced by different technologies in Italy. *J. Appl. Microbiol.* 109(3), 996-1006. doi:10.1111/j.1365-2672.2010.04727.x.

Dhokane, V. S., Hajare, S., Shashiddhar, R., Sharma, A. and Bandekar, J. R. 2006. Radiation processing to ensure safety of minimally processed carrot (*Daucus carota*) and cucumber (*Cucumis sativus*): optimization of dose for the elimination of *Salmonella Typhimurium* and *Listeria monocytogenes*. *J. Food Prot.* 69(2), 444-8. doi:10.4315/0362-028X-69.2.444.

Direction générale de la concurrence, de la consommation et de la répression des fraudes. 2009. Arrête du 27 août 2009 modifiant l'arrêté du 19 octobre 2006 relatif à l'emploi d'auxiliaires technologiques dans la fabrication de certaines denrées alimentaires. NOR: ECEC0911390A. Available at: http://www.legifrance.gouv.fr/

affichTexte.do;jsessionid=?cidTexte=JORFT EXT000021158238&dateTexte=&oldAction=rechJO&categorieLien=id (accessed on 12 June 2011).

Doyle, M. P. and Erickson, M. C. 2008. Summer meeting 2007 - the problems with fresh produce: an overview. *J. Appl. Microbiol.* 105(2), 317–30. doi:10.1111/j.1365-2672.2008.03746.x.

Du, J., Han, Y. and Linton, R. H. 2002. Inactivation by chlorine dioxide gas (ClO_2) of *Listeria monocytogenes* spotted onto different apple surfaces. *J. Food Prot.* 19(5), 481–90. doi:10.1006/fmic.2002.0501.

Du, J., Fu, Y. and Wang, N. 2009. Effects of aqueous chlorine dioxide treatment on browning of fresh-cut lotus root. *LWT Food Sci. Technol.* 42(2), 654–9. doi:10.1016/j.lwt.2008.08.007.

Fan, X. and Sokorai, K. J. B. 2008. Retention of quality and nutritional value of 13 fresh-cut vegetables treated with low-dose radiation. *J. Food Sci.* 73(7), S367–72. doi:10.1111/j.1750-3841.2008.00871.x.

Fan, X. and Sokorai, K. J. B. 2015. Formation of trichloromethane in chlorinated water and fresh-cut produce and as a result of reaction with citric acid. *Postharvest Biol. Technol.* 109, 65–72. doi:10.1016/j.postharvbio.2015.06.009.

Fan, X., Annous, B. A., Keskinen, L. A. and Mattheis, J. P. 2009. Use of chemical sanitizers to reduce microbial populations and maintain quality of whole and fresh-cut cantaloupe. *J. Food Prot.* 72(12), 2453–60. doi:10.4315/0362-028X-72.12.2453.

Fan, X., Sokorai, K. J., Niemira, B. A., Mills, R. S. and Zhen, M. Y. 2012. Quality of gamma ray-irradiated iceberg lettuce and treatments to minimize irradiation-induced disorders. *Hort. Sci.* 47(8), 1108–12. doi:10.21273/HORTSCI.47.8.1108.

Fang, Q., Shang, C. and Chen, G. H. 2006. MS2 inactivation by chloride-assisted electrochemical disinfection. *J. Environ. Eng.* 132(1), 13–22. doi:10.1061/(ASCE)0733-9372(2006)132:1(13).

Fatemi, P. and Frank, J. F. 1999. Inactivation of *Listeria monocytogenes/Pseudomonas* biofilms by peracid sanitizers. *J. Food Prot.* 62(7), 761–5. doi:10.4315/0362-028X-62.7.761.

Fatica, M. K. and Schneider, K. R. 2009. The use of chlorination and alternative sanitizers in the produce industry. *CAB Rev.* 4(52), 1–10. doi:10.1079/PAVSNNR20094052.

Felkey, K., Archer, D. L., Goodrich, R. M., Bartz, J. A. and Schneider, K. R. 2006. Chorine disinfection of tomato surface wounds contaminated with *Salmonella* spp. *HortTechnology* 16, 253–6.

Feng, C. P., Suzuki, K., Zhao, S. Y., Sugiura, N., Shimada, S. and Maekawa, T. 2004. Water disinfection by electrochemical treatment. *Bioresour. Technol.* 94(1), 21–5. doi:10.1016/j.biortech.2003.11.021.

Finch, G. R. and Fairbairn, N. 1991. Comparative inactivation of poliovirus type 3 and MS2 coliphage in demand-free phosphate buffer by using ozone. *Appl. Environ. Microbiol.* 57, 3121–6.

Food and Agriculture Organization of the United Nations and World Health Organization. 2008. Microbial hazards in fruits and vegetables. Available at: http://www.fao.org/ag/agn/agns/files/FFV_2007_Final.pdf (accessed on 22 December 2011).

Food and Drug Administration, Center for Food Safety and Applied Nutrition. 2008. Guidance for Industry: guide to minimize food safety hazards for fresh-cut fruits and vegetables. *Federal Register* 72, 11364–8.

Freer, P. C. and Novy, F. G. 1902. On the formation, decomposition and germicidal action of benzoylacetyl and diacetyl peroxides. *J. Am. Chem. Soc.* 27, 161–93.

Garrett, E. H., Gorny, J. R., Beuchat, L. R., Farber, J. N., Harris, L. J., Parish, M. E., Suslow, T. V. and Busta, F. F. 2003. Microbiological safety of fresh and fresh-cut produce: description of the situation and economic impact. *Comp. Rev. Food Sci. Food Saf.* 2(s1), 13–37. doi:10.1111/j.1541-4337.2003.tb00029.x.

Gereffi, S., Sreedharan, A. and Schneider, K. R. 2015. Control of *Salmonella* cross-contamination between green round tomatoes in a model flume system. *J. Food Prot.* 78, 1–8.

Gil, M. I., Selma, M. V., López-Gálvez, F. and Allende, A. 2009. Fresh-cut product sanitation and wash water disinfection: problems and solutions. *Int. J. Food Microbiol.* 134(1-2), 37–45. doi:10.1016/j.ijfoodmicro.2009.05.021.

Gombas, D., Luo, Y., Brennan, J., Shergill, G., Petran, R., Walsh, R., Hau, H., Khurana, K., Zomorodi, B., Rosen, J., Varley, R. and Deng, K. 2017. Guidelines to validate control of cross-contamination during washing of fresh-cut leafy vegetables. *J. Food Prot.* 80(2), 312–30. doi:10.4315/0362-028X.JFP-16-258.

Gómez-López, V. M. 2012. Chlorine dioxide. In: Gómez-López, V. M. (Ed.), *Decontamination of Fresh and Minimally Processed Produce* (1st edn.). Wiley-Blackwell, Oxford, UK, pp. 165–75.

Gómez-López, V. M., Ragaert, P., Ryckeboer, J., Jeyachandran, V., Debevere, J. and Devlieghere, F. 2007. Shelf life of minimally processed cabbage treated with neutral electrolyzed oxidizing water and stored under equilibrium modified atmosphere. *Int. J. Food Microbiol.* 117, 91–8.

Gómez-López, V. M., Rajkovic, A., Ragaert, P., Smigic, N. and Devlieghere, F. 2009. Chlorine dioxide for minimally processed produce preservation: a review. *Trends Food Sci. Technol.* 20(1), 17–26. doi:10.1016/j.tifs.2008.09.005.

Gómez-López, V. M., Marín, A., Allende, A., Beuchat, L. R. and Gil, M. I. 2013. Post-harvest handling conditions affect internalization of *Salmonella* in baby spinach during washing. *J. Food Prot.* 76(7), 1145–51. doi:10.4315/0362-028X.JFP-12-539.

Gómez-López, V. M., Lannoo, A.-S., Gil, M. I. and Allende, A. 2014. Minimum free chlorine residual level required for the inactivation of *Escherichia coli* O157:H7 and trihalomethane generation during dynamic washing of fresh-cut spinach. *Food Control* 42, 132–8. doi:10.1016/j.foodcont.2014.01.034.

Gonzalez, R. J., Luo, Y., Ruiz-Cruz, S. and McEvoy, J. L. 2004. Efficacy of sanitizers to inactivate *Escherichia coli* O157:H7 on fresh-cut carrot shreds under simulated process water conditions. *J. Food Prot.* 67(11), 2375–80. doi:10.4315/0362-028X-67.11.2375.

Goodburn, C. and Wallace, C. A. 2013. The microbiological efficacy of decontamination methodologies for fresh produce: a review. *Food Control* 32(2), 418–27. doi:10.1016/j.foodcont.2012.12.012.

Gordon, G. and Rosenblatt, A. A. 2005. Chlorine dioxide: the current state of the art. *Ozone Sci. Eng.* 27(3), 203–7. doi:10.1080/01919510590945741.

Grace Ho, K. L., Luzuriaga, D. A., Rodde, K. M., Tang, S. and Phan, C. 2011. Efficacy of a novel sanitizer composed of lactic acid and peroxyacetic acid against single strains of nonpathogenic *Escherichia coli* K-12, *Listeria innocua*, and *Lactobacillus plantarum* in aqueous solution and on surfaces of romaine lettuce and spinach. *J. Food Prot.* 74(9), 1468–74. doi:10.4315/0362-028X.JFP-11-110.

Graham, D. M. 1997. Use of ozone for food processing. *Food Technol.* 51, 72–5.

Greene, S. K., Daly, E. R., Talbot, E. A., Demma, L. J., Holzbauer, S., Patel, N. J., Hill, T. A., Walderhaug, M. O., Hoekstra, R. M., Lynch, M. F. and Painter, J. A. 2008. Recurrent multistate outbreak of *Salmonella* Newport associated with tomatoes from contaminated fields, 2005. *Epidemiol. Infect.* 136(2), 157–65. doi:10.1017/S095026880700859X.

Gunarathna, N. M., Mancl, K. and Kaletunc, G. 2014. Electrochemical disinfection in the fresh-cut produce industry. Ohio State University Extension Service. AEX-322. Available at: https://ohioline.osu.edu/factsheet/AEX-322 (accessed on 13 December 2018).

Gurtler, J. B., Smelser, A. M., Niemira, B. A., Jin, T. Z., Yan, X. and Geveke, D. J. 2012. Inactivation of *Salmonella enterica* on tomato stem scars by sanitizing solutions and vacuum perfusion. *Int. J. Food Microbiol.* 159(2), 84–92. doi:10.1016/j.ijfoodmicro.2012.08.014.

Gurtler, J. B., Bailey, R. B., Jin, T. Z. and Fan, X. 2014. Reduction of an *E. coli* O157:H7 and *Salmonella* composite on fresh strawberries by varying antimicrobial washes and vacuum perfusion. *Int. J. Food Microbiol.* 189, 113–8. doi:10.1016/j.ijfoodmicro.2014.08.005.

Guzel-Seydim, Z., Bever, P. I. and Greene, A. K. 2004. Efficacy of ozone to reduce bacterial populations in the presence of food components. *Food Microbiol.* 21(4), 475–9. doi:10.1016/j.fm.2003.10.001.

Han, Y., Sherman, D. M., Linton, R. H., Nielsen, S. S. and Nelson, P. E. 2000. The effects of washing and chlorine dioxide gas on survival and attachment of *Escherichia coli* O157:H7 to green pepper surfaces. *Food Microbiol.* 17(5), 521–33. doi:10.1006/fmic.2000.0343.

Heaton, J. C. and Jones, K. 2008. Microbial contamination of fruit and vegetables and the behaviour of enteropathogens in the phyllosphere: a review. *J. Appl. Microbiol.* 104(3), 613–26. doi:10.1111/j.1365-2672.2007.03587.x.

Hellström, S., Kervinen, R., Lyly, M., Ahvenainen-Rantala, R. and Korkeala, H. 2006. Efficacy of disinfectants to reduce *Listeria monocytogenes* on precut iceberg lettuce. *J. Food Prot.* 69(7), 1565–70. doi:10.4315/0362-028X-69.7.1565.

Hertrich, S. M., Boyd, G., Sites, J. and Niemira, B. A. 2017. Cold plasma inactivation of *Salmonella* in pre-packaged, mixed salads is influenced by cross-contamination sequence. *J. Food Prot.* 80(12), 2132–6. doi:10.4315/0362-028X.JFP-17-242.

Hilgren, J. D. and Salverda, J. A. 2000. Antimicrobial efficacy of a peroxyacetic/octanoic acid mixture in fresh-cut vegetable process waters. *J. Food Sci.* 65(8), 1376–9. doi:10.1111/j.1365-2621.2000.tb10615.x.

Hilgren, J. D., Swanson, K. M. J., Diez-Gonzalez, F. and Cords, B. 2007. Inactivation of *Bacillus anthracis* spores by liquid biocides in the presence of food residue. *Appl. Environ. Microbiol.* 73(20), 6370–7. doi:10.1128/AEM.00974-07.

Holvoet, K., Jacxsens, L., Sampers, I. and Uyttendaele, M. 2012. Insight into the prevalence and distribution of microbial contamination to evaluate water management in the fresh produce processing industry. *J. Food Prot.* 75(4), 671–81. doi:10.4315/0362-028X.JFP-11-175.

Hricova, D., Stephan, R. and Zweifel, C. 2008. Electrolyzed water and its application in the food industry. *J. Food Prot.* 71(9), 1934–47. doi:10.4315/0362-028X-71.9.1934.

Hua, G. and Reckhow, D. A. 2007. Comparison of disinfection byproduct formation from chlorine and alternative disinfectants. *Water Res.* 41(8), 1667–78. doi:10.1016/j.watres.2007.01.032.

Huang, J. L., Wang, L., Ren, N. Q., Ma, F. and Juli. 1997. Disinfection effect of chlorine dioxide on bacteria in water. *Water Res.* 31(3), 607–13. doi:10.1016/S0043-1354(96)00275-8.

Huang, T. S., Xu, C. L., Walker, K., West, P., Zhang, S. Q. and Weese, J. 2006. Decontamination efficacy of combined chlorine dioxide with ultrasonication on apples and lettuce. *J. Food Sci.* 71(4), M134–9. doi:10.1111/j.1750-3841.2006.00015.x.

Inatsu, Y., Kitagawa, T., Bari, M. L., Nei, D., Juneja, V. and Kawamoto, S. 2010. Effectiveness of acidified sodium chlorite and other sanitizers to control *Escherichia coli* O157:H7 on tomato surfaces. *Foodborne Pathog. Dis.* 7(6), 629–35. doi:10.1089/fpd.2009.0429.

Issa-Zacharia, A., Kamitani, Y., Miwa, N., Muhimbula, H. and Iwasaki, K. 2011. Application of slightly acidic electrolyzed water as a potential non-thermal food sanitizer for decontamination of fresh ready-to-eat vegetables and sprouts. *Food Control* 22(3–4), 601–7. doi:10.1016/j.foodcont.2010.10.011.

Izumi, H. 1999. Electrolyzed water as a disinfectant for fresh-cut vegetables. *J. Food Sci.* 64(3), 536–9. doi:10.1111/j.1365-2621.1999.tb15079.x.

Jeng, D. K. and Woodworth, A. G. 1990. Chlorine dioxide gas sterilization under square-wave conditions. *Appl. Environ. Microbiol.* 56(2), 514–9.

Jeong, J., Kim, J. Y., Cho, M., Choi, W. and Yoon, J. 2007. Inactivation of *Escherichia coli* in the electrochemical disinfection process using a Pt anode. *Chemosphere* 67(4), 652–9. doi:10.1016/j.chemosphere.2006.11.035.

Joshi, K., Mahendran, R., Alagusundaram, K., Norton, T. and Tiwari, B. K. 2013. Novel disinfectants for fresh produce. *Trends Food Sci. Technol.* 34(1), 54–61. doi:10.1016/j.tifs.2013.08.008.

Kim, J. G., Yousef, A. E. and Chism, G. W. 1999a. Use of ozone to inactivate microorganism on lettuce. *J. Food Saf.* 19(1), 17–34. doi:10.1111/j.1745-4565.1999.tb00231.x.

Kim, J. G., Yousef, A. E. and Dave, S. 1999b. Application of ozone for enhancing the microbiological safety and quality of food: a review. *J. Food Prot.* 62(9), 1071–87. doi:10.4315/0362-028X-62.9.1071.

Kim, C., Hung, Y. C. and Brackett, R. E. 2000. Efficacy of electrolyzed oxidizing (EO) and chemically modified water on different types of foodborne pathogens. *Int. J. Food Microbiol.* 61(2–3), 199–207. doi:10.1016/S0168-1605(00)00405-0.

Kim, C., Hung, Y. C., Brackett, R. E. and Lin, C. S. 2003. Efficacy of electrolyzed oxidizing water in inactivating *Salmonella* on alfalfa seeds and sprouts. *J. Food Prot.* 66(2), 208–14. doi:10.4315/0362-028X-66.2.208.

Kingsley, D. H., Perez-Perez, R. E., Boyd, G., Sites, J. and Niemira, B. A. 2018a. Evaluation of 405-nm monochromatic light for inactivation of Tulane virus on blueberry surfaces. *J. Appl. Microbiol.* 124(4), 1017–22. doi:10.1111/jam.13638.

Kingsley, D. H., Pérez-Pérez, R. E., Niemira, B. A. and Fan, X. 2018b. Evaluation of gaseous chlorine dioxide for the inactivation of Tulane virus on blueberries. *Int. J. Food Microbiol.* 273, 28–32. doi:10.1016/j.ijfoodmicro.2018.01.024.

Kitis, M. 2004. Disinfection of wastewater with peracetic acid: a review. *Environ. Int.* 30(1), 47–55. doi:10.1016/S0160-4120(03)00147-8.

Klaiber, R. G., Baur, S., Magel, L., Hammes, W. P. and Carle, R. 2004. Quality of shredded, packaged carrots as affected by different washing treatments. *J. Food Sci.* 69(4), SNQ161–6. doi:10.1111/j.1365-2621.2004.tb06358.x.

Koivunen, J. and Heinonen-Tanski, H. 2005. Inactivation of enteric microorganisms with chemical disinfectants, UV irradiation and combined chemical/UV treatments. *Water Res.* 39(8), 1519–26. doi:10.1016/j.watres.2005.01.021.

Koseki, S., Yoshida, K., Kamitani, Y. and Itoh, K. 2003. Influence of inoculation method, spot inoculation site, and inoculation size of the efficacy of acidic electrolyzed water against pathogens on lettuce. *J. Food Prot.* 66(11), 2010–6. doi:10.4315/0362-028X-66.11.2010.

Lacombe, A., Niemira, B. A., Sites, J., Boyd, G., Gurtler, J. B., Tyrell, B. and Fleck, M. 2016. Reduction of enteric pathogens and potential surrogates using integrated, high intensity 405nm LED light on the surface of almonds. *J. Food Prot.* 79(11), 1840–5. doi:10.4315/0362-028X.JFP-15-418.

Lacombe, A., Breard, A., Hwang, C.-A., Hill, D., Fan, X., Huang, L., Yoo, B. K., Niemira, B. A., Gurtler, J. B. and Wu, V. C. H. 2017. Inactivation of *Toxoplasma gondii* on blueberries using low dose irradiation without affecting quality. *Food Control* 73, 981–5. doi:10.1016/j.foodcont.2016.10.011.

Lemons, K. E. and Taylor Fresh Food, Inc. 2009. Antimicrobial compositions and methods of use thereof. United States Patent #20,090,192,231.

Li, Y., Brackett, R. E., Chen, J. and Beuchat, L. R. 2001. Survival and growth of *Escherichia coli* O157:H7 inoculated onto cut lettuce before or after heating in chlorinated water, followed by storage at 5 or 15°C. *J. Food Prot.* 64(3), 305–9. doi:10.4315/0362-028X-64.3.305.

Li, Y., Brackett, R. E., Chen, J. and Beuchat, L. R. 2002. Mild heat treatment of lettuce enhances growth of Listeria monocytogenes during subsequent storage at 5°C or 15°C. *J. Appl. Microbiol.* 92(2), 269–75. doi:10.1046/j.1365-2672.2002.01530.x.

Li, X. Y., Diao, H. F., Fan, F. X. J., Gu, J. D., Ding, F. and Tong, A. S. F. 2004. Electrochemical wastewater disinfection: identification of its principal germicidal actions. *J. Environ. Eng.* 130(10), 1217–21. doi:10.1061/(ASCE)0733-9372(2004)130:10(1217).

Liao, C. H. 2009. Acidified sodium chlorite as an alternative to chlorine for elimination of *Salmonella* on alfalfa seeds. *J. Food Sci.* 74(4), M159–64. doi:10.1111/j.1750-3841.2009.01125.x.

Liao, L. B., Chen, W. M. and Xiao, X. M. 2007. The generation and inactivation mechanism of oxidation-reduction potential of electrolyzed oxidizing water. *J. Food Eng.* 78(4), 1326–32. doi:10.1016/j.jfoodeng.2006.01.004.

Lin, C.-S., Wu, C., Yeh, J.-Y. and Saalia, F. K. 2005. The evaluation of electrolyzed water as an agent for reducing microorganisms on vegetables. *Int. J. Food Sci. Technol.* 40(5), 495–500. doi:10.1111/j.1365-2621.2004.00947.x.

López-Gálvez, F., Allende, A., Selma, M. V. and Gil, M. I. 2009. Prevention of *Escherichia coli* cross-contamination by different commercial sanitizers during washing of fresh-cut lettuce. *Int. J. Food Microbiol.* 133(1–2), 167–71. doi:10.1016/j.ijfoodmicro.2009.05.017.

López-Gálvez, F., Allende, A., Truchado, P., Martínez-Sánchez, A., Tudela, J. A., Selma, M. V. and Gil, M. I. 2010a. Suitability of aqueous chlorine dioxide versus sodium hypochlorite as an effective sanitizer for preserving quality fresh-cut lettuce while avoiding by-product formation. *Postharvest Biol. Technol.* 55, 53–60.

López-Gálvez, F., Gil, M. I., Truchado, P., Selma, M. V. and Allende, A. 2010b. Cross-contamination of fresh-cut lettuce after a short-term exposure during pre-washing cannot be controlled after subsequent washing with chlorine dioxide or sodium hypochlorite. *Food Microbiol.* 27(2), 199–204. doi:10.1016/j.fm.2009.09.009.

Luo, Y. G. 2007. Fresh-cut produce wash water reuse affects water quality and packaged product quality and microbial growth in Romaine lettuce. *HortSci.* 42(6), 1413–9. doi:10.21273/HORTSCI.42.6.1413.

Luo, Y., Nou, X., Yang, Y., Alegre, I., Turner, E., Feng, H., Abadias, M. and Conway, W. 2011. Determination of free chlorine concentrations needed to prevent *Escherichia coli*

O157:H7 cross-contamination during fresh-cut produce wash. *J. Food Prot.* 74(3), 352–8. doi:10.4315/0362-028X.JFP-10-429.

Luo, Y., Nou, X., Millner, P., Zhou, B., Shen, C., Yang, Y., Wu, Y., Wang, Q., Feng, H. and Shelton, D. 2012. A pilot plant scale evaluation of a new process aid for enhancing chlorine efficacy against pathogen survival and cross-contamination during produce wash. *Int. J. Food Microbiol.* 158(2), 133–9. doi:10.1016/j.ijfoodmicro.2012.07.008.

Mahovic, M., Bartz, J. A., Schneider, K. R. and Tenney, J. D. 2009. Chlorine dioxide gas from an aqueous solution: reduction of *Salmonella* in wounds on tomato fruit and movement to sinks in a treatment chamber. *J. Food Prot.* 72(5), 952–8. doi:10.4315/0362-028X-72.5.952.

Martin, H. and Maris, P. 2012. Synergism between hydrogen peroxide and seventeen acids against six bacterial strains. *Int. J. Appl. Microbiol.* 113(3), 578–90. doi:10.1111/j.1365-2672.2012.05364.x.

Martinez-Huitle, C. A. and Brillas, E. 2008. Electrochemical alternatives for drinking water disinfection. *Angew. Chem. Int. Ed. Engl.* 47(11), 1998–2005. doi:10.1002/anie.200703621.

Milly, P. J., Toledo, R. T., Chen, J. and Kazem, B. 2007. Hydrodynamic cavitation to improve bulk fluid to surface mass transfer in a nonimmersed ultraviolet system for minimal processing of opaque and transparent fluid foods. *J. Food Sci.* 72(9), M407–13. doi:10.1111/j.1750-3841.2007.00518.x.

Min, S. C., Roh, S. H., Boyd, G., Sites, J. E., Uknalis, J., Fan, X. and Niemira, B. A. 2017. Inactivation of *Escherichia coli* O157:H7 and aerobic microorganisms in Romaine lettuce packaged in a commercial polyethylene terephthalate container using atmospheric cold plasma. *J. Food Prot.* 80(1), 35–43. doi:10.4315/0362-028X.JFP-16-148.

Min, S. C., Roh, S. H., Niemira, B. A., Boyd, G., Sites, J. E., Fan, X., Sokorai, K. and Jin, T. Z. 2018. In-package atmospheric cold plasma treatment of bulk grape tomatoes for their microbiological safety and preservation. *Food Res. Internat.* 108, 378–86.

Mintier, A. M. and Folley, D. M. 2006. Electron bean and gamma irradiation effectively reduce *Listeria monocytogenes* populations on chopped romaine lettuce. *J. Food Prot.* 69(3), 570–4. doi:10.4315/0362-028X-69.3.570.

Mohan, A., Dunn, J., Hunt, M. C. and Sizer, C. E. 2009. Inactivation of *Bacillus atrophaeus* spores with surface-active peracids and characterization of formed free radicals using electron spin resonance spectroscopy. *J. Food Sci.* 74(7), M411–7. doi:10.1111/j.1750-3841.2009.01293.x.

Nei, D., Choi, J. W., Bari, M. L., Kawasaki, S., Kawamoto, S. and Inatsu, Y. 2009. Efficacy of chlorine and acidified sodium chlorite on microbial population and quality changes of spinach leaves. *Foodborne Pathog. Dis.* 6(5), 541–6. doi:10.1089/fpd.2008.0237.

Nguyen, C. and Carlin, F. 1994. The microbiology of minimally processed fresh fruits and vegetables. *Crit. Rev. Food Sci. Nutr.* 34(4), 371–401. doi:10.1080/10408399409527668.

Niemira, B. A. 2008. Irradiation compared with chlorination for elimination of *Escherichia coli* O157:H7 internalized in lettuce leaves: influence of lettuce variety. *J. Food Sci.* 73(5), M208–13. doi:10.1111/j.1750-3841.2008.00746.x.

Niemira, B. A. 2018. Irradiation of fruits and fruit products. In: Rosenthal, A., Deliza, R., Welti-Chanes, J. and Barbosa-Canovas, G. V. (Eds), *Fruit Preservation: Novel and Conventional Technologies*. Springer, New York, NY, pp.399–416. Chapter 14.

Niemira, B. A., Fan, X., Sokorai, K. J. B. and Sommers, C. H. 2003. Ionising radiation sensitivity of *Listeria monocytogenes* ATCC 49594 and *Listeria innocua* ATCC 51742 inoculated on endive (*Chichorium endive*). *J. Food Prot.* 66(6), 993–8. doi:10.4315/0362-028X-66.6.993.

Niemira, B. A., Boyd, G. and Sites, J. 2018. Cold plasma inactivation of Escherichia coli O157:H7 biofilms. *Front. Sustain. Food Syst.* 2, 47 doi:10.3389/fsufs.2018.00047.

NORS, CDC National Outbreak Reporting System. 2019. NORS Dashboard, keyword: "lettuce"; date range 2010-2017. Available at: https://wwwn.cdc.gov/norsdashboard/ (accessed on 1 March 2019).

Nygard, K., Lassen, J., Vold, L., Andersson, Y., Fisher, I., Lofdahl, S., Threlfall, J., Luzzi, I., Peters, T., Hampton, M., Torpdahl, M., Kapperud, G. and Aavitsland, P. 2008. Outbreak of *Salmonella* Thompson infections linked to imported rucola lettuce. *Foodborne Pathog. Dis.* 5(2), 165–73. doi:10.1089/fpd.2007.0053.

Olaimat, A. N. and Holley, R. A. 2012. Factors influencing the microbial safety of fresh produce: a review. *Food Microbiol.* 32(1), 1–19. doi:10.1016/j.fm.2012.04.016.

Ölmez, H. and Kretzschmar, U. 2009. Potential alternative disinfection methods for organic fresh-cut industry for minimizing water consumption and environmental impact. *LWT Food Sci. Technol.* 42(3), 686–93. doi:10.1016/j.lwt.2008.08.001.

Ongeng, D., Devlieghere, F., Coosemans, J. and Ryckeboer, J. 2006. The efficacy of electrolyzed oxidizing water for inactivating spoilage microorganisms in process water and on minimally processed vegetables. *Int. J. Food Microbiol.* 109, 187–97.

Orr, R. V. and Beuchat, L. R. 2000. Efficacy of disinfectants in killing spores of *Alicyclobacillus acidoterrestris* and performance of media for supporting colony development by survivors. *J. Food Prot.* 63(8), 1117–22. doi:10.4315/0362-028X-63.8.1117.

Ortega, Y. R., Mann, A., Torres, M. P. and Cama, V. 2008. Efficacy of gaseous chlorine dioxide as a sanitizer against *Cryptosporidium parvum*, *Cyclospora cayetanensis*, and *Encephalitozoon intestinalis* on produce. *J. Food Prot.* 71(12), 2410–4. doi:10.4315/0362-028X-71.12.2410.

Paola, C. L., Rocío, C. V. and Marcela, M. 2005. Effectiveness of electrolyzed oxidizing water for inactivation *Listeria monocytogenes* in lettuce. *Univ. Sci.* 10, 97–108.

Parish, M. E., Beuchat, L. R., Suslow, T. V., Harris, L. J., Garrett, E. H., Farber, J. N. and Busta, F. F. 2003. Methods to reduce/eliminate pathogens from fresh and fresh-cut produce. *Comp. Rev. Food Sci. Food Saf.* 2(s1), 161–73. doi:10.1111/j.1541-4337.2003.tb00033.x.

Park, C.-M., Hung, Y. -C., Doyle, M. P., Ezeike, G. O. I. and Kim, C. 2001. Pathogen reduction and quality of lettuce treated with electrolyzed oxidizing and acidified chlorinated water. *J. Food Sci.* 66(9), 1368–72. doi:10.1111/j.1365-2621.2001.tb15216.x.

Peeters, J. E., Mazas, E. A., Masschelein, W. J., de Maturana, I. V. M. and Debacker, E. 1989. Effect of disinfection of drinking water with ozone or chlorine dioxide on the survival of *Cryptosporidium parvum* oocysts. *Appl. Environ. Microbiol.* 55, 1519–22.

Perez, A. G., Sanz, C., Rios, J. J., Olias, R. and Olias, J. M. 1999. Effects of ozone treatment on postharvest strawberry quality. *J. Agric. Food Chem.* 47(4), 1652–6. doi:10.1021/jf980829l.

Phua, L. K., Neo, S. Y., Khoo, G. H. and Yuk, H. G. 2014. Comparison of the efficacy of various sanitizers and hot water treatment in inactivating inoculated foodborne pathogens and natural microflora on mung bean sprouts. *Food Control* 42, 270–6. doi:10.1016/j.foodcont.2014.02.013.

Prado-Silva, L., Cadavez, V., Gonzales-Barron, U., Rezende, A. C. B. and Sant'Ana, A. S. 2015. Meta-analysis of the effects of sanitizing treatments on *Salmonella*, *Escherichia coli* O157:H7 and *Listeria monocytogenes* inactivation in fresh produce. *Appl. Environ. Microbiol.* 81(23), 8008–21. doi:10.1128/AEM.02216-15.

Predmore, A. and Li, J. R. 2011. Enhanced removal of a human Norovirus surrogate from fresh vegetables and fruits by a combination of surfactants and sanitizers. *Appl. Environ. Microbiol.* 77(14), 4829–38. doi:10.1128/AEM.00174-11.

Rahman, S. M. E., Khan, I. and Oh, D. H. 2016. Electrolyzed water as a novel sanitizer in the food industry: current trends and future perspectives. *Compr. Rev. Food Sci. Food Saf.* 15(3), 471–90. doi:10.1111/1541-4337.12200.

Reina, L. D., Fleming, H. P. and Humphries, E. G. 1995. Microbiological of cucumber hydrocooling water with chlorine dioxide. *J. Food Prot.* 58(5), 541–6. doi:10.4315/0362-028X-58.5.541.

Restaino, L., Frampton, E. W., Hemphill, J. B. and Palnikar, P. 1995. Efficacy of ozonated water against various food-related microorganisms. *Appl. Environ. Microbiol.* 61(9), 3471–5.

Richardson, S. D., Thruston, A. D. J., Caughran, T. V., Collette, T. W., Patterson, K. S. and Lykins, B. W. J. 1998. Chemical by-products of chlorine and alternative disinfectants. *Food Technol.* 52, 58–61.

Richardson, S. D., Thruston Jr., A. D., Caughran, T. V., Chen, P. H., Collette, T. W., Schenck, K. M., Lykins Jr., B. W., Rav-acha, C. and Glezer, V. 2000. Identification of new drinking water disinfection byproducts from ozone, chlorine dioxide, chloramine, and chlorine. *Water Air Soil Pollut.* 123(1/4), 95–102. doi:10.1023/A:1005265509813.

Rico, D., Martin-Diana, A. B., Barat, J. M. and Barry-Ryan, C. 2007. Extending and measuring the quality of fresh-cut fruit and vegetables: a review. *Trends Food Sci. Technol.* 18(7), 373–86. doi:10.1016/j.tifs.2007.03.011.

Rico, D., Martín-Diana, A. B., Barry-Ryan, C., Frías, J. M., Henehan, G. T. M. and Barat, J. M. 2008. Use of neutral electrolyzed water (EW) for quality maintenance and shelf-life extension of minimally processed lettuce. *Innov. Food Sci. Emerg. Technol.* 9(1), 37–48. doi:10.1016/j.ifset.2007.05.002.

Ruiz-Cruz, S., Acedo-Félix, E., Díaz-Cinco, M., Islas-Osuna, M. A. and González-Aguilar, G. A. 2007. Efficacy of sanitizers in reducing *Escherichia coli* O157:H7, *Salmonella* spp. and *Listeria monocytogenes* populations on fresh-cut carrots. *Food Control* 18(11), 1383–90. doi:10.1016/j.foodcont.2006.09.008.

Sadiq, R. and Rodriguez, M. J. 2004. Disinfection by-products (DBP's) in drinking water and predictive models for their occurrence: a review. *Sci. Total Environ.* 321(1–3), 21–46. doi:10.1016/j.scitotenv.2003.05.001.

Sapers, G. M., Miller, R. L. and Mattrazzo, A. M. 1999. Effectiveness of sanitizing agents in inactivating *Escherichia coli* in golden delicious apples. *J. Food Sci.* 64(4), 734–7. doi:10.1111/j.1365-2621.1999.tb15121.x.

Sarig, P., Zahavi, T., Zutkhi, Y., Yannai, S., Lisker, N. and Ben-Arie, R. 1996. Ozone for control for post-harvest decay of table grapes caused by Rhizopus stolonifera. *Physiol. Mol. Pl. Pathol.* 48, 403–15.

Selma, M. V., Ibañez, A. M., Allende, A., Cantwell, M. and Suslow, T. 2008. Effect of gaseous ozone and hot water on microbial and sensory quality of cantaloupe and potential transference of *Escherichia coli* O157:H7 during cutting. *Food Microbiol.* 25(1), 162–8. doi:10.1016/j.fm.2007.06.003.

Seymour, I. J. 1999. Review of current industry practice on fruit and vegetable decontamination. Gloucestershire Review 14. Campden and Chorleywood Food Research Association Group, Chippin Campden.

Shaw, R. K., Berger, C. N., Feys, B., Knutton, S., Pallen, M. J. and Frankel, G. 2008. Enterohemorrhagic *Escherichia coli* exploits espa filaments for attachment to salad leaves. *Appl. Environ. Microbiol.* 74(9), 2908–14. doi:10.1128/AEM.02704-07.

Shen, C. 2014. Evaluation of chlorine efficacy against *Escherichia coli* O157:H7 survival and cross-contamination during continuous produces washing process with water quality change. *Int. J. Food Sci. Nutr. Diet.* 3, 201–7.

Sheng, L., Hanrahan, I., Sun, X., Taylor, M. H., Mendoza, M. and Zhu, M. J. 2018. Survival of *Listeria innocua* on Fuji apples under commercial cold storage with or without low dose continuous ozone gaseous. *Food Microbiol.* 76, 21–8. doi:10.1016/j.fm.2018.04.006.

Simmons, J. L., Ryu, J. H. and Beuchat, L. R. 2006. Comparison of treatment of fresh-cut lettuce and diced tomatoes with sodium hypochlorite and calcium hypochlorite for effects on microbiological and sensory qualities. *Food Prot. Trends* 26, 662–7.

Singla, R., Ganguli, A. and Ghosh, M. 2011. An effective combined treatment using malic acid and ozone inhibits *Shigella* spp. on sprouts. *Food Control* 22(7), 1032–9. doi:10.1016/j.foodcont.2010.12.012.

Skog, L. J. and Chu, C. L. 2001. Ozone technology for shelf life extension of fruits and vegetables. *Acta Hortic.* 553(553), 431–2. doi:10.17660/ActaHortic.2001.553.100.

Slayton, R. B., Turabelidze, G., Bennett, S. D., Schwensohn, C. A., Yaffee, A. Q., Khan, F., Butler, C., Trees, E., Ayers, T. L., David, M. L., Laufer, A. S., Gladbach, S., Williams, I. and Gieraltowski, L. B. 2013. Outbreak of shiga toxin-producing *Escherichia coli* (STEC) O157:H7 associated with romaine lettuce consumption, 2011. *PLoS ONE* 8(2), e55300. doi:10.1371/journal.pone.0055300.

Small, D. A., Chang, W., Toghrol, F. and Bently, W. E. 2007. Comparative global transcription analysis of sodium hypochlorite, peracetic acid, and hydrogen peroxide on *Pseudomonas aeruginosa*. *Appl. Microbiol. Biotechnol.* 76(5), 1093–105. doi:10.1007/s00253-007-1072-z.

Söderström, A., Österberg, P., Lindqvist, A., Jönsson, B., Lindberg, A., Blide Ulander, S., Welinder-Olsson, C., Löfdahl, S., Kaijser, B., De Jong, B., Kühlmann-Berenzon, S., Boqvist, S., Eriksson, E., Szanto, E., Andersson, S., Allestam, G., Hedenström, I., Ledet Muller, L. and Andersson, Y. 2008. A large *Escherichia coli* O157 outbreak in Sweden associated with locally produced lettuce. *Food. Path. Dis.* 5(3), 339–49. doi:10.1089/fpd.2007.0065.

Sreedharan, A., Li, Y., De, J., Gutierrez, A., Silverberg, R. and Schneider, K. R. 2017. Determination of optimum sanitizer levels for prevention of *Salmonella* cross-contamination of mature round tomatoes in a laboratory model flume system. *J. Food Prot.* 80(9), 1436–42. doi:10.4315/0362-028X.JFP-17-032.

Stopforth, J. D., Mai, T., Kottapalli, B. and Samadpour, M. 2008. Effect of acidified sodium chlorite, chlorine, and acidic electrolyzed water on *Escherichia coli* O157:H7, *Salmonella*, and *Listeria monocytogenes* inoculated onto leafy greens. *J. Food Prot.* 71(3), 625–8. doi:10.4315/0362-028X-71.3.625.

Suslow, T. V. 2001. Water disinfection: a practical approach to calculating dose values for preharvest and postharvest applications. Publication 7256. Division of Agriculture and Natural Resources, University of California, Oakland, CA, pp. 1–4.

Suslow, T. V. 2004. Oxidation-reduction potential (ORP) for water disinfection monitoring, control, and documentation. ANR Catalog Publ. No. 8149. University of California, Davis, CA.

Tan, S. Y., Miks-Krajnik, M., Neo, S. Y., Tan, A., Khoo, G. H. and Yuk, H. G. 2015. Effectiveness of various sanitizer treatments for inactivating natural microflora and *Salmonella* spp. on turnip (*Pachyrhizus erosus*). *Food Control* 54, 216–24. doi:10.1016/j.foodcont.2015.02.012.

Taormina, P. J. and Beuchat, L. R. 1999. Comparison of chemical treatments to eliminate enterohemorrhagic *Escherichia coli* O157:H7 on alfalfa seeds. *J. Food Prot.* 62(4), 318–24. doi:10.4315/0362-028X-62.4.318.

U.S. Environmental Protection Agency (USEPA). 2018. Antimicrobial pesticides. Available at: https://www.epa.gov/pesticides/antimicrobial-pesticides (accessed on 11 December 2018).

U.S. Food and Drug Administration. 2014. Methods to reduce/eliminate pathogens from produce and fresh-cut produce. Available at: https://www.fda.gov/Food/FoodScienceResearch/ucm091363.htm (accessed on 19 July 2018).

U.S. Food and Drug Administration. 2018. 21 CFR 173.325. Acidified sodium chlorite solutions. Available at: https://www.accessdata.fda.gov/scripts/cdrh/cfdocs/cfcfr/CFRSearch.cfm?fr=173.325 (accessed on 18 December 2018).

Vandekinderen, I., Devlieghere, F., De Meulenaer, B., Ragaert, P. and Van Camp, J. 2009. Optimization and evaluation of a decontamination step with peroxyacetic acid for fresh-cut produce. *Food Microbiol.* 26(8), 882–8. doi:10.1016/j.fm.2009.06.004.

van Haute, S., Sampers, I., Jacxsens, L. and Uyttendaele, M. 2015. Selection criteria for water disinfection techniques in agricultural practices. *Crit. Rev. Food Sci. Nutr.* 55, 1529–51.

Wang, L., Sokorai, K., Wu, V. C. H. and Fan, X. 2019. Gaseous chlorine dioxide maintained the sensory and nutritional quality of grape tomatoes and reduced populations of *Salmonella enterica* serovar Typhimurium. *Food Control* 96, 299–309. doi:10.1016/j.foodcont.2018.09.023.

Whipps, J. M., Hand, P., Pink, D. A. C. and Bending, G. D. 2008. Human pathogens and the phyllosphere. *Adv. Appl. Microbiol.* 64, 183-221. doi:10.1016/S0065-2164(08)00407-3. Chapter 7.

WHO. 2009. In Benefits and risks of the use of chlorine-containing disinfectants in food production and food processing: report of a joint FAO/WHO expert meeting, Vol. 2013. WHO, Ann Arbor, MI.

Winniczuk, P. P. and Parish, M. E. 1997. Minimum inhibitory concentrations of antimicrobials against microorganisms related to citrus juice. *Food Microbiol.* 14(4), 373–81. doi:10.1006/fmic.1997.0103.

Young, S. B. and Setlow, P. 2003. Mechanisms of killing of *Bacillus subtilis* spores by hypochlorite and chlorine dioxide. *J. Appl. Microbiol.* 95(1), 54–67. doi:10.1046/j.1365-2672.2003.01960.x.

Zagory, D. 1999. Effects of post-processing handling and packaging on microbial populations. *Postharvest Biol. Technol.* 15(3), 313–21. doi:10.1016/S0925-5214(98)00093-3.

Zhang, S. and Farber, J. M. 1996. The effects of various disinfectants against *Listeria monocytogenes* on fresh-cut vegetables. *Food Microbiol.* 13(4), 311–21. doi:10.1006/fmic.1996.0037.

Zhang, G., Ma, L., Phelan, V. H. and Doyle, M. P. 2009. Efficacy of antimicrobial agents in lettuce leaf processing water for control of *Escherichia coli* O157:H7. *J. Food Prot.* 72(7), 1392–7. doi:10.4315/0362-028X-72.7.1392.

Zhou, B., Luo, Y., Turner, E. R., Wang, Q. and Schneider, K. R. 2014. Evaluation of current industry practices for maintaining tomato dump tank water quality during packinghouse operations. *J. Food Process. Preserv.* 38(6), 2201–8. doi:10.1111/jfpp.12200.

Chapter 8

Advances in using heat for disinfection/ disinfestation of horticultural produce

Elazar Fallik and Susan Lurie, Agricultural Research Organization – Volcani Center (ARO), Israel; and Lisa Jamieson and Allan Woolf, The New Zealand Institute for Plant & Food Research Limited, New Zealand

1 Introduction

Worldwide, about one-third of fresh fruit and vegetables are lost as a result of physiological and pathological deterioration, and in light of the increasing world population, this is unacceptable (Jedermann et al., 2014). To decrease our dependency on chemical control in ensuring quality and safety of fresh harvested produce, alternative methods for postharvest decay and pest control are being developed.

Heat treatments (HTs) can be used very effectively in controlling decay-causing agents and insect pests (Sivakumar and Fallik, 2013). The presence of quarantine insect pests in fruit export can impede trade with other countries. Therefore, to reduce the risk of possible quarantine pests in/on exported fruit, a postharvest disinfestation treatment is essential (Heather and Hallman, 2008). In the past, insect distribution was limited to certain geographical regions and the biology of the organism. However, as international trade has increased, insects may be incidentally transported to new areas (Hansen et al., 2011). Therefore, regulatory agencies worldwide have established phytosanitary and quarantine protocols, which include postharvest disinfestation treatments, with the purpose

http://dx.doi.org/10.19103/AS.2019.0055.10

of preventing the introduction of exotic pests through marketing channels and re-infestation during storage (Wang et al., 2010; Hou et al., 2016). HTs are widely employed for disinfestation of some crops, such as fruit fly treatment in mango.

Decay development, primarily from fungi but also in vegetables from bacteria, is the major cause of losses of fresh commodities. These decay organisms are generally controlled by pre- or postharvest chemical treatments. However, because of concerns about impacts on human health, many traditional postharvest treatments are no longer allowed. In addition, over time microorganisms develop resistance to the chemicals, and new means of control need to be developed. HTs are one of these methods.

Pre-storage HTs to control decay development or insect infestations during short or long storage and marketing periods can be applied for a time of several seconds to hours, or even several days (Fallik, 2010). HT can directly control both pathogens and insects that are found on the fresh produce surface or within two to three cell layers under the skin. HT can also control rot development indirectly by inducing defence mechanisms that enable the fresh produce to inhibit the pathogen's growth or the insect's ability to infest the fruit (Fallik, 2010). Amongst these treatments are: hot water dip/immersion (HWT), short hot water rinsing and brushing (HWRB), hot water drenching (HWD), hot air treatment (HAT) and relatively recently microwave (MW) and radio frequency (RF).

Previous studies demonstrated that effects of HTs on fresh fruit and vegetables depend on pre-harvest treatments as well as maturity stage, cultivar, type of HT applied and storage conditions (Lurie, 2008; Fallik, 2010; Schirra et al., 2011; Sivakumar and Fallik, 2013; Caleb et al., 2016; Sui et al., 2016). This chapter will summarize the latest information on HTs, alone, or in combination with other treatments, to control decay development and insect pests after harvest and during prolonged storage. The mode of action of these physical treatments as well as examples of where HTs have been successfully commercialized will also be discussed.

2 Experimental trials and methods of heat treatment

HTs can be applied using several technologies. Drench – the fresh produce is moved on a conveyor either as a single layer or in a bin/crate while hot water is applied at low pressure over the fresh produce. The conveyor's speed is controlled to ensure the required and optimal length of exposure time to the hot water (Fig. 1). HWT – the fresh produce is dipped/immersed in a hot water tank or submerged in a basket on a conveyor for a certain length of time (Fig. 2). These technologies are equipped with water pumps that circulate the hot water to maintain uniform water temperature in the tank during the treatment. HATs are applied using forced air circulation through bins or crates of product (Fig. 3). Treatments are generally defined by a 'ramp time' (time to heat to target

Figure 1 Commercial hot water drench of apples for 2 min at 52°C. Source: Schloffer, used with permission.

Figure 2 Hot water dip system showing recirculating pumps. Photo – A. Woolf.

temperature) and a 'hold time' after target temperature is reached. Relative humidity is either uncontrolled ('dry') or maintained at a saturated level ('vapour heat'). HWRB – the fresh produce is cleaned and disinfected with tap water rinsing (first chamber) and continues into the second chamber with hot water rinsing at high temperature (45–62°C) over revolving brushes for 15–25 s (Fallik, 2010) (Fig. 4a and b). The time and temperature of treatment depends on the commodity and cultivar for all methods of HT.

2.1 Very short heat treatment durations – seconds

Treatment of mango with HWRB technology at 65°C for 15 s and 2.5 kJ m^{-2} of UV-C gave the best results in controlling decay, showing less stem end rot

Figure 3 Hot air ('vapour heat') treatment system. Photo - Van Phong Nguyen.

Figure 4 (a) Commercial hot water rinsing and brushing of persimmon fruit (50°C for 15 s). Photo - E. Fallik. (b) Commercial hot water rinsing and brushing system for melon (58°C for 20 s) and pepper (55°C for 15 s) fruit showing infeed, treatment and drying systems. Photo - E. Fallik.

caused by *Botryosphaeria dothidea* after 18 days of storage in Brazil (Terao et al., 2018). Luria et al. (2014) reported that HWRB for 15–20 s was used commercially to improve mango fruit quality and reduce postharvest disease. This treatment enabled successful storage for 3–4 weeks at 12°C, with improved colour and reduced disease development. A significant reduction in the decay incidence of apple fruit was achieved by using HWRB for 20 or 25 s at 55°C, followed by up to 100 days cold storage at 2°C and 14 days at 18°C (Maxin et al., 2012). Recently, HWRB at 54°C for 15 s was applied very successfully on acorn squash (*Cucurbita pepo* L.) (Chalupowicz et al., 2018). This treatment significantly maintained commodity quality for 3.5 months, as indicated by higher fruit firmness, lower decay incidence and improved retention of green skin colour. This technology has been commercially used in Israel on persimmon, sweet pepper, melon, mango, avocado, citrus and organic citrus fruit (Fallik, pers. comm.).

Based on HWRB technology, a unique steam-based system was developed to control postharvest decay development on carrots (Gan-Mor et al., 2011). This technology used high-temperature heat delivered from above by steam jets, combined with electric steam-drying elements and reflectors. The carrots were exposed for only 3 s while rotating, and then immediately hydro-cooled reducing sensitivity to post-storage soft rots (Gan-Mor et al., 2011).

The optimal HT conditions and HT combined with preservative treatment were investigated for mandarin fruit (*Citrus reticulata* Blanco cv Ponkan) storage. Results indicated that HT in 55°C water for 20 s reduced the decay rate of the fruit, and a combination of HT and 25% of the normal dose used commercially of iminoctadine tris (albesilate), 2,4-dichlorophenoxyacetic acid and imazalil, significantly reduced decay rate without affecting fruit quality (Tang et al., 2017). In a different work, Pazolini et al. (2016) reported that HWT of peach (*Prunus persica* L. Batsch) at 50°C for 30 s combined with canola and Indian mustard oils successfully controlled brown rot development.

Mild heat treatment (HWT) is a thermal process to reduce microbial load and is based on the use of water at temperatures between 45°C and 60°C, for durations ranging from a few seconds to 20 min. Recently, Zudaire et al. (2018) showed that fresh cut onion subjected to 55°C for 60 s presented the highest reduction of aerobic mesophilic bacteria. Therefore, similar HWT can be used successfully to control both fungal and bacterial pathogens.

2.2 Short heat treatment duration - minutes

Dipping organic bananas at 40°C for 20 min maintained fruit appearance, flavour and controlled anthracnose development without any negative effect on the physicochemical and sensory quality (Vilaplana et al., 2018). The influence of HWT was investigated on cucumber (*Cucumis sativus* L.) (Nasef, 2018) where a 5-min, 55°C treatment totally inhibited decay development, while maintaining fruit quality after 2-week storage at 4°C and 2 days at 20°C. HWT of peach fruit at 48°C for 10 min maintained peach quality after 5 weeks at 4°C (Huan et al., 2017). HWT of asparagus (*Asparagus officinalis* L.) spears at 45°C for 4 min or at 50°C for 3 min maintained the highest level of quality after 14 days at 2.5°C (Saltveit, 2017). For Nagpur mandarin (*Citrus reticulate* Blanco), HWT at 50°C for 1.5 min reduced decay incidence without affecting other quality parameters after 2 months storage (Kumar et al., 2018).

Hot water treatment (48°C for 20 min) combined with calcium chloride (CaCl; 1% for 20 min) delayed anthracnose development in papaya inoculated with *Colletotrichum gloeosporioides* during 20 days of storage at 12°C (Ayón-Reyna et al., 2017a,b). HWT of 'Holland' papaya fruit at 42°C for 40 min with salicylic acid reduced the incidence of rot development during storage (Supapvanich and Promyou, 2017). HWT of persimmon fruit at 50°C for 15 min followed by 24 h 1-MCP (1-methylcyclopropene) at 500 nL L^{-1} before storage of 30 days at 1°C plus 20 days at 13°C maintained fruit quality compared with untreated control fruit or fruit dipped at 45°C for 25 min (Rasouli and Khademi, 2018). A sequential treatment with NaOCl at 200 mg L^{-1} and the fungicide imazalil (IMZ) at 50 mg L^{-1}, both at 50°C for 3 min, was as effective as IMZ at 1000 mg L^{-1} at 20°C in controlling *Penicillium* decay in lemon fruit (D'Aquino et al., 2017).

RF and MW heating are disinfestation methods that are electromagnetic based and chemical free. They rely on the interactions between electromagnetic energy and materials. Under electromagnetic exposure, the heating rate and power generation of the insect pests and host materials depend on their dielectric properties, electric field intensity and their thermal properties (Macana and Baik, 2018). Both methods use electromagnetic energy to heat a commodity (and the pest herein) to a temperature that will be lethal to the pest but not damage the commodity. When any material with polarized molecules and charged ions is subjected to an electromagnetic field that rapidly changes direction, heating occurs as polarized molecules and charged ions interact with the alternating electromagnetic field, resulting in frictional losses as they rotate and move. The higher the frequency of the alternating field, the greater the energy imparted to the material being treated. The uses of RF in the food industry have been mainly for drying cookies and crackers (Jiao et al., 2018). Compared with MW, RF have the advantages of greater penetration depth and longer wavelength, which are more suitable for bulk heating. However, heating non-uniformity is a major problem in RF treatments, and this could result in either insect survival or product damage (Birla et al, 2004; Wang et al., 2005, 2007). Due to the intrinsic properties of dielectric material, fringing field formation around compared to inside the commodity is the cause of non-uniform heating, with the outer part being hotter than the centre of the fruit or vegetable (Jiao et al., 2018). The temperatures at the corners and edges of the field are higher than those in the centre (Pan et al., 2012). Therefore, hot air or water for surface heating, sample moving and mixing are commonly used to improve RF heating uniformity (Gao et al., 2010; Sisquella et al, 2013; Wang et al, 2014).

Although the best applications for RF in disinfestation are for grains, nuts and dried fruits because of the insensitivity of dry foods to heat compared with fresh fruits and vegetables, there have been trials on a number of commodities (Hou et al., 2016). Since RF heats the centre of the product first, a combination treatment can be used where hot water heats the outside, followed by RF heating of the centre. Treatment of apples for disinfestation against codling moth combined preheating in hot water followed by core-focussed RF heating. This resulted in a uniform temperature of 47.3 ± 0.7°C. Holding the RF-heated apples in hot water at 48°C for 10 min further reduced the temperature gradient and improved uniformity (Hou et al., 2016). Water preheating of commodities is useful for reducing RF heating time and improving heating uniformity in the fruit, such as persimmons and mangos (Tiwari et al., 2008; Sosa-Morales et al., 2009). Wang et al. (2006) explored the application of RF treatments to control fifth-instar codling moth in apples in hot water at 45°C for 30 min, followed by

RF heating to 48°C and hot water bath holding at 48°C for 15 min. The treatment was found to be the most practical and effective method both for insect control and apple quality. Jiao et al. (2012) used a 27.12 MHz, 6 kW RF unit with a forced hot air system to conduct industrial scale-up studies on disinfesting lentils. To accomplish 100% cowpea weevil mortality, a RF treatment protocol was developed in combination with forced hot air to heat lentils to 60°C for 10 min, followed by forced ambient air cooling for 20 min.

2.3 Medium heat treatment duration - hours

HAT at 38°C for 12 h reduced postharvest disease of cherry tomato fruit after 20 days at 20°C (Wei et al., 2018). HAT of 38°C for 3 h maintained peach (*Prunus persica* L) fruit quality after 5 weeks at 4°C (Huan et al., 2017). The effect of moist HAT applied to broccoli (*Brassica oleracea*) florets was studied in order to maintain quality and phytochemical compounds during postharvest storage at 4°C (Duarte-Sierra et al., 2017). Application of HAT at 41°C for 3 h maintained floret quality for 21 days at 4°C. The effects of individual and combined biological control on Red 'Fuji' apple fruit, using *Pichia guilliermondii* (at 1×10^8 CFU mL^{-1}) and HAT (at 38°C for 4 days) on three major postharvest diseases, *Botrytis cinerea*, *Penicillium expansum* and *Colletotrichum gloeosporioides*, as well as the resulting fruit quality, were investigated. Results suggested that the combined hot air and antagonistic yeast (*P. guilliermondii*) treatment inhibited the infection of apple fruit wounds by the three major postharvest diseases (Zhao and Yin, 2018).

An important postharvest treatment required for long-term storage of onion bulbs is curing. Curing is a drying process intended to dry down the bulb (Maw et al., 2004; Bayat et al., 2010) to prevent the loss of moisture and attack by pathogenic organisms during storage. Standard conditions for curing are blowing hot air at around 38°C with no RH control for several days (Bansal et al., 2018).

HT can be an effective tool for control of insects. Hsu et al. (2018) reported that when the papaya fruit core temperature increased at a heating rate of 0.0925°C min^{-1} from room temperature to 47.2°C in 3 h, fruit flies showed 100% mortality, thus making it a viable quarantine treatment for the fruit prior to export.

The use of microwave-assisted hot water treatment (MW-HWT) was evaluated on grapefruit (*Citrus paradise*) as a quarantine treatment against the Mexican fruit fly (*Anastrepha ludens*) (Soto-Reyes et al., 2018). MW-HWT was shorter (130 min) and resulted in less fruit damage to the grapefruit when compared with HWT only (188 min). Thus, this treatment could be considered as an alternative method against the Mexican fruit fly in grapefruit.

Table 1 summarizes the recent research on the use of different methods of heating on control of insects, decay organisms and quality of the fruit or vegetable.

Table 1 Recent experimental trials of heat treatment to control insects, fungi and commodity quality

Recent trials of commodities for insect control			
Commodity	*Insect*	*Heat treatment*	*Reference*
Grapefruit	Mexican fruit fly (*Anastrepha ludens*)	HWT heated by MW	Soto-Reyes et al. (2018)
Lentils	Cowpea weevil (*Callosobruchus maculatus*)	HAT with RF	Jiao et al. (2012)
Papaya	Fruit fly	HAT	Hsu et al. (2018)
Recent trials of commodities for fungal control			
Commodity	*Fungus*	*Heat treatment*	*Reference*
Acorn squash	General decay	HWRB	Chalupowicz et al. (2018)
Apple	General decay	HWRB	Maxin et al. (2012)
Apple	*Botrytis cinerea, Penicillium expansum* and *Colletotrichum gloeosporioides*	HAT + BC	Zhao and Yin (2018)
Banana	*Colletotrichum musae*	HWT	Vilaplana et al. (2018)
Carrot	Soft rot	Steam spray	Gan-Mor et al. (2011)
Cherry	General decay	HAT	Wei et al. (2018)
Cucumber	General decay	HWT	Nasef (2018)
Dragon fruit	*Fusarium*	HWT	Ngoc et al. (2015)
Lemon	*Penicillium* spp.	HWT + low concentration IMZ	D'Aquino et al. (2017)
Mango	*Botryosphaeria dothidea*	HWRB + UV	Terao et al. (2018)
Mango	Multiple fungi	HWRB	Luria et al. (2014)
Nagpur *Citrus reticulate*	General decay	HWT	Kumar et al. (2018)

Onion	General decay	HAT		Bansal et al. (2018)
Papaya	*Colletotrichum gloeosporioides*	HWT + Ca		Ayón-Reyna et al. (2017b)
Papaya	General decay	HWT + SA		Supapvanich and Promyou (2017)
Peach	*Monilinia fructicola*	HWT + canola, Indian mustard		Pazolini et al. (2016)
Persimmon	General decay	HWT + 1-MCP		Rasouli and Khademi (2018)
Ponkan Citrus reticulata	General decay	HWT		Tang et al. (2017)
Recent trials for commodity quality control				
Commodity		*Heat treatment*	*Purpose of treatment*	*Reference*
Acorn squash		HWRB	Higher firmness, retention of green colour after storage	Chalupowicz et al. (2018)
Apple		HAT + BC	Maintain quality after storage	Zhao and Yin (2018)
Asparagus		HWT	Maintain quality after storage	Saltveit (2017)
Broccoli		HAT	Maintain quality and phytochemical compounds after storage	Duarte-Sierra et al. (2017)
Mango		HWRB	Improve colour after storage	Luria et al. (2014)
Onion		HAT	Maintain quality in storage	Bansal et al. (2018)
Peach		HAT	Quality after storage	Huan et al. (2017)
Persimmon		HWT + 1-MCP	Maintain firmness after storage	Rasouli and Khademi (2018)

Abbreviations: HWRB = hot water rinse and brush; HWT = hot water dip; HAT = hot air treatment; MW = microwave; RF = radio frequency; UV = ultraviolet light; BC = biological control; Ca = calcium; IMZ = imazalil; GRAS = generally recommended as safe; 1-MCP = 1-methylcyclopropene; SA = salicylic acid.

3 Mode of action on fruit and vegetables

Understanding the mechanisms that underlie the induction of heat damage in products, and the mechanisms of mortality/control of the target pathogen or insect better allow us to achieve an effective treatment. For example, understanding underlying biochemistry also points to possible ways of maximizing the treatment window by inducing tolerance in the target product and/or maximizing mortality of the target organism (Waddell et al., 2000; Woolf et al., 2004).

3.1 Product response

A fruit or vegetable is able, to some extent, to tolerate heat stress by altering its metabolism and gene expression. This includes production of compatible solutes that are able to protect proteins and cellular structures, maintain cell turgor by osmotic adjustment and modify the antioxidant system to re-establish the cellular redox balance and homeostasis (Hasanuzzaman et al., 2013). At the molecular level, heat stress causes alterations in expression of genes involved in direct protection from heat stress. These include genes responsible for the expression of osmoprotectants, detoxifying enzymes, transporters and regulatory proteins (Krasensky and Jonak, 2012; Semenov and Halford, 2009). In conditions such as heat stress, modification of physiological and biochemical processes by gene expression changes leads to the development of heat tolerance.

Heat stress affects the stability of various proteins, membranes, RNA species and cytoskeleton structures, and alters the efficiency of enzymatic reactions in the cell, causing metabolic imbalance (McClung and Davis, 2010; Ruelland and Zachowski, 2010; Suzuki et al., 2011). Disrupting the steady-state flux of metabolites can cause the accumulation of toxic compounds, such as reactive oxygen species (ROS). In fact, ROS are generally a component of heat stress and increases in antioxidant processes are part of the heat stress response (Mittler et al., 2012; Konigshofer et al., 2008). As a response to heat stress, plants reprogram their transcriptome, proteome, metabolome and lipidome. These changes establish a new steady-state balance of metabolic processes that can enable the organism to function at the new temperature. This new balance may also be helpful in responding to another stress such as the low temperatures of storage.

Low temperature storage is used to retard product respiration and delay ripening and senescence. Many commodities will develop chilling injury if the temperature is too low or if the cold conditions are maintained for too long. HTs have been found in many cases to delay or prevent the development of chilling injury (Lurie, 1998b). This has been shown in many commodities to be associated with the prolonged presence of heat shock proteins (HSPs) in the tissue and the protective effect they exert (Lurie, 1998a; Zhang et al., 2005; Yi et al., 2006; Sevillano et al., 2010; He et al., 2012). HSPs increase during a heat stress and generally disappear rapidly when a plant is returned to an ambient temperature. Sabehat et al. (1996) were the first to show that if a commodity was placed at 2°C rather than 20°C following a heat stress, the HSPs were not metabolized. Thus, the thermotolerance induced in a heat stress can give protection against cold stress. Woolf et al. (2004) found that HAT of 36°C for 6 h protected avocados from heat damage of HWT of 50°C for 10 min and from chilling injury when stored at 0°C. However, if the time between the HAT and a subsequent treatment was delayed, the protection gradually disappeared. Transcriptomic and metabolic studies of HT of peaches and citrus before cold storage showed the commonality between the response to heat stress and that of cold (Lauxmann et al., 2012, 2014; Lafuente et al., 2017).

Postharvest HTs alter the normal programme of protein synthesis during heat stress. When heat stress is imposed, there is a rapid disassociation of polyribosomes and protein synthesis stops briefly and then resumes with a new set of proteins including the HSPs (Ferguson et al., 1994). The consequence of this switch is that normal ripening processes are inhibited, and, if the commodity is then placed in low temperature, the inhibition persists for some time. Ripening resumes when the fruit or vegetable is rewarmed after storage. Therefore, postharvest HTs can modulate the rate of ripening of commodities (Hernandez et al., 2017), while, in addition, preventing postharvest storage disorders.

The metabolic and proteomic changes induced by the response of fruit or vegetable tissue to high temperature stress also produce compounds that help them to respond to fungal attack. In addition to the damage that high temperature may inflict on the pathogen (Sui et al., 2016), the presence of antifungal compounds in the plant tissue impedes fungal development. These include pathogen-related (PR) proteins induced by various stresses, including heat. Their presence has been found to suppress fungal development in citrus (Pavoncello et al., 2001; Yun et al., 2013).

3.2 Pathogen response

The efficacy of a HT to control a pathogen is measured by the reduction in viability, either in bacterial or fungal spore germination or myceliar growth. The response of a pathogen to heat can be influenced by the moisture content of the spores, metabolic activity of the pathogen, the concentration and age of the inoculum and the chemical composition and water activity of the treatment medium. Even the culture medium on which the fungus is grown after the HT can influence its apparent viability.

When microorganisms are exposed to high temperature and other abiotic stresses, genes specifically associated with ROS are activated; resulting in the accumulation of ROS in a time-dependent manner. Depending on the level of stress tolerance in an organism, the accumulation of high levels of ROS, because of inadequate detoxification systems, will cause oxidative damage resulting in the impairment of many cell functions, which then leads to a reduction in or complete loss of viability. Reductions in spore viability and hyphal growth because of oxidative injury have been reported in *Botrytis cinerea* (Zhao et al., 2014) and *Aspergillus niger* (Abrashev et al., 2008).

Mitochondrial respiration is responsible for the major portion of intracellular ROS. Liu et al. (2012) reported that HT (40°C for either 5 or 10 min) triggered the accumulation of ROS, collapse of mitochondrial membrane potential and a decrease in intracellular ATP in the conidia of *Monilinia fructicola* (the causal pathogen of brown rot in peach), thus inhibiting spore germination and germ tube elongation. The combination of the direct inhibitory effect of HTs on *M. fructicola,* with the activation of an induced resistance response in peach fruit, resulted in good control of brown rot. Similar results were recently obtained in heat-treated melon fruit with control of *Fusarium oxysporum*, the causal fungal pathogen of fusarium rot (Sui et al., 2014).

In addition to mitochondria, the plasma membrane redox system also participates in ROS generation. NADPH oxidases (NOXs, nicotinamide adenine dinucleotide phosphate-oxidases), which are known as respiratory burst oxidases, are a membrane-bound enzyme complex that generates ROS in multicellular eukaryotes. Activation of the *NoxA* gene in *B. cinerea* by HT (43°C, 10–30 min) was reported to be responsible for intracellular ROS accumulation and subsequent oxidative damage to proteins and lipids, which in turn contributed to an inhibition of grey mould (caused by *B. cinerea*) development in pear fruit (Zhao et al., 2014).

An inhibitory effect of HT has also been reported for other postharvest fungal pathogens, including *Dothiorella dominicana*, *Lasiodiplodia theobromae*, *Colletotrichum gloeosporioides*, *Alternaria alternata*, *Fusarium solani*, *Neonectria galligena*, *P. digitatum* and *P. expansum* (Chen et al., 2015; Fallik et al., 2000; Maxin et al., 2012; Porat et al., 2000).

In addition to a direct effect of high temperature on the microorganism, another factor in controlling decay is the compounds induced in the fruit or vegetable as a result of temperature stress. In citrus, for example, a subset of HSPs are pathogenesis-related proteins such as chitinase and α-1,3-glucanases that inhibit fungal development (Pavoncello et al., 2001). Also, levels of low-molecular-weight molecules with antimicrobial activity (phytoalexins) are enhanced following heat shock in wounded citrus fruit (Nafussi et al., 2001). In apple, there are indications of a similarly complex response to heat. In apple fruit cells, a mild heat shock may induce the production of HSPs, which confers resistance to cold temperatures (Wang et al., 2001). Further, low-molecular-weight compounds with antifungal activity have been extracted from the peel of heated apples but not from untreated fruit (Fallik et al., 1995). The duration of antifungal protection following HT was approximately 3–4 months against infections by *P. expansum* (Fallik et al., 2001; Amiri and Bompeix, 2011). This phenomenon has been termed induced resistance, or cross-protection, whereby one stress (temperature) induces resistance against another stress (pathogen).

3.3 Insect response

In insects, thermal tolerance and heat-induced mortality are dependent on time and temperature exposure and involve a complex interaction of physiological and biochemical changes. That said, thermal mortality death responses of fruit fly eggs and larvae follow a logarithmic function, so that the final number of survivors of any time-temperature treatment is dependent on the initial number of insects (Jang, 1991). The time and temperature effects on different stages of fruit fly eggs and larvae under various heating regimes have been measured (Wang et al., 2007; Hsu et al., 2018), as well as other insects such as moths (Tang et al., 2007; Nevin, 1998). These studies on insects in isolation from the fruit help to predict the thermal response of the insects in the fruit. However, the rates of heating in a commodity depend on the heat transfer characteristics of the fruit, size and shape of the fruit, type of container holding the fruit and the characteristics of the HT.

As with fungal pathogens, various insects have different sensitivities to high temperature. Some insects such as the longtailed mealy bug show no difference in response to heat at different life stages (Hansen et al., 1992). However, for most fruit flies, the larvae are less resistant to high temperature than the eggs (Hansen et al., 1990; Hsu et al., 2018). In a study of hot air disinfestation treatment on papaya for three species of fruit fly, Mediterranean fruit fly (*Ceratitis capitata* Wiedemann) required higher temperature for eradication than melon fly (*Dacus cucurbitae* Coquillett) or oriental fruit fly (*Dacus dorsalis* Hendel). In all three species, the first and third instars were more susceptible to heat than eggs (Armstrong et al., 1995). However, if HT is followed by cold storage, the mortality of the insects is increased (Nevin, 1998; Whiting et al., 1999).

Respiration is affected by heat, and as the body temperature of the insect increases, there are increases in both metabolism and respiration up to a critical thermal limit. As with produce and pathogens, exposure to non-lethal elevated temperatures (often around 30°C) has been shown to increase the tolerance of the insect to a subsequent HT that is lethal to non-conditioned insects through the production of HSPs (Beckett and Evans, 1997; Lester and Greenwood, 1997; Yin et al., 2006, Hu et al., 2014; Li et al., 2018). The way this thermal tolerance declines in insects over time may be important to maintain the integrity of a quarantine HT. Brown et al. (2019) modelled acclimation and decline of thermotolerance and proposed that this may be related to the production and decay of two HSPs, for example Hsp70 and Hsp90 which have been found to be elevated in some fruit fly species after heat stress (Hu et al., 2014).

In addition, high temperature causes changes in the nervous and endocrine systems of insects (Nevin, 2000). In HT of insects, there appears to be a threshold between 41°C and 43°C where some major injury takes place. Treatments in this temperature range cause chronic mortality to the insects, and death occurs during metamorphoses. Cell tolerance to heat is dependent on the developmental stage, with the G1 phase (between mitosis and synthetic phase) the most resistant and the M (mitosis) stage the most sensitive (Mackey and Roti, 1992). HTs interfere with the production of proteins essential for development and tissue differentiation, and in this manner cause death. A temperature above 43°C may cause acute mortality with no survivors. Most protein synthesis, including HSP, ceases when insects are held for some hours at temperatures above 42°C or 43°C (Lurie and Jang, 2007). Therefore, most disinfestation protocols involve raising the fruit temperature above 43°C for a period of time, often to 46°C or 47°C.

4 Commercial application of heat treatments

As previously noted in Section 2, there has been a significant amount of research on HTs of various types. While results can appear very promising in experiments in a laboratory setting and the resulting publications, actual commercial uptake of the treatment can be challenging for many reasons. This section outlines key issues in the practical application of HTs followed by a collation of which commercial HT applications to horticultural crops are being used internationally.

4.1 Practical application of heat treatments

There is a wide range of aspects that affect the success of HTs, both biological and practical. Biological factors are generally addressed in scientific publications. Here, we examine a range of practical aspects that are important in the commercial uptake and success of HTs.

Accuracy. A key aspect is the accuracy of the HT. The aim is to kill the insect or decay organism without affecting the quality of the produce. In many cases, there is a small window between efficacy (e.g. killing insect) and reduction in product quality. Indeed often, the HT will result in some reduction in quality, or reduction in quality in some portion of the treated crop. The potential loss of quality because of HT needs to be considered in relation to quality loss from alternative treatments (e.g. methyl bromide), or complete inability to sell and/or export product.

The effect of temperature and duration of treatment is demonstrated in Fig. 5, showing that even small differences in time or temperature result in very large increases in damage. For example, at 49.5°C, doubling the duration results in a damage increase from ~5% to over 30%. More significantly, at a duration of 2 min, an increase in temperature of 1.5°C (49.5–51°C) results in damage increase from ~5% to ~20%, and another 1.5°C to nearly 90% at 52.5°C.

Equipment should be designed to the highest standard of accuracy in terms of sensitivity of the temperature probes (±0.1°C). Calibration of probes should be carried out regularly at the target temperature (i.e. not a simple ice-point check since offsets at 0°C are not necessary the same as at ~50°C). Accuracy of target treatment and changes over time are more important with shorter treatments, such as HWTs (treatment of minutes), and even more so with HWRB which are treatments of seconds.

Most significantly, spatial variability should be very carefully considered since large differences can occur within even a 300-mm distance. Thus, for HATs, there can be very large temperature differences within a vertical stack since air is generally applied from only one direction. For both water and air treatments, it is important to remember that air or water will take the least path of resistance, and thus will flow around a crate of fruit (for example) not through

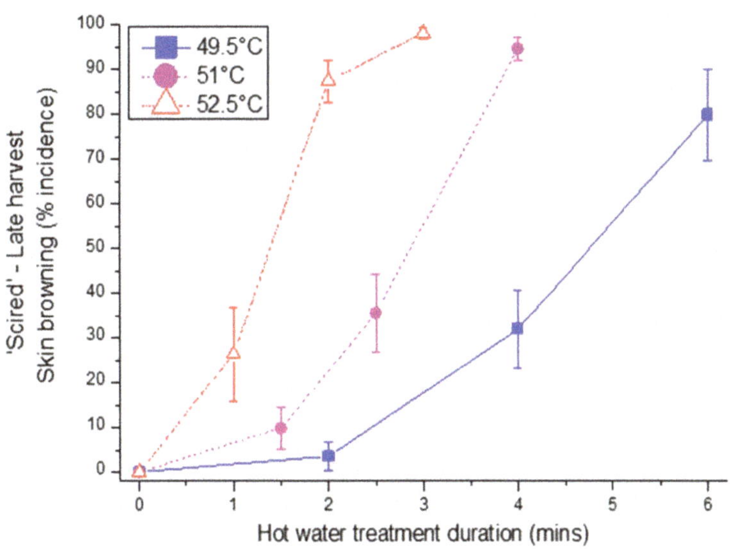

Figure 5 Incidence of heat damage (skin browning) on 'Scired' apples following hot water treatment at 49.5°C, 51°C and 52.5°C for 1, 1.5, 2, 2.5, 3, 4 or 6 min. Damage was rated after storage for 10 weeks at 0.5°C and 7 days at 20°C. Woolf - unpublished data (NZ Apple & Pears Inc. - funded research, used with permission).

it unless it is forced to. Thus, simply placing a 25-kg crate of product in a water bath will result in very poor spatial accuracy of treatment (e.g. the centre of the crate), even if the water bath temperature is accurate. Therefore, engineers experienced in HTs should be used in the system design phase.

Many other variables can affect heating rate, heat absorption by the product and overall evenness of heating. For example, for a given volume of product, small products (such as berryfruit) will absorb heat much faster than larger products because of the high surface area.

4.2 Types of heat treatment

In terms of technologies and temperature variability, HATs are slow (because of low heat transfer, compared with water). Also, because air is difficult to apply evenly to a volume of fruit, temperature variability can be quite significant. However, hot water treatments are relatively easy to apply in a uniform manner as long as high-flow water pumps are used with adequate heating and good design.

For treatments that use hot water, in general, 'pass through' designs will have more accurate heating application compared with 'batch-based' dip systems since they have a more constant heat load. In addition, the peak power requirement (after initial heating) is lower. HWD/shower and hot water brushing systems should, if well designed, be able to apply very even and predictable temperatures because of the nature of treatment. Also, 'pass through' designs have a more constant heat load, which is advantageous in terms of power supply, and thus commercial feasibility.

The challenge of 'tuning' frequencies to achieve accurate heating is one of the key limitations to the use of MW/RF systems to achieve practical outcomes. The field of radiation in both these systems is not spatially even, but is higher in the centre of the chamber than the edges. This is why in most home MWs there is a centre plate that rotates during operation. Therefore, if a static MW or RF system is employed, the produce needs to be moved around in the chamber so that all sides of the product receive the same amount of radiation. Even if a 'pass through' system is used, this difference in field strength needs to be considered regarding how the product is moved.

4.3 Food safety considerations, particularly for water treatment systems

An example of a food safety outbreak/issue in disinfestation systems can be seen in a 1999 *Salmonella* outbreak in the United States that was caused by hot water-treated mangos from Brazil that were cooled in water that was not adequately sanitized (Anon., 1999). Seventy-eight people were ill (confirmed by laboratory testing), and two people died. A study replicating the work found that the most likely cause was internalization (i.e. the pathogen pulled inside the fruit tissue) as fruit was moved from the 47°C HWT to the cooling bath, and most likely occurred through the peduncle (fruit stem), in particular because of the lack of adequate sanitizer dosing and monitoring, and water changes. Thus, with the increasing awareness of food safety issues in relation to horticultural products (Shukla et al., 2018), the potential effects of HTs on food safety outcomes should be carefully considered.

Filtration systems. Where water systems are employed, it is important that water quality is monitored since it will deteriorate over time because of increased organic matter, fruit waxes, insects and agrichemicals. High-quality filtration systems and water dumping and/or continual dilution should be considered. A high-quality, fresh-water rinse system at the end of treatment (recommended to rinse and cool fruit after treatment) might be used for continual top-up and dilution.

Sanitizers. Use of sanitizers with water treatments is recommended, particularly with the increased emphasis on food safety. Depending on the sanitizer, issues such as corrosion of metals and other parts should be considered, as should health and safety of workers (residues/handling). More importantly, off-gassing and other risks, particularly in the HWT tank, should be considered and good ventilation may be required.

Cleaning. All systems developed should be able to be readily cleaned with easy access, and designed to minimize microbe-harbouring areas and biofilm formation. Clean in place (CIP) systems are becoming increasingly common in packhouses and installation should be considered.

Internalization. As observed in the outbreak above, for both hot water and HATs, the issue of human pathogen microbial internalization should be considered where water that is colder than the fruit is applied (Avila-Quezada et al., 2010). This poses the risk of internalizing pathogens into fruit, which does not generally appear to be considered in many applications (e.g. water cooling of fruit after HT of mango).

4.4 Cost of heat treatments

From the perspective of successful commercial uptake, treatments that incur significant capital, energy, maintenance and/or logistic costs are less likely to be adopted, and these issues were noted by a range of researchers around the world in relation to commercialization (see Acknowledgements section).

While capital costs of HTs are significant, they are not outside the general scope of many other postharvest capital expenditures, such as grading equipment and high-pressure washing systems. It is suggested that probably the most significant limitation is the energy cost, particularly if this is based on electricity. Also of significance is that there is an increasing awareness of the carbon cost of electricity or similar heating systems, unless these can be 'greened', that is derived from more sustainable, non-polluting energy sources.

Energy use and options: There is a wide range of options for providing heating, including electrical, gas (natural gas or LPG), diesel or similar oils, solar heating or photovoltaic power generation. Each packhouse will have different issues that drive cost and feasibility of these options, such as location, which can influence proximity to reticulated gas/electricity. Electricity and gas or other petroleum fuels are the most widely used (Edelenbos, pers. comm.). While electricity and industrial-supplied compressed natural gas (CNG) or diesel are the simplest, other energy sources should be explored and developed that are more environmentally acceptable, such as solar-heated water or photo voltaic power generation. Given that most postharvest HTs are closely associated with coolstores, systems that employ heat exchange from coolstore refrigeration systems is an option that would save costs, and this is being used in Southern Germany (Neuwald, pers. comm.). Finally, because water systems generally require some cooling after treatment, heat recovery (using heat exchangers) might prove economic, although this requires some capital input.

Insulation. To reduce energy costs, heat loss should be minimized, using thermal insulation of water tanks and pipework, as well as insulated enclosures over water tanks and conveyors to reduce heat loss to the atmosphere. However, for HATs, it should be noted that while insulation will reduce costs during heating, once the target temperature is achieved, cooling systems will be required if insulation is used since the energy from fans will lead to overheating.

4.5 Case study: Bell pepper

In 1994, several experiments were conducted in Israel where dipping sweet pepper at 50°C for 3 min maintained fruit quality for 2 weeks at 7°C. At the same time, farmers complained that it was difficult to clean field dust from sweet pepper calyx. In 1996, a machine was developed that combined hot water spray and roller brushes, to both clean the peppers and to remove decay organisms. The first prototype was built at the Agricultural Research Center in Israel with longitudinal brushes, a thermostat-controlled hot water tank and a water pump to pressurize and recycle the hot water. In 1997, the first commercial-scale machine was built and installed in packhouses in the south of the country. This machine was used to determine the type of brushes, water flow per nozzle, the number of nozzles, the water pressure and, above all, the best water temperature and time of exposure to the hot water (for pepper, 55°C ± 1°C for 15 s) (Fallik et al., 1999). However, this machine had limited capacity of produce per hour and was very expensive. Therefore, the third version was designed in 1998 with 22 parallel brushes that were divided into two compartments: ten brushes for the non-recycled tap water rinse (TWRB) (for cleaning) and 12 brushes for the recycled hot water rinse (HWRB) and two extra brushes to remove the water. The fruit were then dried in a 6-m long drying tunnel equipped with hot air fans. To meet the food safety requirements, in 2000 the fourth generation of the machine was designed and built with stainless-steel materials and 24 brushes with the following arrangements: 8 + 12 + 4 (TWRB + HWRB + water removal, respectively). In 2005, the fifth generation of the machine was designed to better clean and disinfect the treated fruit by changing the ratio between the TWRB and the HWRB brushes to 12 and 8, respectively, pressurizing the TWRB, leading to recycling of a third of the water during the treatment. Changes were made to the TWRB nozzles to fan-type, and in the HWRB to cone-type nozzles and finally forced-air and regular fans to dry the fruit, in just a 4.5-metre long drying tunnel.

By 2010, there were 250 units in Israel treating pepper, mango, melon, organic citrus, kumquat, litchi, sweet corn and tomato with different hot water temperatures and exposure times (50-62°C and 15-20 s), depending on the produce. This technology helped Israeli farmers by reducing the labour cost of cleaning the fruit manually, extending storability and shelf life of the treated fruit, and allowed sea transport rather than air transport. As a result, losses of pepper were reduced to ~2% from ~15% after 2-3 weeks of storage, thus saving the farmers more than USD70 million/year. This technology reduced product water loss, better maintained fruit firmness, enhanced resistance to fungal infection, reduced sensitivity to chilling injury and improved the overall appearance of the fresh produce.

4.6 Commercial applications

Disinfestation. Disinfestation by HT application is probably one of the most common applications with HATs to control fruit fly and is widespread for tropical crops such as mango. To provide an effective/robust quarantine treatment, the target treatment is that the core (beside the seed, i.e. the coldest part of the fruit) must reach a temperature of ~47°C, although longer times at lower temperatures (44°C for 8 h) can be effective (Table 2 - papaya, Follett). The reason for the widespread use of HTs for tropical crops is most likely because of the lack of alternatives, in particular the fact that cold (low temperature, c. <2°C) disinfestation is not possible for tropical crops because of chilling injury susceptibility (injury at c. <13°C). Control of internal pests of temperate crops (e.g. codling moth in apples) would be of significant commercial advantage, but the thermotolerance of these crops is rarely adequate to tolerate the treatment. A combined treatment of controlled atmosphere temperature treatments (CATTS) was developed to allow for, and lower, treatment temperature. However, even with the lower temperatures of CATTS compared

Table 2 Commercial applications of heat treatments in various countries

Country	Fruit	Purpose	Treatment	Comments	Source
Disinfestation					
Apple/pear[a]	USA	Codling moth, oriental fruit moth	HAT with CA	12°C/h to 46°C, hold 3 h; at 1% O_2, 15% CO_2	Neven
Asparagus[a]	New Zealand	Thrips, aphids, mites	HWT 45–50 °C for 3–10 min	Tip rot can occur	Woolf
Cherry[a]	USA	Codling moth, western cherry fruit fly	HAT with CA	200°C/h to 45°C, hold ¾ h; at 1% O_2, 15% CO_2	Neven
Citrus	Mexico	Mexican fruit fly, (*Anastrepha ludens)* and other fruit flies	Vapour HAT, 44°C within 90 min, hold for 100 min	Treatment can cause injury for delayed harvested grapefruit	Yahia
Mango	Fiji	Fruit fly	HAT 47.2°C at core 20 min	Ramp temperature 5 h	Waddell
Mango	India/Pakistan	Fruit fly	Vapour HAT 47.2°C at core 20 min	Ramp temperature 5 h	Mazhar
Mango	Mexico, Brazil Guatemala, Peru	Mexican fruit fly (*Anastrepha ludens*)	HWT, 46.5°C 65–110 min depending on fruit size		Yahia
Mango	USA, Puerto Rico	Insects	HWT, 52–55°C up to 90 min	Time, temperature depends on cultivar and fruit size; must wax fruit after	Fallik
Papaya	USA (Hawaii)	Fruit fly	HAT to 47.2°C in >4 h		Follet
Papaya	USA (Hawaii)	Fruit fly	Vapour HAT 44.4°C 8¾ h	Cool immediately	Follet
Peach	USA	Codling moth, oriental fruit fly	HAT with CA	12°C/h to 46°C, hold 3 h; at 1% O_2, 15% CO_2	Neven

Persimmon[a]	New Zealand	Mealybug, leafroller caterpillar, thrips	HWT 51°C 20 min	Skin cracking may occur	Woolf
Sweet pepper	Israel	Medfly	HWRB 55°C 15 s	Reducing chilling sensitivity, and can withstand cold quarantine	Fallik
Decay control					
Apple[a]	Denmark	All decay	HWT 54°C 30 s	Organic	Edelenbos
Apple[a]	Israel	*Penicillium expansum*	HWRB 55°C 15 s	Used when CA is not available	Fallik
Apple	Germany (South)	Storage rot, *Neofabraea alba, N. perennans, Gloeosporium* spp.	HWT 48–52°C 2 min	For organic apples, temperature dependent on cultivar	Neuwald
Apple	Germany (North)	Storage rots/ storage scab	HWT 50°C 3 min, 52°C 3 min	For organic apples, temperature dependent on cultivar	Holthusen
Apple	Germany (North)	Storage rots/ storage scab	HWD, 55°C 30 s	Organic apples	Holthusen
Apple	France	Storage rots	HWD, 48–50°C 2–3 min		Lurol
Apple	Switzerland	Storage rots	HWT 50–52°C 2–3 min	Cultivar dependent, some cultivars heat sensitive	Perren
Avocado	Israel	*Colletotrichum gloeosporioides*	HWD 50°C 10 s	Removing and killing spores. Slowing ripening	Fallik
Basil[a]	Israel	*Botrytis cinerea*	Vapour HAT 38°C 8 h	Inactivating spores	Fallik
Citrus	Israel	All fungi, particularly *Penicillium* spp.	HWRB 52–62°C 15–20 s	Organic citrus with 2% SBC, sealing invisible cracks, shining the peel, removing and killing spores, enhancing resistance	Fallik
Citrus	Israel	All fungi, particularly *Penicillium* spp.	HWT 53°C 15–20 s	With 400 ppm imazalil (instead of 1000 ppm), removing and killing spores, enhancing resistance	Porat

(*Continued*)

Table 2 (*Continued*)

Country	Fruit	Purpose	Treatment	Comments	Source
Dragon Fruit	Vietnam	*Colletotrichum, Alternaria, Fusarium, Rhizopus*	HWT 49–51°C 2–3 min		Woolf
Kumquat[a]	Israel	All fungi	HWRB 55°C 20 s	Removing and killing spores. Slowing ripening, removing and killing spores	Fallik
Litchi[a]	Israel	All fungi	HWRB 52°C 15 s	Removing and killing spores. Slowing ripening, retaining skin colour	Fallik
Mango	Israel	All fungi	HWRB 55°C 15 s	Reducing fungicide use (by 50%), improving general, appearance, removing part of the sap, might accelerate ripening, reducing the amount of wax used	Fallik
Mango	Mexico, Australia, South Africa	Anthracnose, (*Colletotrichum gloeosporioides)*	HWT 48–53°C 5–15 min		Yahia
Melon[a]	Israel	All fungi	HWRB 52°C 15 s	Reducing water loss, preventing the use of waxes, sealing invisible cracks, removing and killing spores, enhancing resistance, reducing chilling sensitivity	Fallik
Peach	France	All decay	HWD	58°C 30 s	Lurol
Persimmon	Israel	All fungi	HWD 50°C 10 s	Removing and killing spores	Fallik
Pitaya[a]	Israel	All fungi	HWRB 52°C 15 s	Improving general appearance. Slowing ripening, sealing invisible cracks, removing and killing spores, removing spikes	Fallik

Sweet pepper	Israel	All fungi	HWRB 55°C 15 s	Improving general appearance. Slowing ripening, sealing invisible cracks, removing and killing spores	Fallik
Sweet potato, organic[a]	Israel	Decay, mainly *Rhizopus stolonifer*	HWRB 52°C for 20 s + 0.2% biocontrol agent	Improving general appearance. Sealing invisible cracks, removing and killing spores	Fallik
Sweet potato	USA	Decay primarily *Rhizopus stolonifer*	HAT 29°C 4–7 days	Induce rapid wound periderm formation to reduce decay and moisture loss in storage	Kays, Clark
Round cherry tomato	Israel	*Botrytis cinerea, Alternaria alternata*	HWRB 52°C 10–15 s	Reducing water loss, preventing the use of waxes, sealing invisible cracks, removing and killing spores, enhancing resistance, reducing chilling sensitivity	Fallik

Abbreviations: HWT, hot water dip; HAT, hot air treatment, HWRB, hot water rinse and brush; HWD, hot water drench; CI, chilling injury; CA, controlled atmosphere; RH, relative humidity.

[a]Currently not in commercial use.

with vapour HT, commercially unacceptable damage is still likely. This, along with the technical challenges of applying heat *and* atmospheres (low O_2/high CO_2) has inhibited commercial success of CATTS, although research into this area continues (Verschoor, pers. comm.).

Commercially successful hot water treatments (Table 2) are more often those that are targeting surface-dwelling insects (e.g. thrips, mites, mealybug and some caterpillar species) since HWTs for only a few minutes heat only the outside of the product and for a few millimetres into the flesh. However, HWT for a longer period is also a successful method for treating mangos, which clearly have a high thermotolerance compared with most other crops.

Decay control. The most commercially successful HT system for controlling decay on horticultural products is the application of hot water either as a dip, drench/shower or spray in combination with brushes (Table 2). These treatments are short in duration (seconds to minutes) and thus lend themselves well to in-line systems in a packhouse, making them much easier to instigate than long-duration, batch-based systems such as HAT. The development of hot water rinse and brush systems shows the potential for successful commercial implementation. This method was widespread in Israel where it was used for many commodities as a means of decay control and improvement of appearance of the fruit or vegetable. However, when fuel prices increased, some producers stopped heating the water, and only used the cold rinse. Probably the largest volume of fruit treated with hot water is citrus where commercial control of postharvest decay is clearly successful and is widely used in California, USA (Smilanick, Arpaia pers. comm.). It is likely that the development of fungicide resistance and the general desire for reduced postharvest chemical use drives uptake, although many HWTs are applied in combination with lower doses of fungicides. The successful commercialization of citrus on such a large volume and relatively low-value crop is a significant support for the technology.

In the Netherlands, HWT has been successfully commercialized for the control of insects in flower bulbs (Table 3). This treatment can also be applied to vegetable crops where nematodes and other insects create quarantine issues. The successful treatments tend to be lower temperatures (~43°C for a few hours) and probably reflect the lower thermotolerance of insects associated with the soil (in particular nematodes).

5 Conclusion

HTs of horticultural produce have been examined experimentally for many decades with great success. While HTs can be efficacious for key outcomes (disinfestation, decay control), the high costs and logistical challenges mean that HTs are generally only used in 'niche' applications such as high-value fruit crops; tropical (Table 2 – tropical for fruit fly), bulb crops (Table 3), although

Table 3 Heat treatments used in the Netherlands on flower bulbs to protect against microorganisms and insects

Flower	Purpose	Species	Treatment	Comments
Allium	Nematodes[a]	*Aphelenchoides subtenuis*	HWT 45°C 4 h	Pretreat 1 week 30°C + 24 h soaking
Amaryllis/Hippeastrum	Mite	Various	HWT 43°C 1–2 h or 48°C 30 min	Also for thrips 1–2 h 46°C
Anemone	Insects and rots	*Colletotrichum* spp.	HWT 47.5°C 1.5 h	
Camassia	Nematodes	*Ditylenchus dipsaci*	HWT 45°C 4 h	Pretreat 1–4 weeks 30°C + 24 h soaking
Chionodoxa	Nematodes	*Ditylenchus* spp.	HWT 45°C 4 h	Pretreat 2–4 weeks 30°C + 24 h soaking
Erythronium	Nematodes	*Aphelechoides ritzemabosi, Ditylenchus destructor*	HWT 43.5°C 4 h	Pretreat 1 week 25°C + 24 h soaking
Freesia	Mite	*Rhizoglyphus* spp.	HAT 43°C 24 h	
Gladiolus cormlets	Decay	*Urocystus gladiolicola* *Stromatinia gladioli*	HWT for *Urocystus* 47°C 1 h HWT for *Stromatinia* 50°C 1 h, HWT 53°C 0.5 h for cormlets	Treat cormlets before Jan. 15. May be some growth loss
Gladiolus	Thrips	various	HAT 43°C 24 h	
Hyacinth	Nematodes	*Ditylenchus destructor, Ditylenchus dipsaci*	HWT for *D. destructor* 45°C 2 h, for *D. dipsaci* 45°C 4 h	*D. destructor* 3–6 weeks 30°C then HWT, *D. dipsaci* 3 weeks 30°C + 24 h soaking the HWT
Hyacinth	Bacteria	*Xanthomonas hyacinthi*	HAT 30°C 4 weeks + 38°C 2 weeks + 44°C 3 days	Needs monitoring during 44°C. Heat damage possible

(*Continued*)

Table 3 (*Continued*)

Flower	Purpose	Species	Treatment	Comments
Iris	Nematodes	*Ditylenchus destructor*	HWT planting size 45°C 2 h, small and saleable size 43.5°C 2.4 h	Pretreat 20–23°C 1–4 weeks then 24 h soaking. Treat within 4 weeks from lifting
Iris	Nematodes	*Aphelechoides ritzemabosi*	HWT 43°C 4 h	30°C 1 week within 10 days from lifting
Crocus	Nematodes	*Aphelechoides subtenuis, Ditylenchus destructor*	HWT saleable size 43.5°C 1 h, planting size 45°C 4 h	Pretreat 25–30°C 10 days + 24 h soaking. Treat within 12 days from lifting
Lily	Nematodes	*Aphelechoides fragariae, Aphelechoides ritzemabosi*	HWT 39°C-41°C 2 h	Also reduces mites (*Rhizoglyphus* spp.)
Narcissus/daffodil	Nematodes	*Ditylenchus dipsaci*	HWT 48°C 4 h	
Narcissus/daffodil	Mites	*Steneotarsonemus laticeps*	HWT 45°C 2 h	
Narcissus/daffodil	Insects	Narcissus fly (*merodon equestris)*	HWT 45°C 1 h	
Nerine (Guernsey lily)	Nematodes	*Aphelechoides ritzemabosi*	HWT 43.5°C 2 h	
Tulip	Nematodes	*Ditylenchus destructor*	HWT 43.5°C 2.5 h	Pretreat 3 weeks 30°C + 24 h soaking
Tulip	Nematodes	*Ditylenchus dipsaci*	HWT 48°C 4 h	Pretreat 1 week 36°C + 24 h soaking. Damage possible depending on tulip variety

Abbreviations: HAT, hot air treatment; HWT, hot water treatment; CATT, controlled atmosphere temperature treatment. Information from Martin van Dam (Wageningen University & Research).

[a]Many nematode HWTs start with a dry storage at high temperature to prevent damage to the bulb. This is followed by soaking for ~4 h and then stored wet for 20 h. Soaking activates the nematodes and therefore enhances the effectiveness of the HWT.

decay control in citrus is a notable exception. This is particularly true in cases where other treatments such as low temperature disinfestation for tropical fruit are not possible.

With the general reduction in postharvest chemical treatments and loss of fungicides both pre- and postharvest leading to reduced options, HTs have become more attractive. Organic packhouses in particular have commercialized HTs, and countries with strong organic drivers such as Denmark and Germany engage in more HT research. Citrus is a crop with wide use of a HWT in California, USA, and Israel, as an adjunct to chemical pathogen control. HT of bulbs is a good example of treatments developed because of the removal of methyl bromide use.

Finally, because energy costs are a significant issue, there is an opportunity for innovation in this area with use of systems that recover heat from cooling steps, use of waste heat from cool stores, or use of renewable sources of energy. Any of these will make the treatment more ecologically and economically viable. Therefore, further technical innovations and 'green'/organic drivers are likely to result in increased commercial uptake of HTs worldwide.

6 Acknowledgements

We would like to acknowledge the generous time of many researchers around the world who provided valuable feedback on what commercial treatments were being used in their countries or areas of expertise. Merete Edelenbos (Aarhus University, Denmark), Peter Follett (USDA, Hilo, Hawaii), Elhadi Yahia (Universidad Autónoma de Querétaro, México), Hinrich Holthusen (Chamber of Agriculture, Germany), Sarah Perren (Agroscope, Switzerland), Stan Kays (University of Georgia, USA), Chris Clark (Louisiana State University, USA), Peter Hofman (Department of Agriculture and Fisheries, Australia), Marta Mari (Bologna University), Maria Concetta Strano (Council for Agricultural Research and Economics, Italy), John Bower (Canada), Joseph Smilanick (USA), Mary Lu Arpaia (University of California, USA), Lisa Neven (USDA, Washington), Ron Porat (Volcani, ARO, Israel), Lisa Jamieson (Plant and Food Research, New Zealand), Barbara Waddell (Plant and Food Research, New Zealand), Muhammad Mazhar (Department of Agriculture and Fisheries, Australia), Peter Leach (Department of Agriculture and Fisheries, Australia), Martin van Dam (Wageningen University & Research, Netherlands), Nguyen Van Phong (Southern Horticultural Research Institute, Viet Nam), Nguyen Khanh Ngoc (Southern Horticultural Research Institute, Viet Nam), Jan Verschoor (Wageningen University & Research, Netherlands), Maria Teresa Lafuente (Agroquímica y Tecnología de Alimentos, Spain). For use of figures: – Fig. 1 – Karl Schloffer (Obstweb, Anger, Austria) and Fig. 5 – work was funded by NZ Apple & Pears Inc. and The Ministry of Business, Innovation and Employment through the Apple Futures II programme. Finally, thanks to Samuel Brown, Natalie Page-Weir and Jason Johnston for reviewing this manuscript.

7 Future trends

Many reports have been published during the last two decades about HTs that control postharvest decay development and insect infestation in agricultural products without significantly reducing quality. However, not all fresh and fresh-cut products, especially heat-sensitive produce, can tolerate the HTs required to control decay-causing agents and insects. In addition, energy efficiency and cost requirements to build and operate the technology are two very important considerations in the development of commercially practical and effective HTs for controlling decay and insects in fruits and vegetables (Shahkoomahally and Ramezanian, 2015). The search for safe and effective HTs is still very challenging because many factors, such as genetic background, pre- and postharvest treatment and conditions, fresh produce maturity during harvest impact the results obtained (Sivakumar and Fallik, 2013; Usall et al., 2016). In addition, many technical and engineering factors should be considered in selecting the most appropriate means of heat treating product (e.g. dipping, hot water brushing, vapor-heat and others), and the potential for commercial-scale application. The majority of the problems associated with each HT can be solved, but the solutions may not always be economically or technically feasible.

8 Where to look for further information

Research institutes and universities, and their publications (particularly reviews) about heat treatments.

Energy companies and agricultural machinery manufacturers around the world (USA, Holland, Israel, Spain, New Zealand, etc.).

9 References

Abrashev, R. I., Pashova, S. B., Stefanova, L. N., Vassilev, S. V., Dolashka-Agnelova, P. A. and Angelova, M. B. (2008), Heat-shock-induced oxidative stress and antioxidant response in *Aspergillus niger*. *Canad. J. Microbiol.*, 54, 977–83.

Amiri, A. and Bompeix, G. (2011), Control of *Penicillium expansum* with potassium phosphite and heat treatment. *Crop Prot.*, 30, 222–7.

Anon. (1999), Foodborne illness outbreak database: Brazilian mangoes 1999. Available at: http://www.outbreakdatabase.com/details/brazilian-mangoes-1999/?outbreak=mangoes&organism=Salmonella (accessed on 25 February 2019).

Armstrong, J. W., Hu, B. K. S. and Brown, S. A. (1995), Single-temperature forced hot-air quarantine treatment to control fruit flies (Diptera: Tephritidae) in papaya. *J. Econ. Entomol.*, 88, 678–82.

Avila-Quezada, G., Sánchez, E., Gardea-Béjar, A. A. and Acedo-Félix, E. (2010), *Salmonella* spp. and *Escherichia coli*: survival and growth in plant tissue. *NZ J. Crop Hort. Sci.*, 38, 47–55.

Ayón-Reyna, L. E., González-Robles, A., Rendón-Maldonado, J. G., Báez-Flores, M. E., López-López, M. E. and Vega-García, M. O. (2017a), Application of a hydrothermal-calcium chloride treatment to inhibit postharvest anthracnose development in papaya. *Postharvest Biol. Technol.*, 124, 85–90.

Ayón-Reyna, L. E., López-Valenzuela, J. A., Delgado-Vargas, F., López-López, M. E., Molina-Corral, F. J., Carrillo-López, A. and Vega-García, M. O. (2017b), Effect of the combination hot water-calcium chloride on the in vitro growth of *Colletotrichum gloeosporioides* and the postharvest quality of infected papaya. *Plant Pathol. J.*, 33, 572–81.

Bansal, M. K., Boyhan, G. E. and MacLean, D. D. (2018), Effect of postharvest chemical treatments, heat curing, and refrigerated storage on marketability of short-day onions. *HortTechnology*, 28, 129–35.

Bayat, F., Rezvani, S. and Nosrati, A. E. (2010), Effect of harvesting time and curing temperature on some properties of Iranian white garlic. *Acta Hortic.*, 877, 869–75.

Beckett, S. J. and Evans, D. E. (1997), The effects of thermal acclimation on immature mortality in the Queensland fruit fly *Bactrocera tryoni* and the light brown apple moth *Epiphyas postvittana* at a lethal temperature. *Entom. Exp. Appl.*, 82(1), 45–51.

Birla, S. L., Wang, S., Tang, J. and Hallman, G. (2004), Improving heating uniformity of fresh fruit in radio frequency treatments for pest control. *Postharvest Biol. Technol.*, 38, 205–17.

Brown, S. D. J., Jamieson, L. E., Laidlaw, W. G., De Silva, N. and Waddell, B. C. (2019), Decay of thermal tolerance in Queensland fruit fly eggs (*Bactrocera tryoni*, Diptera: Tephritidae) following non-lethal heat hardening. *J. Econ. Entomol.*, 112(3), 1469–75.

Caleb, O. J., Wegner, G., Rolleczek, C., Herppich, W. B., Geyer, M. and Mahajan, P. V. (2016), Hot water dipping: impact on postharvest quality, individual sugars, and bioactive compounds during storage of 'Sonata' strawberry. *Sci. Hortic.*, 210, 150–7.

Chalupowicz, D., Alkalai-Tuvia, S., Zaaroor-Presman, M. and Fallik, E. (2018), The potential use of hot water rinsing and brushing technology to extend storability and shelf life of sweet acorn squash (*Cucurbita pepo* L.). *Horticulturae*, 4, 19. doi:10.3390/horticulturae4030019.

Chen, H., Cheng, Z., Wisniewsky, M., Liu, Y. and Liu, J. (2015), Ecofriendly hot water treatment reduces postharvest decay and elicits defense response in kiwifruit. *Environ. Sci. Pollut. Res.*, 22, 15037–45.

D'Aquino, S., Dai, S., Deng, Z., Gentile, A., Angioni, A., De Pau, L. and Palma, A. (2017), A sequential treatment with sodium hypochlorite and a reduced dose of imazalil heated at 50°C effectively control decay of individually film-wrapped lemons stored at 20°C. *Postharvest Biol. Technol.*, 124, 75–84.

Duarte-Sierra, A., Forney, C. F., Michaud, D., Angers, P. and Arul, J. (2017), Influence of hormetic heat treatment on quality and phytochemical compounds of broccoli florets during storage. *Postharvest Biol. Technol.*, 128, 44–53.

Fallik, E. (2010), Hot water treatments of fruit and vegetables for postharvest storage. *Hortic. Rev.*, 38, 191–212.

Fallik, E., Grinberg, S., Gambourg, M., Klein, J. D. and Lurie, S. (1995), Prestorage heat treatment reduces pathogenicity of Penicillium expansum in apple fruit. *Plant Pathol.*, 45, 92–7.

Fallik, E., Grinberg, S., Alkalai, S., Yekutieli, O., Wiseblum, A., Regev, R., Beres, H. and Bar Lev, E. (1999), A unique rapid hot water treatment to improve storage quality of sweet pepper. *Postharvest Biol. Technol.*, 15, 25-32.

Fallik, E., Aharoni, Y., Copel, A., Rodov, V., Tuvia-Alkalai, S. and Horev, B. (2000), Reduction of postharvest losses of Galia melon by a short hot-water rinse. *Plant Pathol.*, 49, 333-8.

Fallik, E., Tuvia-Alkalai, S., Feng, X. and Lurie, S. (2001), Ripening characterisation and 280 decay development of stored apples after a short pre-storage hot water rinsing and brushing. *Innov. Food Sci. Emerg. Technol.*, 2, 127-32.

Ferguson, I. B., Lurie, S. and Bowen, J. H. (1994), Protein synthesis and breakdown during heat shock of cultured pear (*Pyrus communis*) cells. *Plant Physiol.*, 104, 1429-37.

Gan-Mor, S., Regev, R., Levi, A. and Eshel, D. (2011), Adapted thermal imaging for the development of postharvest precision steam-disinfection technology for carrots. *Postharvest Biol. Technol.*, 59, 265-71.

Gao, M., Tang, J., Wang, Y., Powers, J. and Wang, S. (2010), Almond quality as influenced by radio frequency heat treatments for disinfestation. *Postharvest Biol. Technol.*, 58, 225-31.

Hansen, J. D., Armstrong, B. K., Hu, S. and Brown, S. A. (1990), Thermal death of oriental fruit fly (Diptera: Tephritidae) third instars in developing quarantine treatments for papayas. *J. Econ. Entom.*, 83, 160-7.

Hansen, J. D., Hara, A. H. and Tenbrink, V. L. (1992), Vapor heat: a potential treatment to disinfect tropical cut flowers and foliage. *HortScience*, 27, 139-43.

Hansen, J. D., Johnson, J. A. and Winter, D. A. (2011), History and use of heat in pest control: a review. *Intl. J. Pest Manag.*, 57, 267-89.

Hasanuzzaman, M., Nahar, K., Alam, M. M., Roychowdhury, R. and Fujita, M. (2013), Physiological, biochemical, and molecular mechanisms of heat stress tolerance in plants. *Intl. J. Mol. Sci.*, 14, 9643-84.

He, L. H., Chen, J. Y., Kuang, J. F. and Lu, W. J. (2012), Expression of three sHSP genes involved in heat pretreatment-inducing chilling tolerance in banana fruit. *J. Sci. Food Agric.*, 92, 1924-30.

Heather, N. W. and Hallman, G. J. (2008), Development of postharvest phytosanitary disinfestation treatments. In: N. W. Heather and G. J. Hallman (Eds), *Pest Management and Phytosanitary Trade Barriers*. CABI, Wallingford, UK, pp. 71-95.

Hernandez, I., Fuentealba, C., Olaeta, J. A., Poblete-Escheverria, C., Defilippi, B. G., Gonzalez-Aguero, M., Campos-Vargas, R., Lurie, S. and Pedreschi, R. (2017), Effects of heat shock and nitrogen shock pre-treatments on ripening heterogeneity of Hass avocados stored in controlled atmosphere. *Sci. Hortic.*, 225, 408-15.

Hou, L., Johnson, J. A. and Wang, S. (2016), Radio frequency heating for postharvest control of pests in agricultural products: a review. *Postharvest Biol. Technol.*, 113, 106-18.

Hsu, Y. L., Chen, S. C., Lin, K. W., Shiesh, C. C., Lin, C. H. and Yeh, W. B. (2018), Quarantine vapor heat treatment of papaya fruit for *Bactrocera dorsalis* and *Bactrocera cucurbitae* (Diptera: Tephritidae). *J. Econ. Entomol.*, 111, 2101-9.

Hu, J.-T., Chen, B. and Li, Z.-H. (2014), Thermal plasticity is related to the hardening response of heat shock protein expression in two Bactrocera fruit flies. *J. Insect Physiol.*, 67(Suppl. C), 105-13.

Huan, C., Han, S., Jiang, L., An, X., Yu, M., Xu, Y., Ma, R. and Yu, Z. (2017), Postharvest hot air and hot water treatments affect the antioxidant system in peach fruit during refrigerated storage. *Postharvest Biol. Technol.*, 126, 1-14.

Jang, E. B. (1991), Thermal death kinetics and heat tolerance in early and late third instars of the oriental fruit fly (Diptera: Tephritidae). *J. Econ. Entom.*, 84, 1298–303.

Jedermann, R., Nicometo, M., Uysal, I. and Lang W. (2014), Reducing food losses by intelligent food logistics. *Phil. Trans. R. Soc. A*, 372, 20130302. doi:10.1098/rsta.2013.0302.

Jiao, S., Johnson, J. A., Tang, J. and Wang, S. (2012), Industrial-scale radio frequency treatments for insect control in lentils. *J. Stored Prod. Res.*, 48, 143–8.

Jiao, Y., Tang, J., Wang, Y. and Kora, T. L. (2018), Radio-frequency applications for food processing and safety. *Annu. Rev. Food Sci. Technol.*, 9, 105–27.

Konigshofer, H., Thromballa, H. W. and Loppert, H. G. (2008), Early events in signalling high temperature stress in tobacco BY2 cells involve alterations in membrane fluidity and enhanced hydrogen peroxide production. *Plant Cell Environ.*, 31, 1771–80.

Krasensky, J. and Jonak, C. (2012), Drought, salt and temperature stress-induced metabolic rearrangements and regulatory networks. *J. Exp. Bot.*, 63, 1593–608.

Kumar, D., Ram, L., Kumar, S. and Khadse, A. (2018), Wet heat treatment of Nagpur mandarin (*Citrus reticulata*) fruit to reduce decay loss. *Indian J. Agric. Sci.*, 88, 112–16.

Lafuente, M. T., Estables-Ortiz, B. and Gonzalez-Candelas, L. (2017), Insights into the molecular events that regulate heat-induced chilling tolerance in citrus fruit. *Front. Plant Sci.*, 8, 1113. doi:10.3389/fpls.2017.01113.

Lauxmann, M. A., Brun, B. and Borsani J. (2012), Transcriptomic profiling during the postharvest of heat-treated Dixiland *Prunus persica* fruits: common and distinct response to heat and cold. *PLoS ONE*, 7, e51052.

Lauxmann, M. A., Borsani, J. and Osorio, S. (2014), Deciphering the metabolic pathways influencing heat and cold responses during post-harvest physiology of peach fruit. *Plant Cell Environ.*, 37, 601–16.

Lester, P. J. and Greenwood, D. R. (1997), Pretreatment induced thermotolerance in lightbrown apple moth (Lepidoptera: Tortricidae) and associated induction of heat shock protein synthesis. *J. Ecol. Entomol.*, 90(1), 199–204.

Li, M., Li, X.-J., Lü, J.-H. and Huo, M.-F. (2018), The effect of acclimation on heat tolerance of *Lasioderma serricorne* (Fabricius) (Coleoptera: Anobiidae). *Journal of Thermal Biol.*, 71(Suppl. C), 153–7.

Liu, J., Sui, Y. Wisniewsky, M., Droby, S., Tian, S. and Norelli, J. (2012), Effect of heat treatment on inhibition of *Monolinia fructicola* and induction of disease resistance in peach fruit. *Postharvest Biol. Technol.*, 65, 61–8.

Luria, N., Sela, N., Yaari, M., Feygenberg, O., Kobiler, I., Lers, A. and Prusky, D. (2014), *De-novo* assembly of mango fruit peel transcriptome reveal mechanisms of mango response to hot water treatment. *BMC Genomics*, 15, 957–71.

Lurie, S. (1998a), Postharvest heat treatments of horticultural crops. *Hort. Rev.*, 22, 91–122.

Lurie, S. (1998b), Postharvest heat treatments. *Postharvest Biol. Technol.*, 14, 257–69.

Lurie, S. (2008), Heat treatment for enhancing postharvest quality. In P. Paliyath, D. P. Murr, A. V. Handa and S. Lurie (Eds), *Postharvest Biology and Technology of Fruit, Vegetables, and Flowers*. Wiley-Blackwell, Ames, IA, pp. 246–59.

Lurie, S. and Jang, E. (2007), The influence of heat shock proteins on insect pests and fruit in thermal treatments. In J. Tang, E. Mitcham, S. Wang and S. Lurie (Eds), *Heat Treatments for Postharvest Pest Control*. CAB International, Wallingford, UK, pp. 269–90.

Macana, R. J. and Baik, O. D. (2018), Disinfestation of insect pests in stored agricultural materials using microwave and radio frequency heating: a review. *Food Rev. Intl.*, 34, 483–510.

Mackey, M. A. and Roti, J. L. (1992), A model of heat induced clonogenic death. *J. Theor. Biol.*, 156, 133–46.

Maw, B. W., Butts, C. L., Purvis, A. C., Seebold, K. and Mullinix, B. G. (2004), High temperature continuous-flow curing of sweet onions. *Appl. Eng. Agric.*, 20, 657–63.

Maxin, P., Weber, R. W. S., Pedersen, H. L. and William, M. (2012), Control of a wide range of storage rots in naturally infected apples by hot-water dipping and rinsing. *Postharvest Biol. Technol.*, 70, 25–31.

McClung, C. R. and Davis, S. J. (2010), Ambient thermometers in plants: From physiological outputs towards mechanisms of thermal sensing. *Curr. Biol.*, 20, 1086–92.

Mittler, R., Finka, A. and Goloubinoff, P. (2012), How do plants feel the heat? *Trends Biochem. Sci.*, 37, 118–25.

Nafussi, B., Ben-Yehoshua, S., Rodov, V., Peretz, J., Ozer, B. K. and D'Hallewin, G. (2001), Mode of action of hot water dip in reducing decay of lemon fruit. *J. Agric. Food Chem.*, 49, 107–13.

Nasef, I. N. (2018), Short hot water as safe treatment induces chilling tolerance and antioxidant enzymes, prevents decay and maintains quality of cold-stored cucumbers. *Postharvest Biol. Technol.*, 138, 1–10.

Nevin, L. G. (1998), Effects of heating rate on the mortality of fifth-instar codling moth (Lepidoptera : Tortricidae). *J. Econ. Entom.*, 91, 297–301.

Nevin, L. G. (2000), Physiological responses of insects to heat. *Postharvest Biol. Technol.*, 21, 103–11.

Ngoc, N. K., Phong, N. V. and Fullerton, B. (2015), Study of post-harvest diseases on dragon fruit (*Hylocereus undatus* Haw.) causing by species of *Fusarium* spp. and the treating of hot water and sanitizer (Chlorine). *Vietnam. J. Plant Prot.*, (4), 42.

Pan, L., Jiao, S., Gautz, L., Tu, K. and Wang, S. (2012), Coffee bean heating uniformity and quality as influenced by radio frequency treatments for postharvest disinfestations. *Trans. ASABE*, 55, 2293–300.

Pavoncello, D., Lurie, S., Droby, S. and Porat, R. (2001), A hot water treatment induces resistance to *Penicillium digitatum* and promotes the accumulation of heat shock and pathogenesis-related proteins in grapefruit flavedo. *Physiol. Plant.*, 111, 17–22.

Pazolini, K., dos Santos, I., Giaretta, R. D., Marcondes, M. M., Reiner, D. A. and Citadin, I. (2016), The use of brassica extracts and thermotherapy for the postharvest control of brown rot in peach. *Sci. Hortic.*, 209, 41–6.

Porat, R., Daus, A., Weiss, B, Cohen, L., Fallik, E. and Droby, S. (2000), Reduction of postharvest decay in organic citrus fruit by a short hot water brushing treatment. *Postharvest Biol. Technol.*, 18, 151–7.

Rasouli, M. and Khademi, O. (2018), Influence of storage temperature, hot water, and 1-MCP treatments on the postharvest quality of 'Karaj' persimmon. *Intl. J. Fruit Sci.*, 18, 320–37.

Ruelland, E. and Zachowski, A. (2010), How plants sense temperature. *Environ. Exp. Bot.*, 69, 225–32.

Sabehat, A., Weiss, D. and Lurie, S. (1996), The correlation between heat-shock protein accumulation and persistence and chilling tolerance in tomato fruit. *Plant Physiol.*, 110, 531–7.

Saltveit, M. E. (2017), Heat-shock treatments increase the chilling tolerance of harvested asparagus spears. *HortScience*, 52, 1563–8.

Schirra, M., D'Aquino, S., Cabras, P. and Angioni, A. (2011), Control of postharvest diseases of fruit by heat and fungicides: Efficacy, residue levels, and residue persistence. A review. *J. Agric. Food Chem.*, 59, 8531–42.

Semenov, M. A. and Halford, N. G. (2009), Identifying target traits and molecular mechanisms for wheat breeding under a changing climate. *J. Exp. Bot.*, 60, 2791–804.

Sevillano, L., Mar Sola, M. and Vargas, A. M. (2010), Induction of small heat-shock proteins in mesocarp of cherimoya fruit (*Annona cherimola* Mill.) produces chilling tolerance. *J. Food Biochem.*, 34, 625–38.

Shahkoomahally, S. and Ramezanian, A. (2015), Hot water combined with calcium treatment improves physical and physicochemical attributes of kiwifruit (*Actinidia deliciosa* cv. Hayward) during storage. *HortScience* 50, 412–15.

Shukla, S., Cho H., Kwon O. J., Chung S. H. and Kim M. (2018), Prevalence and evaluation strategies for viral contamination in food products: risk to human health-a review. *Crit. Rev. Food Sci. Nutr.*, 58, 405–19.

Sisquella, M., Casals, C., Picouet, P., Vinas, I., Toerres, R. and Usall, J. (2013), Immersion of fruit in water to improve radio frequency treatment to control brown rot in stone fruit. *Postharvest Biol. Technol.*, 80, 31–6.

Sivakumar, D. and Fallik, E. (2013), Influence of heat treatments on quality retention of fresh and fresh-cut produce. *Food Rev. Intl.*, 29, 294–320.

Sosa-Morales, M. E., Tiwari, G., Wang, S., Tang, J., Garcia, H. S. and Lopez-Malo, A. (2009), Dielectric heating as a potential post-harvest treatment of disinfesting mangoes. Part II: development of RF-based protocols and quality evaluation of treated fruit. *Biosys. Eng.*, 103, 287–96.

Soto-Reyes, N., López-Malo, A., Rojas-Laguna, R., Gómez-Salazar, J. A. and Sosa-Morales, M. E. (2018), Effects of microwave-assisted hot water treatments designed against Mexican fruit fly (*Anastrepha ludens*) on grapefruit (*Citrus paradisi*) quality. *J. Sci. Food Agric.*, 98, 3659–66.

Sui, Y., Droby, S., Zhang, D., Wang, W. and Liu, Y. (2014), Reduction of Fusarium rot and maintenance of fruit quality in melon using ecofriendly hot water treatment. *Environ. Sci. Poll. Res.*, 21, 13956–63.

Sui, Y., Wisniewski, M., Droby, S., Norelli, J. and Liu, J. (2016), Recent advances and current status of the use of heat treatments in postharvest disease management systems: Is it time to turn up the heat? *Trends Food Sci. Technol.*, 51, 34–40.

Supapvanich, S. and Promyou, S. (2017), Hot water incorporated with salicylic acid dips maintaining physicochemical quality of 'Holland' papaya fruit stored at room temperature. *Emirates J. Food Agric.*, 29, 18–24.

Suzuki, N., Koussevitzki, S. and Mittler, R. (2011), ROS and redox signalling in the response of plants to abiotic stress. *Plant Cell Environ.*, 35, 259–70.

Tang, J., Wang, S. and Johnson, J. A. (2007), Biology and thermal death kinetics of selected insects. In: J. Tang, E. Mitcham, S. Wang and S. Lurie (Eds), *Heat Treatments for Postharvest Pest Control*. CAB International, Wallingford, UK, pp. 133–61.

Tang, D., Lin, Q., Lin, J., Wang, D., Liu, C., Wu, W., Sun, C. and Chen, K. (2017), Effects of combined heat and preservative treatment on storability of Ponkan fruit (*Citrus reticulata* Blanco cv. Ponkan) during postharvest storage. *J. Food Qual.*, 17, 1–7.

Terao, D., de Lima Nechet, K., Toyoko Shiraishi Frighetto, R., de Almeida Anjos, V. D., Benato, E. A. and de Almeida Halfeld-Vieira, B. (2018), Physical postharvest treatments in the control of stem-end rot of mango. *J. Phytopathol.*, 166, 581–9.

Tiwari, G., Wang, S., Birla, S. L. and Tang, J. (2008), Effect of water-assisted radio frequency heat treatment on the quality of 'Fuyu' persimmons. *Biosys. Eng.*, 100, 227-34.

Usall, J., Ippolito, A., Sisquella, M. and Neri, F. 2016, Physical treatments to control postharvest diseases of fresh fruits and vegetables. *Postharvest Biol. Technol.*, 122, 30–40.

Vilaplana, R., Hurtado, G. and Valencia-Chamorro, S. (2018), Hot water dips elicit disease resistance against anthracnose caused by *Colletotrichum musae* in organic bananas (*Musa acuminata*). *LWT-Food Sci. Technol.*, 95, 247–54.

Waddell, B. C., Jones, V. M., Petry, R. J., Sales, F., Paulaud, D., Maindonald, J. H. and Laidlaw, W. G. (2000), Thermal conditioning in *Bactrocera tryoni* eggs (Diptera: Tephritidae) following hot-water immersion. *Postharvest Biol. Technol.*, 21, 113–28.

Wang, C. Y., Bowen, J. H., Weir, I. E., Allan, A. C. and Ferguson, I. B. (2001) Heat-induced protection against death of suspension-cultured apple fruit cells exposed to low temperature. *Plant Cell Environ.*, 24, 1199-207.

Wang, S., Birla, S. L., Tang, J. and Hansen, J. D. (2006), Postharvest treatment to control codling moth in fresh apples using water assisted radio frequency heating. *Postharvest Biol. Technol.*, 40, 89–96.

Wang, S., Monzon, M., Gazit, Y., Tang, J., Mitcham. E. J. and Armstrong, J. W. (2005), Temperature dependent dielectric properties of selected subtropical and tropical fruit and associated insect pests. *Trans. ASAE*, 48, 1873–80.

Wang, S., Tang, J. and Hansen, J. D. (2007), Experimental and simulation methods of insect thermal death kinetics. In: J. Tang, E. Mitcham, S. Wang and S. Lurie (Eds), *Heat Treatments for Postharvest Pest Control*. CAB International, Wallingford, UK, pp. 105–32.

Wang, S., Tiwari, G., Jiao, S., Johnson, J. A. and Tang, J. (2010), Developing postharvest disinfestation treatments for legumes using radio frequency energy. *Biosys. Eng.*, 105, 341-9.

Wang, Y., Zhang, L., Johnson, J., Gao, M., Tang, J. Powers, J. R. and Wang, S. (2014), Developing hot air-assisted radio frequency drying for in-shell macadamia nuts. *Food Bioprod. Technol.*, 7, 278–88.

Wei, Y., Zhou, D., Wang, Z., Tu, S., Shao, X., Peng, J., Pan, L. and Tu, K. (2018), Hot air treatment reduces postharvest decay and delays softening of cherry tomato by regulating gene expression and activities of cell wall-degrading enzymes. *J. Sci. Food Agric.*, 98, 2105-12.

Whiting, D. C., Jamieson, L. E., Spooner, K. J. and Lay-Yee, M. (1999), Combination high-temperature controlled atmosphere and cold storage as a quarantine treatment against *Ctenopseustis obliquana* and *Epiphyas postvittana* on 'Royal Gala' apples. *Postharvest Biol. Technol.*, 16(2), 119–26.

Woolf, A. B., Bowen, J. H., Ball, S., Durand, S., Laidlaw, W. G. and Ferguson, I. B. (2004), A delay between a 38°C pretreatment and damaging high and low temperature treatments influences pretreatment efficacy in 'Hass' avocados. *Postharvest Biol. Technol.*, 34, 143–53.

Yi, S. Y., Sun, A. Q., Sun, Y., Yang, J. Y., Zhao, C. M. and Liu, J. (2006), Differential regulation of Lehsp23.8 in tomato plants: analysis of a multiple stress inducible promoter. *Plant Sci.*, 171, 398–407.

Yin, X., Wang, S., Tang, J. and Hansen, J. D. (2006), Thermal resistance of fifth-instar *Cydia pomonella* (L.) (Lepidoptera: Tortricidae) as affected by pretreatment conditioning. *J. Stored Prod. Res.*, 42(1), 75–85.

Yun, Z., Gao, H., Liu, P., Liu, S., Luo, T., Jin, S., Xu, Q., Xu, J., Cheng, Y. and Deng, X. (2013), Comparative proteomic and metabolomic profiling of citrus fruit with enhancement of disease resistance by postharvest heat treatment. *BMC Plant Biol.*, 13, 44. doi:10.1186/1471-2229-13-44.

Zhang, J. H., Huang, W. D., Pan, Q. H. and Liu, Y. (2005), Improvement of chilling tolerance and accumulation heat shock proteins in grape berries (*Vitis vinifera* cv. Jingxiu) by heat pretreatment. *Postharvest Biol. Technol.*, 38, 80–90.

Zhao, Y. and Yin, J. (2018), Effects of *Pichia guilliermondii* and hot air treatment on the postharvest preservation of Red Fuji apple quality attributes. *J. Food Prot.*, 81, 186–94.

Zhao, W., Wisniewsky, M., Want, W., Liu, J. and Liu, Y. (2014), Heat-induced oxidative injury contributes to inhibition of *Botrytis cinerea* spore germination and growth. *World J. Microbiol. Biotech.*, 63, 89–96.

Zudaire, L., Vinas, I., Abadias, M., Simo, J. and Aguilo-Aguayo, I. (2018), Efficacy of chlorine, peroxyacetic acid and mild-heat treatment on the reduction of natural microflora and maintenance of quality of fresh-cut calçots (*Allium cepa* L.). *LWT-Food Sci. Technol.*, 95, 339–45.

Chapter 9

Advances in the use of irradiation for the market access of fresh horticultural produce

John Golding and Sukhvinder Pal Singh, New South Wales Department of Primary Industries and University of Newcastle, Australia

1 Introduction

Irradiation treatment involves the use of ionizing radiation to facilitate a range of commercial postharvest horticultural outcomes, such as a market access treatment, to maintain product quality and inhibit sprouting in tubers and bulbs. Consuming irradiated food, including fresh fruit and vegetables, is safe (WHO, 1981, 1999; European Food Safety Authority, 2011a) and is allowed in many countries such as the United States, Australia and New Zealand (FSANZ, 2014). Like all postharvest treatments and technologies, irradiation has a range of technical and non-technological benefits and issues which must be managed to allow its integration into commercial horticulture. There is growing interest in the irradiation of fresh fruit and vegetables as a market access treatment. Irradiation has been approved for use in many markets as an approved phytosanitary / quarantine treatment to kill or sterilize quarantine insect pests to allow international trade. This interest for irradiation is reflected in the increase in research papers and reviews in this area, particularly in tropical fruit (Amir et al., 2018; Ashtari et al., 2019; Barkai-Golan and Follett, 2017; Golding et al.,

http://dx.doi.org/10.19103/AS.2019.0055.11

2015; Ramakrishnan et al., 2019; Reddy et al., 2018; Singh and Pal, 2009; Wall, 2015).

Irradiation treatment damages the target cell DNA so that living cells become inactivated, whereby microorganisms (decay), insect gametes (quarantine pests) and plant meristems (tuber / bulb sprouts) are prevented from reproducing, resulting in various preservative effects as a function of the absorbed radiation dose. The effects of irradiation are not selective, where only the target organisms are affected. Irradiation treatment can induce other chemical changes in the fresh fruit or vegetable which can sometimes result in damage to product quality following treatment and storage. This will be discussed in more detail in the '*Product quality*' section.

There has been a broad and comprehensive range of research conducted on irradiation of fresh fruit and vegetables for over 60 years. Perhaps no other food technology has been more studied than irradiation.

Postharvest irradiation treatment can exert fungicidal or fungistatic effects against spoilage microorganisms and has the potential to be used in combination with other treatments to replace chemical treatments. Irradiation is also capable of inhibiting the accumulation of human pathogenic bacteria that frequently contaminate the surface of fresh and fresh-cut fruit and vegetables and are responsible for serious outbreaks. However, the levels of ionizing radiation required for these purposes are often higher than recommended from fresh fruit and vegetables where 'high' levels of radiation can cause irreversible damage to product quality.

Irradiation is used in some countries as an inhibitor of sprouting potato tubers or onion and garlic bulbs to increase their storage and shelf life. Irradiation has also been shown to extend the postharvest life of some fresh fruit and vegetables by retarding the physiological activity, mainly those associated with the ripening and senescence processes. The inhibition of sprouting, ripening and senescence processes underpins the potential of this technology as a tool to minimize food losses and contribute to global food security.

Irradiation of fresh horticultural produce is becoming an increasingly important commercial and market access treatment in some parts of the world. The International Plant Protection Convention (IPPC) has several International Standards for Phytosanitary Measures (ISPM) relating to the use of postharvest irradiation of fresh fruit and vegetables for phytosanitary purposes. This use of irradiation as a market access treatment is currently facilitating the trade of fresh fruit and vegetables around the world without the risk of insect contamination and with limited use of postharvest chemicals or fumigants. This aspect of irradiation is discussed in more detail in the '*Phytosanitary/market access*' section.

The maximum irradiation dose to meet the United States Food and Drug Administration (US FDA) regulations for the preservation and disinfestation of fresh fruit and vegetables is 1000 Gray (Gy) (US FDA, 2008). While the minimum absorbed dose for phytosanitary purposes in fresh fruit and vegetables is 400 Gy, the produce may receive maximum doses of 600–800 Gy irradiation treatment. Therefore, this chapter will focus on the use, application and effect of these low-dose irradiation treatments (<1000 Gy) using citrus fruit as a representative example for other horticultural produce.

2 Irradiation technology

The sources of radiation can either derive from radioactive isotopes or are generated electrically, and are quantified by the amount of energy absorbed by the irradiated product. Dosimetry is the measurement of absorbed radiation dose (commonly referred to as dose) and has units of Gray (Gy). 1 Gy is equal to the transfer of 1 joule of energy per kilogram of material. For the purposes of treating horticultural produce, the following types of ionizing radiations are covered by international standards of the Joint Food and Agricultural Organization (FAO) / World Health Organization (WHO) Codex Alimentarius Commission.

2.1 Gamma irradiation

Gamma rays are emitted spontaneously during the radioactive decay of cobalt-60. Gamma rays with energies of 1.17 and 1.33 MeV (as emitted by the radionuclide cobalt-60) are the standard irradiation treatment for food and fresh horticultural produce. Gamma rays have high penetrating characteristics, thus they can be used to treat horticultural produce even in pallet-size lots. The gamma radiation cannot be switched off and treatment facilities can be operated continuously to make full use of the radioactive material.

2.2 X-rays

X-ray production starts with high-energy electrons which are accelerated at a metallic target (e.g. tantalum, tungsten or gold) and this generates a stream of X-rays. These are generated with machine sources with energies up to 7.5 MeV, which in contrast to gamma irradiation can be switched off when not in use. However similar to gamma rays, X-rays have high penetrating characteristics and are used to treat pallet-size lots. X-ray-based irradiation facilities overcome limitations and risks generally linked to the use and handling of radioactive material.

2.3 E-beam

High-energy electron beams are produced by electron accelerating machines and produce accelerated high-energy electrons with energies up to 10 MeV.

E-beams deliver the dose at a high rate, and in less than a second can impart a dose that would take hours for a gamma irradiation facility to deliver. In contrast to gamma facilities, E-beams can be switched off. However, E-beam has low penetrability, thus the practical usable penetration depth-limit for 10 MeV electrons is a depth of just a few centimetres. This is useful for treating surface contaminated products on a process line but its use in horticulture is limited because treatment of large bulky boxes / pallets is usually required.

Each of the radiation sources (gamma rays, X-rays and E-beams) have different properties and therefore present different technological benefits and disadvantages. However, each type of ionizing radiation initially interacts with the fresh fruit or vegetable to produce similar reactive chemical intermediates that are transient and dissipate rapidly after exposure to ionizing radiation. The nature of these free radicals is quite similar to those produced during any other processing method such as cooking. The effects of irradiation are mainly due to the indirect action of these transient chemicals rather than by the direct effect of the radiation itself. A given dose of either gamma rays, X-rays or E-beam radiation gives rise to a similar effect. For example, irradiation damage is partially caused by the creation of free radicals from oxygen and water which react with and damage nearby molecules.

It is important to note that these radiation treatments do not induce radioactivity in the irradiated food (European Food Safety Authority, 2011a). A significant benefit of irradiation is that treatment causes practically no temperature rise in the product. This is important for the treatment of horticultural produce where the maintenance of the cool chain is critical for quality management. The use of gamma or X-rays, which have excellent penetrating abilities, allow treatment of entire pallets of packed fruit and vegetables without de-stacking of the pallet which also enables the maintenance of the cold chain during treatment.

Irradiation is widely used around the world for the sterilization of medical devices, pharmaceuticals, packaging and so forth. Although most treatment facilities are generally dedicated to one purpose (e.g. medical device sterilization), many treatment facilities are multipurpose and can irradiate a wide range of different products, allowing the possibility of re-alignment to allow treatment of horticultural produce with low doses (<1000 Gy). Thus, the treatment of commercial quantities of irradiated horticultural produce is available at reasonable costs.

3 Safety of irradiated food

No food technology has been as extensively studied with respect to food safety and nutrition as food irradiation. Panels of experts have systematically evaluated data from animal feeding tests and multi-generation tests in animals and the

Joint FAO/IEAE/WHO Expert Committee (JECFI) on the Wholesomeness of Irradiated Food (WHO, 1981, 1999) affirmed that 'Irradiation of any food commodity up to an overall average dose of 10 000 Gy (10 kGy) introduces no toxicological hazard; hence toxicological testing of food so treated is no longer required'. The JECFI also stated that irradiation of food up to a dose of 10 000 Gy introduces no special microbiological or nutritional problems. Further extensive reviews and investigations since this time have continued to support the JECFI's conclusions (European Food Safety Authority, 2011b; FSANZ, 2014) and irradiation of fresh fruit and vegetables is now approved and applied in many countries for different purposes.

4 Uses of irradiation in postharvest horticulture

Irradiation of fresh horticultural produce at low doses has been shown to (1) eliminate or sterilize insect pests, (2) inhibit sprouting and (3) reduce microbial pathogens. Each of these benefits are discussed in the following sections.

4.1 *Irradiation as a phytosanitary/market access treatment*

The primary use of the postharvest application of irradiation around the world is as a market access/phytosanitary treatment. To allow international trade of fresh fruit and vegetables, it is essential to ensure that no living or fertile insect quarantine pests are exported. Ionizing radiation has been shown to break chemical bonds within DNA and other biomolecules, thereby disrupting normal cellular function in the infesting quarantine pest. This accepted use of irradiation is outlined in the ISPM publication 18 '*Guidelines for the use of irradiation as a phytosanitary measure*' and was adopted by IPPC (FAO, 2003). The IPPC is an international plant health agreement which has over 180 contracting parties/countries which aims to protect the environment by preventing the introduction and spread of pests. The ISPM 18 which provides specific procedures for the application of ionizing radiation as a phytosanitary treatment for regulated (quarantine) pests was agreed by all signatories in 2003.

Irradiation as a horticultural market access treatment ensures that no living and viable insect pests can be accidently transported to importing countries. The approved ISPM 18 irradiation treatment does not necessarily have to result in the mortality of the pests, that is the insect could still be living after treatment. It is accepted that irradiation is effective in preventing the pest's successful development (e.g. non-emergence of adults), the inability of the pest to reproduce (e.g. sterility) or its complete inactivation. The use of ISPM 18 as a market access treatment does not include treatments used for the production of sterile organisms for pest control, sanitary treatments (i.e. food safety) or the preservation or improvement of commodity quality (e.g. shelf life extension). This ISPM agreement is only for the use of irradiation to facilitate market access.

'Phytosanitary Treatments for Regulated Pests' (ISPM 28) sets out minimum doses for a range of regulated pests (FAO, 2009). This is a landmark development for the use of irradiation as phytosanitary treatment as it recognized that a minimum 150 Gy as a generic dose will prevent the emergence of all adult Tephritid fruit flies in any host commodity. Tephritid fruit flies, which include the genus *Bactrocera* and *Ceratitis capitata* (Mediterranean fruit fly), are some of the most important quarantine pests around the world and their presence in growing regions can restrict trade of fruit and vegetables.

The US Department of Agriculture (USDA) Animal and Plant Health Inspection Service (APHIS, 2006) approved generic doses of 150 Gy for Tephritid fruit flies and 400 Gy for all insects except pupa and adult Lepidoptera (USDA APHIS PPQ, 2012). Further, APHIS rulings and new rulings by the IPPC have approved new minimum doses for six fruit fly pests and 14 other plant insect pests regardless of the host product, at doses between 60 and 300 Gy. These treatment schedules are now accepted and have facilitated increased world trade of fresh fruit and vegetables using low-dose irradiation treatment (Follett, 2009).

However, if fresh produce is to be treated with a minimum absorbed dose of 400 Gy required for phytosanitary treatment, then the actual maximum doses absorbed by the produce is between 600 Gy and 800 Gy due to the dose uniformity ratios at commercial facilities which are typically 1.5–2.0 (Wall, 2015). It is important that maximum doses remain below 1000 Gy to meet regulations by the US FDA for the preservation and disinfestation of fresh fruit and vegetables with irradiation (US FDA, 2008). This 1000 Gy limit for phytosanitary irradiation has been adopted by other countries. Therefore, this chapter will focus on the effect of product tolerance and quality at or below this 1000 Gy limit.

4.2 Irradiation to inhibit tuber sprouting

Successful long-term storage of potatoes, onions and garlic requires the use of a sprout inhibitor in combination with proper storage management. While chemical treatments such as chlorpropham are regularly used in many countries, low doses of irradiation are also used in some countries to inhibit sprout development. The use of ionizing radiation for sprout inhibition of white potatoes was approved in the United States in 1964, whereas a commercial use of this process began in 1983 in Japan (Barkai-Golan and Follett, 2017).

A minimum dose of 30 Gy is required for sprout inhibition of potato tubers and lower doses could stimulate sprouting while doses above 150 Gy could cause detrimental effects on the tubers (Matsuyama and Umeda, 1983). Among the factors affecting sprout inhibition, Barkai-Golan and Follett (2017) showed that radiation dose, dose rate, time between harvest and irradiation, cultivar, storage temperature and humidity and storage duration affect sprout inhibition. Although different cultivars may differ in their response to ionizing radiation, it has generally been found that a dose of up to 100 Gy can irreversibly inhibit sprouting (Barkai-Golan and Follett, 2017).

4.3 Irradiation to improve microbial safety

Irradiation of fresh fruit and vegetables inactivates microbial pathogens. The mechanism of inactivation is mainly due to the damage to nucleic acids, direct damage or indirect damage which is affected by oxidative radicals originating from the radiolysis of water. The effectiveness of irradiation as a fungicidal or fungistatic treatment in fresh fruit and vegetables depends on the pathogen, its stage of growth and the number of viable fungal cells on or within the tissue (Cia et al., 2010).

The control of postharvest diseases by irradiation depends on the radiation sensitivity of the pathogens. The sensitivity of an organism is correlated with its complexity and the repair mechanisms by which many cells can reverse the damage caused by radiation (Jeong and Jeong, 2018). Viruses show the highest resistance to inactivation by irradiation, whereas insects and parasites are the most sensitive. The lethal effect of irradiation on microbes is measured by the D10 value, which is the dose required to inactivate 90% of a population; for example, the D10 value for blue mould, *Penicillium expansum,* in pear is 230 Gy, which means that irradiation with 230 Gy will kill 90% of a population of this pathogen. Furthermore, some bacteria can form a dense hardy spore, which means they enter a compact and inert hibernation state. Spore forming bacteria are generally the most resistant to irradiation with D10 in the order of 2800 Gy while yeasts are more resistant and viruses are highly radiation resistant (Lacroix and Follett, 2015). In addition, young fungal mycelia are more resistant than mature mycelia. Interestingly, fungal pathogens show different radiosensitivity to various irradiations, indicating that differences in fungal response to radiation reflect species-specific radiosensitivity, which might be linked to their adaptive strategies under radiation stress (Jeong and Jeong, 2018). Postharvest decay responses to irradiation are affected by a range of factors such as the presence of mycelial water content as a natural radioprotector, ability to repair DNA breaks, induction of certain enzymes associated with the recovery from radiation damage and presence of chemical substances suppressing the process of growth (Geweely and Nawar, 2006).

The effects of irradiation on the development of postharvest decay have been well studied (Cia et al., 2010; Jeong and Jeong, 2018). In general, effective control of established infections of *Penicillium* on citrus fruit requires irradiation doses higher than 1000 Gy (Palou et al., 2007). Indeed, the D10 for blue mould, *P. italicum,* in citrus is 1430 Gy, while green mould *P. digitatum* in citrus is 1100 Gy (Jeong and Jeong, 2018). These higher irradiation doses are not allowed as phytosanitary treatments and can also induce fruit damage such as rind pitting in some citrus types and cultivars (see 'Effects of irradiation on fruit quality' section).

The use of combination treatments of irradiation and other physical or chemical treatments (such as heat, modified atmospheres and natural antimicrobials) to improve the efficacy of each treatment is reviewed by Lacroix and Follett (2015), Barkai-Golan and Follett (2017) and Jeong and Jeong (2018). A good example of the effectiveness and promise of combination treatments is the integration of sodium carbonate treatment and X-ray irradiation (at doses of 510 and 875 Gy) on Clemenules clementine mandarins for the control of postharvest green and blue moulds. While a significant but slight reduction of both postharvest disease incidence and severity on treated fruit was found, it was not sufficient for satisfactory disease control under simulated commercial conditions (Palou et al., 2007). Irradiation can reduce postharvest decay, but may not completely control decay for a commercial outcome. More work is required to determine optimum combination treatments for commercial applications.

5 Current trade in irradiated horticultural produce

The advantages of irradiation as a market access treatment over other phytosanitary treatments such as heat, cold and fumigation include general product tolerance to low-dose irradiation (<1000 Gy), the ability to treat all produce within the final packaging and in pallet loads within the current cold chain (i.e. no heating) and the absence of chemical residues within the fruit (Hallman, 2011).

The availability of generic radiation phytosanitary treatments, such as ISPM 28, has stimulated worldwide interest in irradiation as a market access treatment. While the ISPMs which allow irradiation as a market access treatment are signed under the IPPC, the acceptance of irradiation as a treatment is left to the discretion of each country. While approximately 60 countries have approved the use of food irradiation in their health or food regulations for at least one food (e.g. dried spices), its acceptance for fresh fruit is more limited, where over 40 countries allow the trade of irradiated fresh fruit and vegetables. It is worth noting that all countries that default to Codex Alimentarius and do not specifically have irradiation food standards accept the treatment by default, for example Fiji and many of the South Pacific islands. Major exporting countries which are using irradiation as a market access treatment include India, Pakistan, South Africa, United States, Thailand, Vietnam, Peru and Australia.

These countries are exporting a variety of fresh produce with growing world trade of irradiated produce since the approval of generic doses. The fruits which use irradiation as a market access treatment include mangoes (India, Pakistan, Mexico and Australia); guava (Mexico), persimmons (South Africa); mangosteen (Thailand); and dragon fruit, longan and rambutan (Vietnam) (Table 1).

Hallman (2011) identified challenges to the increased use of phytosanitary irradiation as a market access treatment. He identified costs as an inhibitor of use as the commercial use of irradiation as treatment volumes have not yet reached an optimum economy of scale in some parts of the world, the lack of facilities due to the large investment costs to establish commercial facilities and the current inability to feasibly locate them in packing facilities. All treatments which facilitate market access have additional costs (systems approaches, fumigation cold treatment etc.) and the economics of the use of irradiation is slowly changing with increasing use and volumes of produce being treated and traded. However, the current end-point treatment at centralized facilities has ensured that the cost of treatment, exporter access, staffing and auditing of the treatment are efficient. Furthermore, the off-farm nature of the centralized treatment ensures that staff treating the horticultural produce are safe and well trained, and the pathway offers the highest possible integrity. Hallman (2011) also identified the lack of approved treatments for some quarantine pests as an issue with irradiation as a market access treatment, but the generic 400 Gy treatment agreement accepted by the United States, Australia and New Zealand, among others, have largely overcome this issue. Finally, there is the concern about the process by key decision makers, such as packers, shippers and retailers with the use of irradiation as a market access treatment (Hallman, 2011). Almost all countries that permit phytosanitary irradiation require the irradiated produce to be labelled. While there is a perception that consumers will not purchase irradiated produce, studies show that a majority of consumers will buy high-quality irradiated foods and the percentage increases with education about the process and benefits (Barkai-Golan and Follett, 2017). Furthermore, there is a growing market for irradiated horticultural produce around the world (Table 1). In some countries where there are market opportunities for treatment and export, new commercial irradiation facilities (mainly X-ray facilities) are being built to treat commercial-sized pallets of fruit without any interruption to the cold chain. For example, Australia has a dedicated service provider ('Steritech') investing in facilities with specialized produce handling infrastructure and treatment design for uniform absorbed doses with tighter dose uniformity ratios. It is foreseeable that Australia's investment into specialized infrastructure is a key reason why the volume of trade using the treatment continues to grow (Fig. 1), both in diversity of products and

Table 1 Selected approved treatments and world trade of irradiated fresh fruit and vegetables (in 2019)

Exporting country	Fruit/vegetable	Importing country	Dose (Gy)
Australia	Capsicum, lychee, mango, papaya, tomato, table grapes	New Zealand	250–500
	Cherry, plums, table grape, onions, garlic, citrus, zucchini, scaloppini, pumpkin, custard apple, pear, persimmon, peach, nectarine, tomato, melon, passionfruit, avocado, cucumber, apricot, litchi, pomegranate	Indonesia	150
	Mango	Malaysia	300–400
	Mango, lychee	United States	300–400
	Cherry, table grapes	Vietnam	200–400
	Persimmon	Thailand	150 or 400
	Apple, apricot, blueberry, breadfruit, capsicum, carambola, cherry, custard apple, honeydew, litchi, longan, mango, mangosteen, nectarine, papaya, peach, persimmon, plum, raspberry, rambutan, rockmelon, scallopini, strawberry, table grapes, tomato, zucchini	Local internal domestic/ interstate trade	150
Dominican Republic	Mango	United States	400
Granada	Ambarella, yellow mombin, purple mombin	United States	150
Ghana	Eggplant, okra, pepper	United States	400
Hawaii	Abiu, atemoya, banana, breadfruit, Capsicum spp., carambola, citrus, cowpea (pod), Cucurbita spp., curry leaf, dragon fruit (pitahaya, pitaya), eggplant, guava, jackfruit, litchi, longan, mango, mangosteen, melon, moringa pods, papaya, pineapple, rambutan, sapodilla, sweet potato, tomato	US mainland	150 or 400 (mango 300)
India	Mango	Australia, United States	400
Jamaica	Mango	United States	150
Indonesia	Mango	Australia	400
Malaysia	Carambola, jackfruit, papaya, pineapple, rambutan	United States	400

Mexico	Carambola, clementine/mandarin/tangerine (Citrus reticulata), fig, grapefruit (C. paradisi), guava, mango, manzano pepper (Capsicum pubescens), pitahaya/pitaya/dragon fruit, pomegranate, sweet lime (C. limettioides), sweet orange (C. sinensis), tangelo (C. tangelo)	United States	150 or 400
Pakistan	Mango	United States	400[a]
	Mango	Australia	400
Peru	Blueberry, fig, pomegranate	United States	150
Philippines	Litchi, longan, mango, rambutan	United States	150 or 400 (mango 165 or 300)
South Africa	Grape, lychee, persimmon	United States	400[a]
Thailand	Dragon fruit (pitahaya, pitaya), litchi, longan, mango, mangosteen, pineapple, rambutan	United States	400
	Mango	Australia	
United States	Peach	Mexico	250
Vietnam	Dragon fruit (pitahaya, pitaya), litchi, longan, mango, rambutan, star apple	United States	400
	Mango, litchi	Australia	

[a] Treated on arrival in the United States.

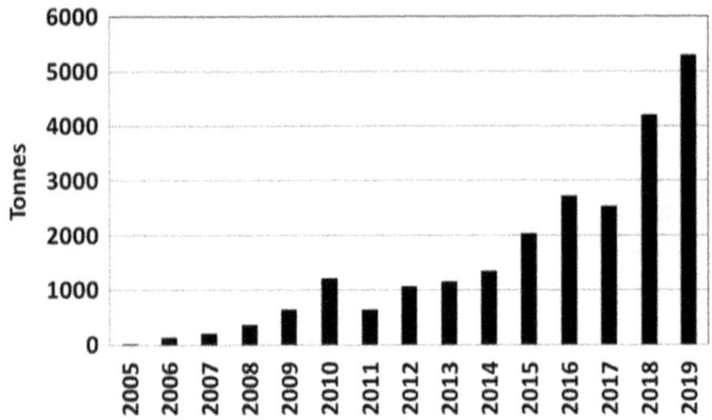

Figure 1 Growth of Australian phytosanitary irradiation treatment (export and domestic volumes)*. *Data supplied by 'Steritech', Australia.

destination markets. Already significant international trade using irradiation as a market access treatment has been established and this trade is growing as industry, regulators, retailers and consumers develop confidence in this treatment.

6 Effects of irradiation on quality

A significant issue with the older scientific literature on the irradiation of fresh horticultural produce is that many studies examined the effects of relatively higher doses (greater than 1000 Gy) on product quality. While it is valid to report these effects, these higher irradiation doses are not commercially used and are not applicable for comparing to current commercial treatments. This section will only source studies with commercial market access treatments, that is doses less than 1000 Gy.

It is impossible to summarize the effect of irradiation on all fresh fruit and vegetables in this chapter. There have been extensive reviews on the effects of irradiation on product tolerance and quality after treatment and storage (Morris and Jessup, 1994; Wall, 2015; Barkai-Golan and Follett, 2017). Rather than give an overview of different fruit and vegetable responses to irradiation treatment, we provide a case study on the selected responses of irradiation on citrus fruit quality. The challenges of irradiation with different citrus types, cultivars, treatment times that affect quality out-turn following treatment are highlighted. This case study is indicative of responses from other horticultural produce responses and focusses on the effects of irradiation on fruit internal quality and how it relates to consumer acceptability, fruit injury and vitamin C content. Each of these factors are key consumer and nutritional concerns and need to be researched and addressed for consumer and commercial acceptance.

6.1 Effects of irradiation on citrus internal fruit quality

Total soluble solids (TSS) is a relative measure of the fruit sugar content and together with titratable acidity (TA) combine to determine overall fruit taste and acceptability. In general, irradiation does not consistently affect TSS or TA in citrus fruit; however, some significant differences have been reported between different cultivars, doses and storage times. For example, O'Mahony and Goldstein (1987) and McDonald et al. (2013) reported a decrease in TSS and TA at irradiation dose levels between 300 Gy and 600 Gy but Miller et al. (2000) did not report changes in TA of oranges treated at irradiation doses of 150–450 Gy. In mandarin fruit, Miller et al. (2000) showed that irradiation (150, 300 and 450 Gy) did not generally affect TSS or TA of five mandarin hybrids ('Fallglo', 'Minneola', 'Murcott', 'Sunburst' and 'Temple'), with the exception of a decrease in 'Sunburst' mandarin at 450 Gy and a reduction in TA of 'Sunburst' and 'Temple' at 450 Gy, but also showed an increase in TA at 300 Gy in 'Murcott'. These differential responses to different cultivar and citrus types highlight the need for more detailed research and applications to different cultivars and maturities and its effects on actual sensory/consumer acceptability. Additional research has shown an interaction between fruit age and irradiation treatment, suggesting that irradiation stresses the fruit and makes it more sensitive to handling. More research is needed.

Consumer acceptability of irradiated fresh produce is the ultimate determinant of commercial acceptability. If the fruit is deleteriously affected by treatment, then commercial irradiation treatments will not succeed. There have been limited studies on the trained panel and consumer acceptability of citrus following irradiation. In general, while there may be some measurable effects of irradiation on fruit quality, either positive of negative on the levels of aroma volatiles, internal drying/granulation, TA and TSS, these have not generally been shown to influence perceived flavour as indicated by the trained sensory panel or consumers (McDonald et al., 2013). Furthermore, the growing commercial trade of irradiated produce around the world highlights the lack of negative commercial effects on fruit taste and consumer acceptability. However, little systematic research has been conducted to examine these important quality parameters.

6.2 Injury and disorders in citrus

The response of citrus fruit to irradiation has been widely studied with the results being highly variable and dependent on species, hybrid and cultivar. Physical injury has been reported at low doses (<1000 Gy) and usually occurs as peel pitting and fruit softening. Miller et al. (2000) treated ten citrus cultivars grown in Florida, including the five orange [*Citrus sinensis* (L.) Osbeck] cultivars,

'Ambersweet', 'Hamlin', 'Navel', 'Pineapple' and 'Valencia', and the five mandarin hybrids (*Citrus reticulata* Blanco), 'Fallglo', 'Minneola', 'Murcott', 'Sunburst' and 'Temple', with low-dose irradiation at 0, 250, 300 and 450 Gy, and stored for 14 days at 1°C or 5°C plus 3 days at 20°C, to determine dose tolerance based on fruit injury. They showed that fruit softening of 'Valencia', 'Minneola', 'Murcott' and 'Temple' was dose-dependent, but that of other cultivars was unaffected. They also showed that only 'Ambersweet', 'Valencia', 'Minneola' and 'Murcott' did not develop peel pitting at 150 Gy or higher, while all the other cultivars were injured. This illustrates the cultivar-specific nature of treatment responses and may also reflect maturity and handling interactions.

However, some of these differences in fruit quality out-turn in the literature may be due to production, treatment infrastructure design and sample handling. The commercial irradiation treatment facility in Australia ('Steritech') routinely holds retention samples from commercial shipments and no commercial impact or quality differences between treated and non-treated produce have been reported (Ben Reilly, Steritech Australia, pers. comm.).

A range of tolerances to irradiation treatment exist, where some cultivars have tolerance to irradiation, while some mandarin types are very sensitive to irradiation. For example, Seedless 'Kishu' mandarins (*Citrus kinokuni mukakukishu*) treated with 150 Gy have been shown to develop peel damage and browning of the calyx (Ornelas-Paz et al., 2017). This is not the classic pitting damage, but a general browning and senescence of the peel which becomes highly susceptible to fungal damage (Fig. 2). The severity of the browning damage was shown to increase with irradiation dose, especially for fruit located on the top layer of the cases and more research is required to investigate this damage, especially in commercial treatments.

6.3 Effects of irradiation on nutrition and vitamin C content

Numerous reviews have been published on the effects of irradiation on the nutritional quality of fresh fruit and vegetables, including citrus, which have shown the health and safety of irradiated produce (WHO, 1981, 1994, 1999; Arvanitoyannis, 2010; European Food Safety Authority, 2011b). In some cases, irradiation has been shown to induce changes in nutrient content but its effects depend on factors such as the irradiation dose, fruit composition, packaging material, ambient temperature and atmospheric oxygen concentration (Diehl, 1995; Kilcast, 1994; WHO, 1994). A relatively small proportion of nutrients are sensitive to irradiation, where higher doses of irradiation are associated with greater nutritional losses (WHO, 1999). Vitamins have been shown to be susceptible to oxidation and breakdown with high levels of irradiation treatment. There is a general hierarchy of vitamin sensitivity to irradiation, with vitamins A, C, E and thiamine being most sensitive (Fig. 3).

Figure 2 Irradiation damage on the peel of kishu mandarins after 3 weeks of storage. Source: adapted from Ornelas-Paz et al. (2017).

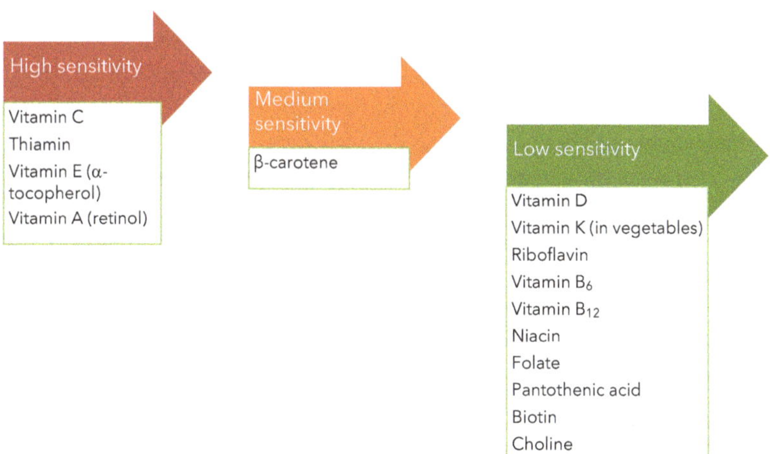

Figure 3 General sensitivity of vitamins in fresh fruit and vegetables to irradiation. Source: modified from FSANZ (2014) and Kilcast (1994).

Citrus is a well known and rich source of ascorbic acid (vitamin C) and is a major dietary contributor to many consumers providing between 5% and 17% of ascorbic acid intake of Australian consumers (FSANZ, 2014). Therefore, any potential effects of irradiation treatment of citrus on the vitamin C content are important. Many reviews have documented the natural variation in levels of vitamin C in citrus fruit (Lee and Kader, 2000; Magwaza et al., 2017). Extensive natural variation occurs in the vitamin C content of citrus, where the main sources of variation are cultivar, season, growing location and orchard management; more than fourfold differences in vitamin C are commonly observed between citrus types (Magwaza et al., 2017). In addition, postharvest treatments and storage time result in decreased levels of vitamin C (Lee and Kader, 2000; Mditshwa et al., 2017).

Vitamin C is widely regarded as a most important water-soluble antioxidant. It includes all compounds exhibiting the biological activity of L-ascorbic acid plus L-dehydroascorbic acid, its oxidation product (Lee and Kader, 2000). However, vitamin C is inherently unstable in solution, with its destruction due to the effects of temperature, light and pH. As such, vitamin C is one of the most sensitive vitamins to irradiation, particularly at higher treatment doses, with the effects of irradiation influenced by exposure to oxygen, storage and temperature, as well as the pH of the food matrix or storage medium (Kilcast, 1994). Irradiation treatment results in some L-ascorbic acid being converted to L-dehydroascorbic acid; however, both forms have vitamin C activity. Therefore, when interpreting findings of irradiation studies on fresh horticultural produce, it is important to consider that losses due to irradiation may be overestimated if only ascorbic acid is reported. Hence, total vitamin C (L-ascorbic acid plus L-dehydroascorbic acid) content is a more reliable indicator of the effects of irradiation.

There have been numerous studies on the effects of irradiation on nutrient composition of orange, mandarin, lemon, lime and grapefruit. In the majority of these cases, the vitamin C content of irradiated fruit remained within the range of natural variation (Magwaza et al., 2017), with different citrus fruit and cultivars responding differently to irradiation, temperature and storage treatments. For example, irradiation has been shown to result in lower vitamin C levels in Navel oranges (Noh et al., 2016), where treatment with 200, 400, 600, 800 and 1000 Gy resulted in lower vitamin C levels in Navel oranges but these effects were variable at different storage times. However, irradiation has also been shown to have no effect on vitamin C content (Mitchell et al., 1992; Boylston et al., 2002; Rojas-Argudo et al., 2012) and in some cases irradiation with 250 and 500 Gy slowed the loss of vitamin C during 6 weeks storage of blood oranges, resulting in higher vitamin C levels in oranges irradiated with 500 Gy (Khalil et al., 2009). If vitamin C loss is a concern, then irradiation treatments could be conducted at low temperatures to minimize any potential losses (Mditshwa et al., 2017).

7 Future trends and conclusion

Irradiation is currently being used for the growing trade in horticultural produce, but issues with product tolerance with some combinations of fruit maturities, cultivars and handling exist which need to be understood and overcome. In an excellent review of the effects of phytosanitary irradiation and fresh fruit quality, Wall (2015) described some of the multiple preharvest factors (cultivar, maturity, temperature, disease pressure) and postharvest conditions that can alter a commodity's tolerance to irradiation stress. In particular, cultivar, maturity and dose must be carefully managed to achieve a beneficial effect from radiation as a quarantine treatment (Wall, 2015). For example, as previously noted, Miller et al. (2000) treated ten citrus cultivars, including five orange cultivars and five mandarin hybrids and showed significant differences in tolerance (fruit softening, peel pitting etc.) between the different types and cultivars, where some were unaffected by irradiation and some were negatively affected. This illustrates the cultivar-specific nature of treatment responses and may also reflect maturity and handling interactions. Indeed, the effect of handling, packing density as well as time of harvest should be evaluated in future studies. For example, Rodriguez et al. (2018) showed that damage in irradiated Navel oranges depends on treatment dose and layer within the treatment pallet, with the top layers showing greater physiological damage and bottom layers showing more physical damage. In addition, with the development and introduction of new and improved cultivars by breeders, this area of postharvest research will be essential for the continued successful application of irradiation as a commercial market access treatment. It is also worthy to compare different market access treatments, such as methyl bromide fumigation, cold treatment and irradiation, on final fruit quality and commercial acceptability. Research is also required to examine preharvest factors affecting final fruit quality after treatment.

There is also the potential of combining traditional market access treatments such as cold and heat treatments with lower dose irradiation treatments to kill quarantine pests (Lacroix and Follett, 2015; Barkai-Golan and Follett, 2017). It is proposed that shorter cold or heat treatment times with even lower irradiation doses would potentially reduce produce damage from both temperature and irradiation treatments. Cold treatment is currently widely used as a market access treatment for the export of citrus fruit. This is where citrus fruit are routinely commercially treated at 1°C for up to 18 days to kill fruit fly eggs and larvae within the fruit. However, this extended cold treatment can result in chilling injury to the fruit and it is a significant limitation of the export of many citrus types such as lemons. It may be possible to combine even lower doses of irradiation (e.g. 50 Gy) with low temperature treatment with shorter treatment times that meet the technical requirements of killing quarantine pests and also minimize any potential irradiation damage or chilling injury. While no combination phytosanitary treatments using irradiation are approved, possibilities such as the lower irradiation dose and cold treatment combinations are possible (Barkai-Golan and Follett, 2017). However, the application of any potential market access treatment must be approved by plant health regulators and governments in both the exporting and importing country and this can take some time. But this is a worthwhile area of research to assist industry develop new market access combination treatments which do not adversely affect fruit quality.

8 Acknowledgements

This chapter is a contribution from the Australian Citrus Postharvest Science Program (CT15010) funded by Horticulture Innovation and NSW Department of Primary Industries. The authors thank Ben Reilly from 'Steritech' (Australia) for ongoing discussions and providing the current and historical treatment volumes of treated horticultural produce. The authors also thank Yves Herron and Guy Hallman for their contributions to Table 1.

9 Where to look for further information

Irradiation technology is a relatively 'old' postharvest technology but is currently in a renascence due to the utilization of internationally accepted market access generic treatments that are starting to transform international trade. Growing international trade is providing a renewed interest in irradiation and is an active area of postharvest horticulture.

The International Atomic Energy Agency (IAEA) is the foremost source of information on food irradiation (www.iaea.org). It is the international organization under the United Nations that seeks to promote the peaceful use of nuclear energy, including food irradiation. The global advocacy of irradiation (including food irradiation) which promotes the safe and beneficial use of irradiation for the radiation processing industry is the International Irradiation Association (https://iiaglobal.com).

There is a plethora of information on food irradiation, including its use on fresh horticultural produce. Several comprehensive books and reviews, such as Barkai-Golan and Follett (2017) and Morris and Jessup (1994), are key sources of irradiation and postharvest horticulture. These provide excellent overviews of the key benefits and challenges of irradiation as a postharvest technology. Jeong and Jeong (2018) also reviewed the recent advances in the use of irradiation to control postharvest diseases in fresh produce, the possible mechanisms underlying pathogen inactivation, radiation efficiency, safety, critical issues and future prospects. Hallman (2011) published an excellent review on the use of irradiation as a phytosanitary measure to overcome quarantine barriers to trade in food. The review outlines the history and present use of the technology as well as challenges for further applications.

There are a range of pure and applied scientific journals dedicated to the publication of original and new irradiation research. For example, '*Radiation Physics and Chemistry*' is a multidisciplinary journal that publishes research and developments involving ionizing radiation in radiation physics, radiation chemistry and radiation processing, including food. More specific applied journals for the treatment of fresh fruit and vegetables include '*Postharvest Biology and Technology*' and '*Journal of Agricultural and Food Chemistry*'.

There are many active research groups around the world which are specializing in assessment and research of irradiation treatments as market access treatments for fresh fruit and vegetables. Chapman University in California, United States, the Canadian Irradiation Centre (INRS – Institut Armand-Frappier) in Canada, Kyungpook National University in South Korea and King Mongkut's University of Technology Thonburi in Thailand are active research centres which have focussed their recent efforts into irradiation on horticultural produce. New South Wales Department of Primary Industries in Australia has also been conducting industry-focussed research to facilitate the development and regulatory approval of irradiation for fresh fruit (Golding et al., 2014, 2015). Chapman University also coordinates an annual phytosanitary forum which is held around the world and gives updates on recent developments in postharvest horticultural uses of irradiation.

The IAEA with the Food and Agriculture Organization of the United Nations has prepared an International Database on Commodity Tolerance (FAO IDCT) which provides access to scientific and other technical information which detail findings on dose tolerance required for phytosanitary irradiation of horticultural commodities (www.nucleus.iaea.org/sites/naipc/IDCT/). The information in the database provides valuable data to determine optimum methods of applying irradiation including pre- and post-treatment handling and if radiation may augment quality or prolong shelf life. The database could also be used to identify gaps and inconsistencies in knowledge that may be explored by researchers.

10 References

Amir, K., Hussain, S., Shuaib, M., Hussain, F., Urooj, Z., Khan, W. M., Zeb, U., Ali, K., Zeb, M. A. and Hussain, F. 2018. Effect of gamma irradiation on okra (*Abelmoschus esculentus* L.). *Acta Ecologica Sinica* 38(5), 368–73. doi:10.1016/j.chnaes.2018.02.002.

APHIS (Animal and Plant Health Inspection Service, U.S. Department of Agriculture). 2006. Treatments for fruits and vegetables. *Federal Register 2006* 71(18), 4451–64.

Arvanitoyannis, I. S. 2010. *Irradiation of Food Commodities: Techniques, Applications, Detection, Legislation, Safety and Consumer Opinion*. Elsevier.

Ashtari, M., Khademi, O., Soufbaf, M., Afsharmanesh, H. and Sarcheshmeh, M. A. A. 2019. Effect of gamma irradiation on antioxidants, microbiological properties and shelf life of pomegranate arils cv.'Malas Saveh'. *Scientia Horticulturae* 244, 365–71.

Barkai-Golan, R. and Follett, P. A. 2017. *Irradiation for Quality Improvement, Microbial Safety and Phytosanitation of Fresh Produce*. Academic Press, London UK. ISBN 9780128110256.

Boylston, T. D., Reitmeier, C. A., Moy, J. H., Mosher, G. A. and Taladriz, L. 2002. Sensory quality and nutrient composition of three Hawaiian fruits treated by X-irradiation. *Journal of Food Quality* 25(5), 419–33. doi:10.1111/j.1745-4557.2002.tb01037.x.

Cia, P., Benato, E. A. and Pascholati, S. F. 2010. Use of irradiation in postharvest disease management: problems and solutions. *Stewart Postharvest Review* 4, 3.

Diehl, J. F. 1995. *Safety of Irradiated Foods*. Marcel Dekker, New York, NY.

European Food Safety Authority. 2011a. EFSA panel on food contact materials, enzymes, flavourings and processing aids (CEF); Scientific opinion on the chemical safety of food irradiation. *EFSA Journal* 9(4), 1930.

European Food Safety Authority. 2011b. Statement summarising the conclusions and recommendations from the opinions on the safety of irradiation of food adopted by the BIOHAZ and CEF Panels. *EFSA Journal* 9, 2107.

FAO (Food and Agriculture Organisation of the United Nations). 2003. International standards for phytosanitary measures. Guidelines for the use of irradiation as a phytosanitary measure. International Standards for Phytosanitary Measures. Publication No. 18. Produced by the Secretariat of the International Plant Protection Convention, 16pp.

FAO (Food and Agriculture Organisation of the United Nations). 2009. Phytosanitary treatments for regulated pests. ISPM No. 28. FAO, Rome.

Follett, P. A. 2009. Generic radiation quarantine treatments: the next steps. *Journal of Economic Entomology* 102(4), 1399–406. doi:10.1603/029.102.0401.

FSANZ (Food Standards Australia New Zealand). 2014. Nutritional impact of phytosanitary irradiation of fruits and vegetables. Food Standards Australia New Zealand, 62pp.

Geweely, N. S. I. and Nawar, L. S. 2006. Sensitivity to gamma irradiation of post-harvest pathogens of pear. *International Journal of Agriculture and Biology* 8, 710–16.

Golding, J. B., Blades, B. L., Satyan, S., Jessup, A. J., Spohr, L. J., Harris, A. M., Banos, C. and Davies, J. B. 2014. Low dose gamma irradiation does not affect the quality, proximate or nutritional profile of 'Brigitta' blueberry and 'Maravilla' raspberry fruit. *Postharvest Biology and Technology* 96, 49–52. doi:10.1016/j.postharvbio.2014.05.002.

Golding, J. B., Blades, B. L., Satyan, S., Spohr, L. J., Harris, A., Jessup, A. J., Archer, J. R., Davies, J. B. and Banos, C. 2015. Low dose gamma irradiation does not affect the quality or total ascorbic acid concentration of 'Sweetheart' passionfruit (*Passiflora edulis*). *Foods* 4(3), 376–90. doi:10.3390/foods4030376.

Hallman, G. J. 2011. Phytosanitary applications of irradiation. *Comprehensive Reviews in Food Science and Food Safety* 10(2), 143–51. doi:10.1111/j.1541-4337.2010.00144.x.

Jeong, M. A. and Jeong, R. D. 2018. Applications of ionizing radiation for the control of postharvest diseases in fresh produce: recent advances. *Plant Pathology* 67(1), 18–29. doi:10.1111/ppa.12739.

Khalil, S. A., Hussain, S., Khan, M. and Khattak, A. B. 2009. Effects of gamma irradiation on quality of Pakistani blood red oranges (*Citrus sinensis* L. Osbeck). *International Journal of Food Science and Technology* 44(5), 927–31. doi:10.1111/j.1365-2621.2008.01669.x.

Kilcast, D. 1994. Effect of irradiation on vitamins. *Food Chemistry* 49(2), 157–64. doi:10.1016/0308-8146(94)90152-X.

Lacroix, M. and Follett, P. 2015. Combination irradiation treatments for food safety and phytosanitary uses. *Stewart Postharvest Review* 3, 4.

Lee, S. K. and Kader, A. A. 2000. Preharvest and postharvest factors influencing vitamin C content of horticultural crops. *Postharvest Biology and Technology* 20(3), 207–20. doi:10.1016/S0925-5214(00)00133-2.

Magwaza, L. S., Mditshwa, A., Tesfay, S. Z. and Opara, U. L. 2017. An overview of preharvest factors affecting vitamin C content of citrus fruit. *Scientia Horticulturae* 216, 12–21. doi:10.1016/j.scienta.2016.12.021.

Matsuyama, A. and Umeda, K. 1983. Sprout inhibition in tubers and bulbs. In: Josephson, E. S. and Peterson, M. S. (Eds), *Preservation of Foods by Ionizing Radiation*. CRC Press, Boca Raton, FL, pp. 159–213.

McDonald, H., Arpaia, M. L., Caporaso, F., Obenland, D., Were, L., Rakovski, C. and Prakash, A. 2013. Effect of gamma irradiation treatment at phytosanitary dose levels on the quality of 'Lane Late' navel oranges. *Postharvest Biology and Technology* 86, 91–9. doi:10.1016/j.postharvbio.2013.06.018.

Mditshwa, A., Magwaza, L. S., Tesfay, S. Z. and Opara, U. L. 2017. Postharvest factors affecting vitamin C content of citrus fruits: a review. *Scientia Horticulturae* 218, 95–104. doi:10.1016/j.scienta.2017.02.024.

Miller, W. R., McDonald, R. E. and Chaparro, J. 2000. Tolerance of selected orange and mandarin hybrid fruit to low-dose irradiation for quarantine purposes. *HortScience* 35(7), 1288–91. doi:10.21273/HORTSCI.35.7.1288.

Mitchell, G. E., McLauchlan, R. L., Isaacs, A. R., Williams, D. J. and Nottingham, S. M. 1992. Effect of low dose irradiation on composition of tropical fruits and vegetables. *Journal of Food Composition and Analysis* 5(4), 291–311. doi:10.1016/0889-1575(92)90063-P.

Morris, S. C. and Jessup, A. J. 1994. Irradiation. In: Paull, R. E. and Armstrong, J. W. (Eds), *Insect Pests and Fresh Horticultural Products: Treatments and Responses*. CAB International, Wallingford, UK, pp. 163–90.

Noh, D.-B., Kim, K.-H.. and Yook, H.-S. 2016. Characteristics of low-dose X-ray irradiated-imported navel oranges during storage at room temperature (20°C). *Journal of the Korean Society of Food Science and Nutrition* 45(1), 109–16. doi:10.3746/jkfn.2016.45.1.109.

O'Mahony, M. and Goldstein, L. R. 1987. Sensory techniques for measuring differences in California Navel oranges treated with doses of gamma-radiation below 0.6K gray. *Journal of Food Science* 52(2), 348–52. doi:10.1111/j.1365-2621.1987.tb06610.x.

Ornelas-Paz, J. J., Meza, M. B., Obenland, D., Rodríguez Friscia, K., Jain, A., Thornton, S. and Prakash, A. 2017. Effect of phytosanitary irradiation on the postharvest quality of Seedless Kishu mandarins (*Citrus kinokuni mukakukishu*). *Food Chemistry* 230, 712–20. doi:10.1016/j.foodchem.2017.02.125.

Palou, L., Marcilla, A., Rojas-Argudo, C., Alonso, M., Jacas, J. A. and del Río, M. Á. 2007. Effects of X-ray irradiation and sodium carbonate treatments on postharvest *Penicillium* decay and quality attributes of clementine mandarins. *Postharvest Biology and Technology* 46(3), 252–61. doi:10.1016/j.postharvbio.2007.05.006.

Ramakrishnan, S. R., Jo, Y., Nam, H.-A., Gu, S.-Y., Baek, M.-E. and Kwon, J.-H. 2019. Implications of low-dose e-beam irradiation as a phytosanitary treatment on physicochemical and sensory qualities of grapefruit and lemons during postharvest cold storage. *Scientia Horticulturae* 245, 1–6. doi:10.1016/j.scienta.2018.09.058.

Reddy, S. V. R., Sharma, R. R. and Gundewadi, G. 2018. Use of irradiation for postharvest disinfection of fruits and vegetables. In: Siddiqui, M. W. (Ed.), *Postharvest Disinfection of Fruits and Vegetables*. Academic Press, pp. 121–36.

Rodriguez, K. C., Ornelas-Paz, J., Ibarra-Junquera, V., Toto, M. C., Jain, A. and Prakash, A. 2018. Effects of fruit position in standard place pack cartons and gamma irradiation on the postharvest quality of 'Barnfield' Navel oranges. *Food and Bioprocess Technology* 11(12), 2153 - 63. doi:10.1007/s11947-018-2174-6.

Rojas-Argudo, C., Palou, L., Bermejo, A., Cano, A., del Río, M. A. and Gonzàlez-Mas, M. C. 2012. Effect of X-ray irradiation on nutritional and antifungal bioactive compounds of 'Clemenules' clementine mandarins. *Postharvest Biology and Technology* 68, 47–53. doi:10.1016/j.postharvbio.2012.02.004.

Singh, S. P. and Pal, R. K. 2009. Ionizing radiation treatment to improve postharvest life and maintain quality of fresh guava fruit. *Radiation Physics and Chemistry* 78(2), 135–40. doi:10.1016/j.radphyschem.2008.09.007.

USDA APHIS PPQ (United States Animal and Plant Health Inspection Service. Plant Protection and Quarantine). 2012. Treatment Manual. Section 3-8. Nonchemical treatments. Irradiation. United States Department of Agriculture, Washington DC.

US FDA (United States Food and Drug Administration). 2008. Foods permitted to be irradiated under FDA regulations (21 CFR 179.26). US FDA, Washington DC.

Wall, M. M. 2015. Phytosanitary irradiation and fresh fruit quality: cultivar and maturity effects. *Stewart Postharvest Review* 3, 6.

World Health Organization. 1981. Wholesomeness of irradiated food. Report of a Joint FAO/IAEA WHO Expert Committee, Technical Report Series No. 659. World Health Organization, Geneva.

World Health Organization. 1994. *Safety and Nutritional Adequacy of Irradiated Food*. World Health Organization, Geneva.

World Health Organization. 1999. High-dose irradiation: wholesomeness of food irradiated with doses above 10 kGy. Report of a Joint FAO/IAEA/WHO Study Group 890. World Health Organization, Geneva.

Chapter 10

Advances in the potential use of non-thermal plasma in postharvest treatment of fresh horticultural produce

Sukhvinder Pal Singh and John Golding, New South Wales Department of Primary Industries and University of Newcastle, Australia

1 Introduction

Increased consumption of fruits, vegetables and tree nuts has been associated with the prevention of chronic diseases and lifestyle disorders. While the health benefits of fresh produce are numerous, the consumption of contaminated horticultural produce can cause serious foodborne illnesses and even death. In addition, this can lead to huge commercial losses to industry, consumer confidence and national productivity. Animal products were historically linked to foodborne illness outbreaks. However, from the mid-1990s, the contribution of fresh horticultural produce to these outbreaks has been on the rise in different parts of the world including the United States, the European Union and Australia (Callejón et al., 2015; Nuesch-Inderbinen and Stephan, 2016; Wadamori et al., 2017; Yeni et al., 2016).

The increase in foodborne illnesses associated with fresh produce in the past 25 years could be attributed to a number of factors such as the increased global trade and complexities of supply chains, increased distance between

http://dx.doi.org/10.19103/AS.2019.0055.18

the sites of production and consumption, better diagnostics and traceability tools, and the increased surveillance by public health and food regulators. The application of whole genome sequencing has also recently opened up new methods for rapid and conclusive trace back investigations. Perhaps, the technological developments in diagnostics and surveillance to protect the public health have outpaced the evolution of food safety systems and risk mitigation tools available to horticultural industries. In various parts of the world, food and health regulators have already recognized this gap and are implementing stricter food safety regulations. As a result, the regulatory landscapes are rapidly evolving. For example, the introduction of 'Food Safety Modernization Act (2011)' in the United States has regulated stringent requirements for plant products. In addition to the regulatory compliance, it is the fundamental right for consumers to access fresh produce which is nutritious and safe. Regulators' requirements and consumers' expectations are thus intertwined uniquely to an extent that leaves no scope for compromise with either.

Microbial pathogens continuously pose a problem for horticultural industries from both a safety and cost standpoint. There are a number of microorganisms involved in fresh horticultural produce-related outbreaks which include bacteria (e.g. *Salmonella*, *Listeria monocytogenes*), viruses (e.g. noroviruses, hepatitis A) and parasites (e.g. *Cryptosporidium*). However, there are significant gaps in our knowledge of food safety in horticultural produce. For example, how to prevent the contamination of fresh produce in the first place and then how the microbial pathogens can be killed or inactivated to make the produce safe for consumption. With the implementation of good agricultural and postharvest practices, microbial food safety risks can be reduced but not be completely eliminated. Postharvest washing and sanitizing often achieve less than 2-log reduction in microbial load on the fresh produce. But current postharvest sanitation practices (e.g. chlorination) have several limitations due to chemical residues, worker safety, potential packing line corrosiveness, limited effectiveness against foodborne pathogens and regulatory environmental issues. A global push towards ultra-low chemical residues and freedom from pathogenic microbes has created a technological gap which needs to be filled with an innovative and cost-effective decontamination process acceptable to the industry, consumers and regulators.

2 Non-thermal plasma (NTP) technology

Plasma is the fourth state of matter that contains partially or wholly ionized gas with a net neutral charge. It can be broadly categorized into two categories - thermal and non-thermal. Thermal plasma has very high temperatures of up to thousands of degrees Celsius and is used in industrial processes such as in casting processes in metallurgical industries. Non-thermal plasma (NTP) with temperatures closer to ambient (contrary to thermal plasma) is more suitable for heat-sensitive products such as horticultural produce. NTP is generated by the application of an electric or electromagnetic field to a gas. The applied energy causes the free electrons in the gas to accelerate and ionizes the gas atoms and molecules. This in turn releases more free electrons which then initiate new ionizations. Consequently, the plasma is constituted of molecules and atoms in an excited state – positive and negative ions, free radicals, electrons, UV radiation and reactive oxygen and nitrogen species (such as ozone, superoxide, hydroxyl radicals, singlet oxygen, atomic oxygen, nitric oxide or nitrogen dioxide). The lifetime of these reactive species varies from a few microseconds to hours, depending on environmental conditions. Interestingly, all these agents have strong biocidal properties against a broad range of microbes such as bacteria, fungi, yeasts and viruses (Niemira, 2012a). The nature and concentration of these reactive species in the plasma depend on the gas used, voltage applied, treatment conditions and plasma generation device. There are different methods of NTP generation for which detailed discussion is out of the scope of this chapter and are thoroughly reviewed by Conrads and Schmidt (2000) and Niemira (2012a).

There are a number of non-food applications of NTP in modification of surface properties including oxidation, etching, altering hydrophobicity, adhesion and inking. Indeed, NTP as a surface treatment has been extensively used for surface modification and polymer functionalization by the packaging and polymer industries for several decades. However, significant research attention and interest in NTP is mainly because of its strong antimicrobial effects at relatively low gas temperatures (Niemira, 2012a). NTP can be generated at atmospheric or under low-pressure conditions. The NTP generated at atmospheric pressure is heterogeneous with localized electron discharges. However, low-pressure NTP is limited to be used as a batch processing method as it is prerequisite to draw partial vacuum before the treatment. This chapter is focussed on the atmospheric pressure of NTP.

3 Mechanism of action of non-thermal plasma (NTP)

The antimicrobial properties of NTP have been known for a long time, but its exact mechanism of microbial inactivation is not known. The cellular envelopes, DNA and proteins are recognized as potential targets of NTP treatment, with the damage occurring as a result of the simultaneous action of the reactive oxidative species and other components in plasma. The dysfunction of macromolecules compromises cellular viability, leading to cell death and inactivation of microorganisms even with a short exposure time. The relative contribution of the different individual components of the plasma to the total lethal effect of the NTP treatment is difficult to understand and estimate.

Depending upon the method of plasma generation, treatment time and the interaction between plasma and commodity, the sanitizing efficacy of plasma is complex due to multiple modes of antimicrobial action. There are several other factors known to influence the microbial inactivation achieved through NTP treatment. Irrespective of the method of plasma generation, the composition of plasma and microbial inactivation potential will depend upon the voltage applied to the feed gas, the power and frequency of the energy supplied and also the composition and flow rate of the feed gas. Due to these variables, it is often difficult to compare the antimicrobial effects achieved in different research studies, but general conclusions can be drawn:

- The levels of microbial inactivation generally increase with the treatment times, which can vary from seconds to a few minutes.
- The increase in voltage increases microbial inactivation of both vegetative and bacterial spores.
- The type of feed gas or mixtures of gases (e.g. air) affects the reactive oxidative species in terms of both qualitative and quantitative effects.
- A wide variety of gases can give rise to plasma capable of achieving microbial inactivation, but there is no general agreement on which one is most effective (López et al., 2019).

The desirable NTP composition to inactivate a specific pathogen for a specific product can be achieved by optimizing various parameters under a given set of conditions. The development of produce-specific postharvest treatment requires extensive research and development effort to achieve the desired outcomes.

4 Methods of non-thermal plasma (NTP) treatment

The application method of NTP is a critical factor in achieving the microbial inactivation on the target products. The treatment can be carried out either directly or indirectly. Direct treatment involves locating the target product (fruit) in the plasma field where plasma is generated and is therefore in direct contact with all the components of plasma. Indirect treatment occurs when the plasma is generated away from the product and is then pushed towards it with a rapid flow of feed gas. There are conflicting results reported on the effectiveness of direct versus indirect treatments. Some studies have shown no differences in the effectiveness between direct and indirect treatments when treating biofilms of *Listeria monocytogenes*, *E. coli* and *Staphylococcus aureus* (Ziuzina et al., 2015). However, Han et al. (2014) reported that indirect treatments were more effective against *Listeria monocytogenes* and *E. coli* than direct treatments. This difference could be attributed to the recombination of reactive species before reaching the target product/pathogen during the indirect treatment giving rise to new chemical species with strong bactericidal effects. Therefore, in this indirect treatment the lethality of NTP treatment is determined by the distance between the plasma generation and the target product. Another form of indirect treatment is the infusion of NTP into water, called plasma-activated water (PAW), which has strong antimicrobial properties. Several studies have shown that PAW can achieve effective decontamination of both surfaces and food products. This type of indirect application is a promising alternative to the chemical disinfectants currently used for cleaning and sanitation of surfaces, equipment, fresh produce and fresh-cut products.

5 Potential applications of non-thermal plasma (NTP) in postharvest horticulture

5.1 Food safety

NTP has a strong decontamination ability for food, packaging materials, equipment and the processing environments. Research on NTP application in food decontamination emerged in the mid-1990s, when *in vitro* efficacy of plasma was investigated and reported by several studies (Surowsky et al., 2015). However, interest in application of plasma in fresh produce has surged in the past 10 years where researchers treated a variety of fresh horticultural produce with major foodborne bacterial pathogens and then treated with different types of NTP to achieve killing or inactivation of the pathogens (Table 1). These studies have demonstrated the effectiveness of plasma to achieve microbial load reduction in the produce using different types of NTP technologies, different types of feed gases including key treatment variables such as voltage, treatment time, direct versus indirect exposure and the initial population of target pathogen. Pignata et al. (2017) reviewed and analysed the published research on the NTP treatment potential for foods and found that 40% of the studies used NTP on fresh fruits and vegetables, 21% on dry fruit, nuts and seeds, 19% on meat and cold cuts, 10% on liquids and 4% on egg shells. The range of plasma technologies used by these studies included plasma jet (34%), dielectric barrier discharge (20%), microwave plasma (14%), low-pressure plasma (12%) and other plasma processes (20%). Furthermore, the most widely used feed gas was air (44%), followed by argon (17%), mixtures of helium/oxygen and argon/oxygen (12%) and pure nitrogen (9%). This wide range of technologies used in experiments combined with a number of other significant variables pose a challenge in interpreting and comparing the published results to draw solid conclusions. Despite these limitations, it has been estimated that the plasma treatments are capable of reducing the microorganisms on fresh produce by an average of 2.7 ± 1.4 log reduction and highlights the potential

Table 1 A summary of non-thermal plasma treatment effects on a range of foodborne pathogens and quality of fresh horticultural products

Commodity	Treatment regime	Effects on microbial population	Effects on product quality attributes	Reference
Fruits				
Apple	'Red Delicious' apples treated with one atmosphere uniform glow discharge plasma, 9 kV, 6 kHz and treatment distance 11.4 cm	>2 log reduction in *Escherichia coli O157:H7* population after 2 min	Not assessed	Critzer et al. (2007)
	'Golden Delicious' apples, gliding arc plasma, 40 L/min, indirect	~3.5 log reduction in *Escherichia coli O157:H7* and *Salmonella* in 2 min	No change in colour and texture	Niemira and Sites (2008)
	'Pink Lady®' apple slices treated with dielectric barrier discharge plasma for 10–30 min	-	Reduced enzymatic browning and polyphenol oxidase activity in treated tissues No adverse effects on soluble solids, titratable acidity and firmness	Tappi et al. (2014)
	Argon (5L/min) + O_2 (0.1%); 10 kV; 8W; 17 mm distance	~4.7 log reduction of *E. coli* in 1 min	Not assessed	Baier et al. (2014)
	Air (20 L/min), microwave plasma, 10 min indirect treatment	~4.6 log reduction	Not assessed	Baier et al. (2015)
	'Pink Lady®' fresh-cut apples treated with double-barrier discharge plasma, 70 mm distance from electrodes, treated for 30 min (15 + 15 min on each side)	-	A slight reduction (10%) in antioxidant capacity No significant difference in total phenolics content but significant decrease in total phenolic index	Ramazzina et al. (2016)

	'Golden Delicious' apples treated with a corona discharge system, 200 Watt, 50 Hz; air-flow rate 17 L min^{-1}, treatment distance 35 mm	*Salmonella* and *E. coli* decreased by >5.0 log in 3 min	Not assessed	Kilongzo-Nthenge et al. (2018)
Blueberry	Atmospheric pressure plasma jet, 47 kHz, 549W, air 4–7 cubic feet/min, 7.5 cm treatment distance for 2 min	Reduced total aerobic count by ~0.9 log in 90 s. and yeast and mould count by ~1.2 log in 2 min	Fruit firmness and anthocyanins decreased in treated fruit with an impact on surface colour	Lacombe et al. (2015)
	DBD, 50 Hz, 60–80 kV, fruit in a polyethylene terephthalate packaged, sealed and treated for 1 or 5 min	-	Decrease in firmness, total phenol, flavonoid and anthocyanins Significant increase in total soluble solids No significant change in acidity	Sarangapani et al. (2017)
Cherry tomato	Dielectric barrier discharge plasma, 70 kV, 50 Hz; 14–16 cm treatment distance	~6.7 log reduction in *L. monocytogenes* in 2 min, ~3.1 log reduction in *E. coli*, and ~6.3 log reduction in *Salmonella Typhimurium*	Not assessed	Ziuzina et al. (2014)
	Dielectric barrier discharge plasma, 100 kV, 150 s treatment time	>5 and 3.5 log reduction in *E. coli* and *L. innocua.* Up to 3.5 log reduction in mesophiles, yeast and mould count	No significant effect on colour, firmness and soluble solids	Ziuzina et al. (2016)
Kiwifruit	Fresh-cut kiwifruit treated with double-barrier discharge plasma, 70 mm distance from electrodes, treated for 30 min (15 + 15 min on each side)	-	Improved colour retention and reduced darkened area formation during storage. No significant changes in colour, vitamin C firmness and antioxidant activity	Ramazzina et al. (2015)

(*Continued*)

Table 1 (*Continued*)

Commodity	Treatment regime	Effects on microbial population	Effects on product quality attributes	Reference
Mandarin	'Satsuma' mandarin treated with a microwave plasma, 2.45 GHz, 900 W, 1 L/min, 0.7 kPa, N_2, He, $N_2 + O_2$ (4:1), 10 min	Inhibition of *Penicillium italicum* on fruit, leading to a significant decrease (84%) in disease incidence	Increased total phenolic content and antioxidant activity of mandarin peel. No significant effect on respiration, weight loss, soluble solids, titratable acidity, ascorbic acid concentration (flesh) or surface colour during storage	Won et al. (2017)
Mango	Mango fruit (cultivar not reported) pericarp treated with atmospheric pressure plasma jet with He + 0.5% O_2 as a feed gas, direct treatment	>3.0 log reduction in *E. coli* in 5 s	Not assessed	Perni et al. (2008a)
	Fresh-cut mango fruit (cultivar not reported) treated with atmospheric pressure plasma jet with He + 0.5% O_2 as a feed gas, direct treatment	2.5 log reduction in *E. coli* and *L. monocytogenes* populations in 30 s	Not assessed	Perni et al. (2008b)
Melon	Cantaloupe rind treated with one atmosphere uniform glow discharge plasma, 9 kV, 6 kHz and treatment distance from plasma source 11.4 cm	Salmonella reduced by >2 log after 1 min and >3 log after 3 or 5 min	Not assessed	Critzer et al. (2007)
	Honeydew melon pericarp treated with atmospheric pressure plasma jet with helium + 0.5% O_2 as a feed gas, direct treatment	> 3.0 log reduction in *E. coli* population in 5 s	Not assessed	Perni et al. (2008a)

	Fresh-cut cantaloupe treated with atmospheric pressure plasma jet with helium + 0.5% O_2 as a feed gas, direct treatment	1.5 log reduction in *E. coli* population in 40 s	Not assessed	Perni et al. (2008b)
	Fresh-cut cantaloupe slices treated with a dielectric discharge, 15 kV for 30 min (15 +15 min on each side) and 60 min (30 + 30 min on each side)	3.4 and 2 log reduction of mesophilic and lactic acid bacteria, respectively	No significant effect on sugars and dry matter content. Inhibition of peroxidase and pectin methyl esterase activities by 17% and 7%, respectively Increase in shelf-life following the 15 + 15 min plasma treatment	Tappi et al. (2016)
	Cantaloupe rind treated with H_2O_2 aerosol aerosolized with cold plasma generating activated H_2O_2 droplets, 17 kV, 45 s treatment time with 30 min dwell time	*S. Typhimurium*, *E. coli* and *L. innocua* reduced by 1.3, 4.9 and 3 log, respectively	Not assessed	Jiang et al. (2017)
Strawberry	Dielectric barrier discharge plasma, 60 kV, 50 Hz; with different feed gas mixtures - air, 90% N_2 + 10% O_2, 65% O_2 + 16% N2 + 19% CO_2	2.4–3.7 log reduction depending upon the feed gas mixture	Fruit respiration rate, colour and firmness not significantly affected by treatment	Misra et al. (2014)
	Dielectric barrier discharge plasma, 70 kV, 50 Hz; 14–16 cm treatment distance	~4.2 log reduction in *L. monocytogenes*, ~3.5 log reduction in *E. coli and* ~3.8 log reduction in *Salmonella Typhimurium* in 5 min	Not assessed	Ziuzina et al. (2014)

(*Continued*)

Table 1 (*Continued*)

Commodity	Treatment regime	Effects on microbial population	Effects on product quality attributes	Reference
Vegetables				
Chicory	Red chicory treated with a dielectric barrier discharge plasma, 15 kV, 12.5 kHz, 15–30 min, air, 1.5 m/s	1.35 log reduction of *E. coli O158:H7* and 2.2 log reduction of *L. monocytogenes*	No significant effect on antioxidant activity and external appearance	Pasquali et al. (2016)
	Red chicory treated with a dielectric barrier discharge plasma, 19.15 kV, 3.15 A, 15 min deionized water	>4 log reduction of L. monocytogenes and >5 log reduction of (VTEC) *E. coli*	No adverse effects on freshness, texture and colour. Overall quality and acceptability slightly decreased during storage	Trevisani et al. (2017)
Lettuce	Iceberg lettuce treated with one atmosphere uniform glow discharge plasma, 9 kV, 6 kHz and treatment distance from plasma source 11.4 cm	*L. monocytogenes* reduced by >3 and >5 log after 3 and 5 min treatment, respectively	Not assessed	Critzer et al. (2007)
	Iceberg lettuce treated with a dielectric barrier discharge plasma indirectly	~2–4 log reduction in *E. coli*, *Salmonella* and *L. monocytogenes* populations in 5 min	Not assessed	Ziuzina et al. (2015)
	Romaine lettuce treated with a dielectric barrier discharge, 42.6 kV, 1.5 A, 10 min with air as feed gas	0.4–0.8 log reduction of *E. coli* O157:H7 in leaf samples in the one, three and five layer configurations and 1.1 log reduction in bulk stacking with seven layers	No significant change in the surface morphology, colour, respiration rate and weight loss	Min et al. (2017)

Nuts				
Almond	Almond kernels treated with atmospheric pressure plasma jet directly, 47 kHz, 549 W, 6 cm distance	*E. coli* O157:H7 and *Salmonella* populations reduced by ~1.2 log in 20 s	No major changes in colour, aroma and surface properties	Niemira (2012b)
	Almond kernels treated with a dielectric barrier discharge plasma, 30 kV, 2 kHz	*E. coli* population reduced by 5 log in 30 s	Not assessed	Deng et al. (2007)
Hazelnut	Atmospheric pressure fluidized bed plasma (APFBP) with stainless steel mesh based on plasma jets. Pressure: 1 atm, dry air; N_2, Flow rate: 50 L/min, 5–10 kV, 25 kHz, 460–665 W, treatment time 1–5 min	~3.8–4.5 log reduction in *Aspergilus flavus* and ~3.8–4.2 log reduction in *A. parasiticus*	Not assessed	Dasan et al. (2016, 2017)

for postharvest decontamination applications (Pignata et al., 2017). Similarly among the tested foodborne bacterial pathogens, *E. coli* appears to be the pathogen that requires the lowest time to achieve kill or inactivation. There are other microbial-related factors which influence the ability of plasma to kill the target pathogens. In general, vegetative cells of bacteria are more sensitive than bacterial spores. Within the vegetative bacterial cell types, Gram-positive bacteria are generally more resistant to NTP treatment than Gram-negative bacteria due to the fact that they have a thicker layer of peptidoglycan in their cellular envelopes enhancing their ability to hinder diffusion of reactive species through the envelopes. Some studies have reported unexpected findings of similar resistance of both bacterial groups to NTP and also less sensitivity of Gram-positive bacteria compared to Gram-negative species as reviewed by López et al. (2019). Microbial resistance to various inactivation treatments is also influenced by the physiological state of bacteria and their growth conditions in the culture media which are seldom reported by many studies.

Among different types of plasma methods, plasma jets are the most effective in achieving human pathogen activation in the shortest time compared with other methods (Pignata et al., 2017). The pathogen inactivation potential of plasma jets could be a great fit into a multi-stage process based on the hurdle approach. The treatment time and distance from the plasma source are the key factors to be considered in optimization of plasma treatment for direct exposure techniques. For a successful in-line NTP application in fresh produce processing and packaging, the rapid killing with uniform application around the entire produce surface are fundamental requirements and can be achieved in current postharvest packing lines.

In addition to the plasma generation and treatment methods, the structural and textural features of the fresh horticultural produce itself influence the variability in reduction of microbial load on the surfaces. A great diversity in the skin texture of fresh fruit, vegetables and nuts, pigmentation and topography poses a significant problem in reproducing and translating the laboratory research across the whole range of different horticultural produce. In general, the antimicrobial efficacy of NTP treatment is very high on smooth peel surfaces such as apples and cherry tomatoes (Table 1), whereas rough, porous and irregular surfaces on fruit such as melons and avocados offer opportunities for microorganisms to avoid the action of plasma (López et al., 2019). Some studies using electron microscopy have shown that inoculated microorganisms are capable of being protected within various food irregularities such as cracks, grooves or gaps in the fruit skin. In addition, bacteria are also capable of forming biofilms which can be become internalized within the produce tissues leading to a reduction in the effectiveness of NTP treatment.

A potential for NTP treatment is for the generation and application of the electric field to the air that is inside packages such as in a salad/lettuce pre-pack bag (Ziuzina et al., 2014). The reactive oxidative species generated by the NTP can decontaminate both the packaging material and fresh produce within the package. This method eliminates the possibility of cross-contamination and recontamination of fresh produce during the packaging step, as the bag is sealed following NTP treatment. Studies have shown that in-package treatment of cherry tomatoes, lettuce and strawberries with NTP generated inside the package can result in a significant reduction of total plate count and foodborne bacterial pathogens such as *L. monocytogenes*, *Salmonella* and *E. coli* O157:H7 (Ziuzina et al., 2014; Misra et al., 2014). The highly oxidative environment surrounding the product inside the package allows adequate contact time to achieve decontamination. After a short period of time, the plasma species in the package recombines and the atmosphere returns to its initial composition. The in-package NTP generation and treatment of fresh produce has a commercial application.

5.2 Shelf-life extension and quality maintenance

NTP treatment has been shown to affect both physical and internal quality attributes of fresh horticultural produce. However, information on the effects of plasma treatment on nutritional quality and sensory attributes of treated produce is scant.

5.2.1 Colour

Loss of colour in strawberries, apples, kiwifruit and cherry tomatoes was unaffected by plasma treatment (Niemira and Sites, 2008; Misra et al., 2014; Ziuzina et al., 2016). However, minor changes in colour after NTP treatment of some fresh produce have been reported previously (Lacombe et al., 2015; Sarangapani et al., 2017). These discrepancies between the different studies are probably due to the variability in the equipment, plasma treatment conditions and the maturity or overall quality of treated produce. For example, plasma-treated blueberries had anthocyanin loss with extended exposure (Lacombe et al., 2015; Sarangapani et al., 2017). The interaction of plasma with the anthocyanin pigments in dark-colour produce such as blueberries is thought to heat the produce, which can further accelerate the loss of heat-sensitive pigments and other phytonutrients. However, no information is available on the undesirable heat production during plasma treatment and on its effects on quality attributes.

5.2.2 Texture

Fruit texture is an important quality attribute that affects the potential postharvest life and consumer acceptance of fresh produce. NTP treatment has been shown to have varying effects on the firmness of treated produce. NTP treatment did not affect the textural properties of strawberry, apple, cherry tomato and melon fruit (Misra et al., 2014; Niemira and Sites, 2008; Tappi et al., 2014, 2016; Ziuzina et al., 2016), while other studies have reported that it increased softening of blueberries (Lacombe et al., 2015; Sarangapani et al., 2017). Softening was mainly attributed to the increase in fruit temperature and the mechanical damage due to the high air-flow rates of the plasma jet. The loss of the waxy bloom on the blueberry fruit surface due to the interaction with plasma could be the first step that initiates a cascade of detrimental effects such as fruit softening and colour loss. The composition of plasma gas can also influence the impact of NTP treatment on the retention of firmness. In another study, NTP-treated strawberries in modified atmosphere packaging containing a high oxygen environment (65% O_2 + 16% N_2 + 19% CO_2) retained better firmness than NPT-treated strawberries treated in a nitrogen-rich environment (90% N_2 + 10% O_2) (Misra et al., 2014). This study demonstrated the effect of plasma gas composition on fruit quality which might have been influenced by high ozone concentrations as a reactive oxygen species in the environment during treatment.

5.2.3 Internal quality

The internal quality of fresh produce can be influenced by plasma treatment, but its effects are dependent on the produce type, maturity, treatment regime and postharvest handling conditions. Since plasma is a complex mixture of reactive oxidative species, its interaction with intact and cut-produce is likely to cause some changes in various chemical attributes which have potential to influence their appearance, flavour and nutritional quality. These changes in chemical attributes are eventually determinants of storage/shelf-life and consumer acceptance of fresh produce. Some studies have shown that NTP treatment did not adversely affect the soluble solids (sugars) and acid contents in apples, cherry tomato, melons and mandarins (Won et al., 2017; Tappi et al., 2014, 2016; Ziuzina et al., 2016). An increase in soluble solids content without a significant change in titratable acidity was observed in blueberries treated with NTP (Sarangapani et al., 2017). The concentrations of soluble solids and their ratio to titratable acidity can greatly influence the overall quality and acceptance of fruits.

5.2.4 Nutritional quality

Little information on the effects of NTP treatment on nutritional components in fruit is available. A few studies have reported the treatment effects on vitamin C (ascorbic acid), which is an important indicator of the potential nutritional loss during a postharvest treatment. Most of the studies on fresh and fresh-cut horticultural produce following NRP treatment have reported no significant loss of vitamin C (Ramazzina et al., 2015; Won et al., 2017; Pankaj et al., 2018). The effects of NTP on the levels of phenolics and antioxidant capacity are also variable depending upon the type of produce and treatment conditions. For example, Ramazzina et al. (2016) reported no significant change in total phenolics content in fresh-cut 'Pink Lady™' apples treated with NTP, but a 10% decrease in antioxidant capacity. No significant effect on the antioxidant capacity of red chicory was observed after NTP treatment (Pasquali et al., 2016), while an increase in the antioxidant capacity of mandarin peel was observed in response to plasma treatment (Won et al., 2017). There are few studies which report the effects of NTP treatment on the important fruit quality enzymes such as polyphenol oxidase and pectin methyl esterase. Tappi et al. (2016) reported an inhibition of peroxidase and pectin methyl esterase activities by 17% and 7%, respectively, in cut melons treated with NTP, which might have contributed to the increase of shelf-life of the treated product. Similarly, in fresh-cut apple slices, inhibition of polyphenol oxidase activity was associated with reduced browning following an NTP treatment (Tappi et al., 2014).

5.2.5 Postharvest disease management

While the major focus on NTP research on fresh produce has been on inactivation of foodborne bacterial pathogens, there is an increasing interest in the application of NTP to manage postharvest fungal diseases in produce and mycotoxins in nuts. The effectiveness of plasma treatments on a range of fungi and mycotoxins in various foods, including fresh produce and nuts, has been reviewed (Misra et al., 2019). In general, fungi and yeasts exhibit higher resistance to NTP treatment than vegetative bacterial cells due to differences between prokaryotic and eukaryotic cells in cellular structure and molecular composition (López et al., 2019). As a result, higher levels of NTP are required to achieve greater log reduction or killing of moulds and these higher levels could potentially affect the fresh produce quality. Some latent fungal infections in tree fruits (e.g. avocado and mango) occur in the field during early stages of fruit development and remain deep seated in the fruit tissues until fruit ripening and senescence, resulting in the development of the fungi and postharvest disease symptoms. The success of the NTP to control these latent infections inside the fruit is limited due to its largely contact-based mechanism of action. However, the effects of surface treatment on the surface yeast and moulds have shown promising results in controlling some postharvest diseases and extending shelf-life of fresh produce.

Lacombe et al. (2015) reported a reduction in yeast and moulds on NTP-treated blueberries by 0.8–1.6 log CFU after 1 day and 1.5–2.0 log CFU reduction after 7 days. Similarly, the in-package treatment of strawberries with NTP for 5 min resulted in 3.3 log cycle reduction of naturally occurring yeasts and moulds on the surface after 24 h of in-pack storage (Misra et al., 2014). In another study, a microwave-powered nitrogen NTP treatment of mandarins at 0.7 kPa pressure for 10 min inactivated *P. italicum* and reduced the disease incidence by 84% (Won et al., 2017). As the plasma power and treatment time increase, a further increase in the inhibition of *P. italicum* was observed. Schnabel et al. (2015) observed inactivation of *Candida albicans* yeast by 6.2 log on apple peel and strawberry following an exposure to microwave plasma. Ouf et al. (2016) reported that treatment of the postharvest wash water with double atmospheric pressure NTP for 7.5 min reduced the population of *Alternaria alternata*, *Aspergillus niger* and *P. italicum*. This study highlights the potential of NTP in mould control in the postharvest wash water and thus minimizing the reliance on chemical sanitizers and fungicides for cross-contamination control.

Postharvest chemical residues are a concern for consumers and regulators all over the world. There is an increasing push to achieve either zero or ultra-low residues. Fungicide residues present on the produce surface or in tissue provide residual protection by inhibiting the germination of spores which are already present or to be introduced during the supply chain. The application of NTP as a supplementary treatment can therefore potentially minimize the initial fungal load on the surface and minimize the fungicide use as a combination treatment. The potential application of NTP in fresh produce has been pitched as an alternative to postharvest fungicide treatments, but it is not practically feasible due to the short-term nature of NTP protection against pathogens.

5.2.6 Postharvest mycotoxins management

Mycotoxins are a serious food safety issue for the storage and consumption of many tree nuts. *Aspergillus* species are universally known for their potential to infect tree nuts and produce aflatoxins - a class of mycotoxins. Several studies have shown the potential of plasma treatment in controlling aflatoxigenic moulds in cereal grains and nuts and these have been extensively reviewed (Misra et al., 2019). Basaran et al. (2008) reported a 5 log decrease for *A. parasiticus* population on hazelnut, peanut and pistachio nut surfaces after a 20 min treatment with low-pressure NTP. While Pignata et al. (2014) reported a 2 log reduction in the population of *A. brasiliensis* on pistachios after 1 min of low-pressure plasma treatment in argon/oxygen (10:1 v/v). A complete destruction of *A. flavus* inoculated on fresh and dried walnut surface was reported following an argon plasma jet exposure of 11 min and 10 min, respectively (Amini and Ghoranneviss, 2016). No changes were observed in phenolic content and antioxidant activity of treated and control walnut after treatment and 30 days of storage (Amini and Ghoranneviss, 2016). Dasan et al. (2016) reported a reduction in *A. flavus* and *A. parasiticus* on hazelnuts by 4.50 and 4.19 log reduction, respectively, after 5 min plasma treatment in dry air. Complete degradation of fumonisin B2 and ochratoxin A produced by *A. niger* inoculated onto date palm slices after 6 min and 7.5 min double argon jet plasma treatments, respectively, has been reported (Ouf et al., 2015). *A. niger* spores inoculated onto date palm fruit slices were also eliminated by the treatment. This research suggested the potential of NTP treatment in degrading mycotoxins in food but also reducing the capability of harmful fungi in mycotoxin production. These studies show that NTP has the potential to achieve mycotoxin control in tree nuts and dates and the effectiveness of the plasma treatment can be managed with the increase in applied voltage and frequency of plasma generation systems. The susceptibility of main species of *Aspergillus* to NTP treatments has been observed to differ widely and needs further investigations. A better understanding of the influence of NTP on genes regulating the biosynthesis of mycotoxins in fungi is also required. The data on nut quality (including lipid stability) and storage life of NTP-treated nuts is limited. More work is required in this field to generate further evidence not only in favour of NTP, but also the benefits it brings to the processors compared with existing technologies such as pasteurization.

6 Conclusion and future trends

NTP is a promising decontamination technology that is capable of killing or inactivating bacteria, moulds, yeasts and viruses on food and non-food surfaces. There are several other interesting applications of NTP in improving quality and safety of foods including fresh produce and nuts. However, the existing information is not very helpful in scaling up and translating the laboratory experiments into a commercial reality. The variability in experimental methodologies and equipment is a major hindrance for verification and validation of results from different parts of the world. It is difficult to draw valid conclusions and develop a comparative account of the existing information. López et al. (2019) recommended the establishment of a set of guidelines for conducting and reporting NTP experiments to help improve the reliability of experimental data and facilitate process optimization for commercial applications. The proposed guidelines should include variables and minimum description for each critical factor. We have modified the general recommendations for foods by López et al. (2019) so as to make them pertinent to the fresh horticultural produce applications.

Process equipment and conditions

- Type of equipment, discharge, electrode configuration, equipment model and vendor name
- Feed gas type, flow rate and moisture content
- Energy supplied - voltage, power and frequency
- Distance between the point of plasma generation and the sample
- Treatment time
- Composition of the generated plasma.

Produce

- Type of fresh produce (cultivar, harvest maturity and supply chain information)
- Storage conditions (temperature and atmosphere conditions)
- Postharvest storage behaviour (respiration and ethylene production rates)
- Effects on water loss, appearance, colour, texture, flavour, nutritional attributes, sensory quality and shelf-life
- Produce temperature (before and after treatment)
- Deleterious effects or observations.

Microbial factors

- Genus, species and strain(s) used and its source
- History of the microbial inoculum
- Procedures for preparation of the inoculum, inoculation (e.g. spot or immersion)
- Initial microbial load and procedure for enumerating microorganism.

It is also important that these treatment parameters are measured with calibrated instruments and not just reported from the instrument display.

Future research is needed to address the critical gaps in information to underpin the effectiveness, efficacy, economic and environmental sustainability of the technology. The above-described set of guidelines should be the minimum standard for information supplied with the experiments for robustness and repeatability. While it is necessary to understand the kinetics of microbial inactivation due to plasma treatment, it is not sufficient to reach the commercial potential of the technology. Therefore, a major research thrust is required on the development of plasma treatment regime assessing the impacts of treatment on fresh horticultural produce quality including nutritional and sensory quality. A systematic optimization process should be followed where treatment performance meets the requirements of inactivation of pathogens while maintaining produce quality in the supply chain.

Research efforts are required to identify the potential of various plasma species responsible for antimicrobial properties. The generation of plasma tailored to the target pathogen and produce would be ideal for effective decontamination within short treatment times. The chemistry of reactive species of plasma and their interaction with various macro and micro-molecules in fresh produce requires a deeper understanding to mitigate the adverse effects of the technology and facilitate its application in highly sensitive products. Despite the tremendous potential of NTP in postharvest horticulture, the application of technology should not be considered in isolation, and other factors such as washing, fungicide treatment should be considered to develop a hurdle-based approach to achieve synergistic effects of other processes. The widespread application of plasma technology can only be adopted if it can be retrofitted into current processing lines in an energy- and cost-effective manner.

7 Acknowledgements

This chapter is a contribution from the Innovative cold plasma for horticultural industries project (HN15000) which is funded by the Hort Frontiers Health, Nutrition and Food Safety Fund, part of the Hort Frontiers strategic partnership initiative developed by Hort Innovation, with co-investment from NSW Department of Primary Industries and contributions from the Australian Government.

8 Where to look for further information

Further information on NTP technology and its applications is scattered in multidisciplinary journals in the fields of applied physics, engineering, food and agriculture. The fundamental research on the advances in understanding of physical properties and methods of plasma generation can be accessed in journals such as *Physics of Plasmas*, *Journal of Plasma Physics*, *Plasma Science and Technology*, *Plasma Chemistry and Plasma Processing*, *Frontiers in Physics*, *Journal of Applied Physics* and so forth. The applications of plasma in foods including fresh horticultural produce are generally published in leading journals of food science, horticulture and postharvest such as *Food Microbiology*, *Journal of Applied Microbiology*, *Postharvest Biology and Technology*, *Journal of Agricultural and Food Chemistry*, *Food Control* and the *Journal of Food Protection*. There are several institutes and leading research groups throughout the world conducting research on various aspects of NTP. Some of these institutes are Culham Science Centre (United Kingdom), Max-Planck-Institut für Plasmaphysik (Germany), European Fusion Development Agreement (Germany), Institute for Plasma Physics Rijnhuizen (the Netherlands), Oakridge National Laboratory (United States), Princeton National Laboratory (United States), Los Alamos

National Laboratory (United States), Lawrence Livermore National Laboratory (United States), National Institute for Fusion Science (Japan), Japan Atomic Energy Research Institute (Japan), Institute of Plasma Physics-Chinese Academy of Sciences (China), Institute for Plasma Research (India) and Commonwealth Scientific and Industrial Research Organisation (Australia). There are several active research groups conducting research on the applications of NTP in fresh horticultural produce such as United States Department of Agriculture – Eastern Regional Research Centre – Philadelphia (United States), Drexel University (United States), Iowa State University (United States), Dublin Institute of Technology (Ireland) and New South Wales Department of Primary Industries, Ourimbah (Australia). In addition, conferences and symposia are held around the world to exchange the advances in plasma science and applications.

9 References

Amini, M. and Ghoranneviss, M. 2016. Effects of cold plasma treatment on antioxidants activity, phenolic contents and shelf life of fresh and dried walnut (*Juglans regia* L.) cultivars during storage. *LWT* 73, 178–84. doi:10.1016/j.lwt.2016.06.014.

Baier, M., Görgen, M., Ehlbeck, J., Knorr, D., Herppich, W. B. and Schlüter, O. 2014. Non-thermal atmospheric pressure plasma: screening for gentle process conditions and antibacterial efficiency on perishable fresh produce. *Innovative Food Science and Emerging Technologies* 22, 147–57. doi:10.1016/j.ifset.2014.01.011.

Baier, M., Ehlbeck, J., Knorr, D., Herppich, W. B. and Schlüter, O. 2015. Impact of plasma processed air (PPA) on quality parameters of fresh produce. *Postharvest Biology and Technology* 100, 120–6. doi:10.1016/j.postharvbio.2014.09.015.

Basaran, P., Basaran-Akgul, N. and Oksuz, L. 2008. Elimination of *Aspergillus parasiticus* from nut surface with low pressure cold plasma (LPCP) treatment. *Food Microbiology* 25(4), 626–32. doi:10.1016/j.fm.2007.12.005.

Callejón, R. M., Rodríguez-Naranjo, M. I., Ubeda, C., Hornedo-Ortega, R., Garcia-Parrilla, M. C. and Troncoso, A. M. 2015. Reported foodborne outbreaks due to fresh produce in the United States and European Union: trends and causes. *Foodborne Pathogens and Disease* 12(1), 32–8. doi:10.1089/fpd.2014.1821.

Conrads, H. and Schmidt, M. 2000. Plasma generation and plasma sources. *Plasma Sources Science and Technology* 9(4), 441–54. doi:10.1088/0963-0252/9/4/301.

Critzer, F. J., Kelly-Wintenberg, K., South, S. L. and Golden, D. A. 2007. Atmospheric plasma inactivation of foodborne pathogens on fresh produce surfaces. *Journal of Food Protection* 70(10), 2290–6. doi:10.4315/0362-028X-70.10.2290.

Dasan, B. G., Mutlu, M. and Boyaci, I. H. 2016. Decontamination of *Aspergillus flavus* and *Aspergillus parasiticus* spores on hazelnuts via atmospheric pressure fluidized bed plasma reactor. *International Journal of Food Microbiology* 216, 50–9. doi:10.1016/j.ijfoodmicro.2015.09.006.

Dasan, B. G., Boyaci, I. H. and Mutlu, M. 2017. Nonthermal plasma treatment of *Aspergillus* spp. spores on hazelnuts in an atmospheric pressure fluidized bed plasma system: impact of process parameters and surveillance of the residual viability of spores. *Journal of Food Engineering* 196, 139–49. doi:10.1016/j.jfoodeng.2016.09.028.

Deng, S., Ruan, R., Mok, C. K., Huang, G., Lin, X. and Chen, P. 2007. Inactivation of *Escherichia coli* on almonds using nonthermal plasma. *Journal of Food Science* 72(2), M62-6. doi:10.1111/j.1750-3841.2007.00275.x.

Han, L., Patil, S., Keener, K. M., Cullen, P. J. and Bourke, P. 2014. Bacterial inactivation by high-voltage atmospheric cold plasma: influence of process parameters and effects on cell leakage and DNA. *Journal of Applied Microbiology* 116(4), 784-94. doi:10.1111/jam.12426.

Jiang, Y., Sokorai, K., Pyrgiotakis, G., Demokritou, P., Li, X., Mukhopadhyay, S., Jin, T. and Fan, X. 2017. Cold plasma-activated hydrogen peroxide aerosol inactivates *Escherichia coli* O157:H7, *Salmonella* Typhimurium, and *Listeria innocua* and maintains quality of grape tomato, spinach and cantaloupe. *International Journal of Food Microbiology* 249, 53-60. doi:10.1016/j.ijfoodmicro.2017.03.004.

Kilonzo-Nthenge, A., Liu, S., Yannam, S. and Patras, A. 2018. Atmospheric cold plasma inactivation of *Salmonella* and *Escherichia coli* on the surface of golden delicious apples. *Frontiers in Nutrition* 5, 120. doi:10.3389/fnut.2018.00120.

Lacombe, A., Niemira, B. A., Gurtler, J. B., Fan, X., Sites, J., Boyd, G. and Chen, H. 2015. Atmospheric cold plasma inactivation of aerobic microorganisms on blueberries and effects on quality attributes. *Food Microbiology* 46, 479-84. doi:10.1016/j.fm.2014.09.010.

López, M., Calvo, T., Prieto, M., Múgica-Vidal, R., Muro-Fraguas, I., Alba-Elías, F. and Alvarez-Ordóñez, A. 2019. A review on non-thermal atmospheric plasma for food preservation: mode of action, determinants of effectiveness, and applications. *Frontiers in Microbiology* 10, 622. doi:10.3389/fmicb.2019.00622.

Min, S. C., Roh, S. H., Niemira, B. A., Boyd, G., Sites, J. E., Uknalis, J. and Fan, X. 2017. In-package inhibition of *E. coli* O157:H7 on bulk Romaine lettuce using cold plasma. *Food Microbiology* 65, 1-6. doi:10.1016/j.fm.2017.01.010.

Misra, N. N., Patil, S., Moiseev, T., Bourke, P., Mosnier, J. P., Keener, K. M. and Cullen, P. J. 2014. In-package atmospheric pressure cold plasma treatment of strawberries. *Journal of Food Engineering* 125, 131-8. doi:10.1016/j.jfoodeng.2013.10.023.

Misra, N. N., Yadav, B., Roopesh, M. S. and Jo, C. 2019. Cold plasma for effective fungal and mycotoxin control in foods: mechanisms, inactivation effects, and applications. *Comprehensive Reviews in Food Science and Food Safety* 18(1), 106-20. doi:10.1111/1541-4337.12398.

Niemira, B. A. 2012a. Cold plasma decontamination of foods. *Annual Review of Food Science and Technology* 3, 125-42. doi:10.1146/annurev-food-022811-101132.

Niemira, B. A. 2012b. Cold plasma reduction of *Salmonella* and *Escherichia coli* O157:H7 on almonds using ambient pressure gases. *Journal of Food Science* 77(3), M171-5. doi:10.1111/j.1750-3841.2011.02594.x.

Niemira, B. A. and Sites, J. 2008. Cold plasma inactivates *Salmonella* Stanley and *Escherichia coli* O157:H7 inoculated on golden delicious apples. *Journal of Food Protection* 71(7), 1357-65. doi:10.4315/0362-028X-71.7.1357.

Nuesch-Inderbinen, M. and Stephan, R. 2016. Fresh fruit and vegetables as vehicles of bacterial foodborne disease: a review and analysis of outbreaks registered by proMEDmail associated with fresh produce. *Journal of Food Safety and Food Quality - Archive for Food Hygiene* 67, 32-9.

Ouf, S. A., Basher, A. H. and Mohamed, A. A. H. 2015. Inhibitory effect of double atmospheric pressure argon cold plasma on spores and mycotoxin production of

Aspergillus niger contaminating date palm fruits. *Journal of the Science of Food and Agriculture* 95(15), 3204–10. doi:10.1002/jsfa.7060.

Ouf, S. A., Mohamed, A. H. and El-Sayed, W. S. 2016. Fungal decontamination of fleshy fruit water washes by double atmospheric pressure cold plasma. *CLEAN - Soil, Air, Water* 44(2), 134–42. doi:10.1002/clen.201400575.

Pankaj, S. K., Wan, Z. and Keener, K. M. 2018. Effects of cold plasma on food quality: a review. *Foods* 7(1), 4. doi:10.3390/foods7010004.

Pasquali, F., Stratakos, A. Ch., Koidis, A., Berardinelli, A., Cevoli, C., Ragni, L., Mancusi, R., Manfreda, G. and Trevisani, M. 2016. Atmospheric cold plasma process for vegetable leaf decontamination: A feasibility study on radicchio (red chicory, *Cichorium intybus* L.). *Food Control* 60, 552–9. doi:10.1016/j.foodcont.2015.08.043.

Perni, S., Liu, D. W., Shama, G. and Kong, M. G. 2008a. Cold atmospheric plasma decontamination of the pericarps of fruit. *Journal of Food Protection* 71(2), 302–8. doi:10.4315/0362-028X-71.2.302.

Perni, S., Shama, G. and Kong, M. G. 2008b. Cold atmospheric plasma disinfection of cut fruit surfaces contaminated with migrating microorganisms. *Journal of Food Protection* 71(8), 1619–25. doi:10.4315/0362-028X-71.8.1619.

Pignata, C., D'Angelo, D., Basso, D., Cavallero, M. C., Beneventi, S., Tartaro, D., Meineri, V. and Gilli, G. 2014. Low-temperature, low-pressure gas plasma application on *Aspergillus brasiliensis*, *Escherichia coli* and pistachios. *Journal of Applied Microbiology* 116(5), 1137–48. doi:10.1111/jam.12448.

Pignata, C., D'Angelo, D., Fea, E. and Gilli, G. 2017. A review on microbiological decontamination of fresh produce with nonthermal plasma. *Journal of Applied Microbiology* 122(6), 1438–55. doi:10.1111/jam.13412.

Ramazzina, I., Berardinelli, A., Rizzi, F., Tappi, S., Ragni, L., Sacchetti, G. and Rocculi, P. 2015. Effect of cold plasma treatment on physico-chemical parameters and antioxidant activity of minimally processed kiwifruit. *Postharvest Biology and Technology* 107, 55–65. doi:10.1016/j.postharvbio.2015.04.008.

Ramazzina, I., Tappi, S., Rocculi, P., Sacchetti, G., Berardinelli, A., Marseglia, A. and Rizzi, F. 2016. Effect of cold plasma treatment on the functional properties of fresh-cut apples. *Journal of Agricultural and Food Chemistry* 64(42), 8010–8. doi:10.1021/acs.jafc.6b02730.

Sarangapani, C., O'Toole, G., Cullen, P. J. and Bourke, P. 2017. Atmospheric cold plasma dissipation efficiency of agrochemicals on blueberries. *Innovative Food Science and Emerging Technologies* 44, 235–41. doi:10.1016/j.ifset.2017.02.012.

Schnabel, U., Niquet, R., Schlüter, O., Gniffke, H. and Ehlbeck, J. 2015. Decontamination and sensory properties of microbiologically contaminated fresh fruits and vegetables by microwave plasma processed air (PPA). *Journal of Food Processing and Preservation* 39, 653–62. doi:10.1111/jfpp.12273.

Surowsky, B., Schlüter, O. and Knorr, D. 2015. Interactions of non-thermal atmospheric pressure plasma with solid and liquid food systems: a review. *Food Engineering Reviews* 7(2), 82–108. doi:10.1007/s12393-014-9088-5.

Tappi, S., Berardinelli, A., Ragni, L., Dalla Rosa, M., Guarnieri, A. and Rocculi, P. 2014. Atmospheric gas plasma treatment of fresh-cut apples. *Innovative Food Science and Emerging Technologies* 21, 114–22. doi:10.1016/j.ifset.2013.09.012.

Tappi, S., Gozzi, G., Vannini, L., Berardinelli, A., Romani, S., Ragni, L. and Rocculi, P. 2016. Cold plasma treatment for fresh-cut melon stabilization. *Innovative Food Science and Emerging Technologies* 33, 225–33. doi:10.1016/j.ifset.2015.12.022.

Trevisani, M., Berardinelli, A., Cevoli, C., Cecchini, M., Ragni, L. and Pasquali, F. 2017. Effects of sanitizing treatments with atmospheric cold plasma, SDS and lactic acid on verotoxin-producing *Escherichia coli* and *Listeria monocytogenes* in red chicory (radicchio). *Food Control* 78, 138–43. doi:10.1016/j.foodcont.2017.02.056.

Wadamori, Y., Gooneratne, R. and Hussain, M. A. 2017. Outbreaks and factors influencing microbiological contamination of fresh produce. *Journal of the Science of Food and Agriculture* 97(5), 1396–403. doi:10.1002/jsfa.8125.

Won, M. Y., Lee, S. J. and Min, S. C. 2017. Mandarin preservation by microwave-powered cold plasma treatment. *Innovative Food Science and Emerging Technologies* 39, 25–32. doi:10.1016/j.ifset.2016.10.021.

Yeni, F., Yavaş, S., Alpas, H. and Soyer, Y. 2016. Most common foodborne pathogens and mycotoxins on fresh produce: a review of recent outbreaks. *Critical Reviews in Food Science and Nutrition* 56(9), 1532–44. doi:10.1080/10408398.2013.777021.

Ziuzina, D., Patil, S., Cullen, P. J., Keener, K. M. and Bourke, P. 2014. Atmospheric cold plasma inactivation of *Escherichia coli*, *Salmonella enterica* serovar Typhimurium and *Listeria monocytogenes* inoculated on fresh produce. *Food Microbiology* 42, 109–16. doi:10.1016/j.fm.2014.02.007.

Ziuzina, D., Han, L., Cullen, P. J. and Bourke, P. 2015. Cold plasma inactivation of internalised bacteria and biofilms for *Salmonella enterica* serovar Typhimurium, *Listeria monocytogenes* and *Escherichia coli*. *International Journal of Food Microbiology* 210, 53–61. doi:10.1016/j.ijfoodmicro.2015.05.019.

Ziuzina, D., Misra, N. N., Cullen, P. J., Keener, K. M., Mosnier, J. P., Vilaró, I., Gaston, E. and Bourke, P. 2016. Demonstrating the potential of industrial scale in-package atmospheric cold plasma for decontamination of cherry tomatoes. *Plasma Medicine* 6(3–4), 397–412. doi:10.1615/PlasmaMed.2017019498.

Chapter 11

Advances in the use of ozone in the disinfection of horticultural produce

Marcin Glowacz, Natural Resources Institute (NRI) – University of Greenwich, UK

1 Introduction

Fresh produce is generally recognized for its health-promoting properties and it is not surprising that the fruit and vegetable market is constantly growing. At the same time, consumption of fresh and fresh-cut fruits and vegetables has been associated with numerous bacterial illnesses and outbreaks (Yeni et al., 2016; Murray et al., 2017) caused by foodborne pathogens, for example *Escherichia coli* and *Listeria* sp. The contamination may come from numerous sources either at the production stage (preharvest) or later (postharvest) within the supply chain due to poor handling/processing of the produce. Thus, from the Food Standards Agency point of view (https://www.food.gov.uk/), the food safety - fresh produce is commonly eaten raw or minimally processed - for example microbial contamination is among the key factors limiting storability of fresh produce and its suitability for consumption. From the consumer's point of view, on the other hand, the marketability of the produce is limited/reduced by any visible signs of deterioration, for example fungal decay, physical injury (bruises and wounds) or colour changes associated with ripening (produce being overripe) or obvious symptoms of physiological deterioration, in particular tissue browning in the case of fresh-cut produce.

Disinfection of horticultural produce (fruits and vegetables) after harvest is an essential step for ensuring fresh-produce consumers' safety and well-being

http://dx.doi.org/10.19103/AS.2019.0055.12
Published by Burleigh Dodds Science Publishing Limited, 2020.

by reducing microbial counts on the produce, and to prevent fungal decay by controlling/reducing disease incidence and fungal growth during storage. Thus, there is clearly a need for technologies that reduce undesired microbial contamination which at the same time do not affect the product's nutritional, visual and textural quality (Allende et al., 2008; Artes et al., 2009). There is a growing body of literature that recognizes the potential of ozone to be used in the disinfection of fresh produce. The interest in using ozone as a sanitizer/postharvest treatment of fruits and vegetables has increased and a few comprehensive reviews have recently been published (Horvitz and Cantalejo, 2014; Glowacz et al., 2015a).

Interest in ozone is associated with its potential to reduce microbial contamination of the produce, without leaving any chemical residues, as it is simply converted to oxygen (Khadre et al., 2001). Ozone has been generally recognized as safe (US FDA, 2001, http://www.fda.gov/), thus being considered as a good alternative to other chemical treatments, such as chlorine. Furthermore, postharvest treatments with ozone have no adverse effect on product quality as long as they are being used at the proper dose, with numerous cases where positive effects of ozone exposure on fresh-produce quality have been reported.

Chemical and physical properties of ozone have been described elsewhere (Guzel-Seydim et al., 2004). Thus, the focus of this chapter is on the antimicrobial efficacy of ozone, mechanisms of action and on the effects of ozone exposure on the fresh-produce quality, defined by the maintenance of texture, visual quality and its nutritional content. The key challenges in postharvest use of ozone with fresh produce are (1) the end users' awareness of its practical limitations and recognition of all the factors affecting its efficiency when applied to fresh produce (Glowacz and Rees, 2016a) and (2) finding optimum conditions and doses for specific commodities and varieties.

2 Ozone as an antimicrobial agent

Ozone (O_3) is a well-known strong oxidizing agent that has been used by the fresh-produce industry as an antimicrobial agent for many years. Glowacz and Rees (2016a) emphasized that the first decision to be made, when considering the use of ozone with fresh produce, is whether to use it in aqueous or gaseous form. The method of ozone application is often informed by the current industrial practice as an aqueous form can only be used as a postharvest treatment prior to storage (sanitizing the produce on the processing line), whereas gaseous form of ozone gives the user more flexibility, that is it could be used either prior to storage and/or continuously/intermittently during subsequent storage to ensure that excessive microbiological contamination is prevented. In some cases, fresh produce might be washed first with a sanitizer with subsequent gaseous ozone treatment to complement existing practice.

Published by Burleigh Dodds Science Publishing Limited, 2020.

The use of gaseous ozone allows better control of the applied concentration, which for safety reasons must be monitored to avoid personnel health risks, for example headaches, and eye, nose, throat and respiratory tract irritation. Thus the Occupational Safety and Health Administration (https://www.osha.gov/) requires a concentration limit exposure of 0.1 μmol mol^{-1} over an 8 h shift.

Numerous studies have assessed the efficiency of ozone against bacteria and fungi immediately after the treatment (simulating the processing line - short-term treatment), whereas others have monitored the microbial counts during storage. The key findings will be discussed here to evaluate the practical potential of ozone for the disinfection of horticultural produce.

2.1 Bacteria

Ozonated (O_3) water efficiently reduces microbial counts on numerous products (Table 1). These include apples, carrot sticks, fresh-cut celery, fresh-cut lettuce, fresh-cut parsley, green bell pepper, red bell peppers and spinach, fitting well with the suitability of this mode of application for minimally processed produce. However, Papachristodoulou et al. (2018) have reported that even though the microbial counts were reduced on the fresh-cut spinach washed with ozonated water, the effect lasted only for 5 days, suggesting that bacteria proliferation was faster on sanitized samples.

On the other hand, application of ozone in the gaseous form (Table 2) was found to be efficient in reducing microbial counts on apples, blueberries, mulberries, raspberries (Bialka et al., 2008), strawberries (Bialka et al., 2008), papaya (Yeoh et al., 2014), peppers (Han et al., 2002; Horvitz and Cantalejo, 2012; Glowacz et al.,2015b), coriander (Wani et al., 2016), fresh-cut parsley (Karaca and Velioglu, 2014), rocket leaves (Wani et al., 2016; Gutierrez et al., 2017), spinach (Shynkaryk et al., 2016; Wani et al., 2016) and tomatoes. Again, in some cases, such as black mulberry fruit, the microbial counts of total aerobic mesophilic bacteria and *Enterobacteriaceae* were reduced by continuous exposure to gaseous ozone at 2.5 μmol mol^{-1} with no benefit at 0.3 μmol mol^{-1} (Tabakoglu and Karaca, 2018). However, the effect lasted only for 4 days, which could be related to the very short shelf-life of this fruit.

From the food-safety perspective, the key finding was that treatment with ozone reduced the levels of foodborne pathogens on the fresh produce, that is *E. coli* (Rahman et al., 2010; Karaca and Velioglu, 2014; Bridges et al., 2018), *Listeria* sp. (Rahman et al., 2010; Alexandre et al., 2011; Karaca and Velioglu, 2014;), *Salmonella* sp. (Selma et al., 2008; Bridges et al., 2018) and *Shigella* sp. (Selma et al., 2007). Bridges et al. (2018) reported much higher reduction in *L. monocytogenes* counts than *E. coli* and *Salmonella* sp. counts on ozone-exposed blueberries, which could be explained by poor attachment of *L. monocytogenes* to smooth surface of the berries. Similarly, when evaluating the

Table 1 Efficiency of ozonated water treatments against microbial populations found on selected fruits and vegetables

Product	Ozone dose	Reduction of microbes	Reference
Apple	21–28 μmol mol^{-1} for 1–5 min	*E. coli* counts by up to 3.7 log CFU	Achen and Yousef (2001)
	3 μmol mol^{-1} for 5 min	*E. coli* O157:H7 below the detection limit from the initial 6 log CFU	Rodgers et al. (2004)
		L. monocytogenes below the detection limit from the initial 6 log CFU	
		Mesophilic bacteria counts by up to 5.0 log CFU	
		Yeast and mould counts by up to 2.5 log CFU	
Carrot (fresh-cut)	10 μmol mol^{-1} for 10 min	Standard plate counts by up to 3.2 log CFU	Chauhan et al. (2011)
		Yeast and mould counts by up to 1.9 log CFU	
		Coliform counts by up to 2.4 log CFU	
Celery (fresh-cut)	0.03–0.18 μmol mol^{-1} for 5 min	Total bacteria count by up to 1.7 log CFU	Zhang et al. (2005)
Lettuce (fresh-cut)	10–20 μmol mol^{-1} for 3 min	Mesophilic bacteria counts by 1.6 log CFU	Beltran et al. (2005)
		Coliform counts by up to 4.9 log CFU	
	4 μmol mol^{-1} for 2 min	Mesophilic bacteria counts by 1.7 log CFU	Akbas and Olmez (2007)
		Psychrotrophic bacteria counts by 1.5 log CFU	
	0.5 μmol mol^{-1} for 5–30 min	Mesophilic bacteria counts by up to 3.1 log CFU	Alexopoulos et al. (2013)
		Coliforms counts by up to 2.5 log CFU	
		Yeast and mould counts by up to 2.3 log CFU	
Parsley (fresh-cut)	12 μmol mol^{-1} for 15 min	*E. coli* counts by 2.2 log CFU	Karaca and Velioglu (2014)
		L. innocua counts by 2.2 log CFU	
Peppers (whole)	0.5 μmol mol^{-1} for 5–30 min	Mesophilic bacteria counts by up to 3.3 log CFU	Alexopoulos et al. (2013)
		Coliform counts by up to 3.7 log CFU	
		Yeast and mould counts by up to 2.0 log CFU	

Product	Ozone dose	Reduction of microbes	Reference
Peppers (fresh-cut)	0.3–2 μmol mol^{-1} for 1–3 min	*L. innocua* counts by up to 2.8 log CFU	Alexandre et al. (2011)
Spinach (fresh-cut)	5 μmol mol^{-1} for 0.5–7 min	Total bacteria counts by 1.1 log CFU	Rahman et al. (2010)
		Yeast and mould counts by 0.9 log CFU	
		E. coli counts by 1.2 log CFU	
		L. monocytogenes counts by 1.4 log CFU	
	12 μmol mol^{-1} for 15 min	*E. coli* counts by 1.7 log CFU	Karaca and Velioglu (2014)
		L. innocua counts by 2.1 log CFU	

efficiency of in-package ozonation at 1000 μmol mol^{-1} in reducing *L. innocua*, *S. enterica* and *E. coli* O157:H7 from tomatoes, Fan et al. (2012) found the highest reduction rate for *Listeria* which was reduced to non-detectable levels within 40 s of the treatment. Alwi and Ali (2014) have also observed higher ozone efficiency in reducing microbial counts of *L. monocytogenes* and *E. coli* O157 than *S. enterica* on fresh-cut bell peppers, and suggested that high resistance of *S. enterica* could be associated with higher density of its cell membrane - higher content of phospholipid components (Heinrichs et al., 1998). All these findings indicate that *Listeria* must indeed be more sensitive to ozone when compared with other pathogens, which would be in agreement with the results of Kim and Yousef (2000).

There were however some reports (Ketteringham et al., 2006; Allende et al., 2007; Bermudez-Aguirre and Barbosa-Canovas, 2013) where no or limited benefits of ozone treatment were described. There are several possible explanations, for example lower efficiency of ozone could be due to the leaching of organic matter from the produce (Ketteringham et al., 2006), which will result in ozone reacting with it rather than fully expressing its antimicrobial potential. Ketteringham et al. (2006) suggested that whole peppers are more suitable for ozone treatment than their fresh-cut counterparts. A similar observation was reported (Rodgers et al., 2004) when comparing the efficiency of ozonated water treatment at 3 μmol mol^{-1} for 5 min in reducing *L. monocytogenes* and *E. coli* counts from whole apples and lettuce with their sliced and shredded counterparts, respectively.

Limited benefits of ozone treatment could also be related to the dose of ozone being too high (Olmez and Akbas, 2009) leading to severe damage. For example, in lettuce the lack of positive effect of gaseous ozone exposure at 5 μmol mol^{-1} for 3–15 min on microbial counts was due to the dose of ozone being too high (Bermudez-Aguirre and Barbosa-Canovas, 2013) leading to tissue damage and in this way compromising the overall quality of the produce.

Table 2 Efficiency of gaseous ozone treatments against microbial populations found on selected fruits and vegetables

Product	Ozone dose	Reduction in microbes	Reference
Apple	50 μmol mol^{-1} for 30 weeks	*L. innocua* counts by up to 2.0 log CFU	Sheng et al. (2018)
Blueberries	30 000 μmol mol^{-1} for 2–64 min	*E. coli* counts by up to 2.2 log CFU	Bialka and Demirci (2007a)
		Salmonella counts by up to 3.0 log CFU	
	0.45–0.9 μmol mol^{-1} for 2.5–5 h	*L. monocytogenes* counts by 1.4 log CFU	Bridges et al. (2018)
		Salmonella counts by up to 1.0 log CFU	
	2.5–4 μmol mol^{-1} for 10 days	*L. monocytogenes* counts by up to 3.0 log CFU	Concha-Meyer et al. (2014)
Mulberries	2.5 μmol mol^{-1} for 6 days	Total aerobic mesophilic bacteria counts by up to 2.0 log CFU	Tabakoglu and Karaca (2018)
		Enterobacteriaceae counts by up to 2.5 log CFU	
Raspberries	30 000 μmol mol^{-1} for 2–64 min	*E. coli* counts by up to 3.8 log CFU	Bialka and Demirci (2007b)
		Salmonella counts by up to 3.6 log CFU	
Strawberries	30 000 μmol mol^{-1} for 2–64 min	*E. coli* counts by up to 3.0 log CFU	Bialka and Demirci (2007b)
		Salmonella counts by up to 2.6 log CFU	
Peppers	0.7 μmol mol^{-1} for 1–5 min	Mesophilic bacteria counts by up to 2.6 log CFU	Horvitz and Cantalejo (2012)
		Psychrotrophic bacteria counts by up to 5.8 log CFU	
		Yeast and mould counts by up to 2.3 log CFU	
Spinach	750–2 000 μmol mol^{-1} for 5 min	*E. coli* counts by up to 5.6 log CFU	Klockow and Keener (2009)
	935 μmol mol^{-1} for 5–30 min	*E. coli* counts by up to 2.4 log CFU	Vurma et al. (2009)
	5–10 μmol mol^{-1} for 3 days	*E. coli* counts by up to 1.8 log CFU	
Tomato	4 μmol mol^{-1} for 30 min every 3 h	Mesophilic bacteria counts by up to 1.1 log CFU	Aguayo et al. (2006)
		Psychrotrophic bacteria counts by up to 1.3 log CFU	
	2 500–9 500 μmol mol^{-1} for 5–20 min	*Salmonella* counts by up to 7 log CFU	Das et al. (2006)
	5 μmol mol^{-1} for 3–15 min	*E. coli* counts by up to 2.2 log CFU	Bermudez-Aguirre and Barbosa-Canovas (2013)
	0.45–0.9 μmol mol^{-1} for 2.5–5 h	*L. monocytogenes* counts by 1.1 log CFU	Bridges et al. (2018)
		Salmonella counts by up to 1.1 log CFU	

Microbial reductions on lettuce treated with ozonated water have indeed been reported by others when lower doses were used, for example 0.5 μmol mol^{-1} for 5–30 min (Alexopoulos et al., 2013), 0.5–4.5 μmol mol^{-1} for 0.5–3.5 min (Olmez and Akbas, 2009), 2.5–4 μmol mol^{-1} for 5 min (Beltran et al., 2005) and 4 μmol mol^{-1} for 2 min (Akbas and Olmez, 2007). In some of these studies, the doses of ozone were also too high (Alexopoulos et al., 2013); however, sensory evaluation of the produce was either not conducted or carried out by a random, untrained panel who nonetheless were not pleased with the quality of lettuce treated with ozone, as indicated by their lack of willingness to purchase the product. Olmez and Akbas (2009) focused on optimization of ozonated water treatment for lettuce and concluded that overall visual quality declines with the concentration of ozone being increased beyond 2.5 μmol mol^{-1} and exposure time extended above 2.5 min.

Another challenge when sanitizing leafy vegetables is the fact that they are typically vacuum cooled after harvest and sanitized later. Vacuum cooling facilitates the transfer of microorganisms into inner leaf structures via cut edges and stomata, making it more difficult for ozone to reduce the bacterial cells counts (Li et al., 2008; Shynkaryk et al., 2016; Yesil et al., 2017).

The reduction of microbial counts is often determined by the type of produce, for example efficiency of ozone in reducing microbial counts on carrots and lettuce (Bermudez-Aguirre and Barbosa-Canovas, 2013) was lower compared to tomatoes and was explained by better attachment of the bacteria to porous and rough surfaces in which they were protected from the ozone action. These authors observed little benefit of exposing carrots to gaseous ozone at 5 μmol mol^{-1} for 3–15 min on *E. coli* counts, which is not surprising as others (Singh et al., 2002; Chauhan et al., 2011) reported that to ensure efficiency in reducing microbial contamination of carrots ozone needs to be applied at 1000 μmol mol^{-1} for 5 min in its gaseous form (Singh et al., 2002), or about 10 μmol mol^{-1} for at least 10 min when used in the aqueous form (Singh et al., 2002; Chauhan et al., 2011). However, ozone exposure at 1000 μmol mol^{-1} for 5 min adversely affects other quality characteristics of carrots (Singh et al., 2002).

Finally, another aspect that needs to be taken into consideration is the initial microbial load on the produce. Alexandre et al. (2011, 2012) found that when the initial microbial counts on strawberries were high (10^7 CFU g^{-1}), reduction by 1.21 log unit was observed on the fruit washed with ozonated water at 0.3 μmol mol^{-1} for 2 min and by 2.30 log unit at 2 μmol mol^{-1} for 3 min, respectively. On the other hand, no benefit of ozone treatment was observed when initial microbial count (10^2–10^4 colony forming units (CFU) g^{-1}) was low (Allende et al., 2007).

To sum up, in general the efficiency of ozone in reducing microbial counts on fresh produce is dose dependent, with the highest microbial reductions being observed at the highest doses of ozone due to differences in the sensitivity to ozone among different commodities and varieties; however, it is crucial to establish the optimum treatment dose for each product to avoid unwanted tissue damage, which among other things would lead to an increased microbial contamination. Numerous studies (e.g. Ketteringham et al., 2006; Kim and Hung, 2012; Alwi and Ali, 2014), however, have only focused on the antimicrobial efficacy of ozone treatment and not on its effect on the sensory characteristics of the produce, which means that these studies are of limited relevance for the industry as some of the doses of ozone that give high reductions in microbial counts would often be detrimental to quality of the produce.

2.2 Fungi

Treatment of fresh produce with ozone has been reported to reduce fungal development, measured as lesion size, on a number of horticultural products. These include broccoli continuously exposed to ozone at 0.2 and 0.7 μmol mol^{-1} for 12 days (Forney et al., 2003), carrots exposed to gaseous ozone at 0.45 μmol mol^{-1} for 2 days (Sharpe et al., 2009), at 1 μmol mol^{-1} for 4 days (Forney et al., 2007) and intermittently (8 h day^{-1}) exposed to ozone at 15 μmol mol^{-1} for 4 weeks (Liew and Prange, 1994) or continuously exposed to low level of ozone at 0.05 μmol mol^{-1} over a 6-month storage period (Hildebrand et al., 2008). Reduced growth of grey mould (*Botrytis cinerea*) and white mould (*Sclerotinia sclerotiorum*) was observed in ozone-exposed carrots (Liew and Prange, 1994; Sharpe et al., 2009), while others (Forney et al., 2007; Hildebrand et al., 2008) reported reduced growth of *B. cinerea* but not *S. sclerotiorum*. Hildebrand et al. (2008) reported an increase in the content of isocoumarin 6-methoxymellein (3-methyl-6-methoxy-8-hydroxy-3,4-dihydro isocoumarin; 6-MM) - compound associated with increased resistance to *B. cinerea* - in carrots continuously exposed to ozone. Even though Liew and Prange (1994) found ozone to be fungistatic, the high dose of ozone at 7.5-60 μmol mol^{-1} (8 h day^{-1}) for 4 weeks was injurious to the produce as indicated by the increased respiration rate, electrolyte leakage and colour bleaching. Furthermore, the positive effect of ozone exposure was reported to be lost during the long-term storage (Forney et al., 2007; Hildebrand et al., 2008).

Similarly, reduced fungal development has been reported for numerous fruits, for example apples exposed to gaseous ozone at 0.45 μmol mol^{-1} for 2 days (Sharpe et al., 2009), kiwifruit exposed to ozone at 0.3 μmol mol^{-1} over 4-month storage period (Minas et al., 2010), papaya exposed to ozone at 0.04, 1.6 and 4 μmol mol^{-1} for 48–144 h (Ong et al., 2013), peaches exposed to ozone at 0.3 μmol mol^{-1} over a 4-week period (Palou et al., 2002) and tomatoes exposed to ozone at 0.2, 1 and 5 μmol mol^{-1} for up to 13 days (Tzortzakis et al.,

Published by Burleigh Dodds Science Publishing Limited, 2020.

2007a,b, 2008, 2011). Boonkorn et al. (2012) found, and documented with a scanning electron microscopy (SEM), that exposure to ozone at 200 μmol mol^{-1} for 4 or 6 h delayed the development of green mould on wound-inoculated (10^6 conidia mL^{-1} of *P. digitatum*) tangerine fruit (Fig. 1), thus reducing disease severity.

No effect, however, was observed in the case of plums exposed to ozone at 0.1 μmol mol^{-1} (Tzortzakis et al., 2007b), suggesting that the dose was too low to prevent the fungal growth. By comparing the efficiency of ozone exposure in reducing fungal development in clementines, plums and tomatoes inoculated with *B. cinerea* spores at low (2×10^3), intermediate (2×10^5) or high (2×10^7) spore concentration, Tzortzakis et al. (2007b) clearly demonstrated that the outcome is more determined by the type of produce than by the spore concentration. Their findings highlight the fact that some commodities would not benefit from exposure to ozone as higher doses that might be necessary may lead to damage/cracks in the fruit epidermis, thus facilitating fungal infection. These findings are further supported by Sharpe et al. (2009) who found blueberries to be far more susceptible to fungal infection compared to apples and grapes.

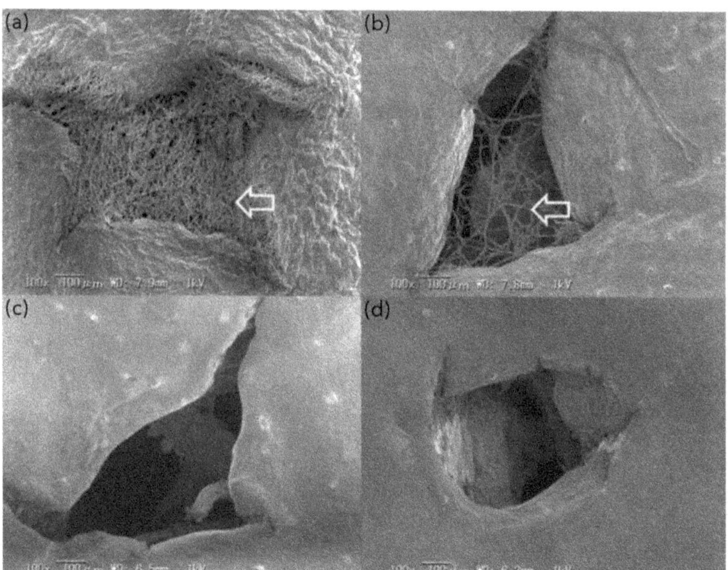

Figure 1 Development of *P. digitatum* conidia on tangerine fruit peel, 72 h after fumigation with 200 μL L^{-1} gaseous ozone for (a) 0, (b) 2, (c) 4 or (d) 6 h. Fruit were wounded, inoculated with 1×10^6 conidia mL^{-1} of *P. digitatum* before ozone treatment and kept for 3 days at 25°C, 75-80% RH. Arrows indicate *P. digitatum* mycelia growth. Source: reprinted from Boonkorn et al. (2012) with permission from Elsevier.

Palou et al. (2002) have reported that continuous exposure to ozone at 0.3 µmol mol^{-1} reduced the mycelial growth and sporulation of *Monilinia fructicola, B. cinerea, Mucor piriformis* and *Penicillium expansum* on wound-inoculated peaches. However, pathogens often survive if they reside within the wounds, thus disease development was only delayed in ozone-exposed fruit. Similar results have been observed for Valencia oranges (Palou et al., 2001), Lanelate oranges (Palou et al., 2003) and Eureka lemons (Palou et al., 2001) wound inoculated with *P. digitatum* and *P. italicum* and subsequently exposed to ozone at 0.3 µmol mol^{-1} for 4 weeks, 0.3 µmol mol^{-1} for 2 weeks and at 0.3 µmol mol^{-1} (only at night) for 9 weeks, respectively. More recently, Garcia-Martin et al. (2018) investigated the effect of continuous ozone exposure (from 0.75 to 30 µmol mol^{-1}) and intermittent exposure to 0.75 µmol mol^{-1} gaseous ozone (only at night) on disease incidence in six citrus varieties: two mandarins - Fortune and Ortanique, and four oranges - Navelate, Lanelate, Salustiana and Valencia. Disease development was delayed in ozone-exposed fruit, but in Lanelate and Navelate oranges ozone exposure led to increase in the incidence of oleocellosis - a physiological disorder provoked by the rupture of peel oil glands. This disorder is associated with mechanical injuries often resulting in high postharvest losses, and these findings suggest that not all citrus varieties are suitable for storage in ozone-enriched air.

Interestingly, disease incidence in kiwifruit was reduced if fruits were exposed to ozone at 0.3 µmol mol^{-1} for 8–144 h prior to inoculation with *B. cinerea* (Minas et al., 2010), whereas the exposure of kiwifruit to ozone post-inoculation had no beneficial effect. This finding suggests that exposure of kiwifruit to ozone may have affected the fruit-pathogen interaction and somehow enhanced their disease resistance. The changes in fruit biochemistry related to defence responses have also been reported, for example in ozone-exposed grapes where synthesis of phytoalexins, that is resveratrol and pterostilbene, was induced by ozone making the berries more resistant to subsequent *Rhizopus stolonifer* infection (Sarig et al., 1996).

Ozone reduced disease incidence (per cent of fruit showing fungal infection) in grapes exposed to 200 µmol mol^{-1} for 15 min (Ozkan et al., 2011) or at 2500 µmol mol^{-1} for 2 h, 5000 µmol mol^{-1} for 1 h and 10 000 µmol mol^{-1} for 30 min (Mlikota Gabler et al., 2010). Similarly, a positive effect of ozone was observed in grapes continuously exposed to the gas at 0.1 µmol mol^{-1} for up to 60 days (Artes-Hernandez et al., 2004), at 0.3 µmol mol^{-1} over 7 weeks (Palou et al., 2002) and in grapes continuously or intermittently (12 h day^{-1}) at 2 µmol mol^{-1} for 72 days (Cayuela et al., 2009). These findings highlight that ozone might be a good alternative method to control grey mould on table grapes (Romanazzi et al., 2012). One must, however, take into account that contamination of grapes with yeasts and moulds would vary among different studies due to different climatic conditions during the grape-growing period, the variety and degree of maturity and also on-farm/vineyard practice including plant-protecting agents (Martins et al., 2014).

Interestingly, Feliziani et al. (2014) reported that ozone exposure at 0.1/0.3 μmol mol^{-1} (day/night cycle) inhibited the spread of aerial mycelia of *B. cinerea* by approximately 65% after 5–8 weeks of storage. However, in contrast to sulphur dioxide, fumigation was not effective against other fungi, such as *Alternaria* sp. and *Penicillium* sp. thereby highlighting its limitation for practical use. The high efficiency against *B. cinerea* might be related to the fact that this pathogen does not grow deep into the berry tissue.

Furthermore, due to numerous benefits associated with reduced microbial counts of various fungi reported for grapes treated with ozone, for example an effective sanitizer for Pignola grapes (Botondi et al., 2015) and Petit Verdot grapes (Bellincontro et al., 2017), there is increasing interest in using ozone during the wine-making process (Morata et al., 2017). This has important industry relevance especially since changes in the microbiota may have consequences for the chemical composition of wine (Cravero et al., 2016, 2018), for example by reducing undesirable apiculate yeasts – *Brettanomyces* sp. – considered to be wine spoilage yeasts due to their ability to produce off-flavours by producing high levels of acetic acid. This aspect of ozone application, however, is beyond the scope of this review.

As evidenced in this chapter, the benefits of using ozone for the disinfection of horticultural produce have been widely reported; however, one must remember that the results from individual experiments are valid only for those specific conditions and in some cases still need to be tested and validated under commercial conditions. For example, ozone efficiency may be affected by the sensitivity of specific fungal species to ozone (Tzortzakis et al., 2008; Ozkan et al., 2011) and by environmental (storage) conditions, such as temperature (Sharpe et al., 2009) and relative humidity (Ozkan et al., 2011). Ozkan et al. (2011) observed that when exposing conidia of *P. digitatum*, *P. italicum* and *B. cinerea* to ozone at 200 μmol mol^{-1} for 15 min at relative humidity of 95%, 75% and 35% respectively, the efficiency of ozone in inhibiting their germination declined with decreasing relative humidity.

Fungal pathogens that cause decay of horticultural produce reproduce by the production of conidia. Ames et al. (2013) have investigated the germinability of conidia of several fungal species (*Alternaria alternata*, *Aspergillus flavus*, *Aspergillus niger*, *P. digitatum*, *P. expansum* and *P. italicum*) over the period of 100 days at 2°C under the ozone-enriched atmosphere at 0.15 μmol mol^{-1}. They observed that the mortality of conidia is rather slow, although taking into account that sometimes produce may spend weeks in transit/storage, being slow does not necessarily mean inefficient. The authors, however, suggested that the primary action of ozone must be to inhibit the aerial mycelial growth rather than to inactivate the conidia. The age of the culture from which the conidia were collected affected their susceptibility to ozone, with conidia coming from younger cultures being more sensitive. Tzortzakis et al. (2007b) reported reduction in *B. cinerea* spore production in wound-inoculated clementines, Thompson seedless grapes, tomatoes and plums stored for 13 days at 13°C under the ozone-enriched atmosphere at 0.1 µmol mol^{-1}, regardless of whether the initial spore concentration was low (2×10^3), intermediate (2×10^5) or high (2×10^7).

Antony-Babu and Singleton (2009) compared the influence of ozone fumigation at three different levels (0.2 µmol mol^{-1} for 12 days, 200 and 300 µmol mol^{-1} for 10 min) on the growth, sporulation and viability of conidia of *Aspergillus nidulans* and *A. ochraceus*. All treatments reduced the mycelial growth compared to the control, with a stronger delay in the case of *A. ochraceus*. However, no significant difference was noticed among the ozone doses in terms of mycelia growth and spore formation. In contrast, spore formation was completely inhibited when *A. nidulans* was continuously exposed to ozone at 0.2 µmol mol^{-1}. Antony-Babu and Singleton (2009) have reported that high dose of ozone reduced spore germination of both fungi. The ability of ozone to lower disease incidence in horticultural produce by reducing spore production and/or viability has been also confirmed by Minas et al. (2010) and Tzortzakis et al. (2011), which is of huge relevance for the fresh-produce industry, as spread of the disease from injured and infected produce can be substantially reduced by introducing gaseous ozone (at the right dose) into the cold store atmosphere. Hua-Li et al. (2018), who investigated the effect of ozone exposure at 1 µmol mol^{-1} for 120 min on fungal development on muskmelon inoculated with *Fusarium sulphureum*, suggested that in addition to reducing spore germination and mycelial growth ozone may alter fungus ultrastructure, in this way reducing its pathogenicity. However, ozone does not prevent fungal development in the infected wounds/cracks as these would often remain protected from the ozone action.

The generally accepted mechanism of ozone action when reducing microbial counts on fresh produce (Khadre et al., 2001; Guzel-Seydim et al., 2004) is related to ozone destroying microorganisms by oxidation of their cellular compounds located on the bacterial cell surface, for example phospholipid and protein molecules in the cell membranes (ruptures the cell wall, leading to cellular disintegration) and damaging the nucleic acids of the microorganisms, thus compromising cellular functions, affecting their ability to proliferate and eventually leading to cell death. In case of fungal pathogens, continuous/intermittent exposure to a low concentration of ozone during transit/storage seems to be beneficial due to the fact that fungi have to go through several growth cycles that are interrupted by ozone, for example spore production and viability. It has been reported by Khadre and Yousef (2001) who examined *B. subtilis* spores treated with ozone with an SEM, that spore coat layer is damaged by ozone making the cortex and core of the spore accessible for the sanitizer. This was further supported by Young and Setlow (2004) who observed that efficiency of killing *B. subtilis* spores by ozone was substantially increased if spores were chemically decoated. Another possibility is that ozone exposure (if the dose is high enough) causes the degradation of mitochondria of fungal spores (Ong and Ali, 2015). These authors observed distorted mitochondrial cristae in the spores of *Colletotrichum gloeosporioides* exposed to ozone at 3.5 $\mu mol\ mol^{-1}$ for 24 h, with mitochondria being almost completely degraded when the ozone concentration was increased to 5 $\mu mol\ mol^{-1}$.

Even though the oxidation of DNA would seem to be the most likely target for inactivation of microorganisms by ozone, the findings reported by Young and Setlow (2004) suggest that killing of *B. subtilis* spores was not due to damage to the spore's DNA but rather to the degradation of the spore coat by ozone. A similar mechanism has been reported by Predmore et al. (2015) as the primary way that ozone disrupts viruses, that is by altering their capsid structure and damaging its proteins.

3 Ozone-induced changes in the produce

Postharvest treatments with ozone would only be considered by the fresh-produce industry if they have the ability to reduce microbial contamination of horticultural produce while at the same time having no adverse effect on its nutritional, visual and textural quality. Otherwise, they cannot be recommended for introduction into the supply chain.

3.1 Oxidative stress

Horticultural produce continues to be metabolically active after harvest and responds to changes in the environmental/storage conditions. Metabolic processes are regulated by signalling molecules, for example phytohormones and reactive oxygen species (ROS) - that change in response to environmental stresses, for example when exposed to ozone their production may increase (Kangasjarvi et al., 2005). ROS include such compounds as superoxide radicals (O_2^-), singlet oxygen ($^1O_2$) and highly reactive hydroxyl radicals (OH^-). SOD catalyses the dismutation of superoxide (O_2^-) to H_2O_2, which is then transformed to H_2O and O_2 by simultaneous action of ascorbate peroxidase and catalase (CAT). To mitigate ROS (Fig. 2), the biosynthesis of antioxidants, for example ascorbic acid (AsA), which is involved in the reduction in ROS through the ascorbate-glutathione cycle (Mittler, 2002), may be induced. Other antioxidants, that is phenolic compounds, may also play an important role as ROS scavengers (Hernandez et al., 2009). Depending on the level of oxidative stress, changes in ROS may be considered as important signalling molecules;

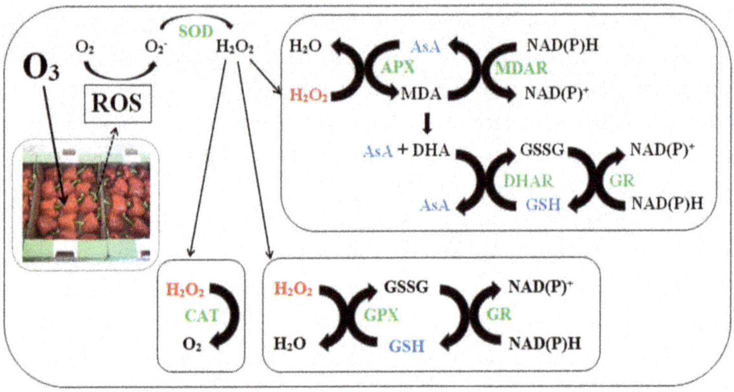

Figure 2 Schematic illustration of how exposure to ozone induces reactive oxygen species (ROS) in fresh produce, and their subsequent scavenging within the produce. APX - ascorbate peroxidase; CAT - catalase; DHA - dehydroascorbic acid; DHAR - DHA reductase; GPX - glutathione peroxidase; GR - glutathione reductase; GSH - reduced glutathione; GSSG - oxidized glutathione; MDA - monodehydroascorbic acid; MDAR - MDA reductase; NAD(P)H - nicotinamide adenine dinucleotide phosphate (reduced form); NAD(P)+ - nicotinamide adenine dinucleotide phosphate (oxidized form); O_3 - ozone; SOD - superoxide dismutase. Source: based on Mittler (2002) and Kangasjarvi et al. (2005).

however, if they are produced in excess they may become toxic to the produce, leading to cell death and senescence/deterioration (Mittler, 2017).

Although the ability of ozone exposure to induce oxidative stress (via ROS) in horticultural produce is well known, with various physiological responses being reported, the changes in the contents of antioxidants (AsA, phenolic compounds) and other secondary metabolites, and associated changes in the antioxidant activity of the produce, thus affecting its biochemistry and nutritional value, are beyond the scope of this chapter and will not be discussed in more depth here.

3.2 Colour

Visual quality of the horticultural produce is of huge importance for the industry because fresh produce with a good appearance is preferred by the customers. In the majority of studies (Glowacz et al., 2015a), ozone exposure had no significant effect on the colour of apples (Sharpe et al., 2009), cantaloupe melons (Toti et al., 2018), cucumbers (Skog and Chu, 2001), mulberry fruit (Han et al., 2017; Tabakoglu and Karaca, 2018), persimmons (Salvador et al., 2006), tangerines (Boonkorn et al., 2012) and tomatoes (Rodoni et al., 2010; Bermudez-Aguirre and Barbosa-Canovas, 2013). On the other hand, others (Zambre et al., 2010; Ali et al., 2014) observed that if ozone is applied at high enough dose, it may interfere with the ripening process, for example delaying red colour development of tomatoes if treated with gaseous ozone at 20–50 µmol mol^{-1} for 10 min (Zambre et al., 2010) or delayed ripening of papaya fruit if exposed to ozone at 2.5 µmol mol^{-1} for 96 h (Ali et al., 2014). Similar effect has recently been reported for oranges (Garcia-Martin et al., 2018), where authors have suggested that ozone may interfere with the synthesis of carotenoids.

In the case of carrots treated with ozone, the effect is dose dependent, that is if the dose is right there would be no difference in colour between control- and ozone-treated carrots (Sharpe et al., 2009; de Souza et al., 2018); however, if the dose of ozone is too high, it will result in bleaching the produce (Liew and Prange, 1994; Bermudez-Aguirre and Barbosa-Canovas, 2013).

Similarly, for leafy vegetables which are very sensitive to ozone action, the dose needs to be properly adjusted. For example, Olmez and Akbas (2009) highlighted that for lettuce, to avoid discolouration, the concentration of ozone should not be higher than 2.5 µmol mol^{-1} and exposure time no longer than 2.5 min. Leaves discolouration has also been reported for lettuce exposed to gaseous ozone at 5 µmol mol^{-1} for 15 min (Bermudez-Aguirre and Barbosa-Canovas, 2013) and spinach (Klockow and Keener, 2009) exposed to gaseous ozone at 0.75 and 2 µmol mol^{-1} for 5 min, and higher doses (Shynkaryk et al., 2016; Pyatkovskyy et al., 2017), respectively.

3.3 Firmness

Firmness changes are often associated with the respiration rates and weight losses. A number of studies have found no significant effect of ozone exposure on firmness of the produce (see Glowacz et al., 2015a), with several papers reporting better firmness retention, for example in cucumbers continuously exposed to ozone at 0.04 µmol mol^{-1} (Skog and Chu, 2001) or 0.1 µmol mol^{-1} (Glowacz et al., 2015b), in kiwifruit continuously exposed to ozone at 0.3 µmol mol^{-1} (Minas et al., 2012), and in tomatoes cyclically exposed to gaseous ozone at 4 µmol mol^{-1} for 30 min every 3 h (Aguayo et al., 2006), exposed to gaseous ozone at 0.05 or 1 µmol mol^{-1} for 6 days (Tzortzakis et al., 2007a), or treated with gaseous ozone at 10 µmol mol^{-1} for 10 min (Rodoni et al., 2010), where softening of the fruit was delayed in ozone-exposed samples. Softening has also been found to be reduced in ozone-exposed citrus fruits (Garcia-Martin et al., 2018).

Better firmness retention of ozone-exposed horticultural produce could be due to the direct effect of ozone on stomata (Kangasjarvi et al., 2005). These authors have suggested that ozone in the surrounding environment will trigger the formation of ROS. This would have a direct effect on stomata functioning by elevating the concentration of Ca^{2+} ions in the cytoplasm of guard cells, which together with the efflux of anions and K^+ ions contributes to the loss of turgor in guard cells, leading to stomata closure (Vahisalu et al., 2010). In our study (Glowacz et al., 2015b), we have observed better firmness retention in courgettes continuously exposed to ozone at 0.1 μmol mol^{-1} and suggested that this was related to the effect of ozone on stomata, as previous reports found stomata to be closed in response to ozone in cucumber (Agrawal et al., 1993) and spinach (Calatayud et al., 2004) leaves. The recent study of Han et al. (2017) confirmed that ozone may lead to stomata shrinkage/closure in mulberry fruits.

Firmness maintenance during storage in ozone-exposed fruits may also be associated with the inhibition of cell wall-degrading enzymes activities, such as polygalacturonase (PG), 1,4-β-glucosidase and pectin methylesterase (PME) as reported in ozone-exposed kiwifruit (Minas et al., 2014) and tomatoes (Rodoni et al., 2010), respectively. However, in the study where cantaloupe melon was exposed to ozone at 0.15/0.3 μmol mol^{-1} (day/night cycle) during storage at 6°C, Toti et al. (2018) observed that the activities of cell wall-modifying enzymes associated with pectin degradation, that is α-arabinopyranosidase, β-galactopyranosidase and PG, were reduced, while the activity of PME was not affected by ozone exposure.

There are few papers reporting negative effects of ozone exposure on firmness maintenance of horticultural produce, unless the dose of ozone being used was too high. Alwi and Ali (2015) reported that exposure of bell peppers to low level of ozone (1–3 μmol mol^{-1}) had no significant effect on the fruit respiration and firmness. However, higher doses, that is 7 and 9 μmol mol^{-1}, were injurious to the produce and led to an increase in the respiration rates, enhanced weight loss and reduced firmness. Water loss from peppers occurs primarily through the cuticle (Kissinger et al., 2005) and the amount of water loss during storage could be affected by its thickness and composition (Parsons et al., 2013; Lara et al., 2014). As previously suggested (Ali et al., 2014), the thick cuticle makes the produce less susceptible to damage caused by a high dose of ozone. Thus, the negative effect of ozone exposure at 7 and 9 μmol mol^{-1} observed by Alwi and Ali (2015) is likely related to the ozone action on the cuticle, resulting in cracks in the skin surface, reduced cell membrane integrity and increased water loss (Lara et al., 2014). Furthermore, due to the fact that some manuscripts do not get published, the possibility of negative results or lack of effect of ozone exposure on the firmness of horticultural produce being not reported cannot be excluded.

4 Practicality aspects

From the practical point of view, it is important for us (researchers) to be fully aware that the industry likes to stay as close as possible to current practice, and is only ready to implement small modifications/amendments within the supply chain. One must remember though that for the technology to be introduced, it must not only be effective in maintaining horticultural produce quality/ reducing quality losses, but ideally (from the industry point of view) should bring some advantage (e.g. being less costly and/or more environmentally friendly) compared to the system that it is going to replace.

As already mentioned (Section 2), the use of the gaseous form of ozone gives the end user more flexibility. We have been approached by several companies from the fresh-produce sector (e.g. Thanet Earth, Mack, Barfoots of Botley Ltd) to investigate the effect of continuous exposure to low levels of ozone (due to health and safety requirements) on the quality changes during the storage/transit of various commodities. These include red bell peppers, cucumbers and courgettes (Glowacz et al., 2015b), green and red chilli peppers (Glowacz and Rees, 2016b), grapes and strawberries (unpublished). Taking into account that the key challenges in postharvest use of ozone with fresh produce are (1) the end users' awareness of its practical limitations and recognition of all the factors affecting its efficiency when applied to fresh produce (Glowacz and Rees, 2016a) and (2) finding optimum conditions and doses for specific commodities and varieties - the key lessons learned through our trials will be described.

Trials with cucumbers and courgettes (Glowacz et al., 2015b) clearly demonstrated that continuous exposure to ozone at 0.1 μmol mol^{-1} resulted in better firmness retention, which we have suggested to be associated with the effect of ozone on stomata. In the same trials, we have learned that ozone has a strong affinity towards green pigments, as the dose of 0.3 μmol mol^{-1} was found to be injurious to the produce, and often resulted in skin colour bleaching (from dark green to light green). Similarly, exposure to ozone even at the lowest dose (0.45 μmol mol^{-1}) tested affected the colour of green chilli peppers which became brighter/lighter (higher L^* value), suggesting that the characteristic dark green colour could indeed be bleached by ozone, due to accelerated chlorophyll degradation (Glowacz and Rees, 2016b). This finding is in agreement with Alwi and Ali (2015) who reported reduced chlorophyll content in green bell peppers exposed to ozone.

Finding optimum conditions and doses for specific commodities is often quite difficult as different produce requires different ranges of ozone concentrations to be tested. The discrepancy among the published studies (sometimes quite substantial) makes it even more difficult (Horvitz and Cantalejo, 2014; Glowacz et al., 2015a). The end user, however, must be aware that the results from individual experiments reported in the literature are only valid for the specific conditions of those particular experiments (ozone dose × commodity and variety being used, maturity, ripeness, growing season/ region), and the applicability of the findings needs to be tested and validated under commercial conditions.

For example, while in the case of cucumbers and courgettes, the level of continuous ozone had to be kept low - 0.1 μmol mol^{-1} (Glowacz et al., 2015b), a dose of 0.3 μmol mol^{-1} was still suitable for red bell peppers and grapes, with doses as high as 0.9 μmol mol^{-1} and higher being found to be efficient in reducing fungal rot development on red chilli peppers (Glowacz and Rees, 2016b) and reducing counts of pathogenic microbes such as *E. coli*, *L. monocytogenes* and *S. enterica* on strawberries, respectively, with no detrimental effect on the produce. This clearly highlights that various commodities differ in their sensitivity to ozone.

What we have also found out is that when scaling up the experimental settings even distribution of ozone within the container(s)/fumigation system is crucial to ensure that horticultural produce in various parts of the container is indeed exposed to the desired concentration of ozone - informed by the small-scale trials and supported by the published literature. This can often be achieved by monitoring the concentration (and thus distribution) of ozone at several locations within a pallet of fresh produce. If this step is ignored, the end user may end up with produce being damaged in some parts of the pallet due to ozone concentration being too high (uneven distribution), which will consequently lead to higher disease incidence, thereby counteracting its beneficial role.

Finally, the efficiency of ozone exposure might be affected not only by the type of produce, dose being used and storage conditions (temperature and relative humidity) but also by the packaging material. It is generally assumed that the diffusion of gases through a material depends on its porosity. Indeed, we found that ozone dose needs to be adjusted according to the packaging material being used, for example cardboard boxes reduced the efficiency of ozone in reducing microbial counts, due to actual concentration of ozone being much lower than the dose supplied/set up on the ozone generator - thus, a higher initial concentration of ozone was required to ensure that the produce is exposed to the desired concentration. Interestingly, the interference of the packaging material with ozone was reduced when corrugated plastic boxes were used. Substantial differences in ozone diffusion into different packaging materials have been reported by others (Palou et al., 2003; Karaca and Smilanick, 2011; Feliziani et al., 2014), which fits with what we have noticed in our trials. These findings also highlight that small-scale experimental trials need to be conducted using industrial settings, that is taking the interaction between ozone and packaging material into account.

5 Future trends

The aim of postharvest science and technology is to provide underpinning information for the fresh-produce industry to ensure the continuous supply of safe and nutritious horticultural produce to the consumer. Future studies on the postharvest use of ozone will most likely include the following: (1) better understanding (with the use of molecular biology tools) of biochemical changes - related to both the nutritional content and resistance to diseases - in ozone-exposed fruits and vegetables to improve quality management within the supply chain; (2) novel packaging including low levels of ozone-releasing sachets and/or made of the material characterized by low interference with ozone; (3) development of technologies for better detection of microorganisms (bacteria and fungi, including spores) and related mycotoxins; (4) potential integration of ozone technology with other technologies, for example UV light, irradiation, hydrogen peroxide, organic acids and so forth (with potential synergistic antimicrobial effect being achieved). Finally, the work will continue on finding the optimum conditions and doses for specific commodities and varieties, especially for those fruits and vegetables where these have not been identified yet.

As briefly mentioned in this chapter, there is already an increased interest in the use of ozone during the wine-making process, while the sanitation of fruits and vegetables would make it an option for the juice industry, especially taking into account an increase in popularity of smoothies. Undoubtedly, due to the size of the beverage and wine industry, and some of the benefits already reported, the potential use of ozone by that sector will be researched further.

6 Conclusion

Disinfection of horticultural produce is an essential step within the supply chain to ensure the food-safety aspect of fresh produce. The use of ozone seems to be among the simple and feasible solutions for the fresh-produce industry. It is well documented in the literature, and reviewed in this chapter, that postharvest treatment of fresh produce with ozone either in its aqueous or gaseous form can reduce microbial contamination of the produce. The fact that each year more and more information regarding the use of ozone with fruits and vegetables is becoming available is of huge importance for the fresh-produce sector, as it provides the necessary information about the 'right doses' to be used for specific commodities, and also improves the end users' awareness of its limitations - scientific papers reporting either the thresholds for ozone doses or reporting little or no benefits to the produce.

It is important to take care when adapting ozone technology, and it is a role of research to establish the optimum conditions and doses for specific commodities and varieties, and subsequently inform the industry/end user about the most up-to-date findings. As a scientific community, we have to make sure that the industry is fully aware that some of the concentrations of ozone that have been reported in the literature as being sufficient in reducing the microbial contamination on the produce, that is 'positive results' are often much higher than ozone concentrations used by others for the produce quality preservation. The doses used by the fresh-produce industry should be informed by the relevant research because if the dose of ozone is too high rather than being beneficial it will cause damage to the produce, for example by oxidizing the cuticle and leading to development of cracks in the skin, in this way increasing the respiration rates of the produce, accelerating weight loss and reducing firmness. High doses of ozone are also likely to result in colour bleaching, thus making the produce unattractive to the consumer.

7 Where to look for further information

A review paper by Guzel-Seydim et al. (2004) titled 'Use of ozone in the food industry' is a good introduction to the subject. A couple of comprehensive reviews on the postharvest use of ozone have been published more recently (Horvitz and Cantalejo, 2014; Glowacz et al., 2015a).

For those interested in ozone, there is an International Ozone Association - Pan American Group and European-African-Asian-Australian Group. It is a not-for-profit educational association involved in information sharing through international symposia, seminars and publications in *Ozone: Science & Engineering* journal. EurO_3zon is another relevant international non-profit association dedicated to promote the use of ozone in Europe. In addition, there is an International Society for Horticultural Science which promotes the research and facilitates knowledge transfer through its symposia and congresses, with 'Postharvest Unlimited Conference' and 'International Postharvest Symposium' series covering most up-to-date findings in postharvest science and technology field, often including sanitation of horticultural produce.

Key journals in the field are Elsevier's *Postharvest Biology and Technology*, *Journal of Food Engineering*, *Food Chemistry* and *Food Control*, and Wiley's *Journal of Food Science* and *Journal of the Science of Food and Agriculture*. There are numerous research centres (see 'References' list) currently working in the field of postharvest use of ozone with horticultural produce.

8 References

Achen, M. A. and Yousef, A. E. 2001. Efficacy of ozone against *Escherichia coli* O157:H7 on apples. *J. Food Sci.* 66(9), 1380–4. doi:10.1111/j.1365-2621.2001.tb15218.x.

Agrawal, M., Krizek, D. T., Agrawal, S. B., Kramer, G. F., Lee, E. H., Mirecki, R. M. and Rowland, R. A. 1993. Influence of inverse day/night temperature on ozone sensitivity and selected morphological and physiological responses of cucumber. *J. Am. Soc. Hortic. Sci.* 118(5), 649–54. doi:10.21273/JASHS.118.5.649.

Aguayo, E., Escalona, V. H. and Artes, F. 2006. Effect of cyclic exposure to ozone gas on physicochemical, sensorial and microbial quality of whole and sliced tomatoes. *Postharvest Biol. Technol.* 39(2), 169-77. doi:10.1016/j.postharvbio.2005.11.005.

Akbas, M. Y. and Olmez, H. 2007. Effectiveness of organic acid, ozonated water and chlorine dippings on microbial reduction and storage quality of fresh-cut iceberg lettuce. *J. Sci. Food. Agric.* 87(14), 2609-16. doi:10.1002/jsfa.3016.

Alexandre, E. M. C., Santos-Pedro, D. M., Brandao, T. R. S. and Silva, C. L. M. 2011. Influence of aqueous ozone, blanching and combined treatments on microbial load of red bell peppers, strawberries and watercress. *J. Food Eng.* 105(2), 277-82. doi:10.1016/j.jfoodeng.2011.02.032.

Alexandre, E. M. C., Brandao, T. R. S. and Silva, C. L. M. 2012. Efficacy of non-thermal technologies and sanitizer solutions on microbial load reduction and quality retention of strawberries. *J. Food Eng.* 108(3), 417-26. doi:10.1016/j.jfoodeng.2011.09.002.

Alexopoulos, A., Plessas, S., Ceciu, S., Lazar, V., Mantzourani, I., Voidarou, C., Stavropoulou, E. and Bezirtzoglou, E. 2013. Evaluation of ozone efficacy on the reduction of microbial population of fresh cut lettuce (*Lactuca sativa*) and green bell pepper (*Capsicum annuum*). *Food Control* 30(2), 491-6. doi:10.1016/j.foodcont.2012.09.018.

Ali, A., Ong, M. K. and Forney, C. F. 2014. Effect of ozone pre-conditioning on quality and antioxidant capacity of papaya fruit during ambient storage. *Food Chem.* 142, 19-26. doi:10.1016/j.foodchem.2013.07.039.

Allende, A., Marin, A., Buendia, B., Tomás-Barberán, F. and Gil, M. I. 2007. Impact of combined postharvest treatments (UV-C light, gaseous O_3, superatmospheric O_2 and high CO_2) on health promoting compounds and shelf-life of strawberries. *Postharvest Biol. Technol.* 46(3), 201-11. doi:10.1016/j.postharvbio.2007.05.007.

Allende, A., Selma, M. V., Lopez-Galvez, F., Villaescusa, R. and Gil, M. I. 2008. Role of commercial sanitizers and washing systems on epiphytic microorganisms and sensory quality of fresh-cut escarole and lettuce. *Postharvest Biol. Technol.* 49(1), 155-63. doi:10.1016/j.postharvbio.2007.12.010.

Alwi, N. A. and Ali, A. 2014. Reduction of *Escherichia coli* O157, Listeria monocytogenes and *Salmonella enterica* sv. Typhimurium populations on fresh-cut bell pepper using gaseous ozone. *Food Control* 46, 304-11. doi:10.1016/j.foodcont.2014.05.037.

Alwi, N. A. and Ali, A. 2015. Dose-dependent effect of ozone fumigation on physiological characteristics, ascorbic acid content and disease development on bell pepper (*Capsicum annuum* L.) during storage. *Food Bioprocess. Technol.* 8(3), 558-66. doi:10.1007/s11947-014-1419-2.

Ames, Z. R., Feliziani, E. and Smilanick, J. L. 2013. Germination of fungal conidia after exposure to low concentration ozone atmospheres. *Postharvest Biol. Technol.* 83, 22-6. doi:10.1016/j.postharvbio.2013.03.008.

Antony-Babu, S. and Singleton, I. 2009. Effect of ozone on spore germination, spore production and biomass production in two *Aspergillus* species. *Antonie van Leeuwenhoek* 96(4), 413-22. doi:10.1007/s10482-009-9355-2.

Artes, F., Gomez, P., Aguayo, E., Escalona, V. and Artes-Hernandez, F. 2009. Sustainable sanitation techniques for keeping quality and safety of fresh-cut plant commodities. *Postharvest Biol. Technol.* 51(3), 287-96. doi:10.1016/j.postharvbio.2008.10.003.

Artes-Hernandez, F., Aguayo, E. and Artes, F. 2004. Alternative atmosphere treatments for keeping quality of 'Autumn seedless' table grapes during long-term cold storage. *Postharvest Biol. Technol.* 31(1), 59-67. doi:10.1016/S0925-5214(03)00116-9.

Bellincontro, A., Catelli, C., Cotarella, R. and Mencarelli, F. 2017. Postharvest ozone fumigation of Petit Verdot grapes to prevent the use of sulfites and to increase anthocyanin in wine. *Aust. J. Grape Wine Res.* 23(2), 200–6. doi:10.1111/ajgw.12257.

Beltran, D., Selma, M. V., Marin, A. and Gil, M. I. 2005. Ozonated water extends the shelf-life of fresh-cut lettuce. *J. Agric. Food Chem.* 53(14), 5654–63. doi:10.1021/jf050359c.

Bermúdez-Aguirre, D. and Barbosa-Cánovas, G. V. 2013. Disinfection of selected vegetables under non-thermal treatments: chlorine, acid citric, ultraviolet light and ozone. *Food Control* 29(1), 82–90. doi:10.1016/j.foodcont.2012.05.073.

Bialka, K. L. and Demirci, A. 2007a. Decontamination of *Escherichia coli* O157:H7 and *Salmonella enterica* on blueberries using ozone and pulsed UV-light. *J. Food Sci.* 72(9), M391–6. doi:10.1111/j.1750-3841.2007.00517.x.

Bialka, K. L. and Demirci, A. 2007b. Utilization of gaseous ozone for the decontamination of *Escherichia coli* O157: H7 and *Salmonella* on raspberries and strawberries. *J. Food Prot.* 70(5), 1093–8. doi:10.4315/0362-028X-70.5.1093.

Bialka, K. L., Demirci, A. and Puri, V. M. 2008. Modeling the inactivation of *Escherichia coli* O157: H7 and *Salmonella enterica* on raspberries and strawberries resulting from exposure to ozone or pulsed UV-light. *J. Food Eng.* 85(3), 444–9. doi:10.1016/j.jfoodeng.2007.08.007.

Boonkorn, P., Gemma, H., Sugaya, S., Setha, S., Uthaibutra, J. and Whangchai, K. 2012. Impact of high-dose, short periods of ozone exposure on green mold and antioxidant enzyme activity of tangerine fruit. *Postharvest Biol. Technol.* 67, 25–8. doi:10.1016/j.postharvbio.2011.12.012.

Botondi, R., De Sanctis, F., Moscatelli, N., Vettraino, A. M., Catelli, C. and Mencarelli, F. 2015. Ozone fumigation for safety and quality of wine grapes in postharvest dehydration. *Food Chem.* 188, 641–7. doi:10.1016/j.foodchem.2015.05.029.

Bridges, D. F., Rane, B. and Wu, V. C. H. 2018. The effectiveness of closed-circulation gaseous chlorine dioxide or ozone treatment against bacterial pathogens on produce. *Food Control* 91, 261–7. doi:10.1016/j.foodcont.2018.04.004.

Calatayud, A., Iglesias, D. J., Talon, M. and Barreno, E. 2004. Response of spinach leaves (*Spinacia oleracea* L.) to ozone measured by gas exchange, chlorophyll a fluorescence, antioxidant systems, and lipid peroxidation. *Photosynthetica* 42(1), 23–9. doi:10.1023/B:PHOT.0000040565.53844.c6.

Cayuela, J. A., Vázquez, A., Pérez, A. G. and García, J. M. 2009. Control of table grapes postharvest decay by ozone treatment and resveratrol induction. *Food Sci. Technol. Int.* 15(5), 495–502. doi:10.1177/1082013209350539.

Chauhan, O. P., Raju, P. S., Ravi, N., Singh, A. and Bawa, A. S. 2011. Effectiveness of ozone in combination with controlled atmosphere on quality characteristics including lignification of carrot sticks. *J. Food Eng.* 102(1), 43–8. doi:10.1016/j.jfoodeng.2010.07.033.

Concha-Meyer, A., Eifert, J., Williams, R., Marcy, J. and Welbaum, G. 2014. Survival of *Listeria monocytogenes* on fresh blueberries (*Vaccinium corymbosum*) stored under controlled atmosphere and ozone. *J. Food Prot.* 77(5), 832–6. doi:10.4315/0362-028X.JFP-13-441.

Cravero, F., Englezos, V., Rantsiou, K., Torchio, F., Giacosa, S., Rio Segade, S. R., Gerbi, V., Rolle, L. and Cocolin, L. 2016. Ozone treatments of post harvested wine grapes: impact on fermentative yeasts and wine chemical properties. *Food Res. Int.* 87, 134–41. doi:10.1016/j.foodres.2016.06.031.

Published by Burleigh Dodds Science Publishing Limited, 2020.

Cravero, F., Englezos, V., Rantsiou, K., Torchio, F., Giacosa, S., Rio Segade, S., Gerbi, V., Rolle, L. and Cocolin, L. 2018. Control of *Brettanomyces bruxellensis* on wine grapes by post-harvest treatments with electrolyzed water, ozonated water and gaseous ozone. *Innov. Food Sci. Emerg. Technol.* 47, 309–16. doi:10.1016/j.ifset.2018.03.017.

Das, E., Gurakan, G. C. and Bayindirli, A. 2006. Effect of controlled atmosphere storage, modified atmosphere packaging and gaseous ozone treatment on the survival of *Salmonella enteritidis* on cherry tomatoes. *Food Microbiol.* 23(5), 430–8. doi:10.1016/j.fm.2005.08.002.

de Souza, L. P., Faroni, L. R. D., Heleno, F. F., Cecon, P. R., Goncalves, T. D. C., da Silva, G. J. and Prates, L. H. F. 2018. Effects of ozone treatment on postharvest carrot quality. *LWT - Food Sci. Technol.* 90, 53–60.

Fan, X., Sokorai, K. J., Engemann, J., Gurtler, J. B. and Liu, Y. 2012. Inactivation of *Listeria innocua*, *Salmonella* Typhimurium, and *Escherichia coli* O157: H7 on surface and stem scar areas of tomatoes using in-package ozonation. *J. Food Prot.* 75(9), 1611–8. doi:10.4315/0362-028X.JFP-12-103.

Feliziani, E., Romanazzi, G. and Smilanick, J. L. 2014. Application of low concentrations of ozone during the cold storage of table grapes. *Postharvest Biol. Technol.* 93, 38–48. doi:10.1016/j.postharvbio.2014.02.006.

Forney, C. F., Song, J., Fan, L., Hildebrand, P. D. and Jordan, M. A. 2003. Ozone and 1-methylcyclopropene alter the postharvest quality of broccoli. *J. Am. Soc. Hort. Sci.* 128(3), 403–8. doi:10.21273/JASHS.128.3.0403.

Forney, C. F., Song, J., Hildebrand, P. D., Fan, L. and McRae, K. B. 2007. Interactive effects of ozone and 1-methylcyclopropene on decay resistance and quality of stored carrots. *Postharvest Biol. Technol.* 45(3), 341–8. doi:10.1016/j.postharvbio.2007.03.006.

Garcia-Martin, J. F., Olmo, M. and Garcia, J. M. 2018. Effect of ozone treatment on postharvest disease and quality of different citrus varieties at laboratory and at industrial facility. *Postharvest Biol. Technol.* 137, 77–85. doi:10.1016/j.postharvbio.2017.11.015.

Glowacz, M. and Rees, D. 2016a. The practicality of using ozone with fruit and vegetables. *J. Sci. Food Agric.* 96(14), 4637–43. doi:10.1002/jsfa.7763.

Glowacz, M. and Rees, D. 2016b. Exposure to ozone reduces postharvest quality loss in red and green chilli peppers. *Food Chem.* 210, 305–10. doi:10.1016/j.foodchem.2016.04.119.

Glowacz, M., Colgan, R. and Rees, D. 2015a. The use of ozone to extend the shelf-life and improve quality of fresh produce. *J. Sci. Food Agric.* 95(4), 662–71. doi:10.1002/jsfa.6776.

Glowacz, M., Colgan, R. and Rees, D. 2015b. Influence of continuous exposure to gaseous ozone on the quality of red bell peppers, cucumbers and zucchini. *Postharvest Biol. Technol.* 99, 1–8. doi:10.1016/j.postharvbio.2014.06.015.

Gutierrez, D. R., Chaves, A. R. and Rodriguez, S. D. C. 2017. Use of UV-C and gaseous ozone as sanitizing agents for keeping the quality of fresh-cut rocket (*Eruca sativa* Mill). *J. Food Process. Preserv.* 41(3), e12968. doi:10.1111/jfpp.12968.

Guzel-Seydim, Z. B., Greene, A. K. and Seydim, A. C. 2004. Use of ozone in the food industry. *LWT - Food Sci. Technol.* 37(4), 453–60. doi:10.1016/j.lwt.2003.10.014.

Han, Y., Floros, J. D., Linton, R. H., Nielsen, S. S. and Nelson, P. E. 2002. Response surface modeling for the inactivation of *Escherichia coli* O157:H7 on green peppers (*Capsicum annuum*) by ozone gas treatment. *J. Food Sci.* 67(3), 1188–93. doi:10.1111/j.1365-2621.2002.tb09475.x.

Han, Q., Gao, H., Chen, H., Fang, X. and Wu, W. 2017. Precooling and ozone treatments affects postharvest quality of black mulberry (*Morus nigra*) fruits. *Food Chem.* 221, 1947–53. doi:10.1016/j.foodchem.2016.11.152.

Heinrichs, D. E., Yethon, J. A. and Whitfield, C. 1998. Molecular basis for structural diversity in the core regions of the lipopolysaccharides of *Escherichia coli* and *Salmonella enterica*. *Mol. Microbiol.* 30(2), 221–32. doi:10.1046/j.1365-2958.1998.01063.x.

Hernandez, I., Alegre, L., Van Breusegem, F. and Munne-Bosch, S. 2009. How relevant are flavonoids as antioxidants in plants? *Trends Plant Sci.* 14(3), 125–32. doi:10.1016/j.tplants.2008.12.003.

Hildebrand, P. D., Forney, C. F., Song, J., Fan, L. and McRae, K. B. 2008. Effect of a continuous low ozone exposure (50 nL L^{-1}) on decay and quality of stored carrots. *Postharvest Biol. Technol.* 49(3), 397–402. doi:10.1016/j.postharvbio.2008.03.012.

Horvitz, S. and Cantalejo, M. J. 2012. Effects of ozone and chlorine postharvest treatments on quality of fresh-cut red bell peppers. *Int. J. Food Sci. Technol.* 47(9), 1935–43. doi:10.1111/j.1365-2621.2012.03053.x.

Horvitz, S. and Cantalejo, M. J. 2014. Application of ozone for the postharvest treatment of fruits and vegetables. *Crit. Rev. Food Sci. Nutr.* 54(3), 312–39. doi:10.1080/10408398.2011.584353.

Hua-Li, X., Yang, B., Raza, H., Hu-Jun, W., Lu-Mei, P., Mi-Na, N., Xiao-Yan, C., Yi, W. and Yong-Cai, L. 2018. Detection of NEO in muskmelon fruits inoculated with *Fusarium sulphureum* and its control by postharvest ozone treatment. *Food Chem.* 254, 193-200. doi:10.1016/j.foodchem.2018.01.149.

Kangasjarvi, J., Jaspers, P. and Kollist, H. 2005. Signalling and cell death in ozone-exposed plants. *Plant Cell Environ.* 28(8), 1021-36. doi:10.1111/j.1365-3040.2005.01325.x.

Karaca, H. and Smilanick, J. L. 2011. The influence of plastic composition and ventilation area on ozone diffusion through some food packaging materials. *Postharvest Biol. Technol.* 62(1), 85–8. doi:10.1016/j.postharvbio.2011.04.004.

Karaca, H. and Velioglu, Y. S. 2014. Effects of ozone treatments on microbial quality and some chemical properties of lettuce, spinach, and parsley. *Postharvest Biol. Technol.* 88, 46–53. doi:10.1016/j.postharvbio.2013.09.003.

Ketteringham, L., Gausseres, R., James, S. J. and James, C. 2006. Application of aqueous ozone for treating pre-cut green peppers (*Capsicum annuum* L.). *J. Food Eng.* 76(1), 104–11. doi:10.1016/j.jfoodeng.2005.05.019.

Khadre, M. A. and Yousef, A. E. 2001. Sporicidal action of ozone and hydrogen peroxide: a comparative study. *Int. J. Food Microbiol.* 71(2–3), 131–8. doi:10.1016/S0168-1605(01)00561-X.

Khadre, M. A., Yousef, A. E. and Kim, J.-G. 2001. Microbiological aspects of ozone applications in food: a review. *J. Food Sci.* 66(9), 1242-52. doi:10.1111/j.1365-2621.2001.tb15196.x.

Kim, C. and Hung, Y. C. 2012. Inactivation of *E. coli* O157:H7 on blueberries by electrolyzed water, ultraviolet light, and ozone. *J. Food Sci.* 77(4), M206–11. doi:10.1111/j.1750-3841.2011.02595.x.

Kim, J.-G. and Yousef, A. E. 2000. Inactivation kinetics of foodborne spoilage and pathogenic bacteria by ozone. *J. Food Sci.* 65(3), 521-8. doi:10.1111/j.1365-2621.2000.tb16040.x.

Kissinger, M., Tuvia-Alkalai, S., Shalom, Y., Fallik, E., Elkind, Y., Jenks, M. A. and Goodwin, M. S. 2005. Characterization of physiological and biochemical factors associated with postharvest water loss in ripe pepper fruit during storage. *J. Am. Soc. Hortic. Sci.* 130(5), 735–41. doi:10.21273/JASHS.130.5.735.

Klockow, P. A. and Keener, K. M. 2009. Safety and quality assessment of packaged spinach treated with a novel ozone-generation system. *LWT - Food Sci. Technol.* 42(6), 1047–53. doi:10.1016/j.lwt.2009.02.011.

Lara, I., Belge, B. and Goulao, L. F. 2014. The fruit cuticle as a modulator of postharvest quality. *Postharvest Biol. Technol.* 87, 103–12. doi:10.1016/j.postharvbio.2013.08.012.

Li, H., Tajkarimi, M. and Osburn, B. I. 2008. Impact of vacuum cooling on *Escherichia coli* O157:H7 infiltration into lettuce tissue. *Appl. Environ. Microbiol.* 74(10), 3138–42. doi:10.1128/AEM.02811-07.

Liew, C. L. and Prange, R. K. 1994. Effect of ozone and storage temperature on postharvest diseases and physiology of carrots (*Daucus carota* L.). *J. Am. Soc. Hort. Sci.* 119(3), 563–7. doi:10.21273/JASHS.119.3.563.

Martins, G., Vallance, J., Mercier, A., Albertin, W., Stamatopoulos, P., Rey, P., Lonvaud, A. and Masneuf-Pomarède, I. 2014. Influence of the farming system on the epiphytic yeasts and yeast-like fungi colonizing grape berries during the ripening process. *Int. J. Food Microbiol.* 177, 21–8. doi:10.1016/j.ijfoodmicro.2014.02.002.

Minas, I. S., Karaoglanidis, G. S., Manganaris, G. A. and Vasilakakis, M. 2010. Effect of ozone application during cold storage of kiwifruit on the development of stem-end rot caused by *Botrytis cinerea*. *Postharvest Biol. Technol.* 58(3), 203–10. doi:10.1016/j.postharvbio.2010.07.002.

Minas, I. S., Tanou, G., Belghazi, M., Job, D., Manganaris, G. A., Molassiotis, A. and Vasilakakis, M. 2012. Physiological and proteomic approaches to address the active role of ozone in kiwifruit postharvest ripening. *J. Exp. Bot.* 63(7), 2449–64. doi:10.1093/jxb/err418.

Minas, I. S., Vicente, A. R., Dhanapal, A. P., Manganaris, G. A., Goulas, V., Vasilakakis, M., Crisosto, C. H. and Molassiotis, A. 2014. Ozone- induced kiwifruit ripening delay is mediated by ethylene biosynthesis inhibition and cell wall dismantling regulation. *Plant Sci.* 229, 76–85. doi:10.1016/j.plantsci.2014.08.016.

Mittler, R. 2002. Oxidative stress, antioxidants and stress tolerance. *Trends Plant Sci.* 7(9), 405–10. doi:10.1016/S1360-1385(02)02312-9.

Mittler, R. 2017. ROS are good. *Trends Plant Sci.* 22(1), 11–9. doi:10.1016/j.tplants.2016.08.002.

Mlikota Gabler, F. M., Smilanick, J. L., Mansour, M. F. and Karaca, H. 2010. Influence of fumigation with high concentrations of ozone gas on postharvest gray mold and fungicide residues on table grapes. *Postharvest Biol. Technol.* 55(2), 85–90. doi:10.1016/j.postharvbio.2009.09.004.

Morata, A., Loira, I., Vejarano, R., Gonzalez, C., Callejo, M. J. and Suarez-Lepe, J. A. 2017. Emerging preservation technologies in grapes for winemaking. *Trends Food Sci. Technol.* 67, 36–43. doi:10.1016/j.tifs.2017.06.014.

Murray, K., Wu, F., Shi, J., Xue, S. J. and Warriner, K. 2017. Challenges in the microbiological food safety of fresh produce: limitations of post-harvest washing and the need for alternative interventions. *Food Qual. Saf.* 1(4), 289–301. doi:10.1093/fqsafe/fyx027.

Olmez, H. and Akbas, M. Y. 2009. Optimization of ozone treatment of fresh-cut green leaf lettuce. *J. Food Eng.* 90(4), 487–94. doi:10.1016/j.jfoodeng.2008.07.026.

Ong, M. K. and Ali, A. 2015. Antifungal action of ozone against *Colletotrichum gloeosporioides* and control of papaya anthracnose. *Postharvest Biol. Technol.* 100, 113-9. doi:10.1016/j.postharvbio.2014.09.023.

Ong, M. K., Kazi, F. K., Forney, C. F. and Ali, A. 2013. Effect of gaseous ozone on papaya anthracnose. *Food Bioprocess. Technol.* 6(11), 2996-3005. doi:10.1007/s11947-012-1013-4.

Ozkan, R., Smilanick, J. L. and Karabulut, O. A. 2011. Toxicity of ozone gas to conidia of *Penicillium digitatum*, *Penicillium italicum*, and *Botrytis cinerea* and control of gray mold on table grapes. *Postharvest Biol. Technol.* 60(1), 47-51. doi:10.1016/j.postharvbio.2010.12.004.

Palou, L., Smilanick, J. L., Crisosto, C. H. and Mansour, M. 2001. Effect of gaseous ozone exposure on the development of green and blue molds on cold stored citrus fruit. *Plant Dis.* 85(6), 632-8. doi:10.1094/PDIS.2001.85.6.632.

Palou, L., Crisosto, C. H., Smilanick, J. L., Adaskaveg, J. E. and Zoffoli, J. P. 2002. Effects of continuous 0.3 ppm ozone exposure on decay development and physiological responses of peaches and table grapes in cold storage. *Postharvest Biol. Technol.* 24(1), 39-48. doi:10.1016/S0925-5214(01)00118-1.

Palou, L., Smilanick, J. L., Crisosto, C. H., Mansour, M. and Plaza, P. 2003. Ozone gas penetration and control of the sporulation of *Penicillium digitatum* and *Penicillium italicum* within commercial packages of oranges during cold storage. *Crop Protect.* 22(9), 1131-4. doi:10.1016/S0261-2194(03)00145-5.

Papachristodoulou, M., Koukounaras, A., Siomos, A. S., Liakou, A. and Gerasopoulos, D. 2018. The effects of ozonated water on the microbial counts and the shelf life attributes of fresh-cut spinach. *J. Food Process. Preserv.* 42(1), e13404. doi:10.1111/jfpp.13404.

Parsons, E. P., Popopvsky, S., Lohrey, G. T., Alkalai-Tuvia, S., Perzelan, Y., Bosland, P., Bebeli, P. J., Paran, I., Fallik, E. and Jenks, M. A. 2013. Fruit cuticle lipid composition and water loss in a diverse collection of pepper (*Capsicum*). *Physiol. Plant.* 149(2), 160-74. doi:10.1111/ppl.12035.

Predmore, A., Sanglay, G., Li, J. R. and Lee, K. 2015. Control of human norovirus surrogates in fresh foods by gaseous ozone and a proposed mechanism of inactivation. *Food Microbiol.* 50, 118-25. doi:10.1016/j.fm.2015.04.004.

Pyatkovskyy, T., Shynkaryk, M., Yousef, A. and Sastry, S. K. 2017. Fresh produce sanitization by combination of gaseous ozone and liquid sanitizer. *J. Food Eng.* 210, 19-26. doi:10.1016/j.jfoodeng.2017.03.031.

Rahman, S. M. E., Ding, T. and Oh, D. H. 2010. Inactivation effect of newly developed low concentration electrolyzed water and other sanitizers against microorganisms on spinach. *Food Control* 21(10), 1383-7. doi:10.1016/j.foodcont.2010.03.011.

Rodgers, S. L., Cash, J. N., Siddiq, M. and Ryser, E. T. 2004. A comparison of different chemical sanitizers for inactivating *Escherichia coli* O157:H7 and *Listeria monocytogenes* in solution and on apples, lettuce, strawberries, and cantaloupe. *J. Food Prot.* 67(4), 721-31. doi:10.4315/0362-028X-67.4.721.

Rodoni, L., Casadei, N., Concellon, A., Alicia, A. R. C. and Vicente, A. R. 2010. Effect of short-term ozone treatments on tomato (*Solanum lycopersicum* L.) fruit quality and cell wall degradation. *J. Agric. Food Chem.* 58(1), 594-9. doi:10.1021/jf9029145.

Romanazzi, G., Lichter, A., Gabler, F. M. and Smilanick, J. L. 2012. Recent advances on the use of natural and safe alternatives to conventional methods to control postharvest gray mold of table grapes. *Postharvest Biol. Technol.* 63(1), 141-7. doi:10.1016/j.postharvbio.2011.06.013.

Salvador, A., Abad, I., Arnal, L. and Martinez-Javega, J. M. 2006. Effect of ozone on postharvest quality of persimmon. *J. Food Sci.* 71(6), S443-6. doi:10.1111/j.1750-3841.2006.00059.x.

Sarig, P., Zahavi, T., Zutkhi, Y., Yannai, S., Lisker, N. and Ben-Arie, R. 1996. Ozone for control of postharvest decay of table grapes caused by *Rhizopus stolonifer*. *Physiol. Mol. Plant* 48, 403-15.

Selma, M. V., Beltran, D., Allende, A., Chacon-Vera, E. and Gil, M. I. 2007. Elimination by ozone of *Shigella sonnei* in shredded lettuce and water. *Food Microbiol.* 24(5), 492-9. doi:10.1016/j.fm.2006.09.005.

Selma, M. V., Ibanez, A. M., Cantwell, M. and Suslow, T. 2008. Reduction by gaseous ozone of *Salmonella* and microbial flora associated with fresh-cut cantaloupe. *Food Microbiol.* 25(4), 558-65. doi:10.1016/j.fm.2008.02.006.

Sharpe, D., Fan, L., McRae, K., Walker, B., MacKay, R. and Doucette, C. 2009. Effects of ozone treatment on *Botrytis cinerea* and *Sclerotinia sclerotiorum* in relation to horticultural product quality. *J. Food Sci.* 74(6), M250-7. doi:10.1111/j.1750-3841.2009.01234.x.

Sheng, L., Hanrahan, I., Sun, X., Taylor, M. H., Mendoza, M. and Zhu, M. J. 2018. Survival of *Listeria innocua* on Fuji apples under commercial cold storage with or without low dose continuous ozone gaseous. *Food Microbiol.* 76, 21-8. doi:10.1016/j.fm.2018.04.006.

Shynkaryk, M. V., Pyatkovskyy, T. I., Yousef, A. E. and Sastry, S. K. 2016. Gaseous ozone treatment of baby spinach within the existing production chain for inactivation of *Escherichia coli* O157:H7. *J. Food Eng.* 191, 10-8. doi:10.1016/j.jfoodeng.2016.06.025.

Singh, N., Singh, R. K., Bhunia, A. K. and Stroshine, R. L. 2002. Efficacy of chlorine dioxide, ozone, and thyme essential oil or a sequential washing in killing *Escherichia coli* O157:H7 on lettuce and baby carrots. *LWT - Food Sci. Technol.* 35(8), 720-9. doi:10.1006/fstl.2002.0933.

Skog, L. J. and Chu, C. L. 2001. Effect of ozone on qualities of fruits and vegetables in cold storage. *Can. J. Plant Sci.* 81(4), 773-8. doi:10.4141/P00-110.

Tabakoglu, N. and Karaca, H. 2018. Effects of ozone-enriched storage atmosphere on postharvest quality of black mulberry fruits (*Morus nigra* L.). *LWT - Food Sci. Technol.* 92, 276-81. doi:10.1016/j.lwt.2018.02.044.

Toti, M., Carboni, C. and Botondi, R. 2018. Postharvest gaseous ozone treatment enhances quality parameters and delays softening in cantaloupe melon during storage at 6°C. *J. Sci. Food Agric.* 98(2), 487-94. doi:10.1002/jsfa.8485.

Tzortzakis, N., Borland, A., Singleton, I. and Barnes, J. 2007a. Impact of atmospheric ozone-enrichment on quality-related attributes of tomato fruit. *Postharvest Biol. Technol.* 45(3), 317-25. doi:10.1016/j.postharvbio.2007.03.004.

Tzortzakis, N., Singleton, I. and Barnes, J. 2007b. Deployment of low-level ozone-enrichment for the preservation of chilled fresh produce. *Postharvest Biol. Technol.* 43(2), 261-70. doi:10.1016/j.postharvbio.2006.09.005.

Tzortzakis, N., Singleton, I. and Barnes, J. 2008. Impact of low-level atmospheric ozone-enrichment on black spot and anthracnose rot of tomato fruit. *Postharvest Biol. Technol.* 47(1), 1-9. doi:10.1016/j.postharvbio.2007.06.004.

Tzortzakis, N., Taybi, T., Roberts, R., Singleton, I., Borland, A. and Barnes, J. 2011. Low-level atmospheric ozone exposure induces protection against *Botrytis cinerea* with down-regulation of ethylene-, jasmonate and pathogenesis-related genes in tomato fruit. *Postharvest Biol. Technol.* 61(2–3), 152–9. doi:10.1016/j.postharvbio.2011.02.013.

US FDA. 2001. Secondary direct food additives permitted in food for human consumption. *Federal Register* 66(123), 33829–30.

Vahisalu, T., Puzorjova, I., Brosche, M., Valk, E., Lepiku, M., Moldau, H., Pechter, P., Wang, Y. S., Lindgren, O., Salojarvi, J., et al. 2010. Ozone-triggered rapid stomatal response involves the production of reactive oxygen species, and is controlled by SLAC1 and OST1. *Plant J.* 62(3), 442–53. doi:10.1111/j.1365-313X.2010.04159.x.

Vurma, M., Pandit, R. B., Sastry, S. K. and Yousef, A. E. 2009. Inactivation of *Escherichia coli* O157:H7 and natural microbiota on spinach leaves using gaseous ozone during vacuum cooling and simulated transportation. *J. Food Prot.* 72(7), 1538–46. doi:10.4315/0362-028X-72.7.1538.

Wani, S., Barnes, J. and Singleton, I. 2016. Investigation of potential reasons for bacterial survival on 'ready-to-eat' leafy produce during exposure to gaseous ozone. *Postharvest Biol. Technol.* 111, 185–90. doi:10.1016/j.postharvbio.2015.08.007.

Yeni, F., Yavas, S., Alpas, H. and Soyer, Y. 2016. Most common foodborne pathogens and mycotoxins on fresh produce: a review of recent outbreaks. *Crit. Rev. Food Sci. Nutr.* 56(9), 1532–44. doi:10.1080/10408398.2013.777021.

Yeoh, W. K., Ali, A. and Forney, C. F. 2014. Effects of ozone on major antioxidants and microbial populations of fresh-cut papaya. *Postharvest Biol. Technol.* 89, 56–8. doi:10.1016/j.postharvbio.2013.11.006.

Yesil, M., Kasler, D. R., Huang, E. and Yousef, A. E. 2017. Efficacy of gaseous ozone application during vacuum cooling against *Escherichia coli* O157:H7 on spinach leaves as influenced by bacterium population size. *J. Food Prot.* 80(7), 1066–71. doi:10.4315/0362-028X.JFP-16-284.

Young, S. B. and Setlow, P. 2004. Mechanisms of *Bacillus subtilis* spore resistance to and killing by aqueous ozone. *J. Appl. Microbiol.* 96(5), 1133–42. doi:10.1111/j.1365-2672.2004.02236.x.

Zambre, S. S., Venkatesh, K. V. and Shah, N. G. 2010. Tomato redness for assessing ozone treatment to extend the shelf-life. *J. Food Eng.* 96(3), 463–8. doi:10.1016/j.jfoodeng.2009.08.027.

Zhang, L., Lu, Z., Yu, Z. and Gao, X. 2005. Preservation of fresh-cut celery by treatment of ozonated water. *Food Control* 16(3), 279–83. doi:10.1016/j.foodcont.2004.03.007.

Chapter 12

Advances in the use of biological control agents in the disinfection of horticultural produce

Samir Droby, Agricultural Research Organization – Volcani Center (ARO), Israel; Michael Wisniewski, USDA-ARS, USA; and Davide Spadaro, University of Torino, Italy

1 Introduction

Food security and safety is at the center of the global debate regarding the development of strategies to ensure a safe and reliable food supply. While increasing food production is and should be a principal focus, there are concerns if this in itself will be enough to meet the need for greater food supplies. Limits to the ability of breeders to continue to increase crop productivity, loss of arable land, increased climate fluctuations, and water shortages, and several other factors, represent significant hurdles to our ability to produce more food. Wilson (2013) has emphasized that the other half of producing food is the ability to safely preserve it after harvest and prevent significant postharvest losses. In this regard, a study by the UN Food and Agricultural Organization (FAO) presented data showing that one-third of the food produced worldwide for human consumption is lost after harvest (Gastavsson et al., 2011).

http://dx.doi.org/10.19103/AS.2019.0055.13

The magnitude of food losses after harvest stated in that report is estimated to be roughly 40–50% of root crops, fruits, and vegetables. These losses can be attributed to a decrease in quality and safety, rendering the food unsuitable for human consumption and or being discarded due to spoilage and the inability to adequately preserve it under appropriate storage conditions.

Postharvest decay is a major component of food losses and can be traced to infections that occur either in the field/orchard from flowering to harvest, or during harvest and subsequent postharvest handling along the supply chain. Infected produce represents a potential health risk since several fungal genera, such as *Penicillium*, *Aspergillus, Alternaria,* and *Fusarium*, produce mycotoxins. For example, *Penicillium expansum*, the causal pathogen of postharvest blue mold in a variety of fruits, produces a number of toxic secondary metabolites including patulin, citrinin, and chaetoglobosins (Andersen et al., 2004). Various *Aspergillus, Alternaria,* and *Fusarium* species are known to produce ochratoxin, alternariol, aflatoxins, and fumonisin (Ismaiel and Papenbrock, 2015). Incidents of contamination of harvested produce by foodborne pathogens (*Salmonella, E. coli,* and *Listeria*) have been increasing, resulting in human fatalities, and requiring the destruction of hundreds of tons of produce.

The management of postharvest pathogens relies mainly on either prevention or eradication strategies. Prevention includes the reduction of pathogen pressure through removal of inoculum sources in the field and in the packinghouse, and the use of treatments (chemical and biological) that inhibit or prevent the pathogens from becoming established. Eradication or curative measures, however, involve the use of treatments (physical, biological, or chemical) that arrest and eradicate infections whether these are latent, incipient, or already active.

The use of synthetic chemical fungicides, alone or in combination, are the general mainstay of the industry. This is because they are, relative to other approaches, easy to use in packing lines, can be employed in an efficient manner, and are cost-effective. The manufacturers of fungicides or other postharvest service companies are also able to provide a strong system (sales representatives and field agents) to support their use. Although the use of synthetic chemical fungicides is still the predominant method of managing postharvest diseases, considerable research has been conducted to identify safe and effective alternatives. This objective has been driven by problems associated with the use of synthetic chemical fungicides, including (1) the development of fungicidal resistance, (2) consumer demand to reduce human and environmental exposure to chemicals, and (3) increased restrictions imposed by governmental regulatory agencies on specific agrochemicals and/or their allowable residues, especially after harvest. The issue of chemical residues and their impact on human health and the environment is especially critical to the search for alternative methods of postharvest disease management. The registration, regulatory statutes, and residue allowances for synthetic fungicides and other agrochemicals have become especially restrictive in Europe (Droby et al., 2016; Wisniewski et al., 2016). Since the economy of the fruit industry is driven by exports, countries that seek to export to Europe are extremely aware of the need to develop safe and effective alternatives to synthetic postharvest fungicides.

This chapter provides an overview of alternative postharvest approaches in managing postharvest diseases of fruit crops that are based on the use of natural plant products (plant- and animal-based materials), as well as microbial biocontrol products. The focus, however, will be on recent advances in the field of biocontrol and future research challenges that need to be addressed. The implementation of an integrated approach will also be discussed in which more than one alternative strategy is used to provide a sufficient, and commercially acceptable level of efficacy in controlling postharvest pathogens.

2 Plant extracts and animal-based materials as postharvest treatments

Dr. Charles Wilson was the first to introduce the concept of an expanded definition of biological control of postharvest diseases (discussed in Wisniewski et al., 2007, 2016; Droby et al., 2009) in which biological control is defined as control of a plant disease by a biological process or the product of a biological process. In addition to the use of microorganisms, this definition included innate and induced resistance and the use of natural antimicrobials, plant and microbial volatile organic compounds (PVOCs, MVOCs), and chitosan were also now seen as biological agents with potential use in disease management. Regardless of factual data, there is a general belief and acceptance by the public that plant-derived natural materials (essential oils in particular) are non-phytotoxic and safe for use as pesticides for crop protection.

The antimicrobial activity of both PVOCs and MVOCs against postharvest diseases has been reviewed in detail by Mari et al. (2011, 2016), Palou et al. (2016), and Sivakumar and Bautista-Baños (2014). Among the wide array of active essential oils (EOs) derived from plants, cinnamaldehyde, citral, eugenol, limonene, hexanal, thymol, carvacrol, and carvone have been widely reported to be the most active against fungal and bacterial pathogens of harvested crops (Spadaro and Gullino, 2014; de Aquino et al., 2015). The use of EOs reduced sensitive pathogens, while favoring less sensitive ones, as recently demonstrated on peaches, where the application of thyme or savory EO favored a reduction of brown rot incidence, caused by *Monilinia fructicola*, but increased gray mold, caused by *Botrytis cinerea* (Santoro et al., 2018). These volatile materials are usually applied as gaseous treatments in a process defined as 'biofumigation' (Regnault-Roger et al., 1993; Mari et al., 2016).

The inhibitory effects of many different EOs against a wide range of postharvest pathogens are well documented in the literature; however, efficacy experiments have been mainly conducted *in vitro* in petri dishes (Palou et al., 2016; Mari et al., 2016). Trials have also been performed on harvested commodities, including grapes, citrus, cherries, peaches, nectarines, banana, apples, pears, tomato, and strawberries. Trials should be performed *in vivo*, as often EOs not only showed antimicrobial activity, but were also able to induce resistance mechanisms in the host fruit, as demonstrated for thyme EO on apple (Banani et al., 2018). In most cases, the concentrations of the EOs, or their active constituents, required to inhibit decay development is much higher than in the levels that are effective in culture media. This may be due to interactions between the active compounds and host plant tissues and/or due to the volatility characteristic of EOs and resulting in a short contact time with pathogen spores. Experiments also showed positive effects of EO application on the reduction of weight loss during storage of sweet cherry (Maghenzani et al., 2018), but also on the preservation of ascorbic acid and carotenoid content in peaches and nectarines (Santoro et al., 2018).

The antimicrobial activity of MVOCs has also been reported in for the control of postharvest diseases. A comprehensive review on this subject was presented by Morath et al. (2012) and discussed by Mari et al. (2016). The most documented example is the VOCs produced by the endophytic fungus, *Muscodor albus* isolated by Strobel et al. (2001) from the bark of *Cinnamomum zeylanicum*. Biofumigation of storage facilities with VOCs of *M. albus* to control postharvest pathogens on various harvested crops has been reported (Mercer and Jimenez, 2004; Mlikota Gabler et al., 2006; Freitas et al., 2005; Braun et al., 2012). The major antimicrobial VOCs produced by *M. albus* are 2-methyl-1-butanol (48.5%), isobutyric acid (14.9%), and ethyl propionate (9.63%) (Mercier and Jimenez, 2004). A commercial company developed a product based on the VOCs of *M. albus* but was denied registration by the EPA in the United States due to the toxicity of the active compounds to animals.

Additional examples of antimicrobial MVOCs are geosmin and phenethyl alcohol (PEA), produced by *Penicillium expansum* and exhibiting inhibitory activity against *B. cinerea, Monilinia* Spp., and *Colletotrichum acutatum* (Rouissi et al., 2013). Yeasts, including *Candida intermedia, Aureobasidium pullulans, Sporidiobolus pararoseus, Pichia anomala, Wickerhamomyces anomalus,* and *S. cerevisiae,* have also been reported to produce a wide array of VOCs with antimicrobial properties. Among these antimicrobial VOCs, the most active are 1,3,5,7-cyclooctatetraene, 3-methyl-1-butanol, 2-nonanone, and PEA (Huang et al., 2012). These compounds, however, exhibited only a limited inhibitory effect against *B. cinerea, C. acutatum,* and *Penicillium* spp. (Mari et al., 2012).

Despite the extensive work that has been done on the use of VOCs of plant and microbial origin, the practical application of these compounds to harvested commodities to control postharvest diseases has not been realized. Dips in water-based emulsions or fumigation have been the most tested methods for application of VOCs and EOs; however, the optimal delivery method will largely depend on the type of commodity and the compatibility of the application process with existing postharvest practices. The application of VOCs could be potentially integrated with the use of controlled atmosphere (CA) storage or modified atmosphere (MA) packaging, where different concentrations of atmospheric gasses and VOCs can be mixed under controlled conditions. New technologies based on the use of encapsulation and methods of slow release (slow release diffusors: Santoro et al., 2018) have also provided opportunities for developing novel formulations of VOCs and EOs for postharvest use. Integration of EOs in waxes and other coatings has also been suggested (Mari et al., 2016; Sivakumar and Bautista-Baños, 2014). However, these latter approaches have still not developed enough to provide a commercial level of postharvest disease protection. The development of formulations that maintain antimicrobial activity for an extended period, while not having negative sensory and phytotoxic effects, remains a challenge. At present, the residual effect of most VOCs and EOs on sensory attributes makes their use as a postharvest treatment problematic and viably difficult. Chitosan, a natural material derived from the shells of crustaceans, insect cuticles, and fungal cell walls, has also been heavily researched as an alternative method of controlling postharvest pathogens (El-Ghaouth et al., 1991; Romanazzi et al., 2009, 2016). Chitosan was reported to have a triple effect when applied to harvested commodities. It exhibits direct antimicrobial activity, induces host resistance, and serves as a film/coating that acts as a physical barrier preventing water loss from harvested produce (Palou et al., 2016). *In vitro* and *in vivo* tests on harvested fruits and vegetables have demonstrated that chitosan has antimicrobial activity against a wide range of bacterial and fungal pathogens, including *C. gloeosporioides*, *Rhizopus stolonifer*, *P. digitatum*, *F. oxysporum*, *P. italicum*, *Botryodiplodia lecanidion*, *Alternaria alternata*, *Monilinia laxa*, and *B. cinerea* (Bautista-Baños et al., 2013; Palou et al., 2016; Romanazzi et al., 2016). Notably, chitosan hydrochloride has been approved in the EU for plant disease management (Reg. EU 494 563/2004) (Romanazzi et al., 2016). Nanotechnologies have been used to develop new formulations of chitosan (Mustafa et al., 2013). Nano-formulations allow for the encapsulation of additional active antimicrobial ingredients, such as EOs and other volatile compounds, that provide inhibitory effects against postharvest pathogens (Mustafa et al., 2013; Ing et al., 2012).

The use of aqueous and organic extracts of various exotic and domesticated medicinal and herbal plants has also been reported as an alternative method of preserving fruits and vegetables after harvest. Reports in the literature describe research conducted on the antimicrobial activity of extracts obtained from a wide range of plants belonging to different botanical families. Inhibitory effects of natural extracts have been reported against several major postharvest pathogens, including *P. digitatum*, *P. italicum*, *Geotrichum citri-aurantii*, *Monilinia laxa*, *Fusarium oxysporum*, *Alternaria alternata*, *C. gloeosporioides*, *C. musae*, *Chalara paradoxa*, *Rhizopus* stolonifer, *Aspergillus niger*, *P. expansum*, *B. cinerea*, *Mucor piriformis*, *Erwinia carotovora* spp. *carotovora*, *Pseudomonas syringae* pv. *tomato*, *Pseudomonas corrugata*, and *Xanthomonas campestris* pv. *vesicatoria* (Bautista-Baños et al., 2013; Palou et al., 2016; Usall et al., 2015; Elsherbiny et al., 2016; Feliziani et al., 2013; Romanazzi et al., 2012; Li Destri Nicosia et al., 2016). As with the exploration of VOCs and EOs, commercial development and application of products based on plant extracts with antimicrobial activity has not been pursued. This is due to various economic, efficacy, and sensory constraints.

Antimicrobial peptides (AMPs) have been widely studied as novel antifungal and antibacterial materials for the control of postharvest pathogens (Jenssen et al., 2006; Keymanesh et al., 2009; Ciociola et al., 2016). In general, these AMPs are less than 30 amino acids in length and can be synthesized. Badosa et al. (2009) designed a peptide designated Bp21 (Ac-FKLFKKILKVL-NH_2) and demonstrated its inhibitory activity against *Fusarium oxysporum*, *Aspergillus niger*, *Rhizopus stolonifer*, and *Penicillium expansum*. In a recent study, Wang et al. (2018) reported that BP21 was also shown to inhibit the growth of *P. digitatum*, *P. italicum*, and *G. citri-aurantii in vitro* and on citrus fruit. Other examples of antifungal peptides that can be used to control postharvest pathogens are iturins and fengycins. These peptides have also been reported to have a role in the mechanism of action of several postharvest bacterial biocontrol agents (e.g., *Bacillus* spp.) (Yánez-Mendizábal et al., 2012; Waewthongrak et al., 2015).

Although inhibitory activity can be demonstrated in *in vitro* and *in vivo* experiments, the challenge is how to overcome the several limitations hindering commercial use of these peptides. These limitations include the cost of production, formulation, delivery, non-specific toxicity, and difficulty in obtaining regulatory approval.

3 The discovery of microbial antagonists to manage postharvest pathogens

The vast majority of postharvest pathogens are necrotrophic and depend on the presence of surface injuries for the infection of harvested produce. The nature of surface wounds may vary from cracks in the cuticle and underlying tissues created during fruit growth, maturation, and ripening, as well as cut surfaces and splitting skin at the stem-end, to mechanical damage during harvesting and subsequent handling of the fruit. Protecting these wounds has been the main target in the development of biological control of postharvest pathogens. The underlying hypothesis has been that there are epiphytic microbial antagonists present on fruit and vegetable surfaces that are capable of colonizing fruit surfaces and wounds, multiply rapidly in wounded tissues, and thus exclude pathogens from an infection site. It was also assumed that isolating antagonistic microbial species and reapplying them in large numbers to the commodity would interfere with spore germination of the pathogen. Wilson and Wisniewski (1989) defined several features required for an effective antagonist, among these, rapid growth and colonization of fresh wounds were the most important. In addition, the antagonist needed to be genetically stable, effective at low concentrations, able to survive adverse environmental conditions, be effective against a wide range of pathogens, not fastidious in its nutrient requirements, amendable to production in inexpensive growth media, non-detrimental to human health, compatible with commercial processing practices, and capable of being produced in a formulation with a long shelf life.

A process for the selection of microbial antagonists that efficiently protect surface wounds in fruits and vegetables against postharvest pathogens was proposed by Wilson et al. (1993). The proposed procedure was rapid, and involved minimal expenditure of time and expense. The proposed procedure has since been used by many researchers conducting postharvest biocontrol research programs. The procedure is based on *in vivo* screening for antagonists and favors the isolation of rapid-growing, antagonistic microorganisms from wounded tissue that are able to inhibit the infection and development of postharvest pathogens (Wilson et al., 1993). Organisms exhibiting antagonistic activity against a broad spectrum of postharvest pathogens on a variety of fruit and vegetable crops are then further characterized using molecular techniques based on conserved regions of the ribosomal DNA or by full genome sequencing (Kurtzman and Droby, 2001; Piombo et al., 2018).

Importantly, the method used for screening microorganisms for biocontrol activity greatly influences the type and properties of the isolated antagonist. Therefore, it is important to define the target pathogen and understand its biology and etiology before selecting a screening strategy. A principal characteristic of the current screening procedure used to identify postharvest biocontrol agents is that in most cases microorganisms will be isolated that rapidly colonize surface injuries, and most likely be a yeast species capable of using sugars more efficiently than other microorganisms. Generally, yeast antagonists provide excellent protection of wounds but lack the ability to stop or inhibit infections that have already been established (curative effect), as well as latent/incipient infections (Droby et al., 1989; El-Ghaouth et al., 2000b). The lack of correlation between laboratory tests evaluating postharvest biocontrol agents and the performance of biocontrol agents/products under highly variable commercial conditions can be explained, at least partially, by the lack of a curative effect (Droby et al., 1993, 2000; Wisniewski et al., 2001).

The use of antagonistic yeasts has been especially emphasized since the production of toxic secondary metabolites (antibiotics) is generally not involved in their inhibitory activity (Wisniewski and Wilson, 1992). The original source (fruit surfaces) for the isolation of yeast antagonists (Wilson and Wisniewski, 1989; Wilson et al., 1993) has expanded to other environments, such as sea water (Wang et al., 2010; Hernández-Montiel et al., 2010) and Antarctic soil samples (Vero et al., 2013). The main intent in exploring these new sources has been to identify yeast that can thrive in stressful environments and to identify isolates with novel modes of action. Harvested agricultural commodities are naturally prone to a wide variety of injuries (bruises, scrapes, broken stems, broken epidermal hairs, etc.). In order for postharvest biocontrol agents to work under commercial conditions, screening methods perhaps need to better reflect the 'real world' where the antagonists will need to function in an efficient and reliable manner.

Several questions have been raised regarding the strategy used in selecting microbial antagonist for postharvest disease control (Droby et al., 2009). The hypothesis and expectation that a single microbial antagonist can provide a consistent commercial level of control may be unrealistic considering the variables involved in biocontrol systems. It has been long believed that a single antagonist isolated from one commodity will be able to provide adequate biocontrol efficacy against a wide range of pathogens on many different commodities. These expectations have perhaps proven to be too ambitious and variables including preharvest history, postharvest treatments, and handling protocols (disinfection, packing, packaging, storage, transit etc.), as well as the physiology of the harvested commodity, and its genetic background, will greatly influence postharvest disease susceptibility and thus the performance of biocontrol agents. Given the knowledge of microbial ecology and plant microbiota that has accumulated in recent years using metagenomic approaches, there is a need for a paradigm shift in our approach to selecting and using microbial antagonists. The relevance of studies on the microbiome of harvested commodities in relation to biocontrol will be discussed further in Section 5.

4 Mode of action of postharvest biocontrol agents

The mechanisms involved in the biocontrol activity of postharvest antagonists have been extensively investigated (Spadaro and Droby, 2016). In general, biocontrol agents may exhibit one or more of the following mechanisms: (i) antibiosis; (ii) competition for limiting nutrients and space; (iii) parasitism or direct interaction with the pathogen; and (iv) induced host resistance (Janisiewicz and Korsten, 2002; Liu et al., 2013; Spadaro and Droby, 2016). For example, strains of the two sister species *Metschnikowia pulcherrima* and *M.*

fructicola can compete for nutrients, release hydrolases, induce host resistance, produce and induce reactive oxygen species (ROS), but can also biodegrade mycotoxins produced by the fungal pathogens (Spadaro et al., 2008, 2013; Macarisin et al., 2010; Hershkovitz et al., 2012, 2013).

A principal mechanism of postharvest antagonists is competition for nutrients and space, resulting from the simultaneous demand for the same resource by the pathogen and the antagonist. In competition for space, yeasts and bacteria have the advantage of rapid multiplication and colonization (Droby et al., 1989; Spadaro et al., 2010), and can also produce extracellular polysaccharide materials that promote adhesion to each other and the fruit surface. The ability to form a biofilm on the surface of wounded cells and tissues, where microorganisms are enclosed in a hydrated matrix, favors competition for nutrients and space (Lutz et al., 2013). Biofilm-forming yeasts are effective against *Penicillium expansum* on apple, but only yeast cells collected from the biofilm phase were able to efficiently colonize the surface of wounded tissues (Ianiri et al., 2013).

Competition for iron frequently occurs in fruit wounds. Iron is essential for fungal growth and pathogenesis, and iron sequestration by non-pathogenic microbes can be exploited in postharvest biocontrol systems. Under iron starvation conditions, fungi exhibit lower catalase activity, and consequently lower resistance to ROS. Rhodotorulic acid, a siderophore produced by *Rhodotorula glutinis,* improved the control of blue mold of apple (Calvente et al., 1999). *Metschnikowia pulcherrima* and *M. fructicola* are able to produce the red pigment pulcherrimin formed from pulcherriminic acid and ferric ions, which plays a role in the control of apple diseases (Saravanakumar et al., 2008; Spadaro and Droby, 2016).

Antibiosis is the inhibition or destruction of a microorganism by diffusible or volatile antibiotics produced by another microorganism. Certain microorganisms begin to produce antibiotics when substrate availability becomes limiting, preventing other microorganisms from using the remaining substrate. Bacteriocins, antibacterial proteins produced by bacteria, induce the formation of pores in the membrane of target cells and deplete the trans-membrane potential, resulting in electrolyte leakage. The antagonist *Bacillus subtilis* produces a range of antimicrobial cyclic lipopeptides, including iturins, fengycins, and surfactins (Ongena and Jacques, 2008; Yánez-Mendizábal et al., 2012). *Aureobasidium pullulans* produces aureobasidin, a cyclic depsipeptide, with antifungal and antibiotic properties. A major concern regarding the use of antibiotic-producing microorganisms in food products, however, is the risk they pose to the development of fruit microorganisms resistant to these compounds and the possible transfer of resistance to human pathogens.

Many fungal species are also known to produce volatile organic compounds (VOCs) with antifungal properties, which can be used for biological fumigation, or biofumigation. As previously mentioned, the endophytic fungus, *Muscodor albus,* is capable of producing a mixture of VOCs that effectively control postharvest diseases of fruit (Mercier and Smilanick, 2005).

Antagonists can also exhibit direct parasitism of other fungi, resulting in the lysis of fungal mycelia and propagules. Mycoparasites use cell-wall-degrading enzymes, such as β-1,3-glucanase, chitinase, and proteases, to dissolve fungal cell walls and penetrate them. *Candida saitoana* cells, when cultivated together with mycelia of *Botrytis cinerea* Pers., attach to the fungal hyphae, degrading the cell wall (El-Ghaouth et al., 1998). The separate and sequential disruption of two exo-ß-1,3-glucanase genes in the yeast antagonist, *Pichia anomala,* resulted in reduced efficacy against *B. cinerea* (Friel et al., 2007). A chitinase gene from *Metschnikowia fructicola* overexpressed in *Pichia pastoris* resulted in significant control of *Monilinia laxa* (Aderh. & Ruhland) Honey and *M. fructicola in vitro* and in peach fruit (Banani et al., 2015). An alkaline serine protease from *A. pullulans*, expressed in *Escherichia coli* (Zhang et al., 2012) and *P. pastoris* (Banani et al., 2014) also controlled several different postharvest pathogens on apples.

Indirect antagonism involves the induction of resistance in host tissues by the presence of beneficial, non-pathogenic microorganisms. Some antagonists interact with host tissue, increasing compartmentalization (healing) of injured and/or infected tissues and lignification of cell walls. Biocontrol agents interact with wounded tissue, inducing β-1,3-glucanase, chitinase, and peroxidase activity, the accumulation of phytoalexins, and the formation of structural barriers. Application of *Candida oleophila* induced phytoalexin production in grapefruit, ethylene biosynthesis, and the accumulation of chitinase and β-1,3-glucanase enzymes in grapefruit (Droby et al., 2002). Application of *Cryptococcus laurentii* to jujube increased β-1,3-glucanase activity and the expression of the *Glu-1* gene (Tian et al., 2007). Notably, however, all of the results on the induction of resistance following antagonist treatment are correlative. Direct evidence for the ability of induced substances to inhibit pathogen infection and development remains to be established.

Oxidative stress plays a crucial role in biocontrol and may be involved in signaling pathways that activate a resistance response (Chan and Tian, 2005). The ability to survive and proliferate in wounded host tissues and to tolerate oxidative stress is believed to be an essential attribute of effective postharvest antagonists (Castoria et al., 2003). Fruit wounding is associated with the accumulation of ROS, which can affect host response, pathogen virulence, and biocontrol efficacy. Macarasin et al. (2010) demonstrated that yeast antagonists have the ability to produce relatively high amounts of superoxide anions. The ability of yeast antagonists to self-generate and possibly stimulate an oxidative response in the host tissue may be an important aspect of biocontrol activity. The application of *M. fructicola* and *C. oleophila* into fruit wounds increased the accumulation of H_2O_2 in fruit tissue (Macarisin et al., 2010). *Penicillium digitatum* suppressed H_2O_2 production in oranges, as well as superoxide dismutase and catalase activity (Torres et al., 2011). In contrast, *Pantoea agglomerans* induced H_2O_2 production and enzymatic activity, which may partially contribute to the prevention of infection of citrus fruit by *P. digitatum* (Macarisin et al., 2007).

Developments in genomics, transcriptomics, proteomics, metagenomics, as well as comparative and functional genomics, is increasing the ability to determine changes in the physiological status of antagonists, and the effect of environmental stress on their physiology and activity (Massart et al., 2015; Sui et al., 2015). The ability to assess changes in the expression of 'biocontrol genes' during mass production, formulation, and storage, or in response to exposure and contact with host plant tissues after application, will greatly contribute to our understanding of the mode of action of biocontrol agents.

5 Microbiome studies and their relevance to postharvest biological control

Advances in metagenomic technologies in relation to characterization of analysis of the plant microbiome is greatly changing our understanding of postharvest biocontrol and offers opportunities to develop more effective biocontrol products. High throughput sequencing (HTS) Meta-omic technologies that characterize the diversity and function of microbial communities present in and on host tissues is revolutionizing our understanding of microbial ecology, plant pathology, and biological control (Abdelfattah et al., 2017). HTS technologies have facilitated the study of microbes *in situ* in their micro-environments and is providing new information on how individual components of the microbial community interact with each other and with their host (Berg et al., 2016). Microbiome studies are providing the understanding and information needed to create a paradigm shift in the way we think about biocontrol systems, biocontrol products, and postharvest biology in general.

Fresh fruits and vegetables contain large diversity of bacteria, filamentous fungi, and yeasts, either as epiphytes or endophytes. Very little is known about the overall diversity and composition of microbial communities on harvested produce and how these communities vary across produce types. The composition and fate of the microbiome on fruits and vegetables over time and in response to postharvest management practices has begun to be explored but the role and function of the microbiome of harvested produce remains to be elucidated. An initial study conducted by Abdelfattah et al. (2016) in which the impact of management practices on the composition of the microflora, and the temporal changes in the microbiome over the course of a growing season, was examined. The composition of the microbiota of apple after harvesting, processing, shipping, and distribution to a local supermarket documented the spatial and compositional diversity of the microbiota on conventional and organic apple fruit. Differences were observed in different portions of the fruit (peel, stem-end, calyx-end) as well as wounded tissues. These findings indicated that the microflora associated with different locations/parts of the apple fruit should be considered when developing biocontrol strategies for the management of postharvest diseases.

The existence and function of endophytic microorganisms within host plants have not been widely studied in harvested commodities. Diskin et al. (2017) provided insights on the taxonomic diversity of endophytic fungal and bacterial families and their dynamics in the stems of mango fruit during storage and ripening. They documented the presence of several stem-end pathogens including *Alternaria* spp., *Colletotrichum* spp., and *Lasiodiplodia* spp., along with other diverse endophytic fungal and bacterial species. The abundance of the different members of the stem-end microbiota was influenced by fruit ripening and by postharvest treatment. It was also shown that stem-ends of fruit that did not develop disease had more diverse microbial communities relative to stem-ends that were diseased. These findings suggested the existence of a natural mechanism of biocontrol in which pathogens have a silent endophytic stage and become active and cause disease when the composition of the microbiota changes in response to host physiological changes and postharvest treatments. This knowledge may influence the way we look at biocontrol and the possible use of endophytes in biocontrol systems.

The development of postharvest biocontrol products have relied on the use of individual microbial species selected from microorganisms cultured from washings of fruit surfaces or isolated from wounds. This approach has resulted in the development of several commercial products that are based on the use of a single antagonist (Droby et al., 2016; Wisniewski et al., 2016). There is a growing realization, however, that a single antagonist may not be able to provide the consistent efficacy and spectrum of activity required commercially against different postharvest pathogens. This raises the question if the utilization of several microorganisms together and their combined function can provide superior and consistent biocontrol under a variety of conditions. In this regard, Massart et al. (2015) promoted the use of microbiota-derived products or the microbiota itself, directly or indirectly, as novel approaches for the protection of plants against pathogens. An initial approach could be the use of a synthetic or a natural consortium (Gopal et al., 2013) that is applied to a harvested commodity to see if the consortium results in better disease control than the use of a single antagonist due to the expression of mechanisms of action expressed by the consortium. Maintaining the right balance and diversity inside a consortium before and after its application, however, may prove to be difficult.

Microbial communities composed of multiple species with a well-defined genetic background, can accomplish a greater number of functions through the interactions that occur among the individual microorganisms making up that community or consortium. Advances in this field of science may lead to a new paradigm in postharvest biocontrol that includes a broader definition of biological control. For example, the biological management of postharvest diseases could utilize specific features of a synthetic microbial community to interfere with the mechanisms associated with the entire disease process, which includes both the response of the pathogen and the host. This synthetic consortium could then be used as a 'biocontrol cocktail' and applied both pre- and postharvest to enhance innate resistance and manage postharvest diseases.

6 Integration of biocontrol agents with non-chemical postharvest treatments

An integrated management strategy for managing postharvest diseases that utilizes a combination of eco-friendly approaches, rather than a single approach, may be needed to create a viable alternative to synthetic, chemical fungicides. The multiple hurdle approach used in the food industry has been a successful approach to managing foodborne pathogens where the cross-contamination of a product is prevented at each stage along the processing line. This approach is based on the use of intervention technologies at critical points throughout the harvesting, handling, storage, shipping, and marketing pipeline. The use of this model should perhaps be considered for the development of effective approaches to prevent postharvest diseases by the use of successive actions/treatments that eventually result in achieving a commercial level (<3%) of decay control in a commodity. In this approach, the different treatments act in an additive or synergistic manner to achieve the necessary level of disease control. Utilizing different approaches at critical points in the harvesting, processing, and delivery pipeline (from the field to the supermarket) may represent the key to success in developing safe, effective, and reliable strategies for managing postharvest pathogens.

Attempts to increase the efficacy of postharvest biocontrol agents have been based on the application of combination of treatments with complimentary modes of actions. The concept behind this approach is based on the ability of the combined treatments to directly or indirectly inhibit or kill the pathogen spores but not the biocontrol agent, enhance the competitiveness of the biocontrol agent, increase the antimicrobial activity of the biocontrol agent in an additive or synergistic manner, or induce resistance responses in the harvested commodity.

New antagonists, new ways to utilize antagonists, and new ways to integrate their use with other alternatives approaches are continually being reported (Wisniewski et al., 2016). In addition to the sanitation of harvesting crates, dip tanks, and other packing house equipment with various disinfecting materials that reduce pathogen inoculum levels, researchers have sought new ways of increasing the efficacy of postharvest biocontrol agents with the use of natural antimicrobials, salts that alter the pH (e.g. sodium carbonate, sodium bicarbonate potassium sorbate), and manipulation of the physical environment (e.g. heat treatment and UV irradiation) (Droby et al., 1993, 2009; Wisniewski et al., 2016). Comprehensive reviews of the use of various physical treatments and natural plant and animal products were published by Usall et al. (2016) and Palou et al. (2016).

The use of calcium and sodium salts, primarily $CaCl_2$ and $NaHCO_3$, in conjunction with biocontrol protocols has been documented (McLaughlin et al., 1990; Wisniewski et al., 1995; Smilanick et al., 1999; Droby et al., 1995). Numerous reports have documented the antimicrobial properties and stimulation of host resistance by chitosan-derived products when used alone or in conjunction with postharvest biocontrol agents (El-Ghaouth et al., 2000a; Meng et al., 2010; Romanazzi et al., 2016) to enhance efficacy. The combination of the yeast, *Candida saitoana*, with the sugar analog, 2-deoxy-D-glucose, glycolchitosan, or with lysozyme, was shown to greatly improve the control of a variety of postharvest diseases of citrus and apple (El-Ghaouth et al., 2000a,b). The integration of UV-C with antagonists (Stevens et al., 1997; Guo et al., 2015) has also been explored. Interestingly, Janisiewicz et al. (2016) demonstrated that following the UV irradiation of plant material with a sufficient dark period greatly enhanced the antimicrobial activity of the UV treatment because it did not allow the microbes to utilize light-activated DNA repair mechanisms to recover from the injury caused by the UV treatment.

The application of postharvest heat treatments (hot air and hot water) with microbial antagonists can provide additive or synergistic enhancement of biocontrol efficacy. Heat treatments are simple techniques that can be easily implemented in packinghouses to reduce postharvest diseases. Plaza et al. (2004) and Casals et al. (2012) reported on the application of a curing treatment (temperatures higher than 30°C and RH higher than 90%) of citrus fruit in conjunction with bacterial biocontrol agents. In these studies, the application of *P. agglomerans* CPA-2 or *Bacillus subtilis* CPA-8 following a curing treatment at 33°C for 65 h for lemon fruit or 50°C for 2 h and 95–99% RH for peaches improved the control of green mold and brown rot, respectively. Importantly, the implementation of the combined treatment completely controlled preexisting infections. Successful control of bitter rot caused by *Colletotrichum acutatum* J.H. Simmonds on apples was also achieved by the combined use of a curing and a biocontrol treatment (Janisiewicz et al., 2003; Conway et al., 2005). The application of strain BIO126 of *M. pulcherrima* after treatment with hot water at 50°C for 3 min improved the biocontrol efficacy of *M. pulcherrima* against *P. expansum* and *Botrytis cinerea* in apple (Spadaro et al., 2004). Zhang et al.

(2007, 2008) demonstrated that a heat treatment at 37°C for 48 h combined with the biocontrol agents *C. laurentii* or *Rhodotorula glutinis* controlled postharvest decay in peaches and pears caused by *P. expansum* and *Rhizopus stolonifer* (Ehrenb.) Vuill., respectively.

A hot air treatment (38–40°C for 24 h) of cherry tomato fruit, combined with the application of the biocontrol agent, *Pichia guilliermondii*, reduced decay caused by B. cinerea, R. stolonifer, and Alternaria alternata (Zhao et al., 2010). A hot water pretreatment of peaches at 55°C for 10 sec also improved the ability of *C. oleophila* to control *B. cinerea* and *P. expansum* (Karabulut and Baykal, 2004).

CA and MA have also been used along with biocontrol agents to improve biocontrol efficacy. The ability of the yeast *C. sake* CPA-1 to control blue mold on apples improved when apples were stored in a CA of 3% O_2 and 3% CO_2 (Usall et al., 2000). Similar results with CA were obtained with the bacterium *P. agglomerans* CPA-2 against most of the major fungal diseases of apples (Nunes et al., 2002).

The use of essential oils has also been reported as a means to improve efficacy of postharvest biocontrol agents. Arrebola et al. (2010) reported that *Bacillus amyloliquefaciens* applied together with oils of lemongrass and thyme gave better control of *B. cinerea, P. expansum,* and *R. stolonifer* on peach fruit than when either agent was used alone. Liu et al. (2010) reported that tea polyphenols inhibited *B. cinerea* and improved the biocontrol efficacy of *Hanseniaspora uvarum*. Phytic acid has been used, due to its antioxidant potential, in combination with biocontrol agents such as *R. glutinis* and *Pichia caribbica*, to control gray mold of strawberry (Zhang et al., 2013) and blue mold of apple (Mahonu et al., 2016).

Plant hormones have also been demonstrated to improve the efficacy of postharvest biocontrol agents. *C. laurentii* combined with N6-benzyladenine suppressed P. *expansum* infection of apple fruit wounds (Yu et al., 2008). Auxins, such as indole-3-acetic acid (IAA), have also been reported to have direct antifungal activity. A combined treatment of *C. laurentii* and IAA was effective in reducing gray mold in apple, and was superior to the effect of each treatment applied alone (Yu et al., 2009). Application of IAA alone greatly reduced gray mold development in apples when IAA was applied 24 h prior to inoculation with *B. cinerea*. The combined use of IAA and *C. laurentii* also increased superoxide dismutase, catalase, and peroxidase activities in apples.

7 Development and commercial application of postharvest biological control products

The majority of postharvest biocontrol agents reported in the literature and most of the antagonists that have been developed into products are yeasts (Droby et al., 2009, 2016). While several products are in development or awaiting registration in the United States or Europe, a number of commercial, biologically based products are also registered for use in controlling postharvest diseases of citrus and other fruits and vegetables. These include Shemer, developed in Israel and based on the yeast *Metschnikowia fructicola;* BioSave, developed in the United States, based on the bacterium *Pseudomonas syringae;* Candifruit, developed in Spain, based on the yeast *Candida sake;* Nexy, developed in Belgium, based on the yeast *Candida oleophila;* Boni-Protect, developed in Germany, based on the yeast *Aureobasidium pullulans;* and Avogreen, developed in South Africa, based on the bacterium *Bacillus subtilis.*

Wide-scale commercial use of postharvest biocontrol products and their general acceptance by industry will largely depend on the ability of the biocontrol products to achieve the same level of disease control as current fungicides in a reliable manner. The development of low-cost, stable formulations that provide consistent control on different crops is also essential. Additionally, technical problems associated with the application of biocontrol product shortly before harvest and/or after harvest on the packing line need be addressed and will require approaches that are dependent on the fruit or vegetable crop being treated.

The commercialization of a biocontrol product involves mass (upscale) production and the development of a formulation with a shelf life of at least 12 months. Evaluating different formulations under commercial conditions is a fundamental step in the process of developing a commercial biocontrol product. Industrial fermentation practices involve conditions that are quite different from those used in shake culture and laboratory-scale production systems. The industrial fermentation process must be cost-effective, rely on industrial by-products as a growth medium, and must be completed within 24–30 h (Hofstein et al., 1994). The type of growth medium selected for large-scale fermentation may impact the final cell count, viability, and stress tolerance of the biocontrol agent. For example, studies with *C. sake* demonstrated that yeast growth in a commercial, molasses-based medium with a lowered water activity (*aw* = 0.98) resulted in high viable cell counts, an increased intracellular concentration of arabitol and glycerol, and increased water stress tolerance than when the yeast were grown in the same medium without the lowered aw (Abadias et al., 2003b). Moreover, the intracellular water potential of cells decreased when grown in the growth medium with the lower *aw* (Abadias et al., 2003a,b).

Production and formulation research programs are often conducted directly or in association with private companies at the last stages of product development, and the outcome of the research is typically protected under secrecy agreements. This has resulted in a lack of published scientific studies on these essential topics. Information on the effect of industrial production practices, as well as formulation technologies themselves, should be investigated early in the development of a commercial biocontrol agent, before the product reaches costly commercial tests.

Downstream processing of the biocontrol agent involves various steps, such as cell separation from the medium, drying, the addition of inert, volumizing materials, adhesives, emulsifiers, and adjuvants. Cell separation is commonly performed by the conventional and well-established method of centrifugation. Two popular types of centrifuges utilized are disc stacks and tubular bowl centrifuges (Jungbauer, 2013). Flotation is another method of cell separation. This process is based on the affinity of cells to attach to air bubbles. The bubbling of air through the medium leads to the accumulation of cells on the surface of the brine where they can be skimmed off and collected (Jungbauer, 2013). The precipitation of cells has also been attempted by the addition of natural and synthetic polymers, inert powders, the lectin Concanavalin A to the brine and the adhesion of the biocontrol agent to fine magnetic particles. These approaches, however, are problematic and generally not allowed when the cultured cells are to be applied to food products.

Different dehydration systems have also been used in the formulation of biocontrol agents. Freeze-drying maintains viability but has a high cost in relation to other systems. The commercial product, BIOSAVE (a strain of the bacterium *Pseudomonas syringae*) was first introduced to the market as freezed paste and later on as a dry powder formulation. Other microorganisms have also been formulated using the freeze-drying technique, including *P. agglomerans* (Costa et al., 2000), *C. sake* CPA-1 (Abadias et al., 2001a,b), *Cryptococcus laurentii* (Li et al., 2008), *Metschnikowia pulcherrima* (Spadaro et al., 2010), and *Pichia anomala* (Melin et al., 2011). Spray-drying is an alternative method for preserving biocontrol agents in a dry state. This approach allows large quantities of cultured cells to be dried in a short time and at a low cost. Few microorganisms, however, are able to survive the high temperatures used during this drying system. Only biocontrol agents able to produce heat-resistant endospores, such as *B. subtilis*, can be formulated by spray-drying (Yánez-Mendizábal et al., 2012). Finally, fluidized bed-drying, a cost-effective method of drying, can be used for heat-sensitive microorganisms because the drying temperatures used are lower than in spray-drying. Fungal biocontrol agents, including *E. nigrum* (Larena et al., 2003) and *P. frequentans* (Guijarro et al., 2006), or the yeast-like fungus, *Aureobasidium pullulans* (Mounir et al., 2007), or the yeast, *C. sake* CPA-2 (Usall et al., 2009), have all been successfully formulated using fluidized bed-drying.

Liquid formulation, which is the simplest way to stabilize microbial cells, involves storing cells in a water- or oil-based solution with different protectants and additives, usually under cold conditions. Isotonic liquid formulations of *C. sake* CPA-1 have been found to be a good alternative to dry formulations (Abadias et al., 2003b; Torres et al., 2003). Liquid formulations of *R. minuta*

(Patiño-Vera et al., 2005), *Cryptococcus laurentii* (Liu et al., 2009), and *P. anomala* (Melin et al., 2011) have also been evaluated.

Very few reports exist in the literature on the formulation of postharvest biocontrol agents. Upscaling, stabilization, and the overall formulation process are often viewed as an art rather than a science. Many researchers have suggested that improvements in the formulation of biocontrol products may represent a key element in achieving reliable and effective performance of these products under commercial conditions.

The prospects for commercial success of postharvest biocontrol products will largely depend on the versatility of their use and the range of the target diseases they can control. A narrow range of activity is considered a limitation that imposes difficulties in marketing the product. One possible approach to expand the uses of a specific biocontrol agent is to develop different formulations for different uses/targets. For example, Boni-Protect, Blossom Protect, and Botector are different formulations of the same biocontrol agent, *A. pullulans*. These different formulations have been introduced for the control of postharvest and preharvest diseases of pome fruit, fire blight on apples, and *Botrytis cinerea* on grapes (Droby et al., 2016).

Another aspect related to the commercialization of biocontrol agents used on agricultural commodities is the health and safety of applying live microorganisms on food products. Although this may represent an obstacle to public acceptance of this technology, the majority of postharvest biocontrol agents were originally isolated from fruit and vegetables and are indigenous to agricultural commodities. Humans are exposed to them daily when consuming fresh vegetables and fruit. Even though these antagonists are introduced in large numbers to the surface of a commodity, they survive and grow only in very restricted sites on the fruit surface (e.g. surface wounds). After their introduction on intact fruit surfaces, antagonist populations usually diminish within a very short period of time to the level of the natural epiphytic microflora.

A major obstacle to the commercialization of postharvest biocontrol products is the registration process required by government regulatory agencies. The registration package must contain a clean record of safety for the biocontrol agent for both humans and the environment, basic toxicological tests on the formulated product (eye and skin irritation, ingestion), detailed data on the mechanisms of action, detailed information about the strain identification and origin, and efficacy data, including semi-commercial and commercial tests using relatively large quantities of fruit treated in a manner similar to commercial conditions. The registration of biocontrol products for postharvest use in some countries (e.g. United States through the Environmental Protection Agency, EPA) has been straightforward and several products have received registration. In Europe, however, the situation until recently has been considerably more complex and will take several years to complete.

8 Conclusions

Postharvest decay caused by fungal and bacterial pathogens is the major causes of food losses along the whole supply chain. Effective management of pathogen infection and subsequent decay development is considered the key for maintaining quality and safety of harvested commodities. The use of synthetic chemical pesticides (fungicides and bacteriocides) has been the major strategy for managing postharvest pathogens; however, their continued use on harvested commodities has been under scrutiny by consumers and regulatory agencies. Several of the fungicides and bacteriocides used for the control of postharvest pathogens have been banned from use due to human and animal health risks. This situation has prompted extensive research to find safer and effective alternative control technologies that are based on materials derived from natural sources and biological control agents based on microbial antagonists.

Wide range of natural material with antimicrobial activity has been reported to inhibit postharvest pathogens in *in vitro* experiments and in trials on fruit. Although significant progress has been reported in identification, production and delivery,, implementation as stand-alone treatment in commercial settings is still very limited. This is mainly due to sensory effects and variable efficacy in large-scale commercial applications.

Several biocontrol products that are based on naturally occurring microbial antagonists have been developed for postharvest use. However, commercial implementation of these products has been very limited and only comprise a very small share of the potential market. The outlook of their use as a viable alternative is very promising; but, a more realistic scenario to combat decay of harvested commodities is the use of an integrated control approach combining biological, natural materials and possibly other means.

The microbiome of harvested agricultural commodities should be considered as an integral component of the host-pathogen system that is influenced by various biotic and abiotic factors affecting the fruit/vegetable after harvest. Characterizing the postharvest microbiome is the first step in exploring the possibility of assembling beneficial microbial consortia that act together in concert to bring about superior biocontrol efficacy.

9 References

Abadias, M., Benabarre, A., Teixidó, N., Usall, J. and Viñas, I. 2001a. Effect of freeze drying and protectants on viability of the biocontrol yeast *Candida sake*. *Int. J. Food Microbiol.* 65(3), 173-82.

Abadias, M., Teixidó, N., Usall, J., Benabarre, A. and Viñas, I. 2001b. Viability, efficacy, and storage stability of freeze-dried biocontrol agent *Candida sake* using different protective and rehydration media. *J. Food Prot.* 64(6), 856-61.

Abadias, M., Teixidó, N., Usall, J. and Viñas, I. 2003a. Optimization of growth conditions of the postharvest biocontrol agent *Candida sake* CPA-1 in a lab-scale fermenter. *J. Appl. Microbiol.* 95, 301-9.

Abadias, M., Usall, J., Teixidó, N. and Viñas, I. 2003b. Liquid formulation of the postharvest biocontrol agent *Candida sake* CPA-1 in isotonic solutions. *Phytopathology* 93(4), 436-42. doi:10.1094/PHYTO.2003.93.4.436.

Abdelfattah, A., Wisniewski, M., Droby, S. and Schena, L. 2016. Spatial and compositional variation in the fungal communities of organic and conventionally grown apple fruit at the consumer point-of-purchase. *Hort. Res.* 3, 16047. doi:10.1038/hortres.2016.47.

Abdelfattah, A., Malacrino, A., Wisniewski, M., Cacciola, S. O. and Schena, L. 2017. Metabarcoding: a powerful tool to investigate microbial communities and shape future plant protection strategies. *Biol. Cont.* 120, 1-10.

Arrebola, E., Sivakumar, D., Bacigalupo, R. and Korsten, L. 2010. Combined application of antagonist *Bacillus amyloliquefaciens* and essential oils for the control of peach postharvest diseases. *Crop Prot.* 29, 369-77.

Badosa, E., Ferré, R., Francés, J., Bardají, E., Feliu, L., Planas, M. and Montesinos, E. 2009. Sporicidal activity of synthetic antifungal undecapeptides and control of Penicillium rot of apples. *Appl. Environ. Microbiol.* 75(17), 5563-9. doi:10.1128/AEM.00711-09.

Banani, H., Spadaro, D., Zhang, D., Matic, S., Garibaldi, A. and Gullino, M. L. 2014. Biocontrol activity of an alkaline serine protease from *Aureobasidium pullulans* expressed in *Pichia pastoris* against four postharvest pathogens on apple. *Int. J. Food Microbiol.* 182-183, 1-8. doi:10.1016/j.ijfoodmicro.2014.05.001.

Banani, H., Spadaro, D., Zhang, D., Matic, S., Garibaldi, A. and Gullino, M. L. 2015. Postharvest application of a novel chitinase cloned from *Metschnikowia fructicola* and overexpressed in *Pichia pastoris* to control brown rot of peaches. *Int. J. Food Microbiol.* 199, 54-61. doi:10.1016/j.ijfoodmicro.2015.01.002.

Banani, H., Olivieri, L., Santoro, K., Garibaldi, A., Gullino, M. L. and Spadaro, D. 2018. Thyme and savory essential oil efficacy and induction of resistance against *Botrytis cinerea* through priming of defense responses in apple. *Foods* 7(2), 11. doi:10.3390/foods7020011.

Bautista-Baños, S., Sivakumar, D., Bello-Pérez, A., Villanueva-Arce, R. and Hernández-López, M. 2013. A review of the management alternatives for controlling fungi on papaya fruit during the postharvest supply chain. *Crop Prot.* 49, 8-20.

Berg, G., Rybakova, D., Grube, M. and Köberl, M. 2016. The plant microbiome explored: implications for experimental botany. *J. Exp. Bot.* 67(4), 995-1002. doi:10.1093/jxb/erv466.

Braun, G., Vailati, M., Prange, R. and Bevis, E. 2012. *Muscodor albus* volatiles control toxigenic fungi under controlled atmosphere (CA) storage conditions. *Int. J. Mol. Sci.* 13(12), 15848-58. doi:10.3390/ijms131215848.

Calvente, V., Benuzzi, D. and de Tosetti, M. I. S. 1999. Antagonistic action of siderophores from *Rhodotorula glutinis* upon the postharvest pathogen *Penicillium expansum*. *Int. Biodeterior. Biodegrad.* 43, 167-72.

Casals, C., Elmer, P. A. G., Viñas, I., Teixidó, N., Sisquella, M. and Usall, J. 2012. The combination of curing with either chitosan or *Bacillus subtilis* CPA-8 to control brown rot infections caused by *Monilinia fructicola*. *Postharvest Biol. Technol.* 64, 126-32.

Castoria, R., Caputo, L., De Curtis, F. and De Cicco, V. 2003. Resistance of postharvest biocontrol yeasts to oxidative stress: a possible new mechanism of action. *Phytopathology* 93(5), 564-72. doi:10.1094/PHYTO.2003.93.5.564.

Chan, Z. and Tian, S. 2005. Interaction of antagonistic yeasts against postharvest pathogens of apple fruit and possible mode of action. *Postharvest Biol.Technol.* 36, 215–23.

Ciociola, T., Giovati, L., Conti, S., Magliani, W., Santinoli, C. and Polonelli, L. 2016. Natural and synthetic peptides with antifungal activity. *Future Med. Chem.* 8(12), 1413–33. doi:10.4155/fmc-2016-0035.

Costa, E., Usall, J., Teixidó, N., García, N. and Viñas, I. 2000. Effect of protective agents, rehydration media and initial cell concentration on viability of *Pantoea agglomerans* strain CPA-2 subjected to freeze-drying. *J. Appl. Microbiol.* 89(5), 793–800.

Diskin, S., Feygenberg, O., Maurer, D., Droby, S., Prusky, D. and Alkan, N. 2017. Microbiome alterations are correlated with occurrence of postharvest stem-end rot in mango fruit. *Phytobiome* 1, 117–27.

Droby, S., Chalutz, E., Wilson, C. L. and Wisniewski, M. E. 1989. Characterization of the biocontrol activity of *Debaryomices hansenii* in the control of *Penicillium digitatum* on grapefruit. *Can. J. Microbiol.* 35, 794–800.

Droby, S., Chalutz, E., Hofstein, R., Wilson, C. L., Wisniewski, M. E., Fridlender, B., Cohen, L., Weiss, B. and Daus, A. 1993. Pilot testing of *Pichia guilliermondiui*: a biocontrol agent of postharvest diseases of citrus fruit. *Biol. Control* 3, 47–52.

Droby, S., Wilson, C., Wisniewski, M. and El Ghaouth, A. 2000. Biologically based technology for the control of postharvest diseases of fruits and vegetables. In: Wilson, C. and Droby, S. (Eds), *Microbial Food Contamination*. CRC Press, Boca Raton, FL, pp. 187–206.

Droby, S., Vinokur, V., Weiss, B., Cohen, L., Daus, A., Goldschmidt, E. E. and Porat, R. 2002. Induction of resistance to *Penicillium digitatum* in grapefruit by the yeast biocontrol agent *Candida oleophila*. *Phytopathology* 92(4), 393–9. doi:10.1094/PHYTO.2002.92.4.393.

Droby, S., Wisniewski, M., Macarisin, D. and Wilson, C. 2009. Twenty years of postharvest biocontrol research: is it time for a new paradigm? *Postharvest Biol. Technol.* 52, 137–45.

Droby, S., Wisniewski, M., Teixidó, N., Spadaro, D. and Jijakli, M. H. 2016. The science, development, and commercialization of postharvest biocontrol products. *Postharvest Biol. Technol.* 122, 22–9.

El-Ghaouth, A., Arul, J., Ponnampalam, R. and Boulet, M. 1991. Chitosan coating effect on storability and quality of fresh strawberries. *J. Food Sci.* 56, 1618–20.

El-Ghaouth, A., Wilson, C. L. and Wishniewski, M. 1998. Ultrastructural and cytochemical aspects of the biological control of *Botrytis cinerea* by *Candida saitoana* in apple fruit. *Phytopathology* 88(4), 282–91. doi:10.1094/PHYTO.1998.88.4.282.

El-Ghaouth, A., Smilanick, J. L. and Wilson, C. L. 2000a. Enhancement of the performance of *Candida saitoana* by the addition of glycolchitosan for the control of postharvest decay of apple and citrus fruit. *Postharvest Biol. Technol.* 19, 103–10.

El-Ghaouth, A., Smilanick, J. L., Wisniewski, M. and Wilson, C. L. 2000b. Improved control of apple and citrus fruit decay with a combination of *Candida saitoana* and 2-deoxy-d-glucose. *Plant Dis.* 84(3), 249–53. doi:10.1094/PDIS.2000.84.3.249.

Elsherbiny, E. A., Amin, B. H. and Baka, Z. A. 2016. Efficiency of pomegranate (*Punica granatum* L.) peels extract as a high potential natural tool towards Fusarium dry rot on potato tubers. *Postharvest Biol. Technol.* 111, 256–63.

Feliziani, E., Santini, M., Landi, L. and Romanazzi, G. 2013. Pre- and postharvest treatment with alternatives to synthetic fungicides to control postharvest decay of sweet cherry. *Postharvest Biol. Technol.* 78, 133–8.

Freitas, P., Suslow, T. and Mercier, J. 2005. Biofumigation with *Muscodor albus* for postharvest control of gray mold rot and *Salmonella* contamination of tomatoes. *Phytopathology* 95, S31.

Friel, D., Pessoa, N. M. G., Vandenbol, M. and Jijakli, M. J. 2007. Separate and combined disruptions of two exo-β-1,3-glucanase genes decrease the efficiency of *Pichia anomala* (strain K) biocontrol against *Botrytis cinerea* on apple. *Mol. Plant-Microbe Inter.* 20, 371–9.

Gopal, M., Gupta, A. and Thomas, G. V. 2013. Bespoke microbiome therapy to manage plant diseases. *Front. Microbiol.* 4, 355. doi:10.3389/fmicb.2013.00355.

Guijarro, B., Larena, I., Melgarejo, P. and De Cal, A. 2006. Effect of drying on viability of *Penicillium frequentans*, a biological control agent against brown rot disease caused by *Monilinia* spp. *Biocont. Sci. Technol.* 16, 257–69.

Gustavsson, J., Cederberg, C., Sonesson, U. and van Otterdijk, R. 2011. *Global Food Losses and Food Waste; Extent, Causes and Prevention*. Swedish Institute for Food and Biotechnology (SIK), Gothenburg, Sweden and FAO, Rome, Italy.

Hernandez-Montiel, L. G., Ochoa, J. L., Troyo-Dieguez, E. and Larralde-Corona, C. P. 2010. Biocontrol of postharvest blue mold (*Penicillium italicum* Wehmer) on Mexican lime by marine and citrus *Debaryomyces hanseii* isolates. *Postharvest Biol. Technol.* 56, 181–7.

Hershkovitz, V., Ben-Dayan, C., Raphael, G., Pasmanik-Chor, M., Liu, J., Belausov, E., Aly, R., Wisniewski, M.. and Droby, S. 2012. Global changes in gene expression of grapefruit peel tissue in response to the yeast biocontrol agent *Metschnikowia fructicola*. *Mol. Plant Pathol.* 13(4), 338–49. doi:10.1111/j.1364-3703.2011.00750.x.

Hershkovitz, V., Sela, N., Taha-Salaime, L., Liu, J., Rafael, G., Kessler, C., Aly, R., Levey, M., Wisniewski, M. and Droby, S. 2013. *De-novo* assemble and characterization of the transcriptome of *Metschnikowia fructicola* reveals differences in gene expression following interaction with *Penicillium digitatum* and grapefruit peel. *BMC Genomics* 14, 168. doi:10.1186/1471-2164-14-168.

Hofstein, R., Fridlender, B., Chalutz, E. and Droby, S. 1994. Large scale production and pilot testing of biocontrol agents of postharvest diseases. In: Wilson, C. L. and Wisniewski, M. E. (Eds), *Biological Control of Postharvest Diseases of Fruits and Vegetables – Theory and Practice*. CRC Press, Boca Raton, FL, pp. 89–100.

Huang, R., Che, H. J., Zhang, J., Yang, L., Jiang, D. H. and Li, G. Q. 2012. Evaluation of *Sporidiobolus pararoseus* strain YCXT3 as biocontrol agent of *Botrytis cinerea* on post-harvest strawberry fruits. *Biol. Control* 62, 53–63.

Ianiri, G., Idnurm, A., Wright, S. A. I., Durán-Patrón, R., Mannina, L., Ferracane, R., Ritieni, A. and Castoria, R. 2013 Searching for genes responsible for patulin degradation in a biocontrol yeast provides insight into the basis for resistance to this mycotoxin. *Appl. Environ. Microbiol.* 79(9), 3101–15. doi:10.1128/AEM.03851-12.

Ing, L. Y., Zin, N. M., Sarwar, A. and Katas, H. 2012. Antifungal activity of chitosan nanoparticles and correlation with their physical properties. *Int. J. Biomater.* 2012, 632698. doi:10.1155/2012/632698.

Ismaiel, A. A. and Papenbrock, J. 2015. Mycotoxins: producing fungi and mechanisms of phytotoxicity. *Agriculture* 5, 492–537.

Janisiewicz, W. J. and Korsten, L. 2002. Biological control of postharvest diseases of fruits. *Ann. Rev. Phytopathol.* 40, 411–41. doi:10.1146/annurev.phyto.40.120401.130158.

Janisiewicz, W. J., Leverentz, B., Conway, W. S., Saftner, R. A., Reed, A. N. and Camp, M. J. 2003. Control of bitter rot and blue mold of apples by integrating heat and

antagonist treatments on 1-MCP treated fruit stored under controlled atmosphere conditions. *Postharvest Biol. Technol.* 29, 129–43.
Janisiewicz, W. J., Takeda, F., Glenn, D. M., Camp, M. J. and Jurick, W. M. 2016. Dark period following UV-C treatment enhances killing of *Botrytis cinerea* conidia and controls gray mold of strawberries. *Phytopathology* 106(4), 386–94. doi:10.1094/PHYTO-09-15-0240-R.
Jenssen, H., Hamill, P. and Hancock, R. E. 2006. Peptide antimicrobial agents. *Clin. Microbiol. Rev.* 19(3), 491–511. doi:10.1128/CMR.00056-05.
Jungbauer, A. 2013. Continuous downstream processing of biopharmaceuticals. *Trends Biotechnol.* 31(8), 479–92. doi:10.1016/j.tibtech.2013.05.011.
Karabulut, O. A. and Baykal, N. 2004. Integrated control of postharvest disease of peaches with a yeast antagonist, hot water and modified atmosphere packaging. *Crop Prot.* 23, 431–5.
Keymanesh, K., Soltani, S. and Sardari, S. 2009. Application of antimicrobial peptides in agriculture and food industry. *World J. Microbiol. Biotechnol.* 25, 933–44.
Kurtzman, C. P. and Droby, S. 2001. *Metschnikowia fructicola*, a new ascosporic yeast effective for biocontrol of postharvest fruit rots. *Syst. Appl. Microbiol.* 24(3), 395–9. doi:10.1078/0723-2020-00045.
Larena, I., De Cal, A., Liñan, M. and Melgarejo, P. 2003. Drying of *Epicoccum nigrum* conidia for obtaining a shelf-stable biological product against Brown rot disease. *J. Appl. Microbiol.* 94(3), 508–14.
Li Destri Nicosia, M. G., Pangallo, S., Raphael, G., Romeo, F. V., Strano, M. C., Rapisarda, P., Droby, S. and Schena, L. 2016. Control of postharvest fungal rots on citrus fruit and sweet cherries using a pomegranate peel extract. *Postharvest Biol. Technol.* 114, 54–61.
Liu, H. M., Guo, J. H., Cheng, Y. J., Liu, P., Long, C. A. and Deng, B. X. 2010. Inhibitory activity of tea polyphenol and *Hanseniaspora uvarum* against *Botrytis cinerea* infections. *Lett. Appl. Microbiol.* 51(3), 258–63. doi:10.1111/j.1472-765X.2010.02888.x.
Liu, J., Sui, Y., Wisniewski, M., Droby, S. and Liu, Y. 2013. Utilization of antagonistic yeasts to manage postharvest fungal diseases of fruit. *Int. J. Food Microbiol.* 167(2), 153–60. doi:10.1016/j.ijfoodmicro.2013.09.004.
Lutz, M. C., Sosa, M. C., Rodriguez, M. E., Lopez, C. A. and Sangorrín, M. P. 2013. Efficacy and putative mode of action of native and commercial antagonistic yeasts against postharvest rots of pear pathogens. *Int. J. Food Microbiol.* 164, 166–72.
Macarisin, D., Cohen, L., Eick, A., Rafael, G., Belausov, E., Wisniewski, M. E. and Droby, S. 2007. *Penicillium digitatum* suppresses production of hydrogen peroxide in host tissue during infection of citrus fruit. *Phytopathology* 97(11), 1491–500. doi:10.1094/PHYTO-97-11-1491.
Macarisin, D., Droby, S., Bauchan, G. and Wisniewski, M. 2010. Superoxide anion and hydrogen peroxide in the yeast antagonist-fruit interaction: a new role for reactive oxygen species in postharvest biocontrol? *Postharvest Biol. Technol.* 58, 194–202.
Maghenzani, M., Chiabrando, V., Santoro, K., Spadaro, D. and Giacalone, G. 2018. Essential oil vapour treatment (*Thymus vulgaris* and *Satureja montana*) on postharvest quality of sweet cherry (cv Ferrovia). *J. Foods Nutr. Res.* 57(2), 161–9.
Mahonu, G. K., Zhang, H., Yang, Q., Zhang, X., Li, D. and Zhou, Y. 2016. Improving the biocontrol efficacy of *Pichia caribbica* with phytic acid against postharvest blue mold and natural decay in apples. *Biol. Control* 92, 172–80.

Mari, M., Neri, F. and Bertolini, P. 2011. Fruit postharvest disease control by plant bioactive compounds. In: Rai, M. and Chikindas, M. (Eds), *Natural Antimicrobial in Food Safety and Quality* (vol. 21). CAB International, Oxford, pp. 242-60.

Mari, M., Martini, C., Spadoni, A., Rouissi, W. and Bertolini, P. 2012. Biocontrol of apple postharvest decay by *Aureobasodium pullulans* strains. *Postharvest Biol. Technol.* 73, 56-62.

Mari, M., Bautista-Baños, S. and Sivakumar, D. 2016. Decay control in the postharvest system: role of microbial and plant volatile organic compounds. *Postharvest Biol. Technol.* 122, 70-81. doi:10.1016/j.postharvbio.2016.04.014.

Massart, S., Perazzolli, M., Höfte, M., Pertot, I. and Jijakli, M. H. 2015. Impact of the omic technologies for understanding the modes of action of biological control agents against plant pathogens. *BioControl* 60, 725-46.

Melin, P., Schnürer, J. and Hakansson, S. 2011. Formulation and stabilisation of the biocontrol yeast *Pichia anomala*. *Antonie van Leeuwenhoek* 99(1), 107-12. doi:10.1007/s10482-010-9522-5.

Mercier, J. and Jimenez, B. 2004. Control of fungal decay of apples and peaches by the biofumigant fungus *Muscodor albus*. *Postharvest Biol. Technol.* 31, 1-8.

Mercier, J. and Smilanick, J. L. 2005. Control of green mold and sour rot of stored lemon by biofumigation with *Muscodor albus*. *Biol. Cont.* 32, 401-7.

McLaughlin, R., Wisniewski, M., Wilson, C. and Chalutz, E. 1990. Effect of inoculum concentration and salt solutions on biological control of postharvest diseases of apple with *Candida* sp. *Phytopathology* 80, 456-61.

Mlikota Gabler, F., Fassel, R., Mercier, J. and Smilanick, J. L. 2006. Influence of temperature, inoculation interval, and dose on biofumigation with *Muscodor albus* to control postharvest gray mold on grapes. *Plant Dis.* 90, 1019-25.

Morath, S. U., Hung, R. and Bennett, J. W. 2012. Fungal volatile organic compounds: a review with emphasis on their biotechnological potential. *Fungal Biol. Rev.* 26, 73-83.

Mounir, R., Durieux, A., Bodo, E., Allard, C., Simon, J. P., Achbani, E. H., El Jaafari, S., Douira, A. and Jijakli, M. H. 2007. Production, formulation and antagonistic activity of the biocontrol like-yeast *Aureobasidium pullulans* against *Penicillium expansum*. *Biotechnol. Lett.* 29(4), 553-9. doi:10.1007/s10529-006-9269-2.

Mustafa, M. A., Ali, A., Manickam, S. and Siddiqui, Y. 2013. Ultrasound-assisted chitosan-surfactant nanostructure assemblies: towards maintaining postharvest quality of tomatoes. *Food Bioprocess Technol.* 7, 2102-11.

Nunes, C., Usall, J., Teixidó, N., Abadias, M. and Viñas, I. 2002. Improved control of postharvest decay of pears by the combination of *Candida sake* (CPA-1) and ammonium molybdate. *Phytopathology* 92(3), 281-7. doi:10.1094/PHYTO.2002.92.3.281.

Ongena, M. and Jacques, P. 2008. *Bacillus* lipopeptides: versatile weapons for plant disease control. *Trends Microbiol.* 16(3), 115-25. doi:10.1016/j.tim.2007.12.009.

Palou, L., Asgar, A., Falik, E. and Romanazzi, G. 2016. GRAS, plant- and animal-derived compounds as alternatives to conventional fungicides for the control of postharvest diseases of fresh horticultural produce. *Postharvest Biol. Technol.* 122, 41-52.

Patiño-Vera, M., Jiménez, B., Balderas, K., Ortiz, M., Allende, R., Carrillo, A. and Galindo, E. 2005. Pilot-scale production and liquid formulation of *Rhodotorula minuta*, a potential biocontrol agent of mango anthracnose. *J. Appl. Microbiol.* 99(3), 540-50. doi:10.1111/j.1365-2672.2005.02646.x.

Piombo, E., Sela, N., Wisniewski, M., Hoffmann, M., Gullino, M. L., Allard, M. W., Levin, E., Spadaro, D. and Droby, S. 2018. Genome sequence, assembly and characterization of two metschnikowia fructicola strains used as biocontrol agents of postharvest diseases. *Front. Microbiol.* 9, 593. doi:10.3389/fmicb.2018.00593.

Plaza, P., Usall, J., Smilanick, J. L., Lamarca, N. and Viñas, I. 2004. Combining *Pantoea agglomerans* (CPA-2) and curing treatments to control established infections of *Penicillium digitatum* on lemons. *J. Food Prot.* 67(4), 781-6.

Regnault-Roger, C., Hamraoui, A., Holeman, M., Theron, E. and Pinel, R. 1993. Insecticidal effect of essential oils from Mediterranean aromatic plants upon Acanthoscelides obtectus Say, Coleoptera, bruchid of kidney bean (*Phaseolus vulgaris* L.). *J. Chem. Ecol.* 19(6), 1233–44. doi:10.1007/BF00987383.

Romanazzi, G., Mlikota Gabler, F. M., Margosan, D. A., Mackey, B. E. and Smilanick, J. L. 2009. Effect of chitosan dissolved in different acids on its ability to control postharvest gray mold of table grape. *Phytopathology* 99(9), 1028-36. doi:10.1094/PHYTO-99-9-1028.

Romanazzi, G., Lichter, A., Mlikota Gabler, F. and Smilanick, J. L. 2012. Recent advances on the use of natural and safe alternatives to conventional methods to control postharvest gray mold of table grapes. *Postharvest Biol. Technol.* 63, 141-7.

Romanazzi, G., Feliziani, E., Bautista-Baños, S. and Sivakumar, D. 2016. Shelf life extension of fresh 1051 fruit and vegetables by chitosan treatment. *Crit. Rev. Food Sci. Nutr.* 57, 579-601.

Rouissi, W., Ugolini, L., Martini, C., Lazzeri, L. and Mari, M. 2013. Control of postharvest fungal 21 pathogens by antifungal compounds from *Penicillium expansum*. *J. Food Protect.* 76, 1879-86.

Santoro, K., Maghenzani, M., Chiabrando, V., Bosio, P., Gullino, M. L., Spadaro, D. and Giacalone, G. 2018. Thyme and savory essential oil vapor treatments control brown rot and improve the storage quality of peaches and nectarines, but could favor gray mold. *Foods* 7(1). doi:10.3390/foods7010007.

Saravanakumar, D., Ciavorella, A., Spadaro, D., Garibaldi, A. and Gullino, M. L. 2008. *Metschnikowia pulcherrima* strain MACH1 outcompetes *Botrytis cinerea*, *Alternaria alternata* and *Penicillium expansum* in apples through iron depletion. *Postharvest Biol. Technol.* 49, 121-8.

Sivakumar, D. and Bautista-Baños, S. 2014. A review on the use of essential oils for postharvest decay 21 control and maintenance of fruit quality during storage. *Crop Prot.* 64, 27-37.

Smilanick, J. L., Margosan, D. A., Mlikota-Gabler, F., Usall, J. and Michael, I. F. 1999. Control of citrus green mold by carbonate and bicarbonate salts and the influence of commercial postharvest practices on their efficacy. *Plant Dis.* 83(2), 139-45. doi:10.1094/PDIS.1999.83.2.139.

Spadaro, D. and Droby, S. 2016. Development of biocontrol products for postharvest diseases of fruit: the importance of elucidating the mechanisms of action of yeast antagonists. *Trends Food Sci. Technol.* 47, 39-49.

Spadaro, D. and Gullino, M. L. 2014. Use of essential oil to control postharvest rots on pome and stone fruit. In: Prusky, D. and Gullino, M. L. (Eds), *Plant Pathology in the 21th Century: Contribution to the 10th International ISPP Congress. Volume 7. Post-harvest Pathology*. Springer, Dordrecht, The Netherlands, pp. 101-10.

Spadaro, D., Garibaldi, A. and Gullino, M. L. 2004. Control of *Botrytis cinerea* and *Penicillium expansum* on apple combining a biocontrol agent with hot water dipping and acibenzolar-S-methyl, baking soda, or ethanol application. *Postharvest Biol. Technol.* 33, 141-51.

Spadaro, D., Sabetta, W., Acquadro, A., Portis, E., Garibaldi, A. and Gullino, M. L. 2008. Use of AFLP for differentiation of *Metschnikowia pulcherrima* strains for postharvest disease biological control. *Microbiol. Res.* 163(5), 523-30. doi:10.1016/j.micres.2007.01.004.

Spadaro, D., Ciavorella, A. A., Lopez-Reyes, J. G., Garibaldi, A. and Gullino, M. L. 2010. Effect of culture age, protectants, and initial cell concentration on viability of freeze-dried cells of *Metschnikowia pulcherrima*. *Can. J. Microbiol.* 56(10), 809-15. doi:10.1139/w10-068.

Spadaro, D., Lorè, A., Garibaldi, A. and Gullino, M. L. 2013. A new strain of *Metschnikowia fructicola* for postharvest control of *Penicillium expansum* and patulin accumulation on four cultivars of apple. *Postharvest Biol. Technol.* 75, 1-8.

Strobel, G. A., Dirkse, E., Sears, J. and Markworth, C. 2001. Volatile antimicrobials from *Muscodor albus*, a novel endophytic fungus. *Microbiology (Reading, Engl.)* 147(11), 2943-50. doi:10.1099/00221287-147-11-2943.

Sui, Y., Wisniewski, M., Droby, S. and Liu, J. 2015. Responses of yeast biocontrol agents to environmental stress. *Appl. Environ. Microbiol.* 81(9), 2968-75. doi:10.1128/AEM.04203-14.

Tian, S. P., Yao, H. J., Deng, X., Xu, X. B., Qin, G. Z. and Chan, Z. L. 2007. Characterization and expression of β-1,3-glucanase genes in jujube fruit induced by the microbial biocontrol agent *Cryptococcus laurentii*. *Phytopathology* 97(3), 260-8. doi:10.1094/PHYTO-97-3-0260.

Torres, R., Usall, J., Teixidó, N., Abadıas, M. and Viñas, I. 2003. Liquid formulation of the biocontrol *agent Candida sake* by modifying water activity or adding protectants. *J. Appl. Microbiol.* 94(2), 330-9.

Torres, R., Teixidó, N., Usall, J., Abadias, M., Mir, N., Larrigaudiere, C. and Viñas, I. 2011. Anti-oxidant activity of oranges after infection with the pathogen *Penicillium digitatum* or treatment with the biocontrol agent *Pantoea agglomerans* CPA-2. *Biol. Cont.* 57, 103-9.

Usall, J., Teixidó, N., Fons, E. and Vinas, I. 2000. Biological control of blue mould by a strain of Candida sake under several controlled atmosphere conditions. *Int. J. Food Microb.* 58, 83-92.

Usall, J., Teixidó, N., Abadias, M., Torres, R., Cañamás, T. and Viñas, I. 2009. Improving formulation of biocontrol agents manipulating production process. In: Prusky, D. and Gullino, M. L. (Eds), *Post-harvest Pathology* (vol. 2). Springer, the Netherlands, pp. 149-70.

Usall, J., Casals, C., Sisquella, M., Palou, L. and De Cal, A. 2015. Alternative technologies to control postharvest diseases of stone fruits. *Stewart Postharvest Rev.* 4(2), 1-6.

Usall, J., Ippolito, A. and Sisquella, M. 2016. Physical treatments to control postharvest diseases of fresh fruits and vegetables. *Postharvest Biol. Technol.* 122, 30-40.

Vero, S., Garmendia, G., Gonzalez, M. B., Bentancur, O. and Wisniewski, M. 2013. Evaluation of yeasts obtained from Antarctic soil samples as biocontrol agents for the management of postharvest diseases of apple (*Malus x domestica*). *FEMS Yeast Res.* 13(2), 189-99. doi:10.1111/1567-1364.12021.

Waewthongrak, W., Pisuchpen, S. and Leelasuphakul, W. 2015. Effect of *Bacillus subtilis* and chitosan applications on green mold (*Penicilium digitatum* Sacc.) decay in citrus fruit. *Postharvest Biol. Technol.* 99, 44-9.

Wang, Y., Wang, P., Xia, J., Yu, T., Luo, B., Wang, J. and Zheng, X. D. 2010. Effect of water activity on stress tolerance and biocontrol activity in antagonistic yeast *Rhodosporidium paludigenum*. *Int. J. Food Microbiol.* 143(3), 103–8. doi:10.1016/j.ijfoodmicro.2010.07.035.

Wang, W., Liu, S., Deng, L., Ming, J., Yao, S. and Zeng, K. 2018. Control of citrus post-harvest green molds, blue molds, and sour rot by the cecropin A-melittin hybrid peptide BP21. *Front. Microbiol.* 9, 2455. doi:10.3389/fmicb.2018.02455.

Wilson, C. L. 2013. Establishment of a world food preservation center. *Agric. Food Sec.* 2(1), 1. doi:10.1186/2048-7010-2-1.

Wilson, C. L. and Wisniewski, M. 1989. Biological control of postharvest diseases of fruits and vegetables: an emerging technology. *Annu. Rev. Phytopathol.* 27, 425–41.

Wilson, C. L., Wisniewski, M., Droby, S. and Chalutz, E. 1993. A selection strategy for microbial antagonist to control postharvest diseases of fruits and vegetables. *Sci. Hortic.* 53, 183–9.

Wisniewski, M. E. and Wilson, C. L. 1992. Biological control of postharvest diseases of fruits and vegetables: recent advances. *HortScience* 27, 94–8.

Wisniewski, M. E., Droby, S., Chalutz, E. and Elam, Y. 1995. Effects of Ca^{2+} and Mg^{2+} on *Botrytis cinerea* and *Penicillium expansum* in vitro and on the biocontrol activity of *Candida oleophila*. *Plant Pathol.* 44, 1016–24.

Wisniewski, M., Wilson, C., El Ghaouth, A. and Droby, S. 2001. Non-chemical approaches to postharvest disease control. *Acta Hort.* 553, 407–12.

Wisniewski, M., Wilson, C., Droby, S., Chalutz, E., El Ghaouth, A. and Stevens, C. 2007. Postharvest biocontrol: new concepts and applications. In: Vincent, C., Goettel, M. S. and Lazarovits, G. (Eds), *Biological Control A Global Perspective*. CABI Publishing, Cambridge, MA, pp.262 - 73.

Wisniewski, M., Droby, S., Norelli, J., Liu, J. and Schena, L. 2016. Alternative management technologies for postharvest disease control: the journey from simplicity to complexity. *Postharvest Biol. Technol.* 122, 3–10.

Yánez-Mendizábal, V., Zeriouh, H., Viñas, I., Torres, R., Usall, J., De Vicente, A., Pérez-García, A. and Teixidó, N. 2012. Biological control of peach brown rot (*Monilinia* spp.) by *Bacillus subtilis* CPA-8 is based on production of fengycin-like lipopeptides. *Eur. J. Plant Pathol.* 132, 609–19.

Yu, T., Wang, L., Yin, Y., Feng, F. and Zheng, X. 2008. Suppression of postharvest blue mould of apple fruit by *Cryptococcus laurentii* and N6-benzyladenine. *J. Sci. Food Agric.* 88, 1266–71.

Yu, T., Chen, J., Lu, H. and Zheng, X. 2009. Indole-3-acetic acid improves postharvest biological control of blue mold rot of apple by *Cryptococcus laurentii*. *Phytopathology* 99(3), 258–64. doi:10.1094/PHYTO-99-3-0258.

Zhang, H., Wang, L., Zheng, X. and Dong, Y. 2007. Effect of yeast antagonist in combination with heat treatment on postharvest blue mold decay and *Rhizopus* decay of peaches. *Int. J. Food Microbiol.* 115(1), 53–8. doi:10.1016/j.ijfoodmicro.2006.10.002.

Zhang, H., Wang, S., Huang, X., Dong, Y. and Zheng, X. 2008. Integrated control of postharvest blue mold decay of pears with hot water treatment and *Rhodotorula glutinis*. *Postharvest Biol. Technol.* 49, 308–13.

Zhang, D., Spadaro, D., Valente, S., Garibaldi, A. and Gullino, M. L. 2012. Cloning, characterization, expression and antifungal activity of an alkaline serine protease of *Aureobasidium pullulans* PL5 involved in the biological control of postharvest pathogens. *Int. J. Food Microbiol.* 153(3), 453–64. doi:10.1016/j.ijfoodmicro.2011.12.016.

Zhang, H., Yang, Q., Lin, H., Ren, X., Zhao, L. and Hou, J. 2013. Phytic acid enhances biocontrol efficacy of *Rhodotorula glutinis* against postharvest gray mold spoilage and natural spoilage of strawberries. *LWT Food Sci. Technol.* 52, 110–5.

Part 3

Monitoring and management

Chapter 13

Monitoring postharvest attributes: instrumental techniques for measuring harvest maturity/fruit quality

Kerry B. Walsh and Nicholas T. Anderson, Central Queensland University, Australia

1 Introduction

Fresh fruit and vegetables (henceforth referred to as 'fruit' for convenience) are perishable products that change between harvest and consumption. In effect, a fruit is a plant organ in senescence from the time of harvest. Postharvest management is about managing this senescence to deliver fruit in an 'optimum' state to the consumer. Such management requires monitoring of fruit attributes relevant to eating quality and storability, typically in the context of specifications set on these attributes.

Attributes of relevance to postharvest management of fruit include storage reserve level (oil, starch or sugars), flesh firmness, skin and flesh colour and extent of internal and external defects, such as internal rots or skin blemishes (for a list of relevant attributes and specification, by commodity, see Florkowski et al., 2014). The specifications on these attributes may vary in the context of the point within the value chain where the assessment is to be made, that is at consumption, retail store, wholesaler, ripening room, storage, packhouse or harvest. Environmental factors, such as temperature and composition of the storage atmosphere, influence the trajectory of the postharvest life of a fruit.

http://dx.doi.org/10.19103/AS.2019.0055.14

Thus, the measurement of both the fruit attributes directly and the environmental attributes related to the development of eating quality parameters is useful to both the understanding of the postharvest process and its management. Postharvest behaviour is also influenced by pre-harvest factors. Ideally, these factors would also be measured to inform postharvest management. Walsh and Wang (2018) provide a review of technologies relevant to the postharvest assessment of mango fruit, and this case study is broadened to a consideration of all fruit in the current chapter.

To generalise, the eating quality of non-climacteric fruit is at a peak at the time of harvest, while for climacteric fruit, eating quality may be improved by ripening after harvest. Thus, for melon, a non-climacteric fruit, attributes relevant to eating quality such as soluble sugar content (SSC) can be assessed at harvest, or any stage postharvest. Other attributes can be monitored as an index of the senescence of the fruit, for example, firmness. In a climacteric fruit such as mango, starch is converted to soluble sugars during ripening, so SSC should be measured for an estimate of eating quality. Alternatively, an attribute that is correlated to the attribute of interest at ripeness can be assessed, such as dry-matter content (DMC). DMC is an index of the sum of starch and soluble sugars in the fruit, and so the DMC of fruit at harvest is correlated to SSC of ripened fruit.

Various actors are at play along the value chain, from the farmer to the consumer, and their ability to use measurement technology and to share data varies. Fruit are also packaged in various formats, from bulk field bins after harvest to a singulated stream on a pack-line, to cartons in storage, transport and marketing. These formats will prescribe the types of measurements possible in commercial practice.

It is practical to assess every fruit in a consignment only when the fruit is singulated, as on a pack-line; otherwise a sampling protocol is required. Pack-lines are typically located in growing districts, either on farms or in packing houses serving several farms. Pack-lines are used to segregate fruit by quality for different markets. Such sorting may occur again in repack centres, for example, in global supply chains operating sea freight over large distances, with repacking occurring in the destination continent. For example, the Netherlands is listed among the top three countries in the world for the value of mango exports, for although it produces no mango fruit, it is a handling point for imports into Europe (http://www.worldsrichestcountries.com/top_mango_exporters.html; accessed 2 September 2019).

In this chapter, various technologies and available instrumentation for the assessment attributes relevant to postharvest characterisation are reviewed, in terms of use for scientific study and commercial implementation, and with attention to the point of use within the value chain. Focus is given to offline instrumentation.

2 Sampling strategies

As a rule, the further upstream in the value chain that a measurement is made, the greater chance for management to impact fruit quality as perceived by consumers. However, along most of the chain only a lot assessment is possible; that is, an estimate of population properties can be achieved by the measurement of a sample taken from the consignment, be that in field after harvest or at receival or distribution points. The issue of sampling and sorting statistics is considered in this section.

The total error in the estimate of a population attribute is the summation of all sources of error, which are principally sampling error (s) and measurement error (e) (Eqn. (1)).

$$\text{Total error} = \sqrt{s^2 + e^2} \quad (1)$$

Sampling error generally far exceeds measurement error. Thus, in any assessment, strong consideration must be given to the sampling regime to be followed. This requires the knowledge of the variation in the attribute of interest - within a single piece of fruit, within a consignment and between consignments. For example, SSC varies within a single peach fruit, with higher levels typically recorded on the blush (sun) side of the fruit. Thus, if the measurement technology does not assess the whole fruit, a standard sampling position should be adopted and the level of variation due to sampling position should be documented and accepted into the acceptable margin of error on a specification.

Fruit SSC also varies between fruit within a tree, with higher levels recorded for fruit from the outer canopy, between trees and between harvests and growing location, with differences in canopy architecture and environmental conditions (e.g. Walsh et al., 2007). A sampling regime is, therefore, required that copes with variation within and between trees, and between harvest consignments, yielding a representative sample of fruit. For example, in the authors' experience, a tomato-processing plant experienced a mismatch in the total solids estimated in 10-kg grab samples taken from the top of 10-tonne trucks arriving at the processing plant, relative to the soluble solids (SS) accounted for in waste streams and final product (tomato paste and sauce; Fig. 1). The issue was traced back to inappropriate sampling of the incoming consignments, with the top layers of fruit having higher SSC than deeper layers. The cause of this inhomogeneity was traced to the loading of the trucks. With growers paid on weight and SSC, incentive existed to layer fruit of higher SSC on the top of the load.

For the example mentioned previously, the error on SCC assessment of a truck consignment can be separated into (i) the instrument measurement error of SSC using a refractometer (typically about 0.2% SSC), (ii) the sampling

Figure 1 Tomato fruit arrive at a processing plant. How should fruit be sampled from these incoming consignments to estimate the average SSC?

measurement error between assessments of SSC from the 10-kg grab sample, with variation due to the level of maceration, and (iii) the sampling position with the truck load. In this case, the last factor was by far the dominant contributor to error. Indeed, as a generalisation, sampling error is the dominant contributor of error in the assessment of population mean, and so this aspect should be the focus of attention in designing a measurement protocol.

Note that, the unit of sampling mentioned previously was a 10-kg grab sample. For fresh fruit, however, the unit of sampling is often an individual fruit.

To generalise, the number of units required for a reliable estimate of the mean of a consignment depends on the variation (standard deviation [SD]) of the attribute between units. If there was no variability, a single sample could be assessed for a reliable estimate, while higher variation requires higher sample numbers for the same level of probability and uncertainty (Eqn. (2)).

$$n = \left(t \cdot \frac{SD}{e} \right)^2 \quad (2)$$

where t is the t statistic for a given probability (e.g. 1.96 for a 95% probability, $n > 30$) and e is the desired level of accuracy. For example, if the SD of SSC for a consignment of fruit is estimated at 1.5% SSC, then for 95% probability and an acceptable error of 0.2% SSC (the typical refractometer measurement error), at least 216 samples should be assessed. If a 0.5% SSC error is accepted, only 35 samples are required. Often the SD of the attribute in past populations is known from previous measurements, and a sample number can be conservatively based on this estimate. Alternatively, Eqn. (2) can be used iteratively as samples are assessed as a guide to the required sample number.

Note that the recommended sample size is independent of the population size; that is, it does not matter if the consignment is of 20, 2000 or 2 000 000 fruit if the population SD is the same, the estimated sample number will be the same. Representative sampling is, however, an issue with large consignments, that is, achieving a validate estimate of SD.

3 Sorting statistics

In some measurement exercises, the desired outcome is not a quantitative estimate of attribute level, but a pass/fail (binary) assessment. The statistics of sorting to a binary criterion is more complex than first meets the eye. Consider the situation of sorting to a criterion on SSC, for example, to a specification of 12% SSC, with consignment rejection if the sample mean is below specification. Note that if the population mean just meets only the specification, then 50% of the population will be below specification, and if attribute SD is high, a proportion of the population will be a good way under specification. Additionally, any measurement has an associated error. For the SSC specification mentioned earlier, as population mean decreases towards a specification, there will be more type I errors than type II (i.e. good fruit considered to be out-of-specification fruit than vice versa), while when population mean is below specification, the opposite is true.

The optimisation of a sorting operation in terms of setting of the cut-off point for acceptance must therefore consider the relative importance of type I and II errors, the specification value and the measurement error. For example, if the specification requires a mean of 12% SSC, the SSC sorting cut-off point could be set above 12% SSC when the population mean is under specification, to decrease the number of defect fruit classified as good fruit.

As another example, consider the specification of a retail chain on internal browning in apple. The specification allows for an incidence level of 2% (of defect fruit in the consignment), with an assessment protocol that involves cutting of 30 fruit, chosen from different boxes of different layers and different pallets of a consignment. In this protocol, if one or more fruit of the sampled 30 fruit reveals internal browning, a second sample of fruit is taken. If this also fails, the consignment is deemed out of specification and can be rejected.

If the defect occurred at a 5% incidence level, a 30-fruit sample would include defect fruit in only 0.78 of sampling events (Eqn. (3)). A consignment with a 5% defect incidence level could thus be accepted in 0.22 of cases using a 30-fruit sample. However, a second sampling event reduces the probability of incorrect acceptance in both events to 0.22 × 0.22 = 0.048.

$$P = 1 - (1 - p)^n \tag{3}$$

where P is the probability of encountering a defect fruit in the sample when n fruit are sampled if the defect is present in a fraction p of the population.

The Excel (Microsoft) binomial distribution function can also be used for this calculation, as:

$$P = 1 - \text{BINOMDIST}(x - 1, n, p, \text{TRUE}) \tag{4}$$

For p = 0.01 and n = 30,

$$P = 1 - (1 - 0.05)^{30} = 0.78 \tag{5}$$

In another example, consider that when sampling lots of 30 fruit from a population with an actual defect incidence level of 1% (i.e. well within specification), the sample lot will reveal one or more defect fruit in 0.26 of sampling events (Eqn. (3)). The probability of this outcome occurring in two independent samplings is 0.26 × 0.26 = 0.068. Thus the proposed sampling regime will result in an inappropriate rejection of the consignment in 6.8% of cases at this incidence level, for a specification that allows 2% of fruit to be defective.

In summary, any postharvest measurement application should give attention to the statistics of the exercise, in context of the aim of the measurement exercise.

4 Currently available technologies for the measurement of different attributes

This section reviews currently available technology for measurement of attributes such as number, weight, provenance, temperature, colour, shape, external blemishes, density, SSC, titratable acidity (TA), firmness, ethylene concentration, volatiles, pigment content and photosystem status.

4.1 Number, weight and lineal size

A basic but primary requirement in postharvest management is knowledge of the amount of product, in terms of number of pieces or weight of fruit. In many value chains, an electronic grader in the packing house provides information on the number of fruit assigned to each grade class, as defined by appearance or size as measured by machine vision and weight as assessed using a load cell. For packhouses without such technology, manual systems such as the count of the number of trays packed at various tray insert sizes can be employed.

A range of technologies are utilised at other points in the value chain. Weight cells measure change in electrical resistance when a load is applied, typically using a Wheatstone bridge circuit. Such cells typically achieve an accuracy of 0.03% of full scale, and approximately 1 g for fruit moving on a finger or cup conveyor (Chen, 2018). Weight cells are used throughout the value chain, from harvest bin weights in orchard through to retailer scales able to apply a price per unit weight to the measured weight to provide actual total price (Fig. 2). In industrialised countries, weight cells used in commerce must be certified. In Australia, the National Measurement Institute of the Federal Government is charged with certification of scales, and can levy fines of up to $170000 on a company and $34000 on an individual shown to be using scales that 'short-measure' (https://www.industry.gov.au/regulations-and-standards/buying-and-selling-goods-and-services-by-weights-and-other-measures; accessed 1 July 2019).

Fruit lineal size is specified for some commodities, rather than weight. Size can be estimated manually using callipers or sizing rings (Fig. 3). Digital versions

Figure 2 Images of retail scales (left) and load bars for larger volume weighing (e.g. harvest bin, right).

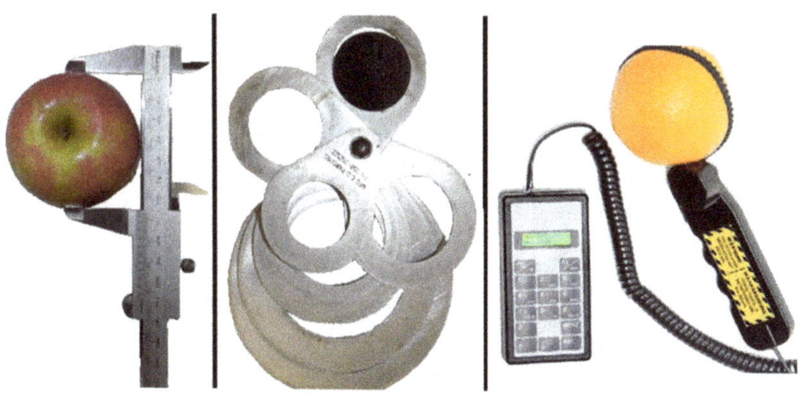

Figure 3 Fruit sizing by calliper (left panel), sizing rings (middle panel) and an electronic fruit size measure (right panel; Guss, Strand, South Africa, http://www.gusstoday.com/datalogger.html; accessed 16 January 2019).

of these measurement devices allow for the transfer of data to a connected tablet (Fig. 3; right). The automation of measurement of lineal dimensions is achieved using camera systems on packing lines, where the camera to fruit distance is fixed. Implementation of camera technology outside of the packing line is not trivial, as variation in the height of the fruit (the distance from the median plane of the fruit to the camera) impacts the calibration of camera pixels to actual distance.

4.2 Temperature

Temperature is a critical environmental variable in the control of postharvest physiological processes. In most value chains, steps are taken to control fruit temperature, for example, placement of fruit harvest bins in shade and refrigeration during storage and transport. A monitoring programme may inform establishment of a practice, but routine monitoring of temperature is still the exception rather than the rule in commercial postharvest practice. In current best practice, temperature measurement and logging is most likely to occur in storeroom and in transport, with potential for alerts on out-of-specification temperature. Multiple temperature sensors can be useful to monitor variation in load, for example, between pallets within a container due to variation in airflow (Moureh and Flick, 2004), or between containers on a ship.

Several technologies are commonly employed in temperature assessment, including liquid expansion thermometers, thermistors, thermocouples and IR sensing devices (Fig. 4). Temperature has long been measured by the coefficient of thermal expansion of a liquid in a confined space. Mercury was commonly used as the liquid, but the toxicity of mercury has seen this technology phased from use, in favour of dyed alcohol. Mercury has a higher freezing and boiling point than alcohol, and so can be used at higher temperatures than alcohol units, while alcohol thermometers can be used to measure lower temperatures than mercury units. Alcohol units can be used for the −10 to 50°C range likely to be encountered in the context of fruit, but care should be taken not to allow the units to overheat, as evaporation can cause the separation of the liquid within the thermometer capillary.

Thermistors (or resistance temperature detectors, RTD) measure the temperature-sensitive electrical resistance of metallic oxides compressed into a coated (epoxy or glass) container. Thermistors can be used across a temperature range of −50 to 250°C, with a typical precision of ±0.2 or 0.1°C and drift of 0.2 or 0.02°C (for epoxy and glass; respectively) per year.

Thermocouples measure the temperature-dependent voltage generated by two dissimilar metal wires attached at the 'hot' junction, which is placed in the sample. The other ends of the wires are connected to a voltage-measuring device. This connection ('cold junction') also involves contact of dissimilar metals, generating another voltage, so the cold junctions should be kept in an ice batch at a known temperature or have their temperature monitored so

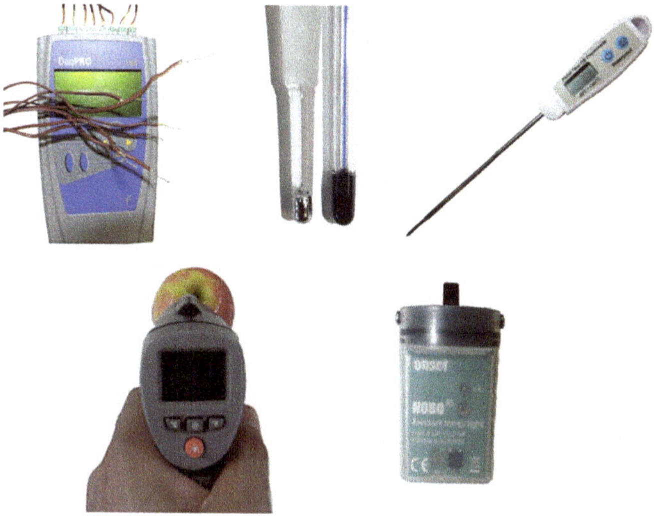

Figure 4 Images of (clockwise from upper left) thermocouples (OMEGA, Norwalk, USA), liquid expansion thermometer (mercury and alcohol), thermistor in insertion probe (HLP Controls, Argyle, Australia), thermistor and data logger in a weather-proof enclosure (Onset, Bourne, USA) and a non-contact IR temp sensor (Dick Smith, Melbourne, Australia).

compensation is possible. Thermocouples are inexpensive and cover a wide range of temperatures (−200 to 1250°C); however, they have less precision than thermistors (1.70–2.20 c.f. 0.05–0.20°C, respectively) and have a higher annual drift per year if used continuously.

Infrared thermometers measure 'blackbody' emissions of infrared wavelengths around 5 μm, at a wavelength at which infrared-absorbing atmospheric gases such as CO_2 and H_2O do not absorb infrared. Blackbody radiation is proportional to temperature. The technique is non-contact and available in imaging as well as point modes. However, the technique involves only surface measurement; that is, the measured emissions are coming from the surface (to a few micrometres) of the sample. Surface temperature can change rapidly with change in ambient conditions, while core temperature lags. Further, the technique requires an 'emissivity' factor to be known for the sample (how well the sample behaves as a black body). Fruit has a high emissivity factor (ε, approximately 0.97), but dry materials, such as a carton, have a lower factor. Thus, an IR temperature device set with $\varepsilon = 0.97$ for the assessment of fruit will give incorrect temperatures if used in the assessment of temperature of other materials, for example, a cardboard carton. Nonetheless, imaging devices can be useful for the assessment of thermal 'leaks', for example, of cold rooms.

The typical technology in use within fruit supply chains is a thermistor sensor with a data logging function, with manual retrieval of data. At present, temperature data are not typically accessed real time in commercial operations. Rather, data are accessed at the end of the journey, for diagnostic uses, such as the identification of issues in cool chain management.

4.3 Colour, shape and blemishes

Consumer visual perception of fruit in terms of colour and blemish greatly influences purchasing. In traditional practice, fruit external (skin) and internal (flesh) colour, shape and blemish level are assessed by eye, ideally with the use of reference colour cards and with consistent lighting and background. Various colour measurement tools exist, and such instruments have been widely adopted in other industries (e.g. paint). However, while the use of such tools in postharvest management occurs for research, uptake within the commercial postharvest industry has primarily been in context of fruit sorting lines. Elsewhere in the value chain, human visual assessment prevails, ideally aided by colour cards spanning the range of colours relevant to a given commodity, and undertaken under consistent lighting.

Several colour spaces have been defined for the instrumental measurement of colour. The Commission Internationale de l'Eclairage (CIE, International Commission on Illumination) established the CIE XYZ colour space in the 1920s, with the tristimulus values representing the response of the three types of receptors ('cones') near the centre of the human eye, which are broadly defined as sensitive to red, green and blue, respectively. However, eye receptor response is actually more complex (Fig. 5), and for the ease of human comprehension other colour scales are often employed.

The most common colour spaces used in postharvest assessment are the CIE L × A × B × (CIE LAB) and the L × C × H × spaces. In CIE LAB, L represents luminosity, a is a red-green space and b is a yellow-blue space (CIE, 1976), In LCH, L stands for lightness, C for chroma (saturation) and H for hue (0° red, 90° yellow, 180° green, 270° blue). The CIE XYZ, LAB and LCH are 'universal colour spaces', representing the colour spectrum visible to the 'average human', and are 'fixed' colour spaces, therefore being 'device-independent'. In contrast, RGB is a variable (device-dependent) space, representing the output of detectors or of screen displays.

Perceived colour is also a function of the spectrum of the illuminating light, so a colour measurement device should also contain a consistent light source, with a correction used to normalise to a standard light source, typically the CIE illuminant D65, which corresponds to the spectral distribution of the midday sun in Western Europe. Further, the colour measurement device should assess diffuse reflection from the sample, not specular reflection. This is typically achieved using an integrating sphere.

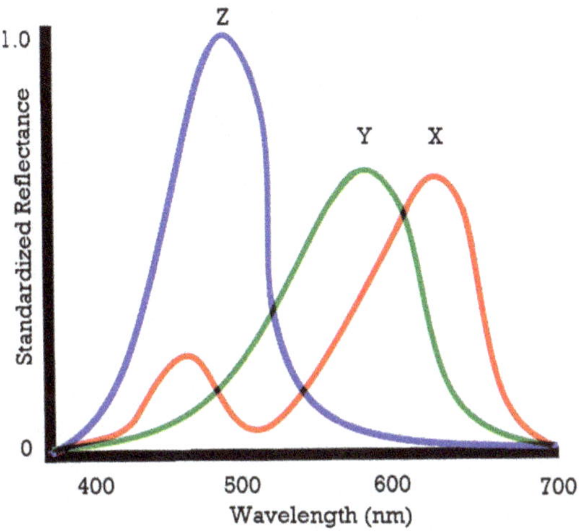

Figure 5 Weighting of CIE X, Y and Z colour spaces (source: adapted from https://www.colourphil.co.uk/xyz_colour_space.shtml).

A spectrometer assessing reflectivity at all wavelengths (e.g. in 10-nm steps) between 380 and 720 nm can be used to estimate the value of these colour spaces of a sample, typically through the calculation of XYZ values (Fig. 5), followed by conversion to other colour spaces. Alternatively, tristimulus colorimeters such as the CR-400 (Minolta, Osaka, Japan; Table 1, Fig. 6) use three filters matched to the CIE X, Y and Z responses and convert these values to other colour spaces. In an attempt to deliver a lower cost solution, several manufacturers now measure reflectance from the sample illuminated by a sequence of specific wavelength LEDS (Table 1, Fig. 6). The assessment of a

Table 1 Examples of handheld chromameters used in the assessment of fruit and vegetables

Instrument	Manufacturer	Light source
Agrocolour[a]	Agrosta, France	LEDs of six wavelengths
Colour Stick[b]	Lincoln University, NZ	Broad spectrum (white) LED
CR400[c]	Minolta, Japan	Xe flash
ProColour[d]	Nix, USA	Broad spectrum (white) LED
STS spectrometer[e]	Ocean Optics, USA	Ambient

[a] https://www.agro-technologies.com/ang/produits/spectrophotometer.htm; accessed on 20 March 2019.
[b] https://www.lincolnagritech.co.nz/capabilities/tech-licensing/colourstick-moist-surface-colorimeter/; accessed on 14 October 2018.
[c] https://sensing.konicaminolta.us/products/cr-400-chroma-meter-colorimeter/; accessed on 14 October 2018.
[d] https://www.nixsensor.com/compare-nixes/; accessed on 20 March 2019.
[e] https://oceanoptics.com/color-light-measurement/; accessed on 20 March 2019.

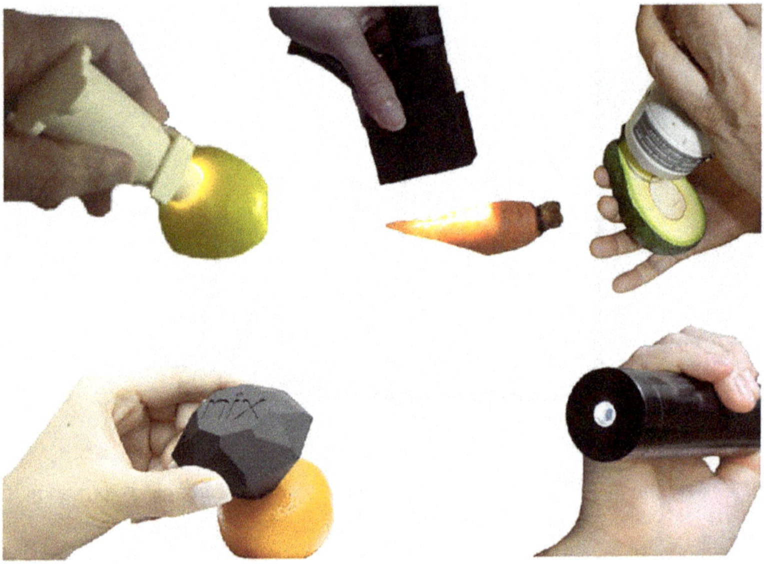

Figure 6 Images of colorimeters (clockwise from upper left) Agrocolour, STS Spectrometer, CR400, Colour Stick and ProColour (as in Table 1).

larger sample area (e.g. 10 cm^2) is also appropriate for the assessment of most fruit, given colour variation across the fruit.

The most common colour space used in commercial practice is RGB, with application occurring in in-line grading since the late 1970s. A typical fruit pack-line utilises multiple RGB cameras within a light box over the pack-line that provides diffuse illumination. Cubero et al. (2011) provide a useful review on machine vision applied to fruit and vegetable evaluation. Colour calibration of the cameras can be carried out, or the operation may simply adjust uncalibrated RGB settings to achieve a desired sorting outcome. Beyond simple estimation of overall colour of produce, images can also be used in the assessment of shape and defects. Traditionally, such assessment involved 'hand-crafted' filters for the segmentation of regions of interest within the image, but increasing use is being made of machine learning techniques (e.g. deep learning algorithms such as convolutional neural networks, Koirala et al., 2019).

4.4 Pigment level and function

Change in fruit chlorophyll, carotenoid and anthocyanin content can occur during fruit maturation and ripening, and so pigment levels can be useful indicators of these processes. Human health benefits have also been claimed for the accessory pigments (e.g. high anthocyanin; Tsuda, 2012). The traditional methods for the assessment of fruit pigment levels involve solvent extracts of fruit tissue, but instrumentation has become available for the proximate assessment of intact fruit (Table 2). Such assessment could use a spectrometer, but a number of field-rugged, lower-cost, dual- or several-wavelength devices have become available. The DA meter (Fig. 7) employs LEDs of peak wavelengths at 680 and 720 nm, with the calculation of a difference index that is related to fruit chlorophyll content. Other devices target anthocyanin estimation. However, while a change in pigment level may be indicative of

Table 2 Examples of dedicated multi-wavelength metres used in the assessment of fruit and vegetables

Instrument	Format	LED peak wavelengths (nm)	Optical geometry	Application
DA meter[a]	Handheld	680, 750	Diffuse reflectance	Chlorophyll
Kiwi meter[a]	Handheld	560, 640, 750	Diffuse reflectance	Red pigments
IDD4[b]	In-line	Four wavelengths	Transmittance	Internal defects
FIORAMA[c]	Handheld	Five wavelengths	Remittance	Chlorophyll, red pigments
Multiplex 330[d]	Handheld	Several wavelengths	Diffuse reflectance and fluorescence	Pigments
SoftRipe[e]	Handheld	Three wavelengths	Diffuse reflectance	Chlorophyll

[a] T.R. Turoni Srl., Forlí, Italy; https://www.trturoni.com/en/fruit-veg-ripeness-quality-control/da-meter/da-meter.5.html; accessed on 20 March 2019.
[b] MAF, Montabaun, France; https://www.maf-roda.com/en/page/electronic-sorting.php; accessed on 20 March 2018.
[c] CP, Potsdam-Golm, Germany; http://cp-info.de/Our-Products/; accessed on 20 March 2019.
[d] Force-A, Paris, France; https://www.force-a.com/en/capteurs-optiques-optical-sensors/multiplex-330-scientific-instrument/; accessed on 20 March 2019.
[e] Frigotec, Landsberg, Germany; https://frigotec.de/; accessed on 20 March 2019.
These units operate by the measurement of reflected or transmitted lights, except the Multiplex which also assesses fluorescence.

Figure 7 Images of pigment analysers from Table 2, left to right: DA meter, Multiplex 330, FIORAMA.

fruit maturation or ripening, the absolute level may be affected by growing conditions (e.g. N availability, temperature). Further, ambient temperature influences LED peak wavelength emission (Hayes et al., 2017). Thus, further work is required to demonstrate measurement reliability in field, and to establish use cases.

Chlorophylls and some anthocyanins exhibit fluorescence (Willstätter and Stoll, 1913; Wolf and Stevens, 1967). This phenomenon can be used to assess the level of these pigments in intact fruit, as claimed for grape (Multiplex 330; Table 2, Fig. 7). Further, the kinetic analysis of chlorophyll fluorescence following brief but intense illumination is used for the assessment of the state of the photosystem, and thus the level of plant stress. This principle is applied in the HarvestWatch (Canada; www.harvestwatch.net/; accessed 15 June 2019) system, with the detection of stress in apple fruit in storage occurring when O_2 levels drop too low.

4.5 Oil gland damage and bruises

Fluorescence can also be used in assessment on certain non-pigment-related attributes. For example, damaged oil glands in citrus fruit exhibit fluorescence under UV excitation (Fig. 8).

Imaging outside the visible region can also be useful in the assessment of certain defects. CCD and CMOS cameras use a silicon detector that is sensitive beyond the visible region, to about 1100 nm. Thus, RGB cameras can be used for far-red imaging, assuming there is no far-red short-pass filter in the camera. Common applications for far-red imaging include bruise detection and improved edge detection of fruit in images.

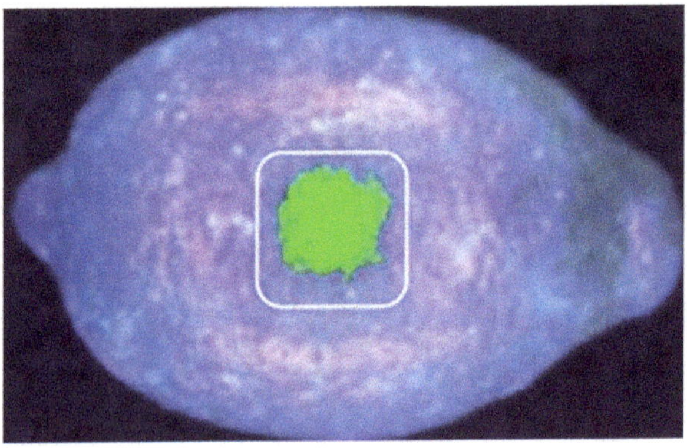

Figure 8 Processed image (segmented on a HSI threshold) of lemon fruit excited by ultraviolet light (390 nm peak), showing the area of mechanical damage to peel (Momin et al., 2013).

4.6 Density

The traditional method for grading fruit for density involved floatation of fruit in a solution adjusted to a desired density by addition of solutes. Sophisticated pack-lines offer 3D reconstruction of imaged fruit, and thus volume estimation. Volume estimation can be combined with weight estimation to allow for density estimation.

X-ray attenuation is a function of path length and average atomic number of the material. This technology is in common use for the detection of foreign objects in postharvest produce, (e.g. glass and metal objects in produce, https://www.mekitec.com/industry-solution/fruits-vegetables/ accessed 3 July 2019; Fig. 9) and can also be used in the detection of voids within produce.

4.7 Soluble sugars, acidity and dry-matter content

Fruit juice is often characterised in terms of 'soluble solids content', also referred to as 'soluble solids' (presented as either g per 100 mL or per 100 g of juice solution, depending on analytical method used in sugar measurement), Brix (assessed using a refractometer, in units of g per 100 g of juice, not water) and Baume (SSC assessed using a hydrometer for measurement of specific gravity, generally in context of grape juice for wine making). The term 'soluble solids' is silent on the form of the dissolved solids. The Baume scale is used for its equivalence to alcohol level after wine fermentation (i.e. a 12 Baume juice should produce a 12% alcohol product; 1.8 Brix = 1 Baume). A Brix value is based on the refractive index of juice extracted from fruit, with the assumption that sucrose is the solute present. Digital refractometers assess the refractive index of the sample based on the angle of total internal reflectance of a monochromatic source of light by the liquid sample (Fig. 10). A Brix reading must be multiplied by the specific gravity of the solution to obtain a value in units of %w/v (i.e. g/100 mL).

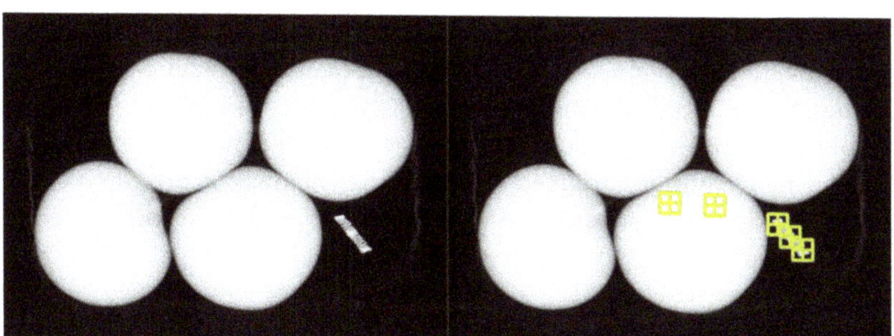

Figure 9 X-ray images of tomatoes with the detection of glass (>2.0 mm) and stainless steel (>0.8 mm) MEKI™ ONE (Mekitec, Oulu, Finland).

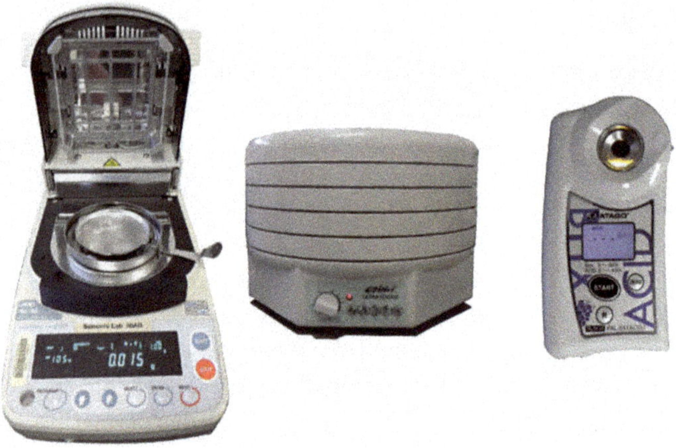

Figure 10 Images (left to right) of ML-50 (A&D Company Ltd, Tokyo, Japan) moisture analyser, FD1000 (Ezidri, Beverley, Australia) forced-air dehydrator and the PAL-BX/ACID2 (Atago, Tokyo, Japan) Brix Acid Meter (refractometer and conductivity meter).

To reiterate this point, the Brix scale is a mass fraction; for example, a 100°Bx solution has 10 g sucrose and 90 g water. At 20°C, 10°Bx has a density of 1.0381 g/mL; that is, a 10°Bx solution is equivalent to 10/1.0381=10.4% w/v (g/100 mL) or 104 g/L. A 20°Bx solution has a density of 1.081; that is, it is equivalent to 10/1.081 = 21.6% w/v (g/100 mL). The difference between Brix and SSC is generally ignored at the concentrations encountered in fruit juices, with terms used interchangeably in common use. SSC is the term in standard usage in postharvest practice.

Refractive index will be influenced by solutes other than sucrose (e.g. salinity level or sucrose level). Thus, if other solutes are present, the Brix measurement is an approximation of the SSC. In lemon juice, the refractometer reading is mostly an assessment of citric acid concentration, with little sucrose present. Juice temperature will also affect refractive index; however, refractometers generally have a temperature compensation mechanism. If the sample is greatly different in temperature to the instrument, a pause before measurement is recommended to allow for correct temperature assessment. Another source of error relates to the settling of particles (particularly starch granules) present in the juice onto the measurement surface. This issue is revealed by change in repeated readings after loading of the sample, and can be avoided by mixing of the sample on the instrument or by reading immediately after sample loading.

Pol (polarity) is commonly assessed in the sugar industry. Glucose and fructose are known as invert sugars as in solution these sugars polarise light in the opposite direction to sucrose. This assessment is rarely used in the fruit industry, despite changing proportions of the soluble sugars during fruit maturation.

TA can be assessed by titration, typically using 0.1 N NaOH and a pH indicator such as phenolphthalein. Alternatively, TA can be assessed by correlation to the conductivity of the solution (e.g. Atago Brix/Acid meter, Fig. 10), given the organic acids are in a de-pronated (charged state; i.e. solution pH is > 7.5). Both methods measure a property of the solution, and do not distinguish between the types of acid or charged solute, respectively, present. An assumption must be made of the dominant moiety present (e.g. citric or malic acid), and results expressed in 'equivalents'.

SSC and TA are expressed in terms of weight per volume of juice, in contrast to DMC, which is expressed relative to whole fruit fresh weight. Consider the example of a fruit that contains 19 g of soluble sugars and 1 g of cell components (cell walls etc.), and thus 80 g of water, per 100 g fresh weight. The DMC is evidently 20% w/w, while the juice SSC is 19 g/80 mL, or 19 g/(19 + 80) g = 19.2°Bx (g/100 g).

DMC is expressed as the percentage ratio of dry to wet weight of a sample. DMC is typically assessed by drying of a sample to a stable weight, typically following 24–48 h of air-forced oven drying, typically at 65°C. Higher temperatures will dry tissue quicker, but at the risk of loss of volatiles (oils etc.) that may impact the DMC estimate. As a low-cost solution, domestic food dehydrators can be used (Fig. 10). Domestic microwave ovens can also be used for tissue drying, but care is required to avoid overheating leading to loss of volatiles or charring. A typical procedure involves operation of the microwave oven in 1-min intervals, with replacement of a beaker of water and reweighing of the sample at each interval, until a stable weight is reached. Laboratory or 'histology' microwave ovens are more expensive than domestic units, but with controlled heating to maintain a set sample temperature. Alternatively, moisture analysers can be used in which a halogen lamp heats the sample to a user set temperature until a stable weight is obtained (e.g. ML-50 moisture analyser; A&D Company Ltd, Tokyo, Japan; Fig. 10). Although the halogen-balance technique can only process one sample at a time, processing time is rapid, at approximately 10 min per sample.

The above-mentioned techniques all require destructive sampling of the fruit sample. Near-infrared spectroscopy (NIR) allows non-invasive and rapid assessment of fruit macro-attributes. The NIR technique is based on the absorption of light, with absorbance bands around 740, 840 and 960 nm associated with O-H bond stretching, and at 910 nm associated with C-H bond stretching. Useful reviews of this technology are provided by Nicolai et al. (2007) and Walsh et al. (2019). The default NIR technology for the fruit application involves the wavelength range of 750–1100 nm and a partial or full transmission optical geometry. The typical root mean square of errors of prediction (RMSEP) is > 1% for DMC, TSS or acidity. For example, Subedi et al. (2007) reported that a multi-cultivar mango model predicted DMC of independent populations of fruit with acceptable accuracy ($R^2 = 0.79$, RMSEP = 0.97%).

There are a number of caveats to the use of the NIR technology with fruit. The technique is not suited to the assessment of constituents present in low concentration, for example, TA of apple. The method is a secondary method, and must be calibrated to the primary method using a multivariate statistical model. As such, a level of quality control is required in the use of such instruments, to confirm calibration performance. The technology suits the rapid assessment of large numbers of samples, as opposed to the sporadic assessment of a small number of samples.

To correct performance issues, a slope and bias correction of results may be required, or a complete multivariate model redevelopment may be required. Calibrations may be based on an indirect assessment (e.g. assessment of an attribute correlated in level to that of the attribute of interest), as long as the association is robust. For example, the Vis–NIR determination of an attribute such as firmness could involve the assessment of a correlated attribute, such as chlorophyll.

There are a number of handheld instruments available (Table 3, Fig. 11), with typical measurement time of a few seconds per sample. Generally, an optical referencing procedure associated with every sample will be required for accurate (low bias) and precise (high R^2) measurements in an environment with a variable ambient light (i.e. in sunlight), a partial transmittance geometry will support more robust calibrations than a reflectance geometry and higher wavelength resolution will support better calibration transfer between instruments.

NIRS technology is also available on fruit packing lines through a number of manufacturers. These systems employ 'point' spectroscopy, but given the use of a partial or full transmission geometry, and fruit moving at up to 1 m s^{-1} (i.e. 25 mm movement during a 25 ms integration time), effectively a large proportion of the fruit is assessed. Matching the physical sampled volume for

Table 3 Examples of commercially available portable Vis–NIR spectrophotometers

Technology	Manufacturer, model	Wavelength range (nm)
Grating and CMOS array-based spectrometer, halogen lamp	SunForest, www.sunforest.kr	790–950
	Felix Instruments, F750, www.felixinstruments.com	300–1100
Grating and Si diode array-based spectrometer, halogen lamp	Sacmi, http://www.sacmi.com	650–1050
Grating and Si diode array-based spectrometer, a fibre-optic bundle ring assemble with halogen lamp	K-BA-100R1, Kubota, jnouki.kubota.co.jp/product/kanren/fruit_selector/	500–1000
Si detector, 20 LEDs of six wavelengths	Atago, Hikari, atago.net/product/	Not available

Figure 11 Images of NIR instruments (left to right) H100-C (SunForest, Icheon, Korea), PAL-HIKARi5 (Atago, Tokyo, Japan) and F750 Produce Quality Meter (Felix Instruments, Camas, USA).

reference analyses to the optically sampled volume can be difficult, and this issue is considered when developing a calibration.

4.8 Firmness

In commercial practice across horticultural value chains, flesh firmness is most commonly assessed using hand feel, with some attempt to standardise assessment by the use of reference latex balls of different firmness (Fig. 12). More rigorous value chains employ instrumental methods to quantify assessment by hand feel ('give' of the fruit) or bite (crunchiness).

The most common tool for fruit firmness assessment in commercial use is the penetrometer. Penetrometers utilise a force gauge to determine the peak force recorded when penetrating into a product at a set speed with a probe of known surface area (e.g. Turoni, Copernico, Italy; https://www.trturoni.com/en). Note that the use of a probe of a different surface area will impact measurement, so it is important to standardise the probe used for an application, and report probe diameter along with peak force. For example, a 6 mm diameter probe is typically used with small fruit, for example, strawberry, while a 13 mm probe is commonly used for the assessment of apple. If an analogue force gauge is used, its maximum range (e.g. 1 or 10 kg) must also be matched to the application. However, handheld penetrometers are subject to operator variation, for example, in speed of insertion (Harker et al., 2006). To reduce this variation, the use of a motorised stand is recommended (Fig. 12).

A texture analyser is a more sophisticated version of a penetrometer in which the profile of force and displacement is recorded as the probe moves into the fruit, for example, Guss (Strand, South Africa, http://www.gusstoday.com/FTAhome.

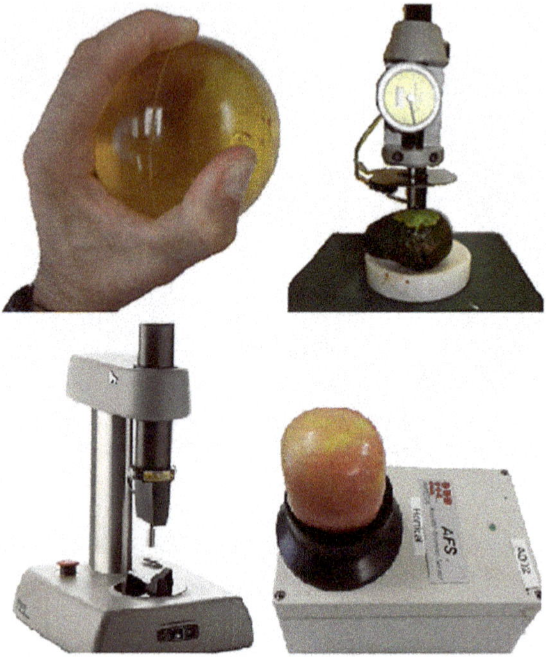

Figure 12 Latex ball for the calibration of hand feel (left top panel), peak force penetrometer on a motorised stand (right top panel), Fruit Texture Analyzer (Guss, South Africa; bottom left panel) and acoustic frequency device (Aweta, Holland; bottom-right panel).

html; accessed 2 September 2019; Fig. 12) and Stable Micro Systems (Surrey, UK; https://www.stablemicrosystems.com, accessed 2 September 2019). Various specialised cells can be used in these texture analysers; for example, Stable Micro Systems offer the 'Ottawa Cell' in which fruit material is extruded through an extrusion plate (https://textureanalysisprofessionals.blogspot.com/2015/03/texture-analysis-in-action-ottawa-cell.html; accessed 2 September 2019).

In practice, the destructive nature of a penetrometer limits sampling effort, given the loss of samples. Non-destructive alternatives exist, although various limitations have restricted postharvest industry adoption. For example, the extent of deformation of a fruit under a set load can be assessed, mimicking the mechanism of hand feel (e.g. the Agrosta14, http://www.agro-technologies.com/ang/produits/crisp_fruits.htm). Fruit acoustic properties also change as flesh firmness (and Young's modulus of the material) changes. These properties can be assessed by either a change in the velocity of a pressure (acoustic) wave travelling through the fruit and past microphones placed at two distances from the impact point or by the assessment of the resonant frequency of the fruit as produced by an impact (Fig. 12; e.g. Subedi and Walsh, 2009). In both systems, a sharp but gentle tap of the fruit produces a pressure wave within the fruit. As this pressure wave reaches the fruit surface, it causes vibration of the surrounding air, detectable by an audio microphone.

4.9 Gases: oxygen, carbon dioxide, water vapour and ethylene

Fruit postharvest physiology is impacted by the level of water vapour, oxygen (O_2) and carbon dioxide (CO_2) levels in the surrounding atmosphere. The assessment of ethylene is also of interest for the postharvest management of climacteric fruit. The measurement of CO_2 and O_2 is routine in controlled atmosphere stores or containers, with several brands of vent control systems available with real-time analysis of O_2/CO_2 (e.g. eAutofresh by Carrier, AFAM$^+$ by ThermoKing and AV$^+$ by MCI). Ethylene is not as routinely assessed in fruit supply chains for want of a sensor technology that is sufficiently sensitive, selective, reliable, robust and cost-efficient.

The classic analytical approach to the analysis of gas composition involves the use of a gas chromatograph (GC). In this technique, an aliquot of sample gas is injected in a stream of an inert carrier gas, with the separation of component gases through differential affinity to a 'stationary phase' packed into a column and detection of the separated component gases on exit from the column. Component gas separation is a function of the stationary phase used, the length of the column, the carrier gas flow rate and temperature. For example, the separation of O_2, CO_2 and C_2H_4 can be achieved using Molecular Sieve 5A in a 60- to 80-cm column, using a thermal conductivity detector (TCD) as a detector. Injection of the gas aliquot can be automated, but the requirement for a carrier gas limits the portability of this instrumentation. Instrumentation cost is also reasonably high, and thus the technology has not been adopted for monitoring in containers or CA stores.

Several other types of detection systems are available for each gas, each with advantages and disadvantages around cost, operational life and detection level. Detection limit is a function of sensitivity and stability and is usually defined as three times the SD of a background/blank measurement.

4.9.1 Oxygen

Oxygen (O_2) is a paramagnetic gas and will move under a magnetic field. In a paramagnetic cell, the movement of O_2 molecules under the influence of a magnetic field induces a shift in an assembly in the gas stream, which is measured by the angle of reflection of a light beam. The technique has a low limit of detection, but it is sensitive to dust. More commonly, O_2 is assessed with an electrochemical galvanic cell, commonly employing a ceramic sensor containing zirconium oxide. These sensors suffer interference from combustible gases (e.g. ethylene), and their operation involves oxidation of the anode, limiting effective life (to several years).

4.9.2 Carbon dioxide

Carbon dioxide (CO_2) levels are commonly assessed based on the absorption of infrared radiation around 4300 nm (2325 cm^{-1} in Fig. 13), based on non-dispersive infrared absorption or a photo-acoustic sensor (e.g. Ntron, https://www.ntron.com/).

4.9.3 Water vapour

Water vapour content of an atmosphere can be assessed using infrared absorption (~2600 nm or 3850 cm^{-1}, Fig. 13), electronic transducers, the temperature difference of a dry and moistened probe or the estimation of dew point temperature.

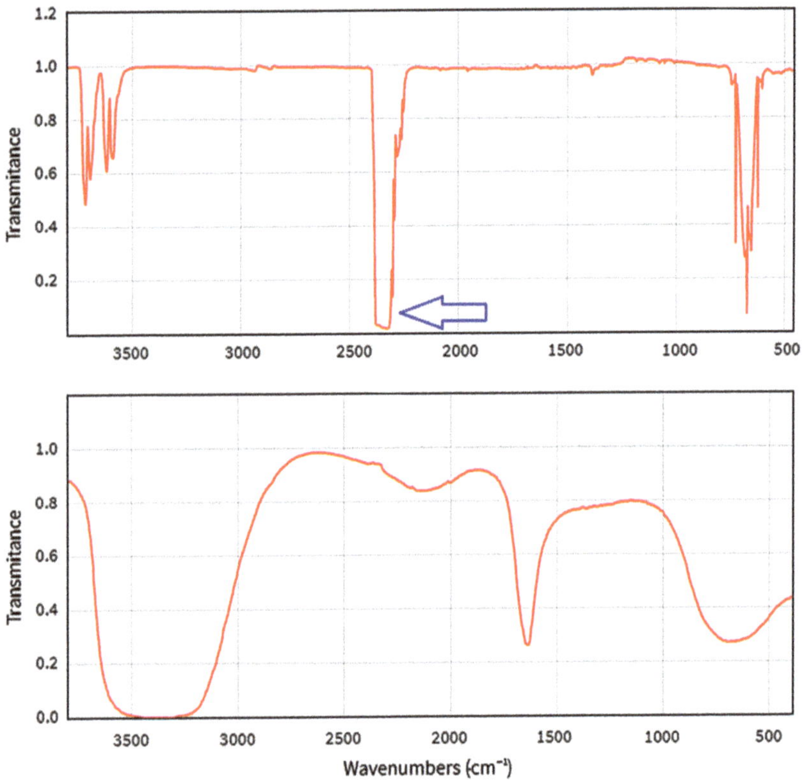

Figure 13 Transmission spectrum of CO_2 (top) and H_2O vapour (bottom; NIST https://webbook.nist.gov; accessed 1 July 2019). CO_2 can be assessed by the measurement of absorption at 2325 cm^{-1} (arrow) at which wavelength interference from water is minor.

4.9.4 Ethylene

Ripening rooms commonly use ethylene levels of around 50–100 ppm for a 24-h period. Ripening in climacteric fruit is triggered by levels around 1 ppm. However, the detection of levels as low as 50 ppb is useful for assessment at pre-climacteric stages. Useful reviews of ethylene detection technologies are provided by Cristescu et al. (2013) and Janssen et al. (2017), as summarised in the following text.

In commercial postharvest practice, ethylene concentrations in an atmosphere are commonly assessed using the Diels-Alder reaction, which provides a coloured end product (Dewar et al., 1986). Practical implementation involves gas flow through a sampling tube, for example, the Kitagawa ethylene detector tube (e.g. https://qasupplies.com/air-gas-analysis/ethylene-testing/ethylene-sampling-tubes-kits/).

Ethylene measurement using a GC was reported for stored apples, first reported by Meigh (1960). However, as noted previously, the need for a carrier gas and relatively expensive instrumentation has restricted the use of this technology in commercial postharvest practice. Attempts have been made to 'ruggedise' the technology to suit commercial practice, including the use of air as the carrier gas; for example, Zaidi et al. (2016) reported the detection of ethylene evolved by bananas to a resolution of 12 ppbv using a micro-GC and a pre-concentrator to trap ethylene molecules, thus separating the ethylene signal from that of H_2O.

Several other technologies have been employed in the instrumentation for the detection of ethylene in postharvest applications: non-dispersive infrared spectroscopy (NDIR), including miniaturised GC, infrared absorption, electrochemical measurement and photo-ionisation. These techniques are described in the following paragraphs.

Measurement by NDIR absorption spectroscopy is based on the absorption of specific wavelengths of IR (e.g. around 10.6 µm) by ethylene, with gas concentration calculated from IR attenuation following the Lambert-Beer law. The SmartGAS system (Mikrosensorik GmbH, Heilbronn, Germany) is based on the NDIR principle intended for postharvest use, with a detection limit of 20 ppm, as opposed to a myriad of systems intended for the detection of the lower explosion limit of ethylene, and thus a detection limit of approximately 100 ppm.

Photo-acoustic spectroscopy is the technique with the lowest detection limit for ethylene. It provides the highest sensitivity values to ethylene gas, and systems can be relatively small and mobile. Example systems include the Sensor Sense EDT-300 (Sensor Sense B.V., Nijmegen, the Netherlands) and the Gasera F10 (Gasera, Turku, Finland), with sensitivity of 0.3 and 800 ppb, respectively. Typically, a CO_2 laser beam is polarised and chopped at a specific frequency.

Gas in the measurement chamber is heated through the absorption of radiation-inducing pressure changes, which are measured using a microphone.

Chemoresistive or electrochemical sensors are based on the change of the resistivity of a sensor induced by the binding of the gas molecules. For example, in a sensor based on a copper complex mixed with single-walled carbon nanotubes placed between two gold electrodes, ethylene is selectively bound by the copper complex, resulting in a change in the resistivity of the nanoparticles. The F950 (Felix Instruments, Camas, WA, USA) is an example of an instrument used in postharvest applications that uses an electrochemical sensor, with a detection limit of 0.15 ppm C_2H_4 reported. The Easi-1 (Absorger SAS, Les Barthes, France) is based on an electrocatalytic sensor in which ethylene is oxidised on an anode consisting of a Nafion membrane covered with gold nanoparticles, with a weak sulphuric acid solution acting as an electrolyte. The ethylene double bond is oxidised by the gold nanoparticles, creating CO_2, water and two protons. The movement of the protons produces a current which is proportional to the ethylene concentration, with a detection limit of 10 ppb claimed. Another technique is based on the change of a capacitor caused by ethylene, with SnO_2 used as an active surface which will negatively ionise oxygen that can react with the ethylene. However, current generations of these detectors display neither high sensitivity or selectively. Yet another approach involves the use of a photo-ionisation detector in which UV photons striking the sample molecules create charged ions that can be detected (e.g. FreshView, Brisbane, Australia; sensitivity of 0.1 ppm).

The development of technologies for the detection of ethylene is ongoing, with results of practical relevance in addressing the typical compromise of cost and sensitivity expected over the next few years.

4.10 Aroma

The aroma of a fruit is created by a number of gaseous compounds created by the fruit. Separation of these gases by GC and compound identification by ion mobility spectrometry provide a characterisation of the aroma. Such technology is used in the FlavourSpec (G.A.S., Dortmund, Germany, https://www.gas-dortmund.de/, accessed 16 June 2019).

Alternatively, electronic noses have been available since the early 1980s, which recognise odours by the use of a sensor array with elements (typically electrochemical cells) sensitive at various levels to specific volatile gases, providing a 'fingerprint' of the odour without the identification of the component compounds. This principle mimics the function of the human nose and tongue. Indeed, olfactory receptors (proteins that bind to volatiles) can be used in bio-electronic noses. For example, Lebrun et al. (2008) used an electronic nose to successfully discriminate between hard-green and ripened mangoes. The

Cyranose 320 (www.sensigent.com/; accessed 16 June 2019) is arguably the benchmark product in this field. A range of other devices also target this area, including low-cost consumer-targeted devices such as the FoodSniffer (http://www.myfoodsniffer.com/; accessed 16 June 2019; Fig. 14).

Potential and limitations of these technologies are usefully reviewed by Baietto and Wilson (2015). In brief, while these technologies are used in research studies, they have yet to become established in commercial practice. The GC-ion mobility technology is too expensive and too slow for commercial use, while the e-noses are prone to sensor drift over time.

Attempts have been made to use low-cost aldehyde sensitive colorimetric strips, for example, based on the Cannizzaro reaction (Kim et al., 2018) for the detection of volatiles associated with the ripening of fruit. For example, RipeSense of New Zealand employed a clamshell (to concentrate emitted volatiles) with a colorimetric volatiles sensor (Fig. 15) in the mid-2000s.

Figure 14 Example e-nose (FOODsniffer, Redwood City, CA, USA; http://www.myfoodsniffer.com/), placed on a lemon.

Figure 15 RipeSense volatile sensor (Auckland, New Zealand), with colour change as fruit ripen (https://www.kiwinet.org.nz/Portfolio/ViewItem/Project/RipeSense-Labels; accessed 1 July 2019).

4.11 Provenance

In developed economies since 1990, fresh produce typically carries a sticker applied at the packhouse that carries branding and a price lookup (PLU) number (www.plucodes.com; accessed 28 June 2019). The PLU number is for the convenience of the retailer, with the entry of the PLU number at the checkout relieving the checkout operator of the task of identifying the type and cultivar of the fruit. In recent years, an extra digit has been added to the PLU number, carrying information on whether the product is organic or genetically modified.

Bar codes (Fig. 16) carry additional information on provenance. These codes can be applied as stickers or via laser etching of the fruit surface (https://www.troteclaser.com/en-au/trotec-laser-applications/food/; accessed 15 June 2019). For in-house purposes, an operation can invent its own codes, but for universal recognition, codes must be purchased from a managing organisation. The GS1 DataBar (formerly known as Reduced Space Symbology, RSS) is the current industry standard for fruit labelling (https://ww.gs1.org/; accessed 1 July 2019). This labelling allows for the identification of the packhouse and indeed the orchard block that produced the fruit. Such information is useful for fruit management (e.g. imposition of storage and ripening regimes to fruit of common orchard management blocks) and for food safety/biosecurity purposes (tracing fruit to origin). Such stickers also allow for consumer information, for example, using a barcode scanner app on a smart phone to access information about the consignment, and are fundamental to the application of blockchain technologies.

4.12 Conclusion

Technologies that are sufficiently 'mature' to be currently employed in postharvest management have been reviewed. Some of these technologies are in widespread use, while some are in relatively limited use. The next section undertakes a

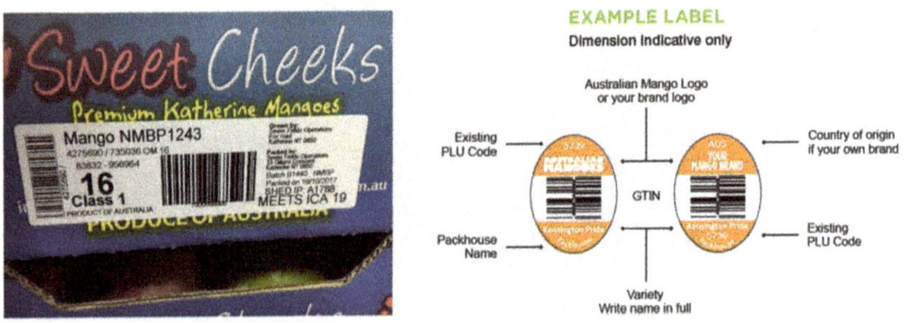

Figure 16 Example bar coding of fruit trays and Australian mango fruit sticker incorporating GS1 DataBar (https://www.industry.mangoes.net.au/resource-collection/2018/7/3/gs1-databar-plu-labels-for-mangoes). https://www.gs1au.org

consideration of technologies that hold promise for application to postharvest management and research but have seen only pilot or no application, as yet.

5 Future technologies for the measurement of different attributes

A wide range of measurement technologies are under development, largely driven by investments in the defence, medical and industrial sectors (e.g. consider the list of sensors available at https://au.mouser.com/Sensors/_/N-5gej; accessed 1 July 2019). Application opportunities to other areas, including postharvest, develop as device cost decreases and reliability increases. One development trend is that of device miniaturisation and the use of multiple sensors within a given device. For example, an iPhone contains a microphone, a proximity sensor, a touch (capacitance) screen, an ambient light sensor, a camera (with face ID capacity), an accelerometer, a gyroscope, a compass, a Touch ID fingerprint sensor, a barometer, a moisture sensor and a temperature sensor, and hosts near field communication (NFC), wifi and cellular communications capacity. Some Android phones carry a 3D camera and a spectrometer. These sensors can be put to a range of 'unintended' uses. For example, the phone app 'Touchscale' uses the capacitance screen of the phone to quantify applied force, and this weight of an object (with capacitance properties) placed on the phone screen.

Of course, the mere availability of a technology does not warrant its uptake by postharvest industries (so-called push factor); rather technology will be rapidly adopted when it addresses a particular need of the industry (the 'pull' factor). In the following sections, key new technologies are outlined and discussed in terms of issues in postharvest practice. These emerging technologies are likely to support unexpected applications in postharvest practice, as has occurred with other technologies and applications (e.g. consider the rise of Facebook on the back of the creation of the World Wide Web).

5.1 Advances in spectroscopy

In most cases, 'point spectroscopy' is adequate for fruit assessment. In some cases, however, knowledge of the distribution of an attribute is useful. In this case, hyperspectral imaging is the relevant technology. As hyperspectral imaging technology matures and decreases in price, commercial applications will develop, for example, for the detection of 'sugar-nose' in potato (Fig. 17).

Near-infrared, mid-infrared (MIR) and Raman spectroscopy are differentiated by the types of molecular vibrations detected (Bureau et al., 2019). Molecular vibrational features in the MIR are sharper and less overlapped than in the NIR, making this region more useful for chemical identification. However,

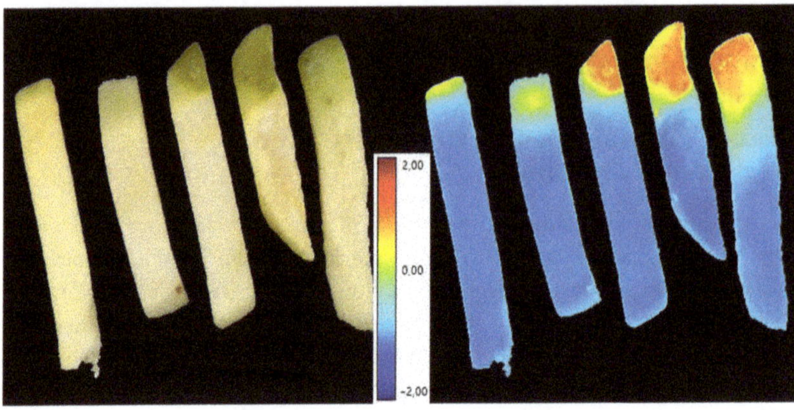

Figure 17 Hyperspectral image of potato strips, showing sugar concentration (https://videometer.com/News/potato-strip-quality-control; accessed 1 July 2019).

water absorption is also very strong in this region, such that the effective depth of penetration of a liquid sample (or a fruit) is only micrometres, compared to centimetres for NIR. MIR spectroscopy is finding use for the assessment of compounds in fruit juice samples, enjoying simple sample preparation and rapid analysis (approximately a minute per sample) relative to HPLC. Limitations include that the technique can only detect molecules with IR active bonds, and so elements are not detected per se, and, as a secondary method, regression models must be maintained. Increasing use by the postharvest industry is likely as instrumentation advances decrease price and increase portability. The application of MIR spectroscopy to address postharvest needs has been reviewed by Bureau et al. (2019).

The application of Raman spectroscopy based on visible light excitation has been limited by the strong fluorescence exhibited by fruit pigments. The advent of a 1064 laser excitation source in Raman spectroscopy has alleviated this limitation, and the technique has been reported in the assessment of fruit pigments and maturity, and in the detection of pathogen and pesticides. The detection of low levels of pesticides using 'surface enhanced Raman spectroscopy' (SERS) holds promise for greater use in the future, given the speed and relative ease of assessment compared to conventional techniques. The application of Raman spectroscopy for postharvest needs has been reviewed by Qin et al. (2019).

5.2 X-ray and magnetic resonance imaging (MRI)

Magnetic resonance imaging (MRI) and X-ray computed tomography (X-ray CT) are technologies that have found widespread use in diagnostic medicine. The technologies can also be used with fruit to reveal internal structure and defects of fruit, but adoption is limited due to high cost, high operator skill requirement and low speed. Both X-ray CT and MRI were utilised by Lammertyn et al. (2003) for the detection and measurement of the area percentage of internal browning and cavity formation of cold stored pears.

MRI detects electromagnetic energy released from protons reorientating after being influenced by strong magnetic and radio frequencies. The speed of the energy release is dependent on the chemical environment of the proton. MRI can be used to map proton distribution within an object, and thus effectively map water within a fruit (Fig. 18). To achieve 3D visualisation of an object, the technique requires the movement of a magnetic field around a stationary object, with data acquisition typically requiring a period of minutes. Aspect Imaging P/L of Israel has attempted a commercial deployment of this technology, with the assessment of boxes of fruit on a stop-start conveyor (https://www.aspectimaging.com/industrial-solutions/agricultural-mri/; accessed 16 May 2019).

X-ray CT detects the differences of mass density from multiple viewpoints around an object, allowing the construction of a 3D image with internal detail. van Dael et al. (2016) applied X-ray CT for the detection of granulation and endoxerosis disorders in citrus, reporting the correct classification of fruit as defect or normal in 95.7% and 93.6% of cases for a population of oranges and lemons, respectively. Porosity mapping is also possible with micro-X-ray CT, as shown by Nugraha et al. (2019) on apple, eggplant, pear and turnips.

However, the acquisition of images from multiple viewpoints requires time, and thus the technique is too slow for commercial fruit assessment. A 2017 patent reports on the use of a 3D representation of fruit from RGB camera

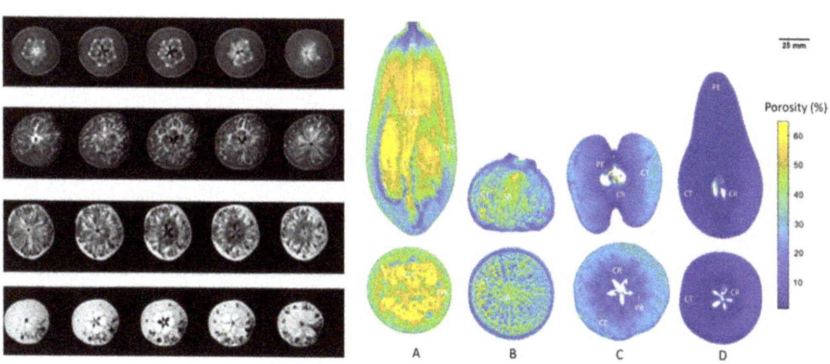

Figure 18 MRI (left panel) of four levels of apple watercore (no watercore to strong watercore, top to bottom) and X-ray CT (right panel) mapping of fruit porosity, sourced from Melado-Herreros et al. (2013) and Nugraha et al. (2019), respectively.

imagery to correct for pathlength differences in a transmission X-ray image (Nicolai et al., 2017), thus avoiding the need for rotation of the X-ray source around a stationary object. van Dael et al. (2019) report on the use of this method, noting detection with an accuracy of >99% of internal defects and voids in fruit larger than 3.5 and 1.5 mm, respectively. This technology appears appropriate for commercial adoption.

5.3 Deep learning in machine vision technology

Deep learning involves the use of several layers of artificial neural networks (ANN) used in conjunction to perform a task (classification, quantitative prediction). These models can process large amounts of data (e.g. for image analysis) by recognising simple patterns in early layers and performing feature extraction (edges, shape etc.), and then undertaking prediction analysis in later layers. For example, in a network searching for blemishes on a fruit image, the task of the first layer might be to identify pixels in the image that are associated to fruit edges, allowing the area of the image that is processed in later steps to be decreased. The process of feature extraction before analysis allows real-time prediction applications. Thus, machine vision techniques for fruit quality measurements that were once too computationally expensive to be commercially viable have become feasible. Koirala et al. (2019) provide a useful review of this topic, summarising developments in this field and the move to architectures that allow detection and counting in a single pass, for example, You only look once (YOLO), that can be implemented real time and on modest computing hardware, for example, a mobile phone.

Example applications at farm include in-field fruit counting and automated harvesting (Koirala et al., 2019), in-field assessments of fruit quality, using images of fruit in field bins (e.g. https://www.croptracker.com/product/harvest-quality-vision.html; accessed 17 June 2019). At the other end of the value chain, retail stores are trialling deep learning machine vision systems to recognise the items chosen by the customer, enabling stores without checkouts (www.amazon.com/b?ie=UTF8&node=16008589011; accessed 16 June 2019).

5.4 Data access and management

Data collected about a consignment of fresh produce in its journey from farm to retail would be valuable to the management of that consignment. An attribute as basic as temperature holds great value in predicting the future quality of the consignment, while assessments of attributes such as SSC or pesticide levels add to the credentials of a consignment. Such information has been managed through a paper trail of documentation of data from sensors accessed on site. However, technology now exists for remote and automated data collection. For example, Escavox (www.escavox.com; accessed 28 June 2019) offer 'smart analytics' around the temperature profile of a consignment, as acquired by a temperature-recording device equipped with data storage and 3G network communications capacity. The device is intended to be placed into a pallet of produce on farm, and real time tracked through to the retailer.

Unfortunately, horticultural production areas and transport pathways are often poorly covered by (3G, 4G) broadband cellular networks, limiting capacity for remote interrogation of sensors or control of processes. Several communication technologies advances could change this scenario in the next decade. For example, Googles' Project Loon (https://loon.com/; accessed 1 July 2019) progresses, redesigning the components of a cell to be carried by balloons at an altitude of 20 km. Other technologies are ground based. For example, the Internet of things (IoT) extends Internet connectivity to physical devices such as sensors, with collected data used to create value in some way. The backbone technology to IoT is a low-power-wide-area-network (LPWAN), enabling the transmission of small data packets. Long-range WAN (LoRaWAN) is the most common protocol (https://lora-alliance.org/; accessed 1 July 2019). The open protocol involves transmission across a range of radio frequencies to decrease the chance that the transmission will be lost by interference. SigFox (https://www.sigfox.com/en; accessed 1 July 2019) utilises a patented slow modulation for transmission and can achieve greater ranges (e.g. to 30 km line of sight); however, a decreased amount of data can be transmitted.

The management of the transient and physically scattered information associated with a fruit consignment is an issue that can be addressed by the distributed ledger system of 'Block chain'. Block chain is a proof-of-work (POW) protocol that obtained notoriety for use in cryptocurrency transactions. Each block is filled with data created by all users in the network, creating a unique 256-digit ID called a hash. If any data are manipulated or changed, the hash is also altered. All blocks in the chain are connected by the previous and next hash ID. POW protocols are meant to decentralise the validation of ledgers. A block chain could allow for rapid sharing of information throughout a horticultural supply chain. Wal-Mart is requiring all leafy green vegetable suppliers to upload their data to the IBM Food Trust Solution blockchain by September 2019. This system is built on Hyperledger Fabric (open-source digital ledger technology) running on IBM Cloud (https://www.hyperledger.org/resources/publications/walmart-case-study; accessed 16 June 2019). The example is given that a mango fruit in store previously took approximately 7 days to trace to its source, while the task was accomplished in 2.2 s using the blockchain.

5.5 Decision support

This chapter has covered the creation and transmission of data related to the measurement of fruit attributes. Turning that data into information, that information to knowledge and that knowledge to understanding is a broader challenge. A produce manager must be able to utilise a range of information about a multitude of consignments in a timely fashion, and thus the information must be 'digested' and presented in a visual format that enables quick comprehension.

Measurement data of a fruit consignment can be used to feed models of fruit consignment behaviour within a decision support system (DSS). For example, the propensity of a consignment of apples to develop flesh browning is impacted by preharvest tree nutrition and temperature regime, and postharvest temperature and gaseous environment (Jobling et al., 2007). At present, such models are based on a set of 'rules' codifying human understanding of the phenomenon, that is, a mechanistic model. As greater amounts of data are collected per consignment, artificial intelligence/deep learning approaches are likely to be used (Fig. 19).

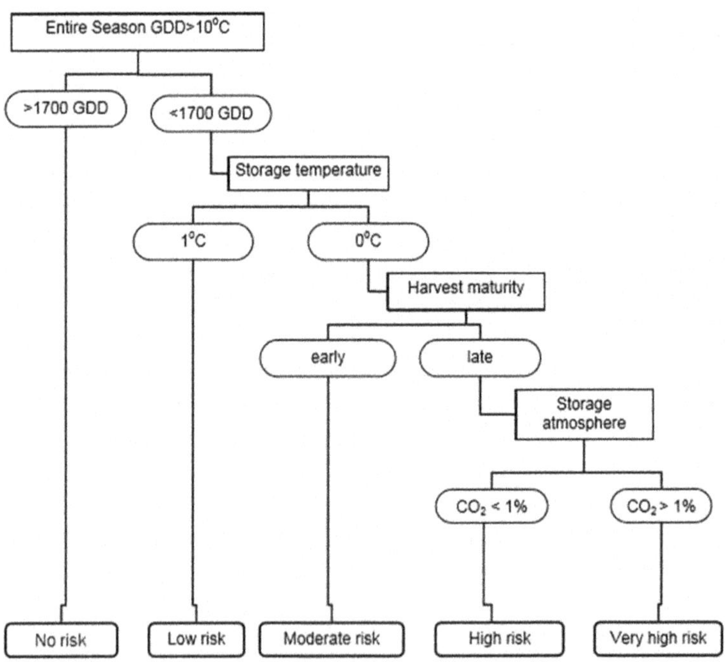

Figure 19 Risk factors for the development of radial flesh browning of 'Cripps Pink' apples during storage related to GDD from Jobling et al. (2007). GDD refers to 'growing degree days', that is, sum of daily (max-min temperature)/2-10°C.

6 Conclusion

This chapter has explored measurement techniques for the assessment of a range of attributes relevant to postharvest quality of fruit, in terms of both fruit attributes and attributes of the fruit's environment. Several areas of likely advances in the near future for measurement instrumentation have been identified. An overview of sampling and sorting statistics has been presented. Further, technologies for the communication, management and interpretation of measurement data have been discussed.

7 Where to look for further information

The reviews of Bureau et al. (2019), Qin et al. (2019) and Walsh et al. (2019) on the application of MIR, NIR and Raman spectroscopy to fruit, and the review by Koirala et al. (2019) on the use of deep learning on machine vision assessment of fruit, are recommended.

8 References

Baietto, M. and Wilson, A. (2015). Electronic-nose applications for fruit identification, ripeness and quality grading. *Sensors*, 15, 899-931. 10.3390/s150100899.

Bureau, S., Cozzolino, D. and Clark, C. J. (2019). Contributions of Fourier-transform mid infrared (FT-MIR) spectroscopy to the study of fruit and vegetables: A review. *Postharvest Biology and Technology*, 148, 1-14, https://doi.org/10.1016/j.postharvbio.2018.10.003.

Chen, G. (2018). *Advances in Agricultural Machinery and Technologies*. CRC Press, Boca Raton, USA.

Cristescu, A. M., Mandon, J., Arslanov, D., De Pessemier, J., Hermans, C. and Harren, F. J. M (2013). Current methods for detecting ethylene in plants. *Annals of Botany*, 111, 347-60.

Cubero, S., Aleixos, N., Moltó, E., Gómez-Sanchis, J. and Blasco and JJF & Technology (2011). Advances in machine vision applications for automatic inspection and quality evaluation of fruits and vegetables. *Food and Bioprocess Technology*, 4, 487-504.

Dewar, M. J., Olivella, S. and Stewart, J. J. (1986). Mechanism of the Diels-Alder reaction: reactions of butadiene with ethylene and cyanoethylenes. *Journal of the American Chemical Society*, 108, 5771-9.

Florkowski, W. J., Shewfelt, R. L., Brueckner, B. and Prussia, S. E. (2014). *Postharvest Handling a Systems Approach*, 3rd edn., Academic Press, San Diego, USA, pp. 167-211.

Harker, F. R., White, A., Gunson, F. A., Hallett, I. C. and De Silva, H. N. (2006). Instrumental measurement of apple texture: A comparison of the single-edge notched bend test and the penetrometer. *Postharvest Biology and Technology*, 39, 185-92.

Hayes, C. J., Walsh, K. B. and Greensill, C. V. (2017). Light-emitting diodes as light sources for spectroscopy: sensitivity to temperature. *Journal of Near Infrared Spectroscopy*, 25, 416-22.

Janssen, S., Schmitt, K., Blanke, M., Bauerfeld, M. L., Wollenstein, J. and Land, W. (2017). Ethylene detection in fruit supply chains. *Philosophical Transactions Series A Mathematical, Physical and Engineering Sciences*, 372: 20130311.

Jobling, J., James, H., Tanner, D., Wilkinson, I., Brown, G., Portman, T., Morris, S., Tustin, S., Mitcham, B. and Zanella, A. (2007). Continuation of the flesh browning of pink lady apple project to validate recommendations. Retrieved from Apple & Pear Australia Ltd: https://apal.org.au/wp-content/uploads/2013/11/AP04008-Continuation-of-the-flesh-browning-of-pink-lady-apple-project-to-validate-recommendations-ext-of-AP02008..pdf.

Kim, Y. H., Yang, Y. J., Kim, J. S., Choi, D. S., Park, S. H., Jin, S. Y. and Park, J. S. (2018). Non-destructive monitoring of apple ripeness using an aldehyde sensitive colorimetric sensor. *Food Chemistry*, 267, 149–56.

Koirala, A., Walsh, K. B., Wang, Z. and McCarthy, C. (2019). Deep learning-method overview and review of use for fruit detection and yield estimation. *Journal of Computers and Electronics in Agriculture*, 162, 219–34.

Lammertyn, J., Dresselaers, T., Van Hecke, P., Jancsók, P., Wevers, M. and Nicolaï, B. M. (2003). MRI and x-ray CT study of spatial distribution of core breakdown in 'Conference' pears. *Magnetic Resonance Imaging*, 21, 805–15.

Lebrun, M., Plotto, A., Goodner, K., Ducamp, M-N. and Baldwin, E. (2008). Discrimination of mango fruit maturity by volatiles using the electronic nose and gas chromatography. *Postharvest Biology and Technology*, 48, 122–31.

Meigh, D. F. J. (1960). Use of gas chromatography in measuring the ethylene production of stored apples. *Journal of the Science of Food Agriculture*, 11, 381–5.

Melado-Herreros, A., Muñoz-García, M-A., Blanco, A., Val, J., Fernández-Valle, M. E. and Barreiro, P. (2013). Assessment of watercore development in apples with MRI: Effect of fruit location in the canopy. *Postharvest Biology and Technology*, 86, 125–33.

Momin, M. A., Kuramoto, M., Kondo, N., Ido, K., Ogawa, Y., Shiigi, T. and Ahmad, U. (2013). Identification of UV-fluorescence components for detecting peel defects of lemon and yuzu using machine vision. *Engineering in Agriculture, Environment and Food*, 6, 165–71.

Moureh, J. and Flick, D. (2004). Airflow pattern and temperature distribution in a typical refrigerated truck configuration loaded with pallets. *International Journal of Refrigeration*, 27, 464–74.

Nicolai, B., Van Dael, M., Verboven, P., Dhaene, J., Van Hoorebeke, L. and Sijbers, J. (2017). https://patents.google.com/patent/CA2979932A1/en?q=fruit&q=Xray&inventor=nicolai.

Nicolai, B. M., Beullens, K., Bobelyn, E., Peirs, A., Saeys, W., Theron, K. I. and Lammertyn, J. (2007). Nondestructive measurement of fruit and vegetable quality by means of NIR spectroscopy: A review. *Postharvest Biology and Technology*, 46, 99–118.

Nugraha, B., Verboven, P., Janssen, S., Wang, Z. and Nicolaï, B. M. (2019). Non-destructive porosity mapping of fruit and vegetables using X-ray CT. *Postharvest Biology and Technology*, 150, 80–8.

Qin, J., Kim, M. S., Chao, K., Dhakal, S., Cho, B-K., Lohumi, S., Mo, C., Peng, Y. and Huang, M. (2019). Advances in Raman spectroscopy and imaging techniques for quality and safety inspection of horticultural products. *Postharvest Biology and Technology*, 149, 101–17.

Subedi, P. and Walsh, K. B. (2009). Non-invasive techniques for measurement of fresh fruit firmness. *Postharvest Biology and Technology*, 51, 297–304.

Subedi, P. P., Walsh, K. B. and Owens, G. (2007). Prediction of mango eating quality at harvest using short-wave near infrared spectrometry. *Postharvest Biology and Technology*, 43, 326–34.

Tsuda, T. (2012). Dietary anthocyanin-rich plants: Biochemical basis and recent progress in health benefits studies. *Food Research*, 56, 159-70.

van Dael, M., Lebotsa, S., Herremans, E., Verboven, P., Sijbers, J., Opara, U. L., Cronje, P. J. and Nicolaï, B. M. (2016). A segmentation and classification algorithm for online detection of internal disorders in citrus using X-ray radiographs. *Postharvest Biology and Technology*, 112, 205–14.

van Dael, M., Verboven. P., Zanella, A., Sijbers, J. and Nicolai, B. (2019). Combination of shape and X-ray inspection for apple internal quality control: In silico analysis of the methodology based on X-ray computed tomography. *Postharvest Biology and Technology*, 148, 218–27. ISSN 0925-5214, https://doi.org/10.1016/j.postharvbio.2018.05.020.

Walsh, K. B. (2018). Advances in Agricultural Machinery and Technologies, in Chen, G. (Ed.), *Fruit and Vegetable Packhouse - Technologies for Assessing Fruit Quality and Quantity*. CRC Press, Boca Raton, pp. 367–95.

Walsh, K. B. and Wang, Z. (2018). Achieving Sustainable Cultivation of Mangoes, in Victor Galan Sauco and Ping Lu (Eds), *Monitoring Fruit Quality and Quantity in Mangoes*. Burleigh Dodds Science Publishing Limited, Cambridge, UK, pp. 313–38.

Walsh, K. B., Long, R. L. and Middleton, S. G. (2007). Use of near infra-red spectroscopy in evaluation of source-sink manipulation to increase the soluble sugar content of stonefruit. *The Journal of Horticultural Science and Biotechnology*, 82, 316–22.

Walsh, K. B., Blasco, J., Zude-Sasse, M. and Sun, X. (2019). Visible-NIR 'point' spectroscopy in postharvest fruit and vegetable assessment. *Postharvest Biology and Technology* (in process).

Willstätter, R. and Stoll, A. (1913). *Untersuchungen über Chlorophyll. Methoden und Ergebnisse*. Berlin: Springer.

Wolf, F. T. and Stevens, M. V. (1967). The fluorescence of carotenoids. *Photochemistry and Photobiology*, 6, 597–9.

Zaidi, N. A., Tahir, M., Vinayaka, P., Lucklum, F., Vellekoop, M. and Lang, W. J. P. E. (2016). Detection of ethylene using gas chromatographic system. *Procedia Engineering*, 168, 380–3.

Chapter 14

Postharvest handling of organically produced specialty crops

Penelope Perkins-Veazie, Chris Gunter and Marlee Trandel, North Carolina State University, USA

1 Introduction

The annual demand for organically produced food continues to increase globally, and was $16.5 billion for produce in the United States in 2017 (McNeil, 2018). Organic farming is a method of crop (and livestock) production that involves a holistic approach to food production, to optimize crop productivity within an agroecosystem encompassing soil microorganisms, plants, and people, for a safe and environmentally harmonious system.

The organic production system routinely incorporates crop rotations, cover crops, and compost/manures, and encourages balanced host/predator relationships. Preventative insect and disease control tactics are implemented, with a heavy focus on soil building and conservation. Postharvest handling completes this production system with factors that are unique to storage and marketing of certified organic produce and specialty crops. The organic integrity of a crop must be maintained from the farm to the consumer, including no contamination with prohibited substances in the packhouse, pack line, packaging, or comingling with nonorganic produce (Plotto and Narcisco, 2006). The primary postharvest challenges for organic crops are determining what is acceptable for extending shelf life, where steps or chemicals used differ in application or amount from conventional postharvest, and how practices in transport, export and import differ from those of nonorganic specialty crops.

http://dx.doi.org/10.19103/AS.2019.0055.16

Materials considered of synthetic origin are not necessarily excluded from use in organic systems, and those considered GRAS (generally recognized as safe) are not automatically allowed in organic systems. The Organic Materials Review Institute (OMRI) provides a database listing materials and allowed uses. In the United States, the National Organic Program (NOP) provides the regulations for materials allowed in organic systems. Petitions for use of a substance can be submitted to NOP and OMRI for consideration and must include selection of the use (such as crop protection or processing) and how it is used (vapor, direct contact with product, additive etc.).

A useful tutorial on US organic regulations and how substances are approved is provided by McEvoy (2012). Regulations of allowed, disallowed, or amended chemicals and the types of application change rapidly for organic produce. The US sources to check for updates include the NOP list of allowed and prohibited substances, the OMRI website, and the Federal Register.

Postharvest application covers direct contact with produce and often includes sanitizers used to clean equipment as well as the water involved in cleaning produce. Fumigant and phytosanitary treatments such as vapor, hot air, atmospheric alterations, light, ionizing radiation or chemicals may be applied to extend shelf life and/or to reduce possible pest/pathogen infestation. Waxes or coatings that are not petroleum based may be applied directly to produce to reduce moisture loss, decay, or contact injury, and to maintain freshness. Packaging and storage facilities may incorporate atmospheric scrubbers, delivery of antimicrobial substances, or altered gas atmosphere ratios of nitrogen, oxygen, carbon dioxide.

2 Differences in postharvest organic and nonorganic approaches

Standard postharvest management practices apply to both organic and conventional systems. Produce that has been grown using nonorganic methods cannot be labeled as organic even if organic postharvest steps are used.

General postharvest operations at the packhouse can differ between the two systems. The packhouse steps of bin dumping, washing, cooling, sorting, waxing, and storing have sanitizers and coatings that can be used for both systems. Systems may also differ in storage, transit, and ripening steps. Allowed substances for postharvest use in organic systems are given in Table 1.

There are currently no organic postharvest practices that are forbidden for conventional postharvest use while substances used in postharvest organic systems may differ in application and/or allowed amount compared to those for nonorganic application.

Additionally, the use of chemicals in postharvest handling and storage of fruits and vegetables is far less than during the production phase. Chemical

Table 1 List of allowed synthetic substances for use in US organic crop postharvest handling under NOP and OMRI

	Use	Restrictions
Copper sulfate	Fungicide and algicide in food processing houses	Application rates limited to those that do not increase baseline soil tests for copper before use
Lignin sulfonate	Dust suppressant	Flotation aid in bin dumping
Ethylene gas	Regulation of pineapple flowering	Cannot use gas produced by catalytic generators
Sodium silicate	Postharvest handling for tree fruit processing	Only allowed as flotation aid for produce
Waxes and emulsions	Postharvest handling	Must be nonpetroleum based
Soda ash	Postharvest handling for tree fruit processing	Flotation aid for fruit

uses are primarily applied for prevention of postharvest disorders (such as antioxidants to prevent surface scald in apple), or to prevent sprouting in potatoes and onions.

2.1 Cleaning and sanitation of packing lines, facilities and produce

Bacterial contaminants can occur in both organic and conventional systems. Food safety and product quality deteriorating organisms are the same regardless of production system. The purpose of cleaning and sanitizing packing areas is to reduce the chances for produce quality to deteriorate and reduce the spread of contamination and foodborne illness causing organisms (Owens, 2017).

The key goals for a sanitation process are to remove soil, kill remaining bacteria, and remove and/or prevent biofilm formation to maintain a clean and sanitary environment for packaging produce. Cleaning refers to the use of chemical and mechanical processes to remove soil from a surface. Sanitizing is the application of a specific chemical sanitizer or treatment to a previously clean surface, in order to kill bacteria. One critical concept of this cleaning and sanitation step is to provide a clean break, separating one packing 'Lot' from the next. This is useful for inventory purposes, and also provides a physical separation of production units in the event of an unforeseen contamination event.

Pest management must adhere to organic standards, including preventative measures, such as removal of pest harborage and food sources, screening or physical plugging of gaps, cracks, and holes that could allow pests to enter. Mechanical methods (for example traps), lures or repellents used must be approved on the Organics Material list for NOP (OMRI, 2001; NOP, 2018) or country equivalent. It is important to check with your certifier to be sure that the method being used meets the approved materials/methods before you use it.

State, provincial, and/or national laws and guidelines should be followed for water quality requirements for handling and packaging fresh produce in the country of origin and for export. In the United States, water used for food contact surfaces in packing houses must meet the requirements outlined in the Produce Safety Rule of the Food Safety Modernization Act (FSMA), such as zero detectable generic *E. coli*. All cleaning and sanitizing materials must labeled for their intended use and be listed in the approved list for organic produce. If there is a question on a specific product, check with your certifier before beginning use. Sanitizing agents approved for use on organic produce in the United States and Canada include peroxyacetic acid (PAA), chlorine, chlorine dioxide, and ozone (Table 2).

Approved chlorine materials may be used, if approved by the respective certifying agency, up to the maximum label rates for disinfecting or sanitizing food contact surfaces. Follow the product label for recommendations on their use. An approved testing method must be used to check the strength of sanitizers used. Test strips, or other methods are available for each sanitizer. If there is a question about the testing procedure, refer to the manufacture guidelines for testing solution concentration.

After a sanitation step, it is important to verify the effectiveness of cleaning and sanitizing. Visual inspection of surfaces following cleaning and sanitizing and the use of swabbing techniques are best management steps to ensure cleanliness and sanitation have occurred (Plotto and Narcisco, 2006). Adenosine triphosphate (ATP) testing swabs and a handheld meter, for example, can be used for rapid feedback of sanitation effectiveness.

2.2 Materials allowed for produce contact in postharvest handling, storage and shipment

2.2.1 Bin dumping, wash and rinse steps

Bin dumping is commonly used for large production operations, especially for apples, stone fruits, and tomatoes. Produce can be released directly into the water or floated from bins. Flotation aids (solids) approved for organic use are listed in Table 1. Both bin dumping systems require some means of water sanitation. Chlorine is most commonly used as a water sanitizer although ozonation of water being used for sanitation is also allowed (Graham, 1997; Suslow, 2000). Wash steps with chlorine include application of a final rinse with water at approved residual chlorine levels of 1–4 mg/L (Extension, 2016).

Table 2 Allowed substances for use in organic systems for food or surface sanitation (Canada and US NOP rules)

Product	Use	Restrictions
Acetic acid i.e. vinegar	Antimicrobial agent used rinse or clean organic produce if from a non-synthetic source.	
Ethanol	Sanitize fruit and vegetable wash products. Effective concentration of 60-90% in water. Activity drops sharply when diluted below a 50% concentration.	Must be from a certified organic source
Isopropyl alcohol	Effective disinfectant against bacteria, viruses, and fungi. Not effective against spores. Concentration at 60-90% in water. Activity drops sharply when diluted below a 50% concentration.	
Alcohol (organic sources)	Effective against fungi, vegetative bacteria (i.e. bacteria in their growing states), Mycobacteria and some viruses, including norovirus but not effective against spores. Most effective at 60-90% in water. Activity drops sharply when diluted below a 50% concentration.	
Ascorbic acid (vitamin C)	Used as sterilizing and sanitation agent and to lower pH in processing.	Must be from non-synthetic source
Chlorines Sodium hypochlorite Calcium hypochlorite Chlorine dioxide	Used to kill bacteria and disinfect, pH must be maintained between 6.5 and 7.5 to optimize effectiveness. Used to treat cooling water, washing fruit, vegetables, and disinfect food contact surfaces.	Discharge level must be at or below 4 ppm (mg/L)
Citric acid	Used as a component of fruit and vegetable wash products.	Non-synthetic sources made from fruit concentrates
Glycerol	Used as a stabilizer of cleaning products that contain a mixture of water and oils.	
Hydrogen peroxide	Food-grade hydrogen peroxide at 3% concentration is the most common for use as a sanitizing agent.	Must be food-grade quality, free of non-NOP adjuvants
Ozone gas	A potent oxidant/disinfectant. Approved only to purify drinking water, water used for pack lines and irrigation cleaner only.	Must be generated on site
Peracetic acid	Effect sanitizer and disinfectant of equipment.	Cannot exceed concentration of 6%
Peroxyacetic acid	Antimicrobial agent due to its high oxidizing potential. Effective against a broad spectrum of coliforms, bacteria, yeast, and molds, at 5-40°C and to a pH maximum of 8.0. Used most commonly in fresh-cut, further processed, and postharvest fruit and vegetable flume and wash water systems.	

(Continued)

Table 2 *(Continued)*

Product	Use	Restrictions
Potassium and sodium bicarbonate	Effective scouring agent when mixed with water.	
Sodium carbonate	Used as a component of cleaning products for disinfectant, grease cutting, and water softening properties.	
Sodium citrate	Allowed as buffering agent in the formulation of cleaning products for postharvest crop handling.	
Sodium hydroxide	Alkaline industrial cleaning agent that dissolves grease, fats, and protein-based deposits for cleaning counters, processing equipment, or areas of food handling.	
Steam	Disinfecting agent.	
Waxes and coatings	Used to coat fruits and vegetables to prevent weight loss and preserve appearance. Includes wood rosin, carnauba, orange shellac, carrageenan, alginates, beeswax.	Must be from natural sources (nonpetroleum based) but can be of nonorganic origin, with exceptions[a]

[a] Nonorganic orange shellac can only be used when an organic form is not commercially available. Only organically derived beeswax can be used with foods labeled organic. Chitosan is not allowed as a coating for organic label (Currier, 2015).

2.2.2 Waxes, coatings and essential oils

Pack line application of approved mold inhibiting products, non-treated tissue wraps, and natural waxes are allowed (Table 2). While chitosan has been widely studied as a coating for fruits and vegetables (Sharif et al., 2018) and is allowed as an adjuvant under the pesticide category in the United States (OMRI), it is not approved under the processing category (where substances having contact with food are placed). Although chitosan originates from the natural chitin present in arthropod shells (such as crab or shrimp), the de-acylation step involves use of chemicals considered synthetic, which classifies chitosan as a synthetic material. A petition for its use as a coating has not been submitted to OMRI.

2.2.3 Storage rooms and storage atmosphere treatments

Controlled or modified atmosphere is allowed for organic produce, but should not be done in rooms where nonorganic produce practices may have left substances that could contaminate organic produce. For instance, diphenylamine (DPA) is a synthetic chemical that has been used for many years for pome fruits such as apple and pear to prevent superficial or surface scald that can form when fruit become stressed. Residues have been reported in storage rooms, which can in turn cause contamination of organic apples (Robatscher et al., 2012).

Controlled atmospheres are created by flushing sealed storage rooms with nitrogen gas to reduce oxygen levels to less than 5% and maintaining good air movement to prevent hypoxia.

Dynamic controlled atmosphere (DCA) is a modification of controlled atmosphere, allowing for fine tuning of oxygen levels in storage rooms (Thewes et al., 2015). DCA was developed for use in recalcitrant apple cultivars to aid in maintaining firmness and reducing scald and has quickly evolved into an effective and organically allowed treatment for apples and pears (Prengamen, 2018).

2.2.4 Transportation

Transportation of organic and conventional produce in mixed loads has to be accomplished in a way that will not cause contamination of the organic produce. This means that product should not be comingled (package separately and insure that loads will not have direct contact) and that nonorganic produce does not have residues that could outgas into the transport container. To avoid comingling, segregate organic and nonorganic products via sealed boxes or crates, or in shrink-wrapped pallets, or use separate trucks or compartments in trucks. In addition, cleaning of trucks, railcars, or containers, and so on are required to prevent contamination by prohibited substances or previous shipments. Shared loading/unloading equipment must also be cleaned before handling organic product. The handler of organic postharvest crops must implement necessary measure to prevent comingling of products. The following are prohibited for use: (1) packaging materials, storage containers or bins that contained synthetic crops, fungicides, preservatives or fumigants and (2) the use or reuse of any bag or container that has been in contact with any substance that compromises the organic integrity of any organically produced produce.

2.2.5 Record keeping and documentation

Good record keeping at all steps is essential with organic product. Documentation before loading must identify produce as organic includes bin tags, labels, scale tickets, and lot control documents. This will prevent misidentification or comingling of organic and nonorganic produce. Shipping documentation must include the product name, organic status, lot number(s) and contact information of the responsible party. The shipping documentation packet should include copies of clean truck affidavits (signed by the truckers), which identify the previous two loads and the cleaning methods that were used (Plotto and Narcisco, 2006).

2.2.6 Use of ethylene and ethylene antagonists

Management of the plant hormone ethylene is a postharvest consideration for quality maintenance during storage and transportation of many specialty crops. Ethylene has been widely applied in postharvest uses to help ripen specialty crops. Under the USDA NOP, ethylene can be applied as a gas only for ripening of tropical fruits and if naturally produced. For instance, addition of synthetic ethylene gas to a room of tomatoes would not be allowed but ethylene generated by the tomatoes is allowed.

In contrast, removal of ethylene gas does not require special approval. This is because there is no direct contact of the products with the crop and the gas is being removed rather than applied (Currier, 2017). Ethylene removal sachets can be used in packages and scrubbing filters such as potassium permanganate can be used in storage facilities.

As with nonorganic produce, ethylene-producing crops (apple, banana, tomato, other subtropical fruit) should not be stored with ethylene-sensitive fruits or vegetables. 1-MCP (1-methylcyclopropene) is used for postharvest management to slow ethylene effects. Because it is not considered a naturally occurring substance, and has direct contact with produce, 1-MCP is not allowed under the organic label.

2.2.7 Anti-sprouting agents

Although sprouting of potatoes and onions can be delayed by holding at low temperature, sprout and sprout growth shorten shelf life quality once dormancy periods have ended (Paul et al., 2016). CIPC [chloropropham or Isoproyl N-(3-chlorophenyl) carbamate] - which is not approved for organic use - is applied via fogging or aerosols to suppress sprouting. Use of essential oils such as clove, peppermint, spearmint, and L-carvone have shown success when applied at sprout emergence (Frazier et al., 2004; Eshel, 2011; Paul et al., 2016). Clove oil applied as an aerosol has been approved as an anti-sprouting treatment in organic potatoes in the United States and Canada.

2.3 What can be done if no approved organic postharvest treatments are allowed for a commodity?

In the absence of available organic postharvest alternatives, good management and practice selection are critical. The general preharvest factors considered important for postharvest shelf life are similar between organic and conventional systems. For instance, genetic selection of material with desired attributes for high yield yet longer shelf life is common to both systems, but disease resistance is of increased importance for organic systems as approved pesticides are limited. In contrast, use of genetically modified organism (GMO) material is not allowed in organic systems. As fertility choices are mostly naturally based rather than synthetic for organic produce, proper plant nutrition is critical before and during crop growth.

Good agricultural practices (GAP) and good handling practices (GHP), for reduced food safety risks, apply to both systems. Prompt removal of field heat, maintenance of the cold chain, and good temperature management from packing house to point of sale remain the most important way to extend shelf life regardless of production system. Temperature control slows respiration, ripening, postharvest disease, and physiological disorders such as chilling injury. Refrigeration, vacuum cooling, room cooling, or forced air cooling do not have restrictions for the organic production system. Removal of field heat via water-based cooling systems (icing, hydrocooling, hydrovacuum cooling) will require a form of water sanitation that meets materials and standards for organic produce as outlined above.

Nutrient and phytochemical content of organically produced crops appears to differ slightly from nonorganic crops in the amount of dry weight, nitrate and nitrite, and phenolic compounds. Ceglie et al. (2016) compared journal publications and concluded that dry weight of organic produce is often higher than that of nonorganic produce, possibly from the longer duration of available soil nutrients. Ultimately, higher dry weight, lower nitrogen content, and higher amounts of secondary metabolites/phytochemicals would argue for an improved postharvest shelf life of organic produce through lower weight loss and better anti-decay mechanisms. However, organic produce may be harvested closer to full ripeness which may offset advantages in nutrient and phytochemical content (Ilic et al., 2018). While per capita postharvest loss of fruit and vegetables is estimated at 12 kg (Xue et al., 2017), the difference in losses between organic or nonorganic produce is not known.

3 Comparison of organic postharvest regulations worldwide

3.1 Organic programs and importation

A number of countries have fully implemented organic practices in Europe, North America, Africa, Asia, and the Pacific Rim (Table 3). Some countries and regions share a common approach, such as the NOP that is aligned across Canada, Mexico, and the United States. India has recently developed their own NOP (INDOCERT, 2001). If an organic program is fully implemented, an

Table 3 Countries with a fully implemented or recognized organic program (https://globalorganictrade.com/country-list)[a]

Austria	Estonia	Lithuania	Slovak Republic
Belgium	Finland	Luxembourg	South Korea
Brazil	France	Macedonia	Spain
Bulgaria	Germany	Malta	Sweden
Canada	Greece	Mexico	Switzerland
Chile	Honduras	Moldova	Taiwan
China	Hungary	Montenegro	The Netherlands
Croatia	Iceland	Norway	Tunisia
Czech Republic	Ireland	Poland	Turkey
Denmark	Italy	Portugal	United Kingdom
Ecuador	Japan	Romania	United States

[a] All European countries follow the European Council Regulations for Organic Production Program (Council of Regulation, 2007). Organic production and labeling of organic productions and repealing Regulation. No. 2092/91. *Official Journal of European Union* 834. https://eur-lex.europa.eu/legal-content/EN/TXT/?uri=CELEX:32007R0834.

equivalency agreement can be utilized. For instance, Canada can label and sell produce as organic in both Canada and the United States without extra certification as long as produce meets the organic guidelines. Documentation with the produce must state 'Certified in compliance with the terms of the US Canada Organic Equivalency Arrangement.' For packaged retail products, labels or stickers must state the name of the USDA authorized certifier and may use the USDA organic seal and/or the Canada organic logo (https://www.ams.usda.gov/services/organic-certification/international-trade/Canada).

Recognition agreements involving the United States are in place with several countries, such as New Zealand and India. These agreements allow a foreign government to accredit their certifying agents to adherence with the USDA organic standards. The certifying agents are authorized to certify organic farms and processing facilities, ensuring that USDA organic products meet or exceed all USDA organic standards. These products can then be imported for sale in the United States.

If a country has a fully implemented program, importation into the United States requires USDA certification that organic standards under NOP were followed. Countries that do not have an implemented organic program can have produce accepted as organic imports to the United States if the specific grower follows the US NOP and obtains agent certification. USDA certifiers determine which plant products are allowed entry into the United States. If an outside operation or grower is audited and found to not comply with NOP organic processes, significant noncompliance will be reported to both countries and enforcement actions may be pursued under the respective country's regulations. The US NOP has started education programs for importers to clarify what is acceptable in postharvest handling (Tucker, 2018).

3.2 Disinfection practices allowed for maintenance of an organic label

In the United States, the US Custom Service and USDA APHIS Plant Protection and Quarantine (PPQ) are the federal agencies responsible for documenting the arrival and disposition of every agricultural shipment entering the country. To bring plants or plant products into the United States, importers must first submit PPQ form 587 which identifies the name and quantity of the commodities being imported, the country of origin, and the intended port of entry. Typically, importers provide Customs and PPQ a computerized manifest of the commodities they are carrying approximately one week prior to arrival of the shipment. This enables quarantine officials to determine which portions of the overall shipment need to be placed on hold. PPQ staff place an automatic hold on all agricultural commodities on the manifest as well as some nonagricultural goods to determine if a potential pest infestation has spread. Upon arrival of the shipment, the importer provides PPQ with a bill of laden (for ground and sea shipments) or an airway bill (for air freight) which lists the commodities being held. Attached to the bill of laden or airway bill are invoices correlated to the individual commodities. After review and inspection, PPQ staff determine whether individual commodities can be released, require a quarantine treatment as a condition of entry, or must be refused entry and destroyed or re-exported.

One of the more frustrating issues for producers, importers and exporters is how to keep the organic label intact as produce moves in or out of countries. Many countries have disinfestation regulations to prevent movement of unwanted pests or pathogens. Some countries still only allow treatments that will negate an organic label. Methyl bromide, which has been used for some years as a treatment for fruit flies, if used on produce that has been grown and treated organically will negate the organic label and can result in fraudulent label charges if produce continues into the country with the organic label. Fortunately, there are nonchemical alternatives for disinfestation. Efforts are ongoing to better educate importers and exporters regarding this important step.

3.3 Certification practices and fumigant treatments for imports/exports of produce

Most countries with national organic standards do not allow chemical postharvest treatments. In contrast, nonchemical practices are usually accepted.

Conventional chemical treatments, including fungicides, aerosols, dips, dusts, and sprays, are not allowed in organic systems if originating from synthetic products for postharvest pest control. A few chemical substances are allowed for sanitation purposes (not for direct food application) (Table 2). In recent years, imports of organic produce to the United States have been carefully scrutinized and NOP standards are refined. Under these standards, if a crop was organically produced, its organic status is nullified if subjected to fumigant treatments such as methyl bromide, sulfuryl fluoride, ionizing radiation or phosphine (USDA-AMS, 2018). Additionally, some countries will not accept organic fumigation treatments and require use of methyl bromide prior to importation, which negates the organic status of the crop. This underscores the importance of checking country import/export regulations for fumigation if considering out of country shipment of organic produce.

3.4 Fumigant treatment issues

Approved quarantine practices must be treatment and commodity specific with a detailed set of conditions applied to a particular fruit, vegetable, grain, or other plant product. Approval of a quarantine treatment option is the joint responsibility of the USDA APHIS PPQ, who sets product performance qualification, and the USDA-Agricultural Research Service (ARS), who reviews data from the exporting countries in support of applications for approval of new treatments. Use of light treatment, including pulsed white light, UVA, and UVB is allowed but ionizing radiation, such as gamma irradiation is not permitted under the organic label.

3.5 Organic regulations and treatment facility approval

Organic handlers and importers must rely upon the same cold storage and heat regulation facilities which serve the nonorganic market. For example, mangos entering the United States (including those from Hawaii and Puerto Rico) must receive hot water immersion, high temperature forced air, or vapor heat treatment for fruit fly disinfestation. Because both organic and nonorganic commodities require identical treatments and the number of approved facilities is limited, individual plants will process both types of fruit. Some certifying agents require separate production runs for organic commodities to ensure that water left over from treatments of nonorganic produce is not reused. Importers of organic commodities may not have sufficient volume to fill a segregated shipping container and therefore have to share space with noncertified goods. Certifying agents may impose additional handling restrictions to preserve the identity of certified product.

Table 4 Types of treatments approved for organic products used to meet quarantine/disinfestation requirements

Treatment	Description	Uses
Vapor heat treatment	Used as kill step for Mediterranean and oriental fruit flies, incorporates relative humidity and temperature of 47.2–50°C	Fruits, vegetables, nursery stock
Forced hot air treatment	Hot air at 47.2–50°C used to kill insects	Seeds, fruit, vegetables, nursery stock
Controlled atmosphere storage	Increased carbon dioxide and low oxygen	Fruits, vegetables, nursery stock
Steam treatment	Steam at 100–125°C to destroy pathogenic microorganisms	Soil, hay, packing materials, and other nonedibles
Hot water immersion	Immersion in water at 50°C for specific time to kill pests	Fruits, vegetables, nursery stock
Cold treatment	Chill pulp close or below freezing (2–0°C) for 14–40 days	Fruits, vegetables
Holistic or system approach	Combined cold treatment with controlled atmosphere	Fruits, vegetables

Source: USDA Treatment manual, nonchemical treatments. 2016. pp.3-1-1 to 3-7-1 https://www.aphis.usda.gov/import_export/plants/manuals/ports/downloads/treatment.pdf.

Treatments that can be applied to organic produce with minimal loss of quality include hot water immersion, vapor heat treatment, forced hot air treatment, controlled atmosphere, or a holistic approach (Table 4). Facilities for hot water immersion, vapor heat treatment, and forced hot air treatment are available. The holistic approach is not currently approved for quarantine standards in the United States and there are no US quarantine-approved controlled atmosphere treatments for agricultural imports.

Grant et al. (2015) analyzed the reported application of phytosanitary treatments in US fruits and vegetables exports from 1999–2009. Use of vapor heat, cold treatment, or quick freeze accounted for 20–80% of the sanitary/phytosanitary measures in fruits and vegetables exported from the United States. While the information does not indicate use of organic labeling, it does show that vapor heat and cold treatments can be successfully used to replace traditional fumigants such as methyl bromide.

4 Conclusion

The postharvest handling of organically produced specialty crops needs to incorporate decisions made long before planting. Facilities, pack lines, sanitation, treatments, storage, and transit systems and substances need to be determined and appropriate records, labeling, and documentation maintained at all steps. Final destination, if out of country, needs to be thoroughly checked to make sure that the appropriate disinfestation treatment is used to maintain the organic label.

5 Future trends

The organic market continues to show steady growth, which will drive research into areas formally considered uneconomical or toward approaches that are less efficacious than those currently used. Postharvest challenges, especially in pests and diseases, are common in fresh produce and are even more challenging with organic produce. Ways to address these challenges include rapid development of organically approved substances and postharvest technologies and long term changes in plant environment or genetics.

Good postharvest management practices greatly reduce the need for postharvest pesticides or antioxidants. Incorporation of system preharvest to screen for ripeness, such as use of portable infrared spectrophotometers, as well as packhouse sorting will minimize the amount of under or overripe, soft, or injured fruit. Adaption of DCA will further reduce postharvest disorders.

Addition of treatments such as UV light or pulsed light that produce that stimulate bioactive compounds may be an organic option to reduce microbial loads and fungal growth in crops such as berries where a wash step cannot be used (Umagiliyage and Choudhary, 2018). Rapid and even application of light with minimal heating, plus evidence of significant reduction of microbial or fungal pathogens will be needed.

Incorporation or encapsulation of plant bioactive compounds may prove to be an organically approved and viable means of suppressing fungal growth in fresh produce (Mari et al., 2007; Eshel, 2011). Essential oils used as dips, combined in coatings, or applied as vapors have been found to be effective at suppressing fungal growth but do not kill fungi (Ilic et al., 2018), requiring a constant presence. These oils include clove, thyme, oregano, mints, lemon grass, and cinnamon. Incorporation of plant bioactive compounds such as isothiocynates or essential oils into postharvest systems as a vapor delivery or sachet in the packaging offers a means of fungistatic control while avoiding off taste.

Continued development of pesticides that utilize naturally derived and ecologically friendly substances or microorganisms will benefit all production systems. Microbial community interaction on fresh produce may stimulate or suppress fungal growth (Droby et al., 2016). This area of research has been challenging but better technologies to follow identification and function may lead to alternative sanitization, storage, and packaging approaches.

6 Where to look for further information

Several websites contain a wealth of information and a gateway into additional resources. These include:

- *Canadian Organic Growers*. https://www.cog.ca/standard-guide/8maintaining-organic-integrity-during-cleaning-preparation-and-transportation/ (accessed on 01 December 18).

  This website explains the organic program of Canada, known as Canada Organic Regime standards. Any organic product that is labeled organic (including food consumption for human and livestock) is inspected by the Canadian Food Inspection Agency and must comply with Food and Drugs Act and Regulations and the Consumer Packaging and Labelling Act and Regulations. Postharvest cleaning, preparation and transport are covered under

- *OMRI (Organic Materials Review Institute)*. https://www.omri.org/about (accessed on 22 May 2019).

  This website covers materials under US and Canada organic program. It has a search function for materials, as well as a subscription for updates from OMRI. Extremely useful if considering use of a product and not sure of its status under organic rules.

- *USDA-AMS Good Agricultural Practices (GAP) & Good Handling Practices (GHP)*. https://www.ams.usda.gov/services/auditing/gap-ghp (accessed 01 December 2018).

  This website provides information for producers on compliance and regulations for US GAP and GHP and audit services.

- *USDA-AMS Organic standards*. https://www.ams.usda.gov/grades-standards/organic-standards (accessed 14 July 2019).

  This website provides links to the organic food production act, USDA organic regulations, and the NOP handbook. Essential for checks on approved new products, discontinuation of previously approved products (including last date can be used), as well as changes in regulations.

- *European Commission Agriculture and Rural Development Organic Farming*. https://ec.europa.eu/agriculture/organic/index_en (accessed 14 July 2019)

  This website provides entry into EU legislation, policy, label rules for produce grown organically in the European Union.

- *Organic Farming Practices: Postharvest Handling*. University of California at Davis. https://vric.ucdavis.edu/pdf/postharvest_organic_handling.pdf (accessed on 14 July 2019)

  This website provides references for organic postharvest handling for specific crops.

- *Production Guide for Storage of Organic Fruits and Vegetables*. Cornell University Cooperative Extension. http://hdl.handle.net/1813/42885 (accessed on 14 July 2019)

  This website provides information on management, practices, and chemicals allowed for postharvest storage and handling of organic crops and provides specific commodity information.

7 References

Cardi, T. (2016), Cisgenesis and genome editing: Combining concepts and efforts for a smarter use of genetic resources in crop breeding. *Plant Breeding*, 138: 139–47.

Ceglie, F., Amodio, M. L. and Colelli, G. (2016), Effect of organic production systems on quality and postharvest performance on horticultural produce. *Horticulturae*, 2: 4. doi:10.3390/horticulturae202004.

Currier, D. (2015), Fruit and vegetable coatings. *Organic Materials Review Institute*. Available at: https://www.omri.org/fruit-and-vegetable-coatings (accessed on 13 July 2019).

Currier, D. (2017), Ethylene filters. *Organic Materials Review Institute*. Available at: https://www.omri.org/ethylene-filters (accessed on 13 July 2019).

Droby, S., Wisniewski, M., Teixido, N., Spadarod, D. and Jikaklie, M. H. (2016), The science, development, and commercialization of postharvest biocontrol products. *Postharvest Biology and Technology*, 122: 22–9.

EFSA (European Food Safety Authority) Panel on Genetically Modified Organisms (GMO). (2012), Scientific opinion addressing the safety assessment of plants developed through cisgenesis and intragenesis. *EFSA Journal*, 10: 2561.

Eshel, D. (2011), Non-chemical approaches for postharvest quality management of underground vegetables. *Stewart Postharvest Solutions*, 1: 6. doi:10.2212/spr.2011.1.6.

European Commission Agriculture and Rural Development Organic Farming Council of Regulation. (2007), Organic production and labeling of organic productions and repealing Regulation. No. 2092/91. *Official Journal of European Union* 834. Available at: https://eur-lex.europa.eu/legal-content/EN/TXT/?uri=CELEX:32007R0834 (accessed on 13 July 2019).

Extension. (2016), Approved chemicals for use in organic postharvest systems. *Organic Agriculture*. Available at: https://articles.extension.org/pages/18355/approved-chemicals-for-use-in-organic-postharvest-systems (accessed on 13 July 2019).

Frazier, M. J., Olsen, N. and Keinkopf, G. (2004), *Organic and Alternative Methods for Potato Sprout Control in Storage*. University of Idaho Extension.

Graham, D. M. (1997), Use of ozone for food processing. *Food Technology*, 51: 72–5.

Grant, J. H., Peterson, E. and Ramniceanu, R. (2015), Assessing the impact of SPS regulations on U.S. fresh fruit and vegetable exports. *Journal of Agricultural and Resource Economics*, 40: 144-63.

Ilic, Z. S., Fallik, E., Manojlovic, M., Kevresan, Z. and Mastilovic, J. (2018), Postharvest practices for organically grown products. *Contemporary Agriculture*, 67: 71-80.

INDOCERT. (2001), Certification. Available at: https://www.indocert.org/ (accessed on 13 July 2019).

Mari, M., Neri, F. and Bertolini, P. (2007), Novel approaches to prevent and control postharvest disease of fruits. *Stewart Postharvest Solutions* 6: 4. doi:10.2212/spr.2011.1.6.

McEvoy, M. (2012), Organic 101: Allowed and prohibited substances. *USDA*. Available at: https://www.usda.gov/media/blog/2012/01/25/organic-101-allowed-and-prohibited-substances.

McNeil, M. (2018), Maturing U.S. organic sector sees steady growth of 6.4 percent in 2017. *Organic Trade Association*, 18 May 2018. Available at: https://ota.com/news/press-releases/20236.

NOP (National Organic Program) (2018), Part 205 National organic program, electronic code of federal regulations, 20 December2018. Available at: https://www.ecfr.gov/cgi-bin/text-idx?SID=a6a0935ddf00e166695f4c2138bd58d8&mc=true&node=pt7.3.205&rgn=div5#sp7.3.205.a.

NPOP (National Program for Organic Products) (2015), India's National Program for Organic Production Reference. Available at: http://apeda.gov.in/apedawebsite/organic/Organic_Products.htm.

OMRI (Organic Materials Review Institute) (2001), OMRI for farmers and organic professionals. Available at: https://www.omri.org/omri-for-farmers.

Owens, E. (2017), *Listeria monocytogenes* Intervention and control workshop. Available at: https://www.pma.com/~/media/pma-files/food-safety/listeria/lm-control-workshop/owens-sanitation-best-practices-11-july-2017.pdf?la=en.

Paul, V., Ezekiel, R. and Pandey, R. (2016), Sprout suppression on potato: Need to look beyond CIPC for more effective and safer alternatives. *Journal of Food Science and Technology*, 53: 1-18.

Plotto, A. and Narcisco, J. A. (2006), Guidelines and acceptable postharvest practices for organically grown produce. *HortScience*, 41: 287-91. Available at: https://naldc.nal.usda.gov/download/16229/PDF.

Prengamen, K. (2018), The challenges of storing organic apples. *Good Fruit Grower*, 24 January 2018. Available at: https://www.goodfruit.com/the-challenges-of-storing-organic-apples/.

Richardson, S. D., Thruston, A. D. Caughran, T. V., Collette, T. W., Patterson, K. S. and Lykins, B. W. (1998), Chemical byproducts of chlorine and alternative disinfectants. *Food Technology*, 52: 58-61.

Robatscher, P., Eisenstecken, D., Sacco, F., Hannes, P., Berger, J., Zanella, A. and Oberhuber, M. (2012), Diphenylamine residues in apples caused by contamination in fruit storage facilities. *Journal of Agriculture and Food Chemistry*, 60: 2205-11.

Sharif, R., Mujataba, M., Ur Rahman, M., Shalmani, A., Ahmad, H., Anwar, T. and Wang, X. P. (2018), The multifunctional role of chitosan in horticultural crops; a review. *Molecules*, 23: 872-89.

Suslow, T.V. (2000), Postharvest handling for organic crops. Organic Vegetable Production in California Series. Pub. 7254. University of California Davis. Available at: http://ucanr.org/freepubs/docs/7254.pdf (accessed on 13 July 2019).

Thewes, F. R., Both, V., Brackmann, A., Weber, A. and Anese, R. D. (2015), Dynamic controlled atmosphere and ultralow oxygen storage on 'Gala' mutants quality maintenance. *Food Chemistry*, 188: 62–70.

Tucker, J. (2018), Impact of fumigation and irradiation requirements on organic imports. United States Department of Agriculture, National Organic Program. Rev01, pp. 1–18.

Umagiliyage, A. L. and Choudhary, R. (2018), Postharvest ultraviolet light treatment of fresh berries for improving quality and safety. *International Journal of Food Science and Nutrition Engineering*, 8: 60–71.

USDA-AMS (2018), United States National Organic Program Joint letter to PACA Licensees, organic labels must be removed from fumigated imports, 18 July 2018. Available at: https://www.ams.usda.gov (accessed on 01 December 2018).

Xue, L., Liue, G., Parfitt, J., Liu, X., van Herpen, E., Stenmarck, A., O'Connor, C., Ostergren, K. and Cheng, S. (2017), Missing food, missing data? A critical review of global food losses and food waste data. *Environmental Science Technology*, 51: 6618–33. doi:10.1021/acs.est.7b00401.

United States National Organic Program (2001a), Agricultural Marketing Service. International Trade Partners. Available at: https://www.ams.usda.gov/services/organic-certification/international-trade; http://postharvest.tfrec.wsu.edu/pages/PC98U (accessed on 01 December 2018).

United States National Organic Program (2001b), Agricultural Marketing Service. USDA-AMS Organic standards. Available at: https://www.ams.usda.gov/grades-standards/organic-standards (accessed on 01 December 2018).

Chapter 15

Smart distribution to maintain shelf life of horticultural produce

J. K. Brecht, University of Florida, USA; I. Uysal and M. C. N. Nunes, University of South Florida, USA; J. P. Emond, The Illuminate Group, USA; S. Mercier, Décathlon Canada, Canada; and U. McCarthy, Waterford Institute of Technology, Ireland

1 Introduction

Recognition of the magnitude of food waste and loss, particularly for fresh horticultural crops, has come to the fore in recent years (Gustavsson et al., 2011; Muth et al., 2011). 'Food waste' comprises the edible food that is either discarded, mostly at the consumer level, or not sold, and therefore not consumed, because of oversupply, or because the product did not meet minimum quality standards at some point during distribution. 'Food loss' consists of food that is rendered inedible or is consumed by decay organisms or animal or insect pests. It is estimated that the total of waste and loss for horticultural food crops is around 40–50%, which is the case for both the more developed countries of North America, Europe, and Pacifica as well as the less developed countries of Africa and Asia (Gustavsson, 2011). The difference is that, in the former countries, the majority is edible food that is discarded, while in the latter, the majority is due to outright loss of food.

The concept of 'shelf life' for fresh produce refers to a time period after harvest until the product becomes unmarketable or inedible due to negative quality changes reaching an unacceptable condition. Shelf life is therefore closely related to the waste portion of the food waste plus loss equation. When

http://dx.doi.org/10.19103/AS.2019.0055.17

the quality of fresh horticultural crops in the food distribution system falls below some minimum level required for acceptability, that product has reached the end of its shelf life. In practice, an entire load of produce may be rejected when only a small portion of the items fail to meet quality standards. This is because most government and commercial grade or quality standards specify allowable percentages for each possible defect or quality factor; loads that fail to meet any one of those must be re-worked (re-sorted) to remove the offending items, which is often considered to be economically infeasible.

Fresh horticultural crops are perishable, losing quality over time even when handled under the very best conditions. To best retain quality and thus extend shelf life, the distribution systems for these products are designed to avoid physical injuries, quickly and uniformly cool the products, strictly manage temperature and humidity levels, and quickly move the products from their origin to their destination. Temperature is well known to play the most critical role in regulating postharvest quality changes of fresh horticultural crops. Each individual crop has an optimal postharvest temperature, which is the lowest safe temperature to avoid either freezing or chilling injuries. Above the optimum temperature, the rate of quality deterioration typically increases by two to four fold for each 10°C (the Q_{10} relationship). Breaks in the 'cold chain' during distribution must be strictly avoided, because not only does exposure to a higher-than optimal temperature accelerate quality deterioration, but also because it is almost impossible to re-cool products while in the distribution system once warming occurs. Although the majority of losses and waste of fresh horticultural crops in developed countries occurs at the consumer level, much of that is influenced by previous temperature and humidity management, which influences the amount of shelf life remaining at the point of purchase. Nunes et al. (2009) conducted the first extensive study on the impact of retail conditions on the quality and waste of several fresh fruits and vegetables. Results from the study showed that improper temperature and humidity was the major cause of produce 'shrink' (i.e. waste and loss).

The purpose of this chapter is to highlight how optimizing management of produce distribution systems can reduce waste and loss and extend shelf life by minimizing quality deterioration, resulting in the best possible quality produce being consistently available to consumers. We believe that achieving this must involve monitoring of conditions during distribution and linking the resulting data to shelf-life modeling. This approach will enable 'smart distribution' practices to be implemented that result in products being further distributed according to their remaining shelf life, resulting in delivery of more consistent quality products, reduced waste and loss, and increased consumer satisfaction.

2 Smart distribution

Smart distribution, intelligent food logistics, and similar terms applied to the food distribution system imply that information about the history and quality of food is collected and used to inform decisions on how, when, and where discrete lots of food should be distributed in order to maximize efficiency and product quality (Jederman et al., 2014). Integral to smart distribution is the concept of first-expired-first-out (FEFO) logistics, which was introduced in the 1980s as 'shortest-remaining-shelf-life' (SRSL) or 'least-shelf-life-first-out' (LSFO; Wells and Singh, 1998). The FEFO approach stands in contrast to the more typical first-in-first-out (FIFO) approach. With FIFO, products are moved as quickly as possible through a distribution system, which does make a certain amount of sense for highly perishable products such as fresh horticultural crops. However, FEFO has the more advanced purpose of ensuring that when and where each item is shipped is matched to its remaining shelf life and the shipping options that are available in terms of scheduling and distance or duration. Applied properly, and with access to reliable, relevant information, FEFO logistics should reduce product waste and more consistently provides higher quality products to the ultimate consumers (Emond, 2006; Emond and Nicometo, 2006; Emond et al., 2006; Jedermann et al., 2008).

For smart distribution to be possible in a distribution cold chain for fresh horticultural crops, the initial quality of the product at harvest must be known. There must also be a means to predict the remaining, quality-based shelf life or, more specifically, the quality level of the product at each possible destination. That requires an accurate and robust shelf-life model that takes into account the complexity of the quality changes that potentially limit shelf life in horticultural crops under different constant and fluctuating temperature regimes. Creating such a shelf-life model in turn requires extensive knowledge of how product quality factors are affected by environmental conditions, primarily temperature, during distribution. Finally, to properly calculate the remaining shelf life, it is necessary to understand customer quality expectations related to customer satisfaction to know when a product's quality changes from 'acceptable' to 'unacceptable'.

Shelf-life modeling of perishable food products can be approached in several different ways, but always involves an end point that is related to a limit of acceptability related to some quality aspect of the product in question. Changes in sensorial quality factors like color, glossiness, texture, and flavor may occur over a range in which acceptability-unacceptability varies according to the market condition or the person making the decision. Other quality factors, especially negative factors such as decay and off-odors, have very low thresholds for limiting shelf life, that is, any occurrence at all renders the product unacceptable. Kinetic modeling is used to follow a reaction in which a desirable product (e.g. sugar) declines or an undesirable product increases until becoming apparent (e.g. browning). Microbial growth models calculate the proliferation rate of human pathogens or decay organisms under different conditions to predict when the microbial load reaches an unacceptable level. Empirical shelf-life models use experimental observations of changes in quality factors affecting acceptability and therefore shelf life during the postharvest period. Due to the complex nature of quality and quality changes in fresh horticultural crops after harvest, it is necessary to include multiple quality factors in produce shelf-life models, with all of those factors needing to be included among the initial measurements used to determine the initial quality index, and with different quality factors predominating in different temperature ranges.

Furthermore, it is necessary to have hardware and software systems to monitor conditions and recommend actions. As described in Hertog et al. (2014), sensor applications in a fresh produce supply chain include first-order, second-order, and third-order logistics (Fig. 1). First-order logistics involves using collected data to establish initial conditions, determine that conditions remain within desired limits, and to calculate the initial quality index for a product sample. Second-order logistics involve processing the environmental and quality data using the shelf-life model to calculate remaining shelf life under different projected scenarios. Third-order logistics involve using the calculated remaining shelf life and quality indices for the existing inventory at any moment during distribution to make intelligent supply chain decisions that maximize product quality for all customers.

Numerous types of fresh horticultural crops are produced in various regions of the world during different times of the year. Thus, produce quality, demand, supply, and price may fluctuate drastically throughout the year. This means that what is acceptable and unacceptable at any given time changes as market conditions change. When different production regions are producing

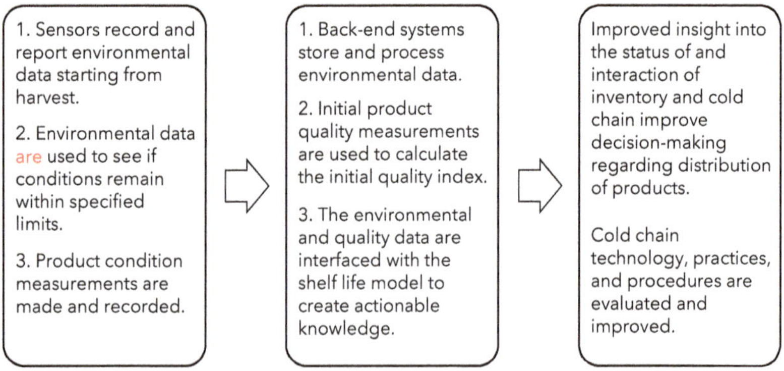

Figure 1 First-, second-, and third-order logistics in a monitored food distribution cold chain. Source: adapted from Hertog et al. (2014).

the same crop at the same time, this further complicates logistical decisions aiming to maximize product quality.

3 Logistics and supply chain management

In his landmark textbook 'Logistics and Supply Chain Management', Christopher (2016) draws a well-referenced distinction between the two, look-alike concepts of logistics and supply chain management. Whereas logistics is defined as the strategical management of movement and storage of materials within a single entity, like an organization, supply chain management builds upon this as a form of coordinated logistics both within and between organizations, which naturally represents a wider and more sophisticated concept. Compared to logistics, for which the academic, commercial, and military past goes back centuries, supply chain management is a relatively new idea. Oliver and Webber (1982) wrote: '...the traditional approach of seeking trade-offs among the various conflicting objective key functions - purchasing, production, distribution and sales - along the supply chain no longer worked well. We needed a new perspective and, following from it, a new approach: Supply chain management.' Although its definition has evolved throughout the years since it was first introduced, a widely accepted, contemporary, definition involves a network approach in which the supply chain is described as a connected network of organizations cooperating to manage the flow of both tangible (materials) and intangible (information) goods from suppliers to consumers (Hugos, 2018).

Fresh horticultural crop supply chain management requires coordination between multiple stakeholders from farmers to packers to shippers to retailers [including both storage facility, distribution center (DC), and store managers]. Shelf life, as the defining qualitative and/or quantitative quality of perishable products, has had a commanding role in the development of best distribution practices over the years for horticultural commodities (Hertog et al., 2014). As the single most important environmental factor affecting the shelf life of perishable products, management of temperature is a critical component, and ultimately resulted in a modified and renamed version of the supply chain called 'the cold chain' for which temperature has to be monitored and regulated (Trienekens and Zuurbier, 2008). The starting point of the cold chain is the harvesting of fresh horticultural crops and the precooling process that brings produce temperatures down to the appropriate ranges for crop-specific storage and transportation. The end point of the cold chain is the consumer's refrigerator or any other temperature-controlled storage unit at home. However, it is what happens between these two points that defines an appropriately managed cold chain with the optimum amount of preservation of produce quality and shelf life *versus* an ineffectively managed one that conversely results in unnecessary product loss and lessened consumer satisfaction. When it comes to cold chain inventory management, however, most companies still employ traditional inventory control methods that rely mostly on supply and demand and the time of arrival (FIFO) rather than data-driven distribution practices (Alamri and Syntetos, 2018).

The first step in the cold chain is the precooling process, for which cut-to-cool time is critical for removal of the residual or 'field' heat from the produce after harvest (Mercier et al., 2019). Precooling facilities are crucial for reducing the produce temperatures to the ideal storage and transportation range since the refrigeration used in storage, and especially transportation, systems are usually designed only to maintain the temperature of the produce rather than reduce it (James et al., 2006). Forced-air cooling, hydrocooling, room cooling, and vacuum cooling constitute different kinds of precooling available for different commodities (Gross et al., 2016; McDonald and Sun, 2000; Reina et al., 1995; Thompson et al., 2008). Ultimately, the choice of the cooling method is based upon the product's physical and biological properties. Once the product temperatures are brought down to a desirable range, the products may be stored, but are eventually always shipped, via different modes of transportation including land, air, sea, rail or multimodal, which may include a chain of more than one form of transportation. Although it is well accepted that proper precooling, storage, and transportation practices are critical for a well-managed cold chain, perhaps the most significant portion of inventory control, in terms of both volume and cost within the cold chain framework, takes place in the DCs (Lee and Billington, 1992).

DCs are commonly owned and operated by the retail companies, including large grocery chains such as Costco, Kroger, and Walmart. They accept purchased goods from the suppliers and store them temporarily until they are shipped to the retail stores. Horticultural crops are typically held at one of the few compromise temperatures and humidities in refrigerated storage rooms (Thompson and Crisosto, 2002). For standard inventory control of horticultural crops, FIFO is the most common method used to manage purchased inventory (Wei et al., 2015). FIFO inventory control requires the first items to arrive at the DC to be the first ones shipped to the stores connected to that warehouse. FIFO is the preferred method for distribution when the products are highly perishable, which is the case for horticultural produce.

Inventory management using FIFO relies on static information (time of arrival at the DC). FEFO is an alternative inventory control method that relies on more dynamic information, such as time-temperature history of the cold chain to serve as an intelligent alternative to FIFO (McCarthy et al., 2018). In FEFO, the products are shipped to the connected stores based on their algorithmically predicted expiration dates, or remaining shelf lives, instead of their arrival times at the DC. Since the application of a FEFO-managed cold chain requires knowledge of the produce's remaining shelf life, a FEFO-enabled DC requires both hardware and software support, including wireless sensors (primarily temperature) embedded throughout the cold chain and the necessary algorithms and cloud services for actionable information processing. Wireless monitoring devices, including radio-frequency-identification (RFID) tags, Bluetooth, and Wi-Fi-enabled sensors, and even sensory loggers with cellular connectivity, have seen groundbreaking advances in accuracy, convenience, and cost-effectiveness. As a result, it is not only feasible but also more sensible to use ubiquitous temperature monitoring technologies for more efficient and data-driven cold chain management.

With regard to FEFO-based perishable inventory management, quantifiable impacts on both the level and consistency of the quality of fresh produce have been shown (Nunes et al., 2014). In a recent study comparing FIFO *versus* FEFO for planning the distribution strategy for pallets of fresh strawberries with different temperature histories shipped to the same DC, Brecht et al. (2014a) showed that the minimum observed quality at the retail stores for products distributed using FEFO strategy was higher compared to FIFO strategy, which meant that the consumers on average had access to better-quality products with longer remaining shelf lives. Furthermore, the standard deviation in quality for FEFO was also less than FIFO, which indicates more consistent quality for the consumers. Simulation-based studies on other perishable commodities support similar conclusions. A FEFO management system implemented at a DC for seafood was shown to reduce the incidence of unacceptable quality at consumption from 7% to 0% when the products were shipped to local retailers and from 30% to 13% when the products were shipped to distant retailers (Koutsoumanis et al., 2002). It is important to note, however, that the majority of studies conducted on FEFO inventory management involve stand-alone retailers where the communication takes place between DCs and stores owned and operated by the same company. At a global scale, the advantages of FEFO management would become even more apparent as demonstrated by a simulation-based study on the banana cold chain between Costa Rica and Europe (Dittmer et al., 2012). A significant food waste reduction of up to 22% was observed, which would have a cascading effect on reducing CO_2 emissions due to smaller numbers of spoiled shipments being replaced.

To realize the true benefits of a well-managed cold chain, coordination between all stakeholders is critical and often categorized into two key management challenges: (1) product management and (2) information management. While these are often viewed and addressed independently of one another in terms of improvement strategies, it is important not to lose sight of one when addressing the other. A key to successful cold chain management is transparency. Transparency involves an ability to view all occurrences (planned and unplanned) that occur during transit in real or near real time. An ability to view (or be aware) of all these events facilitates understanding and possibly predicting future occurrences, thus preventing any unplanned events from occurring again in the future. It is also important to note that effective supply chain management involves the transport of both product and information in unison (Badia-Melis et al., 2016a).

Cold chain management requirements involved in the management, coordination, dispatching, and/or receiving of produce in the current perishable commodity business environment are significantly different from what they were 20 years ago. External to the immediate internal (business processes) and product-based (process/storage requirements) business requirements involved, there is also significantly more societal and environmental awareness with respect to the product and its environmental impact from a variety of stakeholders including consumers and governmental agencies who wish to know the 'story' of products in terms of origin, destination, and means of transport. Advancements in transport, trade, product, and processing treatments and packaging have all served to positively impact global trade and have successfully aided in establishing and maintaining global supply networks. A description of the requirements of both product and information management are outlined below.

3.1 Coordinated product management

Coordinated product management requires precise coordination, control, and monitoring of product to ensure that it reaches its destination in as high a quality as possible relative to its remaining shelf life. This process requires a delicate balance of a series of economic and financial dynamics to ensure this is carried out as efficiently as possible. It involves the management of the product as it is prepared for dispatch and ensuring that all conditions during dispatch are compatible with the safe storage and transit requirements of the product.

3.2 Coordinated information management

Coordinated information management involves the gathering, processing, and interpretation of data specific to the product. In many instances, this information is acquired via a series of environmental sensors traveling with the product. Environmental parameters of significance include temperature, relative humidity, light, and acceleration. Once gathered, these various data sources must be processed and interpreted, which, when combined with referenced data, can paint a 'virtual story' of the product. This virtual story (dataset) can then be referenced against a baseline (ideal storage conditions) dataset and any instances of outlier data points can be labeled as representative of product abuse. Through the use of predictive algorithms, it is also possible to predict the shelf life of the product relative to its transit history (Badia-Melis et al., 2018).

A key challenge, however, is the standardization of this data to ensure that it can be analyzed and interpreted anywhere in the globe and mean the same thing to all parties, which is to say that all data must be stored and processed in a standardized fashion irrespective of location on the globe (McCarthy et al., 2013).

Correctly coordinated, efficient, and well-maintained cold chain management has direct positive impact both on consumer food quality and global food security as well as waste and loss (McCarthy et al., 2018). In combination, both product and information management form the basis of fast, flexible, and responsive global supply networks and facilitate many economic and financial efficiencies as well as positive environmental and societal impacts for the betterment of society.

4 Shelf-life modeling of fresh produce

Shelf life necessarily encompasses all the variable handling, cooling, storage, transport, and display conditions encountered by the produce throughout the supply chain. There are numerous quality factors that can potentially determine the end of shelf life, depending on the environmental factors experienced by the produce and whether or not the produce has been physically damaged, not to mention the presence of decay-causing microorganisms and the susceptibility of the produce to their attacks. The potential shelf life of produce items varies widely, with some, such as tree fruits like apple, and underground storage organs such as sweetpotatoes, possessing the potential to be stored for many months, while others, such as leafy greens, immature fruits such as okra, and fruits harvested fully ripe like strawberries, have a potential shelf life that may be measured by days or weeks. Within the potential shelf-life limits possessed by each produce type, temperature is the main factor affecting the rate of quality deterioration and thus the shelf life (Nunes, 2008a,b).

'Quality' in the context of this discussion refers to all those characteristics of various produce items that determine their attractiveness to consumers. Quality generally declines over time. This differs from the usage in some areas of produce handling and transport, such as claims and inspections, in which quality is distinct from 'condition'. In those situations, quality describes permanent characteristics of the product. Examples include shape, scarring, or internal cavitation. In contrast, condition refers to the characteristics that progress over time, such as decay, softening, wilting, bruising symptoms, or color changes. Visual quality factors are most important in handling fruits, vegetables, and ornamental crops. Visual quality encompasses all aspects of appearance, including size, shape, color, gloss and other aspects of surface appearance, wilting or shriveling, and decay. They are usually of primary importance in determining market value and consumer acceptance. However, it is often said that consumers 'buy with their eyes, but repeat purchases depend on flavor'. Flavor quality factors include both taste and aroma. Taste includes sweetness and sourness and their balance, as well as bitterness and various off-flavors. Aroma includes both the characteristic volatile aromas produced by each produce item and the off-odors that signal spoilage. Texture can be included with flavor as an important part of the overall sensory quality of fresh produce, encompassing crispness, juiciness, fibrousness, melting, and other aspects of mouthfeel. Other quality attributes that are not perceived by consumers, but are equality important, are the chemical and nutritional attributes of produce. These include for example, water content, sugars, vitamins, and bioactive compounds, which also decline as produce senesces and loses quality.

Flavor and texture can often limit produce shelf life prior to the appearance of visual quality defects. However, unfortunately, for many years, shelf life was determined exclusively in terms of loss of visual quality. This has been particularly unfortunate with regard to tropical and subtropical crops that are subject to chilling injury. The recommended lowest safe temperatures for those crops (i.e. the chilling threshold temperature) often result in loss of flavor, because inhibition of aroma volatile synthesis is usually the first symptom of chilling injury to appear. Potential storage life expectations for temperate crops are similarly often overstated for similar reasons since postharvest life based on flavor is usually only 1/2 to 2/3 the shelf life based on visual appearance (Kader, 2003).

Loss of moisture from fruits and vegetables is a major cause of deterioration during handling and distribution and often results from produce being handled at suboptimal levels of humidity (below 85%) combined with temperature abuse. Loss of substantial amounts of water may result in important quality and economic losses, and even when the consequences of weight losses are subtle, the visual, compositional, and eating quality of the produce may be impaired (Chilson et al., 2011; Nunes and Dea, 2016; Nunes and Emond, 2007).

Nutritional components of fresh fruits and vegetables are often quite labile. For example, loss of vitamin C has been shown to be very sensitive to temperature and water loss (Nunes et al., 1998). Although there have been only a few studies involving use of chemical or compositional measurements to model or estimate fresh produce shelf life (Amodio et al., 2013; Odriozola-Serrano et al., 2009; Oms-Oliu et al., 2009), we know that changes in some quantitative parameters such as weight loss, L* value, analytical texture, anthocyanin, ascorbic acid, and chlorophyll contents are highly correlated with subjective attributes such as for example, visual color, subjective texture, wilting, or shriveling, and thus can very likely be used to estimate shelf life (Nunes, 2015).

Shelf-life modeling is used to describe the quality changes that occur in food products and to predict the remaining shelf life under any defined future conditions. Being able to predict future quality changes within the food distribution system has powerful implications for managing logistical operations related to transportation, distribution, and retailing to maximize quality and shelf life and consequentially improve consumer satisfaction (Brecht et al., 2014a; Kelly et al., 2019; Nunes et al., 2014).

Perhaps not surprisingly, since microbial spoilage is the primary cause of end of shelf life for many foods, much shelf-life modeling research has been focused on microbial growth models (Caleb et al., 2013; Ramos et al., 2013; Zwietering et al., 1991). If visible decay can be equated with a specific number of microorganisms per unit mass of food, then the exponential growth rate of the microbes under various conditions can be used to predict shelf life (Fu and Labuza, 1993). A number of different models have been developed or utilized to describe and predict microbial growth, including the exponential model and the Arrhenius model (Anderson, 2010). However, since predictive microbiology modeling is usually based on the response of the typical microflora of a specific food, they are only valid under a specific range of conditions outside which other microorganisms may also cause spoilage (Pin and Baranyi, 1998). Other limitations of applying predictive microbiology to estimate shelf life of fresh fruits and vegetables are the contribution of non-microbial factors to produce deterioration (McMeekin and Ross, 1996; Verhaelen et al., 2012). Besides, such models may not always be accurate in determining the shelf life of fresh fruits and vegetables because deterioration of sensory quality often limits their shelf life before microbial growth exceeds acceptable values (Nunes et al., 2014).

Kinetic modeling is one approach in which it is intended that what is known about the chemical changes that occur during food quality degradation can be used to model its shelf life (van Boekel, 2008). If a particular chemical or biochemical (enzymatic) reaction in a food product is primarily responsible for quality degradation, then kinetic modeling is an attractive approach for modeling the shelf life at different temperatures. An example is the nucleotide catabolism that occurs rapidly postmortem in fresh seafood (Ehira and Uchiyama, 1987). Similarly, the enzymatic reactions of spoilage causing microorganisms lead to the accumulation of a number of volatile compounds, particularly ammonia by the action of bacterial decarboxylases on amino acids and other nitrogenous compounds (Cheng et al., 2015).

A common feature of both microbial and kinetic growth models is that they are applicable only to relatively simple processes and chemical reactions. However, produce shelf life is a complicated biological system, with many different changes occurring, often simultaneously, that may affect quality perceptions. The concept of using different sensory quality attributes simultaneously to model shelf life of fresh fruits and vegetables was first proposed by Nunes et al., who have extensively documented for numerous fruits and vegetables that not only can almost any quality factor be responsible for the end of shelf life, depending on the specific produce item, but which quality factor is responsible for limiting the shelf life changes according to the environmental conditions (Nunes et al., 2001, 2003a,b, 2004, 2007, 2012, 2013; Proulx et al., 2001, 2005, 2010; Chilson et al., 2011). Specifically, the shelf-life limiting quality factors for any produce item are often different at different temperatures. Additionally, loss of shelf life can occur without the occurrence of visible changes, such as with loss of the green pigment chlorophyll, which can easily reach 50% or more with no visible clue. In the produce distribution system, this results in products that appear to be in good condition but in reality have very little remaining shelf life being accepted or forward-routed incorrectly.

Quality can be considered as an absolute standard by which all samples of something can be judged. However, to expect all people to agree on an absolute standard of quality is impractical. The alternative is to treat quality as a relative concept–in modeling the shelf life of fresh produce during distribution and marketing it makes most sense to identify the product quality relative to the initial, at-harvest, quality of each sample. Assuming that the variety possesses intrinsically acceptable quality (or, acceptable *potential* quality, such as with a fruit that will be ripened after harvest), then this approach treats all samples of that type of fruit or vegetable as initially similar in quality.

The situation with fresh horticultural crops thus favors the use of empirical models that relate (primarily) temperature to the unique set of quality changes found in each type of fruit, vegetable, or ornamental crop.

4.1 Empirical respiration rates and quality modeling

Emond et al. (1999) used empirical respiration rate data found in the literature to develop elementary models to predict shelf life based on a quality index. A quality index is the ratio of the product quality at any point during its shelf-life period relative to its initial (i.e. at-harvest) quality for any one or more of the product's quality characteristics. Respiration rates of many produce items at different temperatures can be easily found in literature such as USDA Handbook 66 (Gross et al., 2016). Using the principle of Q_{10}, namely that the rates of chemical reactions roughly double with an increase of 10°C, and data on optimal shelf life, it is possible to develop equations that can be used to predict overall quality index depletion over time at different temperatures. Assuming that quality index depletion follows a natural logarithmic trend such as the Arrhenius equation, it is possible to plot quality index curves at different temperatures as shown in Fig. 2.

One of the major weaknesses of this approach is the nature of the data found in the literature. Shelf life as well as respiration rates come from multiple sources and most of the times are done using different cultivars, which create significant discrepancies for specific applications. Nunes et al.

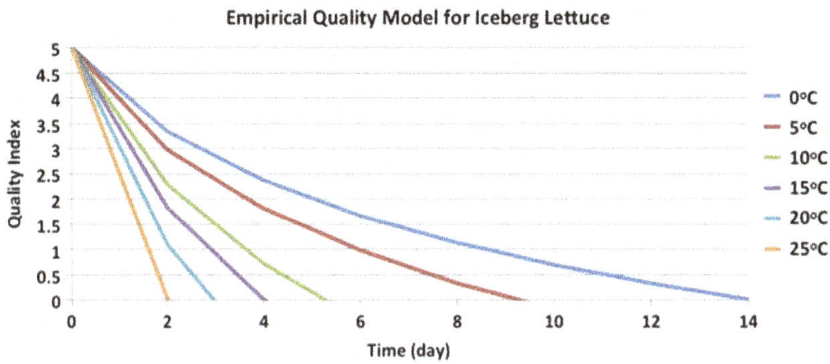

Figure 2 Empirical quality index curves for Iceberg lettuce based on respiration rates found in the literature. Source: adapted from: Emond et al. (1999).

(2004) employed this method in the study of blueberry quality by producing curves from experimental and predicted quality data at varying temperatures. The predicted shelf life was found to be longer than the experimental data below 5°C and shorter above 5°C. Inconsistencies were attributed to varying environmental factors and low reliability of the correlation between quality and respiration rates found in the literature.

4.2 Shelf-life model using multiple quality parameters

As discussed previously, a quality index for fresh horticultural crops can be described in terms of all relevant quality characteristics for that crop: color, texture, taste, and many other attributes. Using an overall quality index does not help for managing logistical operations related to transportation, distribution, and retailing since it does not tell the whole story on how the whole distribution chain can be improved. Nunes et al. (2004) developed quality curves at different temperatures for multiple quality parameters of blueberries, including color, taste, firmness, and aroma. As shown in Fig. 3, some quality parameters decline more rapidly than others.

Multiple quality parameter models can be developed using polynomial trend lines that give the strongest correlations between temperature and quality changes. Empirical quality curves can be obtained from experimental data and converted into a series of curves for multiple temperatures. Emond et al. (2001) developed algorithms to predict residual shelf life when products are exposed to fluctuating temperatures during transit. Based on experimental data, algorithms were developed to integrate changes over time in quality attributes of fresh fruits and vegetables. The algorithms iterate each quality factor with specific time increments. In many cases, only tables for different constant temperatures are available, listing the changes of physical properties

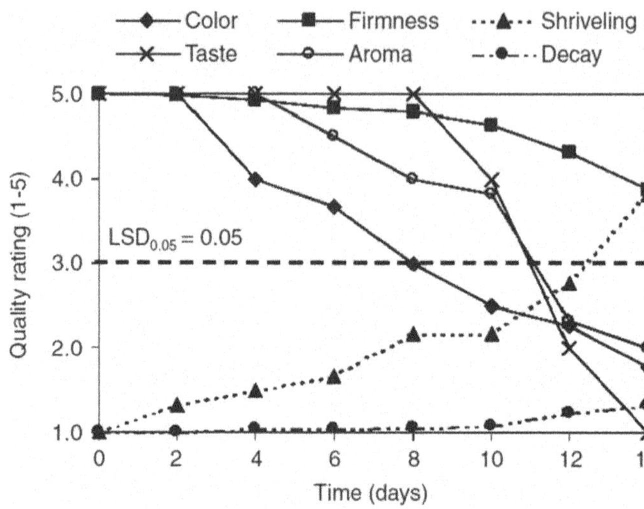

Figure 3 Quality characteristics of 'Patriot' blueberries stored at 0°C; objectionable taste and aroma were the limiting quality factors (the dotted line corresponds to the limit of acceptability before the quality of the fruit became objectionable). Source: adapted from: Nunes et al. (2004).

over time. Examples of differential equations from these measurements can be found in Nunes et al. (2003a, 2004).

Using recorded temperature during transit, it is possible to predict the remaining shelf life at any time by continuously updating the quality index value for designated time steps. Different quality curve equations are used for each designated time step. Intermediate temperatures for which quality curves have not been determined are handled by interpolation. Periods during which product warming or cooling occur are handled by averaging the temperatures during specified time steps. The residual shelf life at any point in time is predicted by projecting either continuation of the current temperature or any selected future temperature scenario. Figure 4 shows the color-based quality curves for iceberg lettuce at 1°C, 4°C, and 8°C and the process to determine the residual shelf life when the transit temperature changes from 1°C to 4°C on Day 5. Considering that the end of shelf life occurs when the color quality index reaches '1', the increase of temperature from 1°C to 4°C on Day 5 resulted in a reduction of the shelf life by 10 days (from 21 days to 11 days).

Simultaneously running multiple quality parameter models will provide at any time the limiting quality parameter responsible for the shortest residual shelf-life duration. For example, a strawberry shelf-life prediction model proposed by Nunes and Emond (unpublished) that included multiple potential shelf-life-limiting quality factors was tested with retailers under real-world conditions with great success. The mathematical model predicted that the shelf

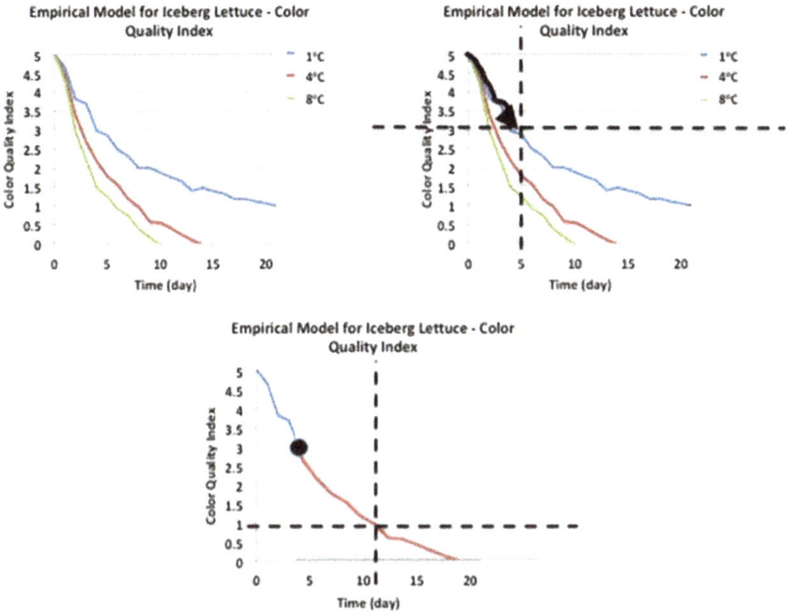

Figure 4 Dynamic prediction of color quality index of iceberg lettuce during transit. Color-based quality curves are determined for constant exposure at 1°C, 4°C, or 8°C (upper left). In the case that the temperature changes from 1°C to 4°C on Day 5, the portions of the 1°C curve up to Day 5 and the 4°C curve after Day 5 are selected (upper right); when combined, the resulting curve (lower left) allows the time when the color quality index will reach '1' (i.e. end of shelf life) to be predicted. Source: adapted from: Emond et al. (1999).

life of strawberries arriving at the DC would reach the limits of salability within approximately 8 h of the time determined by quality evaluations (Nunes et al., 2014). Similar model developments are described for sweetcorn (*Zea mays* L.) in Anderson (2010) and for military rations in Brecht et al. (2011, 2014b).

5 Remote sensing of food quality and the environmental factors that influence food quality

Constant monitoring of fresh horticultural products is critical to the optimization of the cold chain. Temperature is generally considered to be the most important factor to monitor (Hertog et al., 2014; Nunes et al., 2014). Each crop has a product-specific temperature range for the optimal preservation of its shelf-life, below which the risk of cold injuries increases, and above which accelerated loss of quality occurs (Mercier et al., 2017). The loss in quality can be reflected by organoleptic and compositional deterioration of the product, or can be the result of the growth of spoilage microorganisms (Giannakourou and Taoukis,

2003; Gwanpua et al., 2015; Hertog et al., 2014). Knowledge in real time of temperature can often be translated into an estimation of the product's remaining shelf life–this estimate can then be used to improve the supply chain through management systems such as FEFO inventory management, dynamic expiry dates, and dynamic pricing (Dittmer et al., 2012; Mercier et al., 2017; Nunes et al., 2014; Tromp et al., 2012; Wang and Li, 2012). Temperature monitoring along the supply chain can also enable corrective actions, such as changing the destination of a shipment in the event of a temperature deviation (Dittmer et al., 2012). In addition to the temperature, relative humidity, pressure, microbial concentration, and the composition of surrounding gases are additional factors that can be monitored to extend the shelf life (Hertog et al., 2014; Nunes et al., 2014).

Visual inspection of the product (after harvest or at a storage facility or DC) can provide inaccurate estimates of the remaining shelf life (Nunes et al., 2014; Jedermann et al., 2014). The vast majority of produce loads delivered to DCs are accepted, but we know that all those accepted products have actually aged (deteriorated) since harvest. Although the changes associated with aging of produce are mostly not visible, including such things as early stages of chlorophyll degradation and water loss, texture changes, loss of vitamin C and other nutrients, and loss of aroma, it is possible to correlate subjectively evaluated quality factors with environmental measurements and physicochemical characteristics of produce (Nunes, 2015). As such, monitoring of the product quality and environmental factors can be performed using a system of wireless sensors. Radio-frequency-identification (RFID) sensors, built upon an Internet-of-Things (IoT) framework, are popular choices given their increasing monitoring capabilities and cost-efficient results (Badia-Melis et al., 2014).

It is well known that the distribution of quality and temperature is not uniform within a shipment of fresh produce - therefore, a network of sensors is required to reach sufficient spatial resolution. Sometimes, multiple sensors are required even at the scale of a single pallet (Mercier et al., 2017). To limit the cost of the sensor network, the combination of a minimal number of sensors with a predictive mathematical model has been suggested (Badia-Melis et al., 2016a, 2018; Jedermann and Lang, 2009; Jedermann et al., 2011). Mathematical models can be based on machine learning approaches or can be physics-based, that is, developed from mass, momentum, and heat balance equations (Badia-Melis et al., 2016b; Mercier and Uysal, 2018). Machine learning models built using feedforward or recurrent neural networks are more flexible. However, they require a large set of historical data for model training. In contrast, physics-based models do not need a large set of historical data, but do require a good knowledge of the product's thermal properties, and a different model is required for each product, packaging system, and stage along the supply chain (Mercier and Uysal, 2018). While the commercial implementation of a sensor network combined with a predictive model remains challenging because the proper number of sensors to use and their optimal location are often unclear, it represents a promising approach to extend the preservation of fresh horticultural products (Mercier and Uysal, 2018).

Smart distribution of fresh horticultural products to maintain consistent quality and shelf life is fundamentally a data-driven process. The data, in this case, consists of any number of sensory measurements along the cold chain to subsequently enable predictive tools and algorithms that ultimately allow informed decision-making. The front-end of a cutting-edge smart distribution system consists of remote sensing devices that can reliably monitor and report the environmental factors that impact food quality, such as temperature and humidity. As both wireless communication technologies and sensory measurement hardware have significantly improved over the last decade, a wide range of monitoring tools have become available for all cold chain stakeholders, from growers to processers to shippers to retailers (McCarthy et al., 2018). Today, a system integrator can choose from an independent combination of the two technologies, sensory and communications hardware, to meet the exact application requirements for their unique implementation. For instance, a temperature sensor chip can be coupled with an RFID-powered device to allow passive and/or active communications on an existing RFID infrastructure. On the other hand, if humidity is also critical for a certain commodity and the RFID infrastructure may be unavailable, the same sensory circuit can be augmented with a humidity sensor and instead coupled with either a Wi-Fi or Bluetooth low energy (BLE) module to utilize the distribution system facilities' existing wireless router network (in the case of Wi-Fi) or the mobile phones of shipping and warehouse personnel (in the case of BLE).

Even though the implementation details are different between these two scenarios, a common thread exists within the context of 'cloud computing' empowered by the IoT. The IoT concept, although gaining mainstream attention only recently, was introduced as far back as in 1991 in Weiser's renowned article '*The Computer for the 21st Century*', in which he famously claimed, 'Specialized elements of hardware and software, connected by wires, radio waves and infrared, will be so ubiquitous that no one will notice their presence' (Weiser, 1991). Considered by some as a mere re-description of computing devices connected to the Internet, IoT examples are widespread (Li et al., 2015). For instance, a wireless sensor network monitoring the cold chain of a highly perishable horticultural product, such as strawberry, can easily be considered an IoT framework as long as at least one sensor node has access to a remote server (Dargie and Poellabauer, 2010). Recent developments in the RFID, BLE, Wi-Fi, and cellular communication technologies and their wide-scale adoption in the retail supply and cold chains are modern applications of IoT, where Internet-connected RFID readers, BLE-enabled mobile phones, wireless routers, and cell towers communicate with stand-alone sensor devices with unique identifiers attached to the objects in the environment (such as produce cartons, flats, pallets, containers, etc.) (Welbourne et al., 2009). Defined as Web 3.0, which signifies the next generation of ubiquitous computing web, IoT presents remarkable challenges and opportunities at the same time (Gubbi et al., 2013).

6 Conclusion

Approximately half of all the horticultural crops in developed countries are wasted postharvest due to inadequate management of the supply chain and reduced quality below market or consumer requirements. Smart distribution management via ubiquitous monitoring of environmental conditions such as temperature and humidity can extend shelf life and minimize quality loss throughout the cold chain. Sensory data can be used for predictive modeling of the product shelf life as well as transportation and handling quality, which can subsequently be used to intelligently control product flow out of the DCs. Informative and data-driven models such as FEFO can be used to manage fresh horticultural inventory based on quality and remaining shelf life instead of FIFO arrival time logistics. Industry and academy alike have demonstrated quantifiable benefits in terms of both more consistent and higher quality of products delivered to consumers of fresh horticultural produce.

Enabling smart logistics requires both the hardware and software tools to remotely monitor environmental conditions, associate those conditions with product quality, predict the remaining shelf life, and recommend actions. Of the possible shelf-life modeling approaches available, multiple quality parameter shelf-life models using polynomial trend lines give the strongest correlations between temperature and quality changes for fresh horticultural crops. Knowing the remaining shelf life of each inventory item, FEFO logistics dictates that the items with the least remaining shelf life should be distributed first and to the closest destinations, while those items with longer remaining shelf life can either be distributed later or sent to more distant destinations. The result is that consumers are less likely to encounter fresh horticultural crops at retail that are of unacceptable quality or that possess too little remaining shelf life to remain acceptable in their homes until they are consumed. This will both improve consumer satisfaction while reducing food waste.

Nevertheless, the promise of a well-managed, sensor-enabled, and data-driven cold chain is tempered by the significant task of effective collaboration between all the industry stakeholders, from growers to processers to shippers to retailers to even consumers, for successful implementation of all the necessary management systems. Two of the biggest challenges that constrain the so-called success stories are (1) the up-front costs of sensory and communications infrastructure and (2) how data are shared between different companies that operate within the same supply chain information framework. Fortunately, while the groundbreaking advances in wireless communications, such as BLE, help reduce the initial burden of implementation, other technologies such as Blockchain could help in how data are stored and shared between companies. Global and regional regulations, best manufacturing and handling practices, the interoperability of sensor and communication technologies, and standards and data sharing services and agreements all need to work harmoniously to truly realize the potential of smart distribution management. In the end, however, the sustainability of the cold chain has grown only more important with the growing world population and associated food insecurity all across the globe. Low-income households, for instance, benefit disproportionately from reduced fresh fruit and vegetable waste, which helps promote lower prices, and provides consistently better quality and nutritional value for the most financially vulnerable populations.

7 References

Alamri, A. A. and Syntetos, A. A. 2018. Beyond LIFO and FIFO: exploring an Allocation-In-Fraction-Out (AIFO) policy in a two-warehouse inventory model. *International Journal of Production Economics* 206, 33–45.

Amodio, M. L., Derossi, A. and Colelli, G. 2013. Modelling sensorial and nutritional changes to better define quality and shelf life of fresh-cut melons. *Journal of Agricultural Engineering* 44, 38–45.

Anderson, K. 2010. Development of a simulator for sweetcorn cold chain distribution. M.Sc. thesis. University of Florida, Gainesville, FL, 97p.

Badia-Melis, R., Garcia-Hierro, J., Ruiz-Garcia, L., Jiménez-Ariza, T., Villalba, J. I. R. and Barreiro, P. 2014. Assessing the dynamic behavior of WSN motes and RFID semi-passive tags for temperature monitoring. *Computers and Electronics in Agriculture* 103, 11–6.

Badia-Melis, R., McCarthy, U. and Uysal, I. 2016a. Data estimation methods for predicting temperatures of fruit in refrigerated containers. *Biosystems Engineering* 151, 261–72.

Badia-Melis, R., Qian, J. P., Fan, B. L., Hoyos-Echevarria, P., Ruiz-García, L. and Yang, X. T. 2016b. Artificial neural networks and thermal image for temperature prediction in apples. *Food and Bioprocess Technology* 9, 1089–99.

Badia-Melis, R., McCarthy, U., Ruiz-Garcia, L., Garcia-Hierro, J. and Villalba, J. R. 2018. New trends in cold chain monitoring applications - a review. *Food Control* 86, 170–82.

Brecht, J. K., Emond, J. P., Nunes, M. C. N., Sims, C. A., Obadasi, A. and Uysal, I. 2011. Remote environmental monitoring and diagnostics in the perishables supply chain: Phase I. Technical Report Natick/TR-12/009. Prepared for U.S. Army Natick Soldier Research, Development and Engineering Center, Natick, MA. Available at: www.dtic.mil/cgi-bin/GetTRDoc?AD=ADA554037.

Brecht, J. K., Loayza, F. E., Nunes, M. C. N., Emond, J. P., Uysal, I., Badia, R., Wells, J. and Saenz, J. 2014a. Reducing strawberry waste and losses in the postharvest supply chain via intelligent distribution management. *Acta Horticulturae* 1120, 253-60.

Brecht, J. K., Uysal, I., Nunes, M. C. N., Sims, C. A., Obadasi, A., Pliakoni, E., Emond, J. P., Bennett, G. and Wells, J. L. 2014b. Remote environmental monitoring and diagnostics in the perishables supply chain: Phase II. Technical Report. Prepared for U.S. Army Natick Soldier Research, Development and Engineering Center, Natick, MA. Available at: www.dtic.mil/cgi-bin/GetTRDoc?AD=ADA601491.

Caleb, O. J., Mahajan, P. V., Al-Said, F. A. and Opara, U. L. 2013. Modified atmosphere packaging technology of fresh and fresh-cut produce and the microbial consequences - a review. *Food and Bioprocess Technology* 6, 303-29.

Cheng, J. H., Sun, D. W., Zeng, X. A. and Liu, D. 2015. Recent advances in methods and techniques for freshness quality determination and evaluation of fish and fish fillets: a review. *Critical Reviews in Food Science and Nutrition* 55(7), 1012-225. doi:10.1080/10408398.2013.769934.

Chilson, D., Delgado, A. and Nunes, M. C. N. 2011. Shelf life of vine tomato (*Lycopersicum esculentum*) stored at non-chilling temperature and different relative humidity levels. *Proceedings of the Florida State Horticultural Society* 124, 246-55.

Christopher, M. 2016. *Logistics & Supply Chain Management*. Pearson, UK.

Dargie, W. and Poellabauer, C. 2010. *Fundamentals of Wireless Sensor Networks: Theory and Practice*. John Wiley & Sons, Chichester, West Sussex, UK, 311p.

Dittmer, P., Veigt, M., Scholz-Reiter, B., Heidmann, N. and Paul, S. 2012. The intelligent container as a part of the Internet of Things. In: *2012 IEEE International Conference on Cyber Technology in Automation, Control, and Intelligent Systems (CYBER)*. IEEE, pp. 209-14.

Ehira, S. and Uchiyama, H. 1987. Determination of fish freshness using the K value and comments on some other biochemical changes in relation to freshness. In: Kramer, D. E. and Liston, J. (Eds), *Seafood Quality Determination*. Elsevier, Amsterdam, pp. 185-93.

Emond, J. P. 2006. The future of the RFID-enabled supply chain and research needed to drive it forward. *RFID Journal Live*, 1 May 2006, Las Vegas, NV.

Emond, J. P. and Nicometo, M. 2006. Shelf-life prediction and FEFO inventory management with RFID. In: *Cool Chain Association Workshop: Temperature Measurements: When, Where and How?* 13-14 November 2006, Knivsta, Sweden.

Emond, J. P., Nunes, M. C. and Laniel, M. 1999. Effects of the distribution chain on quality of perishables: quality curves as a function of storage temperature. Technical Report. ETR Group AB, Laval University, April 1999, 151p.

Emond, J. P., Nunes, M. C., Molinari, A., Laurin, E. and Proulx, E. 2001. Quality curves as a function of fluctuating temperatures. Technical Annual Report. ETR Group AB, Laval University, May 2001, 148p.

Emond, J. P., Casas, T. and Crawford, T. 2006. Case study: publix tracks fresh produce with RFID. *RFID Journal Live*, 3 May 2006, Las Vegas, NV.

Fu, B. and Labuza, T. P. 1993. Shelf-life prediction: theory and application. *Food Science and Nutrition* 4, 125-33.

Giannakourou, M. C. and Taoukis, P. S. 2003. Application of a TTI-based distribution management system for quality optimization of frozen vegetables at the consumer end. *Journal of Food Science* 68, 201-9.

Gross, K. C., Wang, C. Y. and Saltveit, M. 2016. The commercial storage of fruits, vegetables, and florist and nursery stocks. Agriculture Handbook No. 66. USDA Agricultural Research Service, Washington DC.

Gubbi, J., Buyya, R., Marusic, S. and Palaniswami, M. 2013. Internet of Things (IoT): a vision, architectural elements, and future directions. *Future Generation Computer Systems* 29, 1645–60.

Gustavsson, J., Cederberg, C., Sonesson, U., Otterdijk, R. and Meybeck, A. 2011. *Global Food Losses and Food Waste: Extent, Causes and Prevention*. Food and Agriculture Organization of the United Nations, Rome, 29p.

Gwanpua, S. G., Verboven, P., Leducq, D., Brown, T., Verlinden, B. E., Bekele, E., Aregawi, W., Evans, J., Foster, A., Duret, S., Hoang, H. M., van der Sluis, S., Wissink, E., Hendriksen, L. J. A. M., Taoukis, P., Gogou, E., Stahl, V., El Jabrii, M., LePage, J. F., Claussen, I., Indergård, E., Nicolai, B. M., Alvarez, G. and Geeraerd, A. H. 2015. The FRISBEE tool, a software for optimising the trade-off between food quality, energy use, and global warming impact of cold chains. *Journal of Food Engineering* 148, 2–12.

Hertog, M. L. A. T. M., Uysal, I., McCarthy, U., Verlinden, B. M. and Nicolaï, B. M. 2014. Shelf life modelling for first-expired-first-out warehouse management. *Philosophical Transactions of the Royal Society Series. A* 372(2017)(2017), 20130306. doi:10.1098/rsta.2013.0306.

Hugos, M. H. 2018. *Essentials of Supply Chain Management* (4th edn.). John Wiley & Sons, Hoboken, NJ, 355p.

James, S. J., James, C. and Evans, J. A. 2006. Modelling of food transportation systems – a review. *International Journal of Refrigeration* 29, 947–57.

Jedermann, R. and Lang, W. 2009. The minimum number of sensors – interpolation of spatial temperature profiles in chilled transports. In: *European Conference on Wireless Sensor Networks*. Springer, Berlin and Heidelberg, pp. 232–46.

Jedermann, R., Emond, J. P. and Lang, W. 2008. Shelf life prediction by intelligent RFID – technical limits of model accuracy. In: Kreowski, H. J., Scholz-Reiter, B. and Haasis, H. D. (Eds), *Dynamics in Logistics*. Springer, Berlin and Heidelberg, pp. 231–8.

Jedermann, R., Palafox-Albarrán, J., Barreiro, P., Ruiz-Garcia, L., Robla, J. I. and Lang, W. 2011. Interpolation of spatial temperature profiles by sensor networks. In: *SENSORS, 2011 IEEE*. IEEE, pp. 778–81.

Jedermann, R., Nicometo, M., Uysal, I. and Lang, W. 2014. Reducing food losses by intelligent food logistics. *Philosophical Transactions of the Royal Society Series. A* 372(2017), 20130302. doi:10.1098/rsta.2013.0302.

Kader, A. A. 2003. A perspective on postharvest horticulture. *HortScience* 38, 1004–8.

Kelly, K., Madden, R., Emond, J. P. and Nunes, M. C. N. 2019. A novel approach to determine the impact level of each step along the supply chain on strawberry quality. *Postharvest Biology and Technology* 147, 78–88.

Koutsoumanis, K., Giannakourou, M. C., Taoukis, P. S. and Nychas, G. J. E. 2002. Application of shelf life decision system (SLDS) to marine cultured fish quality. *International Journal of Food Microbiology* 73(2–3), 375–82.

Lee, H. L. and Billington, C. 1992. Managing supply chain inventory: pitfalls and opportunities. *Sloan Management Review* 33(3), 65–74.

Li, S., Da Xu, L. and Zhao, S. 2015. The internet of things: a survey. *Information Systems Frontiers* 17, 243–59.

McCarthy, U., Uysal, I., Laniel, M., Corkery, G., Butler, F., McDonnell, K. P. and Ward, S. 2013. Sustainable global food supply networks. In: Tiwrim, B. K., Norton, T. and Holden, N. M. (Eds), *Sustainable Food Processing*. Wiley, Oxford, UK, pp. 521-38.

McCarthy, U., Uysal, I., Melis, R. B., Mercier, S., Donnell, C. O. and Ktenioudaki, A. 2018. Global food security - issues, challenges and technological solutions. *Trends in Food Science and Technology* 77, 11-20.

McDonald, K. and Sun, D. W. 2000. Vacuum cooling technology for the food processing industry: a review. *Journal of Food Engineering* 45, 55-65.

McMeekin, T. A. and Ross, T. 1996. Shelf life prediction: status and future possibilities. *International Journal of Food Microbiology* 33(1), 65-83.

Mercier, S. and Uysal, I. 2018. Neural network models for predicting perishable food temperatures along the supply chain. *Biosystems Engineering* 171, 91-100.

Mercier, S., Marcos, B. and Uysal, I. 2017. Identification of the best temperature measurement position inside a food pallet for the prediction of its temperature distribution. *International Journal of Refrigeration* 76, 147-59.

Mercier, S., Brecht, J. K. and Uysal, I. 2019. Commercial forced-air precooling of strawberries: a temperature distribution and correlation study. *Journal of Food Engineering* 242, 47-54.

Muth, M. K., Karns, S. A., Nielsen, S. J., Buzby, J. C. and Wells, H. F. 2011. Consumer-level food loss estimates and their use in the ERS loss-adjusted food availability data. USDA Econ. Res. Serv. Tech. Bull. No. 1927. USDA, Washington DC.

Nunes, M. C. N. 2008a. *Color Atlas of Postharvest Quality of Fruits and Vegetables*. Wiley-Blackwell, Ames, IA, 463p.

Nunes, M. C. N. 2008b. Impact of environmental conditions on fruit and vegetable quality. *Stewart Postharvest Review* 4(4), 14.

Nunes, M. C. N. 2015. Correlations between subjective quality and physicochemical attributes of fresh fruits and vegetables. *Postharvest Biology and Technology* 107, 43-54.

Nunes, M. C. N. and Dea, S. 2016. Bioactive compounds in strawberry fruit exposed to optimal and suboptimal relative humidity. *Acta Horticulturae* 1117, 231-6.

Nunes, M. C. N. and Emond, J.-P. 2007. Relationship between weight loss and visual quality of fruits and vegetables. *Proceedings of the Florida State Horticultural Society* 120, 235-45.

Nunes, M. C. N. and Emond, J.-P. 2013. Influence of environmental conditions on the quality attributes and shelf life of 'Goldfinger' bananas. *Postharvest Biology and Technology* 86, 309-20.

Nunes, M. C. N., Brecht, J. K., Morais, A. M. M. B. and Sargent, S. A. 1998. Controlling temperature and water loss to maintain ascorbic acid levels in strawberries during postharvest handling. *Journal of Food Science* 63, 1033-6.

Nunes, M. C. N., Emond, J.-P. and Brecht, J. K. 2001. Temperature abuse during ground and in-flight handling operations affects quality of snap beans. *HortScience* 36, 510 (Abstract).

Nunes, M. C. N., Emond, J.-P. and Brecht, J. K. 2003a. Predicting shelf life and quality of raspberries under different storage temperatures. *Acta Horticulturae* 628, 599-606.

Nunes, M. C. N., Proulx, E., Emond, J.-P. and Brecht, J. K. 2003b. Quality characteristics of 'Horn of Plenty' and 'Medallion' yellow summer squash as a function of the storage temperature. *Acta Horticulturae* 628, 607-17.

Nunes, M. C. N., Emond, J.-P. and Brecht, J. K. 2004. Quality curves for highbush blueberries as a function of the storage temperature. *Small Fruits Review* 3, 423–40.

Nunes, M. C. N., Emond, J.-P., Brecht, J. K., Dea, S. and Proulx, E. 2007. Quality curves for mango as a function of the storage temperature. *Journal of Food Quality* 30, 104–20.

Nunes, M. C. N., Emond, J., Rauth, M., Dea, S. and Chau, K. 2009. Environmental conditions encountered during typical consumer retail display affect fruit and vegetable quality and waste. *Postharvest Biology and Technology* 51, 232–41.

Nunes, M. C. N., Delgado, A. and Emond, J.-P. 2012. Quality curves for green bell pepper (*Capsicum annuum* L) stored at low and recommended relative humidity levels. *Acta Horticulturae* 945, 71–8.

Nunes, M. C. N., Nicometo, M., Emond, J. P., Badia, R. M. and Uysal, I. 2014. Improvement in fresh fruit and vegetables logistics quality: berry logistic field studies. *Philosophical Transactions of the Royal Society Series. Part A* 372, 20130307.

Odriozola-Serrano, I., Soliva-Fortuny, R. and Martin-Belloso, O. 2009. Influence of storage temperature on the kinetics of the changes in anthocyanins, vitamin C, and antioxidant capacity in fresh-cut strawberries stored under high-oxygen atmospheres. *Journal of Food Science* 74(2), C184–91. doi:10.1111/j.1750-3841.2009.01075.x.

Oliver, R. K. and Webber, M. D. 1982. Supply-chain management: logistics catches up with strategy. *Outlook* 5, 42–7.

Oms-Oliu, G., Odriozola-Serrano, I., Soliva-Fortuny, R. and Martin-Belloso, O. 2009. Use of Weibull distribution for describing kinetics of antioxidant potential changes in fresh-cut watermelon. *Journal of Food Engineering* 95, 99–105.

Pin, C. and Baranyi, J. 1998. Predictive models as means to quantify the interactions of spoilage organisms. *International Journal of Food Microbiology* 41(1), 59–72.

Proulx, E., Nunes, M. C. N., Emond, J.-P. and Brecht, J. K. 2001. Quality curves for tomato exposed at chilling and non-chilling temperatures. *HortScience* 36, 509 (Abstract).

Proulx, E., Nunes, M. C. N., Emond, J.-P. and Brecht, J. K. 2005. Quality attributes limiting papaya postharvest life at chilling and non-chilling temperatures. *Proceedings of the Florida State Horticultural Society* 118, 389–95.

Proulx, E., Yagiz, Y., Nunes, M. C. N. and Emond, J.-P. 2010. Quality attributes limiting snap bean (*Phaseolus vulgaris* L.) postharvest life at chilling and non-chilling temperatures. *HortScience* 45, 1238–49.

Ramos, B., Miller, F. A., Brandao, T. R. S., Teixeira, P. and Silva, C. L. M. 2013. Fresh fruits and vegetables-an overview on applied methodologies to improve its quality and safety. *Innovative Food Science and Emerging Technologies* 20, 1–15.

Reina, L. D., Fleming, H. P. and Humphries, E. G. 1995. Microbiological control of cucumber hydrocooling water with chlorine dioxide. *Journal of Food Protection* 58(5), 541–6. doi:10.4315/0362-028X-58.5.541.

Thompson, J. F. and Crisosto, C. H. 2002. Handling at destination markets. In: Kader, A. A. (Ed.), *Postharvest Technology of Horticultural Crops*. UC Publication 3311. University of California, Division of Agriculture and Natural Resources, Berkeley, CA, pp. 271–7.

Thompson, J. F., Mitchell, F. G., Rumsey, T. R., Kasmire, R. F. and Crisosto, C. H. 2008. Commercial cooling of fruits, vegetables, and flowers. UC Publication 21567. University of California, Division of Agriculture and Natural Resources, Berkeley, CA.

Trienekens, J. and Zuurbier, P. 2008. Quality and safety standards in the food industry, developments and challenges. *International Journal of Production Economics* 113, 107–22.

Tromp, S. O., Rijgersberg, H., da Silva, F. P. and Bartels, P. 2012. Retail benefits of dynamic expiry dates–simulating opportunity losses due to product loss, discount policy and out of stock. *International Journal of Production Economics* 139, 14–21.

van Boekel, M. A. J. S. 2008. Kinetic modeling of food quality: a comprehensive review. *Comprehensive Reviews in Food Science and Food Safety* 7, 144–58.

Verhaelen, K., Bouwknegt, M., Lodder-Verschoor, F., Rutjes, S. A. and Husman, A. M. R. 2012. Persistence of human norovirus GII.4 and GI.4, murine norovirus, and human adenovirus on soft berries as compared with PBS at commonly applied storage conditions. *International Journal of Food Microbiology* 160(2), 137–44. doi:10.1016/j.ijfoodmicro.2012.10.008.

Wang, X. and Li, D. 2012. A dynamic product quality evaluation based pricing model for perishable food supply chains. *Omega* 40, 906–17.

Wei, L., Qin, H., Zhu, W. and Wan, L. 2015. A study of perturbation operators for the pickup and delivery traveling salesman problem with LIFO or FIFO loading. *Journal of Heuristics* 21, 617–39.

Weiser, M. 1991. The computer for the 21st century. *Scientific American* 265(3), 94–5, 98–104.

Welbourne, E., Battle, L., Cole, G., Gould, K., Rector, K., Raymer, S., Balazinska, M. and Borriello, G. 2009. Building the internet of things using RFID: the RFID ecosystem experience. *IEEE Internet Computing* 13(3), 48–55.

Wells, J. H. and Singh, R. P. 1998. Quality management during storage and distribution. In: Taub, I. A. and Singh, R. P. (Eds), *Food Storage Stability*. CRC, Boca Raton, FL, pp. 369–86.

Zwietering, M. H., De Koos, J. T., Hasenack, B. E., De Wit, J. C. and van't Riet, K. 1991. Modelling of bacterial growth as a function of temperature. *Applied and Environmental Microbiology* 57, 1875–81.

Index

Milton Keynes UK
Ingram Content Group UK Ltd.
UKHW020611041123
431885UK00001B/11

9 781786 762887